Nik Bessis and Fatos Xhafa (Eds.)

Next Generation Data Technologies for Collective Computational Intelligence

Studies in Computational Intelligence, Volume 352

Editor-in-Chief
Prof. Janusz Kacprzyk
Systems Research Institute
Polish Academy of Sciences
ul. Newelska 6
01-447 Warsaw
Poland
E-mail: kacprzyk@ibspan.waw.pl

Further volumes of this series can be found on our homepage: springer.com

Vol. 330. Steffen Rendle
Context-Aware Ranking with Factorization Models, 2010
ISBN 978-3-642-16897-0

Vol. 331. Athena Vakali and Lakhmi C. Jain (Eds.)
New Directions in Web Data Management 1, 2011
ISBN 978-3-642-17550-3

Vol. 332. Jianguo Zhang, Ling Shao, Lei Zhang, and Graeme A. Jones (Eds.)
Intelligent Video Event Analysis and Understanding, 2011
ISBN 978-3-642-17553-4

Vol. 333. Fedja Hadzic, Henry Tan, and Tharam S. Dillon
Mining of Data with Complex Structures, 2011
ISBN 978-3-642-17556-5

Vol. 334. Álvaro Herrero and Emilio Corchado (Eds.)
Mobile Hybrid Intrusion Detection, 2011
ISBN 978-3-642-18298-3

Vol. 335. Radomir S. Stankovic and Radomir S. Stankovic
From Boolean Logic to Switching Circuits and Automata, 2011
ISBN 978-3-642-11681-0

Vol. 336. Paolo Remagnino, Dorothy N. Monekosso, and Lakhmi C. Jain (Eds.)
Innovations in Defence Support Systems – 3, 2011
ISBN 978-3-642-18277-8

Vol. 337. Sheryl Brahnam and Lakhmi C. Jain (Eds.)
Advanced Computational Intelligence Paradigms in Healthcare 6, 2011
ISBN 978-3-642-17823-8

Vol. 338. Lakhmi C. Jain, Eugene V. Aidman, and Canicious Abeynayake (Eds.)
Innovations in Defence Support Systems – 2, 2011
ISBN 978-3-642-17763-7

Vol. 339. Halina Kwasnicka, Lakhmi C. Jain (Eds.)
Innovations in Intelligent Image Analysis, 2010
ISBN 978-3-642-17933-4

Vol. 340. Heinrich Hussmann, Gerrit Meixner, and Detlef Zuehlke (Eds.)
Model-Driven Development of Advanced User Interfaces, 2011
ISBN 978-3-642-14561-2

Vol. 341. Stéphane Doncieux, Nicolas Bredeche, and Jean-Baptiste Mouret(Eds.)
New Horizons in Evolutionary Robotics, 2011
ISBN 978-3-642-18271-6

Vol. 342. Federico Montesino Pouzols, Diego R. Lopez, and Angel Barriga Barros
Mining and Control of Network Traffic by Computational Intelligence, 2011
ISBN 978-3-642-18083-5

Vol. 343. Kurosh Madani, António Dourado Correia, Agostinho Rosa, and Joaquim Filipe (Eds.)
Computational Intelligence, 2011
ISBN 978-3-642-20205-6

Vol. 344. Atilla Elçi, Mamadou Tadiou Koné, and Mehmet A. Orgun (Eds.)
Semantic Agent Systems, 2011
ISBN 978-3-642-18307-2

Vol. 345. Shi Yu, Léon-Charles Tranchevent, Bart De Moor, and Yves Moreau
Kernel-based Data Fusion for Machine Learning, 2011
ISBN 978-3-642-19405-4

Vol. 346. Weisi Lin, Dacheng Tao, Janusz Kacprzyk, Zhu Li, Ebroul Izquierdo, and Haohong Wang (Eds.)
Multimedia Analysis, Processing and Communications, 2011
ISBN 978-3-642-19550-1

Vol. 347. Sven Helmer, Alexandra Poulovassilis, and Fatos Xhafa
Reasoning in Event-Based Distributed Systems, 2011
ISBN 978-3-642-19723-9

Vol. 348. Beniamino Murgante, Giuseppe Borruso, and Alessandra Lapucci (Eds.)
Geocomputation, Sustainability and Environmental Planning, 2011
ISBN 978-3-642-19732-1

Vol. 349. Vitor R. Carvalho
Modeling Intention in Email, 2011
ISBN 978-3-642-19955-4

Vol. 350. Thanasis Daradoumis, Santi Caballé, Angel A. Juan, and Fatos Xhafa (Eds.)
Technology-Enhanced Systems and Tools for Collaborative Learning Scaffolding, 2011
ISBN 978-3-642-19813-7

Vol. 351. Ngoc Thanh Nguyen, Bogdan Trawiński, and Jason J. Jung (Eds.)
New Challenges for Intelligent Information and Database Systems, 2011
ISBN 978-3-642-19952-3

Vol. 352. Nik Bessis and Fatos Xhafa (Eds.)
Next Generation Data Technologies for Collective Computational Intelligence, 2011
ISBN 978-3-642-20343-5

Nik Bessis and Fatos Xhafa (Eds.)

Next Generation Data Technologies for Collective Computational Intelligence

Springer

Professor Nik Bessis
School of Computing & Maths
University of Derby
Derby, DE22 1GB
United Kingdom (UK)
E-mail: n.bessis@derby.ac.uk

Dr. Fatos Xhafa
Professor Titular d'Universitat
Dept de Llenguatges i Sistemes Informàtics
Universitat Politècnica de Catalunya
Barcelona, Spain
E-mail: fatos@lsi.upc.edu

ISBN 978-3-642-20343-5 e-ISBN 978-3-642-20344-2

DOI 10.1007/978-3-642-20344-2

Studies in Computational Intelligence ISSN 1860-949X

Library of Congress Control Number: 2011925383

Typeset & Cover Design: Scientific Publishing Services Pvt. Ltd., Chennai, India.

Printed on acid-free paper

9 8 7 6 5 4 3 2 1

springer.com

Foreword

It is a great honor to me to write a foreword for this book on "Next Generation Data Technologies for Collective Computational Intelligence". With the rapid development of the Internet, the volume of data being created and digitized is growing at an unprecedented rate, which if combined and analyzed through a collective and computational intelligence manner will make a difference in the organizational settings and their user communities.

The focus of this book is on next generation data technologies in support of collective and computational intelligence. The book distinguish itself from others in that it brings various next generation data technologies together to capture, integrate, analyze, mine, annotate and visualize distributed data – made available from various community users – in a meaningful and collaborative for the organization manner.

This book offers a unique perspective on collective computational intelligence, embracing both theory and strategies fundamentals such as data clustering, graph partitioning, collaborative decision making, self-adaptive ant colony, swarm and evolutionary agents. It also covers emerging and next generation technologies in support of collective computational intelligence such as Web 2.0 enabled social networks, semantic web for data annotation, knowledge representation and inference, data privacy and security, and enabling distributed and collaborative paradigms such as P2P computing, grid computing, cloud computing due to the nature that data is usually geographically dispersed and distributed in the Internet environment.

This book will be of great interest and help to those who are broadly involved in the domains of computer science, computer engineering, applied informatics, business or management information systems. The reader group might include researchers or senior graduates working in academia; academics, instructors and senior students in colleges and universities, and software developers.

Dr. Maozhen Li
Brunel University, UK

Preface

Introduction

The use of collaborative decision and management support systems has evolved over the years through developments in distributed computational science in a manner, which provides applicable intelligence in decision-making. The rapid developments in networking and resource integration domains have resulted in the emergence and in some instances to the maturation of distributed and collaborative paradigms such as Web Services, P2P, Grid and Cloud computing, Data Mashups and Web 2.0. Recent implementations in these areas demonstrate the applicability of the aforementioned next generation technologies in a manner, which seems the panacea for solving very complex problems and grand challenges. A broad range of issues are currently being addressed; however, most of these developments are focused on developing the platforms and the communication and networking infrastructures for solving these very complex problems, which in most instances are well-known challenges. The enabling nature of these technologies allows us to visualize their collaborative and synergetic use in a less conventional manner, which are currently problem focused.

In this book, the focus is on the viewpoints of the organizational setting as well as on the user communities, which those organizations cater to. The book appreciates that in many real-world situations an understanding – using computational techniques – of the organization and the user community needs is a computational intelligence itself. Specifically, current Web and Web 2.0 implementations and future manifestations will store and continuously produce a vast amount of distributed data, which if combined and analyzed through a collective and computational intelligence manner using next generation data technologies will make a difference in the organizational settings and their user communities. Thus, the focus of this book is about the methods and technologies which bring various next generation data technologies together to capture, integrate, analyze, mine, annotate and visualize distributed data – made available from various community users – in a meaningful and collaborative for the organization manner.

In brief, the overall objective of this book is to encapsulate works incorporating various next generation distributed and other emergent collaborative data technologies for collective and computational intelligence, which are also applicable in various organizational settings. Thus, the book aims to cover in a comprehensive manner the combinatorial effort of utilizing and integrating various next generation collaborative and distributed data technologies for computational intelligence in various scenarios. The book also distinguishes itself by focusing on assessing whether utilization and integration of next generation data technologies can assist in the identification of new opportunities, which may also be strategically fit for purpose.

Who Should Read the Book?

The content of the book offers state-of-the-art information and references for work undertaken in the challenging area of collective computational intelligence using emerging distributed computing paradigms. Thus, the book should be of particular interest for:

Researchers and doctoral students working in the area of distributed data technologies, collective intelligence and computational intelligence, primarily as a reference publication. The book should be also a very useful reference for all researchers and doctoral students working in the broader fields of data technologies, distributed computing, collaborative technologies, agent intelligence, artificial intelligence and data mining.

Academics and students engaging in research informed teaching and/or learning in the above fields. The view here is that the book can serve as a good reference offering a solid understanding of the subject area.

Professionals including computing specialists, practitioners, managers and consultants who may be interested in identifying ways and thus, applying a number of well defined and/or applicable cutting edge techniques and processes within the domain area.

Book Organization and Overview

The book contains 22 self-contained chapters that were very carefully selected based on peer review by at least two expert and independent reviewers. The book is organized into four parts according to the thematic topic of each chapter.

Part I: Foundations and Principles

The part focuses on presenting state-of-the-art reviews on the foundations, principles, methods and techniques for collective and computational intelligence. In particular:

Chapter 1 illustrates the space-based computing paradigm aiming to support and facilitate software developers in their efforts to control complexity regarding concerns of interaction in software systems.

Chapter 2 presents a state-of-the-art review on ant colony optimization and data mining techniques and focus on their use for data classification and clustering. They briefly present related applications and examples and outline possible future trends of this promising collaborative use of techniques.

Chapter 3 offers a high-level introduction to the open semantic enterprise architecture. Because of its open nature it is free to adopt and extend, yet retains a root commonality to ensure all participating agents can agree on a common understanding without ambiguity, regardless of the underlying ontology or logic system used.

Chapter 4 discusses and evaluates techniques for automatically classifying and coordinating tags extracted from one or more folksonomies, with the aim of building collective tag intelligence, which can then be exploited to improve the conventional searching functionalities provided by tagging systems.

Chapter 5 provides an overview of the current landscape of computational models of trust and reputation, and it presents an experimental study case in the domain of social search, where it is shown how trust techniques can be applied to enhance the quality of social search engine predictions.

Part II: Advanced Models and Practices

The part focuses on presenting theoretical models and state-of-the-art practices on the area of collective and computational intelligence. These include but not limited to the application of formal concept analysis; classifiers and expression trees; swarm intelligence; channel prediction and message request; time costs and user interfaces. In particular:

Chapter 6 presents the formal concept analysis; a proposed data technology that complements collective intelligence such as that identified in the semantic web. The work demonstrates the discovery of these novel semantics through open source software development and visualizes data's inherent semantics.

Chapter 7 focuses on constructing high quality classifiers through applying collective computational techniques to the field of machine learning. Experiment results confirm gene expression programming and cellular evolutionary algorithms when applied to the field of machine learning, can offer an advantage that can be attributed to their collaborative and synergetic features.

Chapter 8 deals with the load-balancing problem by using a self-organizing approach. In this work, a generic architectural pattern has been presented, which allows the exchanging of different algorithms through plugging. Although it possesses self-organizing properties by itself, a significant contribution to self-organization is given by the application of swarm based algorithms, especially bee algorithms that are modified, adapted and applied for the first time in solving the load balancing problem.

Chapter 9 presents a new scheme for channel prediction in multicarrier frequency hopping spread spectrum system. The technique adaptively estimates the channel conditions and eliminates the need for the system to transmit a request message prior to transmit the packet data.

Chapter 10 discusses a theory on process for decision making under time stress, which is common among two or bilateral decision makers. The work also proposes a formula on strategic points for minimizing the cost of time for a certain process.

Chapter 11 presents a model for amplifying human intelligence, utilizing agents technology for task-oriented contexts. It uses domain ontology and task scripts for handling formal and semiformal knowledge bases, thereby helping to systematically explore the range of alternatives; interpret the problem and the context and finally, maintain awareness of the problem.

Part III: Advanced Applications

The part focuses on presenting cutting-edge applications with a specific focus on social networks; cloud computing; computer games and trust. In particular:

Chapter 12 investigates the use of a proposed architecture for continuous analytics for massively multi-play online games, to support the analytics part of the relevant social networks. The work presents the design and implementation of the platform, with a focus on the cloud-related benefits and challenges.

Chapter 13 studies feature extraction and pattern classification methods in two medical areas, Stabilometry and Electroencephalography. An adaptive fuzzy inference neural network has been applied by using a hybrid supervised/unsupervised clustering scheme while its final fuzzy rule base is optimized through competitive learning. The proposed system is based on a method for generating reference models from a set of time series.

Chapter 14 analyzes a service oriented architecture based next generation mobility management model. In this work, a practical case, e.g., a "mobile messaging" application showing how to apply the proposed approach is presented.

Chapter 15 creates a set of metrics for measuring entertainment in computer games. Specifically, the work here uses evolutionary algorithm to generate new and entertaining games using the proposed entertainment metrics as the fitness function. A human user survey and experiment using the controller learning ability is also included.

Chapter 16 investigates the problem of knowledge extraction from social media. Specifically, the work here presents three methods that use Flickr data to extract different types of knowledge namely, the community structure of tag-networks, the emerging trends and events in users tag activity, and the associations between image regions and tags in user tagged images.

Chapter 17 presents an anonymity model to protect privacy in large survey rating data. Extensive experiments on two real-life data sets show that the proposed slicing technique is fast and scalable with data size and much more efficient in terms of execution time and space overhead than the heuristic pair-wise method.

Part IV: Future Trends and Concepts

Finally, this part focuses on presenting future concepts and trends using either real or realistic scenarios. In particular:

Chapter 18 focuses on the next generation network and how underlying technologies should evolve and be used to help service providers remain competitive. Within this context, a migration strategy is proposed and explored enabling the development of a capable concept of how the structuring of networks must be changed, and in doing so taking into consideration the business needs of diverse service providers and network operators.

Chapter 19 discusses how next generation emerging technologies could help coin and prompt future direction of their fit-to-purpose use in various real-world scenarios including the proposed case of disaster management. Specifically, it reviews their possible combination with intelligence techniques for augmenting computational intelligence in a collective manner for the purpose of managing disasters.

Chapter 20 presents novel technologies for exploiting multiple layers of collective intelligence from user-contributed content. The exploitation of the emerging results is showcased using an emergency response and a consumers social group case studies.

Chapter 21 offers a review of mobile sensing technologies and computational methods for collective intelligence. Specifically, the work presented discusses the application of mobile sensing to understand collective mechanisms and phenomena in face-to-face networks at three different scales: organizations, communities and societies. Finally, the impact that these new sensing technologies may have on the understanding of societies, and how these insights can assist in the design of smarter cities and countries is discussed.

Chapter 22 outlines the key social drivers for dataveillance and illustrate some of the roles emerging technology plays in it. Within this context, the work presents a social ecological model of technology cooption. The proposed model provides a middle range theory for empirical analysis by identifying the key elements of technology cooption and their proposed links and the role of the stakeholders in such cooption.

Professor Nik Bessis
University of Derby, UK
University of Bedfordshire, UK

Professor Fatos Xhafa
Universitat Politècnica de Catalunya, Spain

Acknowledgements

It is with our great pleasure to comment on the hard work and support of many people who have been involved in the development of this book. It is always a major undertaking but most importantly, a great encouragement and definitely a reward and an honor when experiencing the enthusiasm and eagerness of so many people willing to participate and contribute towards the publication of this book. Without their support the book could not have been satisfactory completed.

First and foremost, we wish to thank all the authors who, as distinguished scientists despite busy schedules, devoted so much of their time preparing and writing their chapters, and responding to numerous comments and suggestions made from the reviewers. We have also been fortunate that the following (in no particular order) have honored us with their assistance in this project: Dr Gorell Cheek, University of North Carolina, USA; Pierre Lévy, University of Ottawa, Canada; Dr Maozhen Li, Brunel University, UK; Professor Udai Shanker, M.M.M Engineering College Gorakhpur, India and Professor Gio Wiederhold, Stanford University, USA. Special gratitude goes also to all the reviewers and some of the chapters' authors who also served as referees for chapters written by other authors. The editors wish to apologize to anyone whom they have forgotten.

Last but not least, we wish to thank Professor Janusz Kacprzyk, Editor-in-Chief of "Studies in Computational Intelligence" Springer series, Dr Thomas Ditzinger, Ms Heather King and the whole Springer's editorial team for their strong and continuous support throughout the development of this book.

Finally, we are deeply indebted to our families for their love, patience and support throughout this rewarding experience.

Professor Nik Bessis
University of Derby, UK
University of Bedfordshire, UK

Professor Fatos Xhafa
Universitat Politècnica de Catalunya, Spain

Contents

Part IV: Future Trends and Concepts

Part I
Foundations and Principles

Chapter 1
Coordination Mechanisms in Complex Software Systems

Richard Mordinyi and Eva Kühn

Abstract. Software developers have to deal with software systems which are usually composed of distributed, heterogeneous and interacting application components representing higher-level business goals. Although the message-passing paradigm is a common concept allowing application components to interact with each other in an asynchronous manner, the technology is not entirely suitable for complex coordination requirements since the processing and state of coordination have to be handled explicitly by the application component. Data-driven frameworks support the coordination of application components, but have a limited number of coordination policies requiring from the software developer to implement coordination functionality that is not directly supported by the coordination framework. We illustrate the Space-Based Computing (SBC) paradigm aiming to support and facilitate software developers efficiently in their efforts to control complexity regarding concerns of interaction in software systems. Major results of the evaluation in this context are improved coordination efficiency accompanied with reduced complexity within application components.

1 Introduction

Complex systems are systems [1, 2] whose properties are not fully explained by an understanding of their single component parts. Complex systems (e.g., financial markets, bacteria life cycles) usually consist of a large number of mutually interacting, dynamically interwoven, and indeterminably dis- and reappearing component parts. The understanding often is that the complexity of a system emerges by interaction of a (large) number of component parts, but cannot be explained by looking at the parts alone. Software systems, especially software-intensive systems [3], can be interpreted as complex systems as well, because they usually interact with other software, systems, devices, sensors and people. Over time these systems become more distributed, heterogeneous, decentralized and interdependent, and are operating more often in dynamic and frequently unpredictable

---

Richard Mordinyi · Eva Kühn
Vienna University of Technology, Institute of Computer Languages,
Space-Based Computing Group, Argentinierstrasse 8, 1040 Vienna, Austria
e-mail: {richard.mordinyi,eva.kuehn}@tuwien.ac.at

N. Bessis, F. Xhafa (Eds.): Next Generation Data Technologies for CCI, SCI 352, pp. 3–30.
springerlink.com

environments. Therefore, software developers have to deal with issues like heterogeneity and varying size of components, variety of protocols for interaction with internal and external components, or with a number of potential incidents, like crashed or unreachable components in distributed environments. In the course of developing distributed software systems, software developers cannot avoid coping with these complexity issues. Today's software systems typically consist of mainly distributed application components representing higher-level business goals and a middleware technology abstracting the complexity concerns related to network and distribution. However, software developers still have to deal with the interaction of application components.

The message-passing paradigm is a common concept allowing application components to communicate with each other. The message-oriented middleware [4, 5], prominent representative is the Enterprise Service Bus (ESB) [6], provides synchronous and asynchronous message-passing properties and promises to interconnect application components in a loosely coupled manner. Since message-oriented middleware is only capable of transmitting and transforming messages between application components, it lacks support for complex interaction requirements which involve the participation of several application components for decision making, like in the telecommunication domain [7]. The software developers have no other choice but to take into account both, the application logic representing the business goal and additional coordination logic needed to fulfill the specific coordination requirement. Such logic for instance may contain implementation matters related to synchronization problems. Furthermore, it may include logic for the management and supervision of the latest state of the coordination process itself otherwise the application may get "lost" and the common business goal cannot be reached. Additional management is needed in case the coordinating component crashes and after recovery the failed application component still wants to be part of the running coordinating process [8]. These additional issues introduce potential sources of error, decrease the efficiency of the system, and increase the cognitive complexity [9, 10] of the application component. However, the responsibility of the software developer should focus on the application's business goals and not on concerns related to distribution or coordination.

A framework that has been explicitly designed for coordination purposes is the so called tuple space, based on the Linda coordination model of David Gelernter [11]. It is a data-centered, blackboard based, architectural style that describes the usage of a logically shared memory, the tuple space, by means of simple operations as interaction mechanisms. The approach promotes a clear separation between the computation model, used to express the computational requirements of an algorithm, and the coordination model, used to express the communication and synchronization requirements. The state of coordination is not embedded in the coordinating process itself but in the space [12]. The state of the coordination information on the blackboard determines the way of execution of the process. By means of this coordination model the application may entirely focus on business goals since the model "gives application builders the advantage of ignoring some of the harder aspects of multi-client synchronization, such as tracking names (and addresses) of all active clients, communication line status, and conversation status" [13]. The Linda coordination model uses template matching with random, non-deterministic tuple access to coordinate processes (see Sect. 0) supporting one coordination model only that can be considered a restriction limiting the benefit of using such a communication abstraction. Therefore, with respect to more complex coordination requirements the software developer still needs to implement coordination functionality not directly supported by the coordination framework within application components. Consequently, this increases the complexity of the application component, decreases performance due to additional implementation logic, and leads once again to an unclear separation between computation model and coordination model.

Taking into account the previously mentioned issues regarding interaction in software systems, we illustrate the so called Space-Based Computing (SBC) paradigm [14] supporting and facilitating software developers efficiently in their efforts to control complexity concerns in software systems. SBC extends and strengthens the clear separation between computation and coordination logic by allowing the selection and injection of scenario specific coordination models. From the application's point of view the SBC paradigm is comparable to the blackboard architectural style, orientated on the Linda coordination language. In contrast to traditional Linda coordination frameworks, we define the SBC paradigm to extend the Linda coordination model by introducing exchangeable mechanisms for structuring data in the space using special ordering characteristics and reducing dependencies between application components and coordination models. SBC explicitly embeds sophisticated coordination capabilities in the architectural style, and thus makes the style itself dynamic with respect to the scenario's coordination problem statement. This means that SBC is capable of abstracting coordination requirements and changes from the application. Since coordination requires and thus inherently consists of communication, consequently the abstraction of coordination also means that SBC abstracts communication requirements as well.

We evaluate the Space-Based Computing (SBC) paradigm using an industrial use case from an assembly workshop of a production automation system. The evaluation will demonstrate SBC's coordination and recovery capabilities focusing on aspects like feasibility, effort, robustness, performance, and complexity. Major results of the evaluation are higher coordination efficiency accompanied with minimized complexity within application components.

The remainder of this chapter is structured as follows: Section 2 summarizes related work on coordination models and platforms. Section 3 describes the industrial use case while Section 4 concentrates on the conceptual details of the Space-based Computing paradigm. Section 5 presents the evaluation whereas Section 6 discusses the advantages and limitations of SBC. Finally, Section 7 concludes the chapter and provides further research issues for future work.

2 Related Work

Since significant characteristics of complex systems refer to the interaction between components of complex systems, coordination between these components is an important issue to be investigated. This section summarizes related work with emphasis on coordination theory by giving a definition of coordination, describing coordination models, and presenting technologies built for supporting coordination.

2.1 Coordination Theory

Coordination [15] is the additional organizing activity (like information processing) that is needed in case multiple actors pursue the same goal, that a single actor would not perform. In a more general perspective [16], coordination refers to "*the act working together harmoniously*". However, it can be derived that coordination itself consists of different components, like actors performing some activities which are directed to a goal. Therefore, the definition implies that activities are not independent and thus coordination can be seen as "*the act of managing interdependencies between activities performed to achieve a goal*". Later, Malone and Crowston [17], the founders of interdisciplinary science of coordination theory, describe their definition in a refined form just as "*managing dependencies between activities*". It has to be pointed out that coordination makes only sense if tasks are interdependent. If there are no interdependencies, there is nothing to coordinate either. Given the unavoidable existence of dependencies, a detailed characterization of different sorts of dependencies is given in [17, 18].

2.2 Coordination Models

A coordination model [19] is either a formal or a conceptual framework to model the space of interaction. A formal framework expresses notations and rules for the formal characterization of coordinated systems, as used in the frameworks listed in [20] and [21]. A conceptual framework is required by software developers to manage inter-component interactions, since it provides abstraction mechanisms. In general, the emphasis is on the expressiveness of the abstraction mechanism of the coordination model, and on its effectiveness helping software developers in managing interactions.

From a functionality point of view distributed systems are typically divided into the following three concerns:

- Computational logic (i.e. business logic) performs calculations representing the main intention of the system (i.e. business specific goals)
- Communication responsible for sending and receiving data between components to be further processed.
- Coordination or dependency management responsible to execute tasks in a way where no dependencies are violated and the common coordination goal is achievable.

Sancese et. al. [22] argue that a clear separation of the three parts leads to a reduction of complexity of the entire systems also enabling a reliable and more stable implementation. The process of coordination follows a certain coordination model for which Ciancarini [19] defines a generic coordination model as a triple of

{E, M, L}. In the model, {E} stands for either physical or logical entities to be coordinated. These can be software processes, threads, services, agents, or even human beings interacting with computer-based systems. {M} represents the coordination media (i.e. communication channels) serving as a connector between the entities and enables communication, which is a mandatory prerequisite for coordination [18, 23]. Such coordination media may be message-passing systems, pipes, tuple spaces [11] etc. {L} specifies the coordination laws between the entities defining how the interdependencies have to be resolved and therefore, semantically define the coordination mechanisms. According to [12], existing variations of coordination models and languages can be mainly divided into two categories: control-driven (or task- or process-oriented) or data-driven coordination models, as described in the following sections.

2.2.1 Control-Driven Coordination

In control-driven coordination models [12] processes are treated as black boxes and any data manipulated within the process is of no concern to other system processes. Processes communicate with other processes by means of well defined interfaces, but it is entirely up to the process when communication takes place. In case processes communicate, they send out control messages or events with the aim of letting other interested processes know their interest, in which state they are, or informing them of any state changes.

From a stylistically perspective, in the control-driven coordination model it is easy to separate the processes into two components, namely purely computational ones and purely coordination ones. The reason is that "*the state of the computation at any moment in time is defined in terms of only the coordinated patterns that the processes involved in some computation adhere to*" [12] and that the actual values of the data being manipulated by the processes are almost never involved enabling a coordination component written in a high-level language. Usually, a coordinator process is employed for executing the coordination code. The computations are regarded as black boxes with clearly defined input and output interfaces which are plugged into the coordination code, i.e. they are executed when the program reaches a certain part of the coordination code. In which way (e.g., RPC [24], RMI [24], messaging [5, 6], publish/subscribe [25-28]) events are transmitted to the consumers is up to the middleware technology used in the given context. Examples for control-driven coordination languages include WS-BPEL [29], Manifold [30], CoLaS [31], or ORC [32].

2.2.2 Data-Driven Coordination

In contrast to control-driven coordination models, the main characteristic of the data-driven coordination model is the fact that "*the state of the computation at any moment in time is defined in terms of both the values of the data being received or sent and the actual configuration of the coordinated components*" [12]. This means that a coordinated process is responsible for both examining and manipulating data as well as for coordinating either itself and/or other processes by invoking the coordination mechanism each language provides. A data-driven coordination language typically offers some coordination primitives which are mixed within the computational code implying that processes cannot easily be distinguished as either coordination or computational processes.

Carriero and Gelernter define in [33] that "*a coordination model is the glue that binds separate activities into an ensemble*". They express the need for a clear separation between the specification of the communication entities of a system and the specification of their interactions or dependencies; i.e. a clear separation between the computation model, used to express the computational requirements of an algorithm, and the coordination model, used to express the communication and synchronization requirements. They explain that these two aspects of a system's construction may either be embodied in a single language or, as they prefer, in two separate, specialized languages. Such a coordination language is e.g., the Linda coordination model (see Sect. 0). In this data-driven coordination model, processes exchange information by adding and retrieving data from a so called shared dataspace.

2.3 Linda Coordination Frameworks

The Linda coordination model [11] was developed in the mid-1980's by David Gelernter at Yale University. It describes the usage of a logically shared memory, called tuple space, together with a handful of operations (*out, in, rd, eval*) as a communication mechanism for parallel and distributed processes. In principal, the tuple space is a bag containing tuples with non-deterministic *rd* and *in* operation access. A tuple is built-up of ordered fields containing a value and its type, where unassigned fields are not permitted, e.g. a tuple with the three fields <"index", 24, 75> contains "index" of type string and 24 resp. 75 of type integer.

The defined operations allow placing tuples into the space (*out*) and querying tuples from the space (*rd* and *in*). The difference between *rd* and *in* is that *rd* only returns a copy of the tuple, whereas *in* also removes it from the tuple space. Both operations return a single tuple and will block until a matching tuple is available in the tuple space. There are also non-blocking versions of the *rd* and *in* operation, called *rdp* and *inp*, which return an indication of failure instead of blocking, when no matching tuple is found [34]. The *eval* operation is like the *out* operation, but the tuple space initiates a single or several threads and performs calculations on the tuple to be written. The result of these calculations is a tuple that is written into the space after completed evaluation and that can then be queried by other processes.

The Linda model requires the specification of a tuple as an argument for both query operations and thus supports associative queries, similar to query by example [35]. In such a case, the tuple is called template that allows the usage of a wildcard as a field's value. A wildcard declares only the type of the sought field, but not its value, e.g. the operation *rd("index" ?x, ?y)* returns a tuple, matching the size, the type of the fields and the string "index". A tuple containing wildcards is called an anti-tuple. If a tuple is found, which matches the anti-tuple, the wildcards are replaced by the value of the corresponding fields. The non-deterministic *rd* and *in* operation semantics comes from the fact that in case of several matching tuples a random one is chosen.

Implementations that support the exact tuple matching of the Linda coordination model are: Blossom [36], JavaSpaces [37], LIME [38, 39], MARS [40, 41], and TuCSon [42-44]. Although both MARS and TuCSoN enable the modification of the operations' semantics by adding so called reactions, they cannot influence the way how tuples are queried. JavaSpaces adds subtype matching to the exact tuple matching mechanism to query objects from the space.

The drawback of exact tuple matching is that all collaborating processes must be aware of the tuple's signature they use for information exchange. Hence, there are several tuple space implementations that offer additional queries mechanisms, such as TSpaces [13, 45, 46], XMLSpaces.Net [47, 48] and eLinda [34, 49-51]. TSpaces offers the possibility to query tuples by named fields or by specifying only the field's index and a value or wildcard. Furthermore, TSpaces allows the definition of custom queries by introducing the concept of factories and handlers. Both TSpaces and XMLSpaces.Net support the use of XML-documents in tuple fields and therefore enable the use of several XML query languages such as XQL or XPath. In addition, XMLSpaces.Net uses an XML-document like structuring for its space, which allows the utilization of sophisticated XML queries on the space. eLinda enables the usage of more flexible queries, via its Programmable Matching Engine (PME), such as maximum or range queries. Beside these queries the PME also provides aggregated operations that allow the summary or aggregation of information from a number of tuples, returning the result as a single tuple. The PME allows, like TSpaces with its concept of custom factories and handlers, the simple definition of custom matchers [49].

The Linda coordination model exhibits the problem that access to local tuples is tied to the built-in associative mechanisms of tuple spaces. This implies that any non-directly supported coordination policy, like automatically reading several tuples, has to be charged upon the coordinating processes. This means that processes have to be made aware of the coordination laws increasing the complexity of the application design and so breaking the separation between coordination and business issues. The LuCe framework (stands for Logic Tuple Centres [52-54] and is further development of MARS and TuCSoN) introduces the concept of tuple centres as an extended tuple space, which can work as a *programmable coordination medium*. Beside normal tuples, information about the behavior of the centre is stored in the so called *specification tuples*. The main difference between a tuple space and a tuple centre is that the former supports only Linda coordination while the latter can be programmed to bridge between different representations of information shared by coordinated processes to provide new coordination mechanisms. Such mechanisms are realized by reactions allowing the extension of effects from the execution of communication operations as needed. Reactions map a logical operation onto one or more system operations. Furthermore, the results of an operation can be made visible to the coordinating processes as a single transition.

LuCe extends the Linda coordination model by a dynamic coordination behavior realized by means of reactions. This allows LuCe to satisfy complex coordination requirements, like handling of ordered tuples. Reactions are limited to Linda primitives only. Therefore, they are only capable of handling coordination requirements which do not need the integration of other components for interaction than tuple spaces themselves. Beside the fact that reactions cannot perform blocking operations, to the best knowledge they introduce accidental complexity into the coordination framework due to missing structuring and separation of concern mechanisms. For instance, aggregation and ordering logic has to be implemented in one reaction. Furthermore, in case tuples need to be sorted according to a specific requirement, they have to be extended with additional information representing the current position of the tuple. This implies that every operation performed on the space has to be adapted to the new structure of tuples decreasing the overall performance of the system.

3 Use Case Description

The SAW (Simulation of Assembly Workshop) research project [8] investigates coordination requirements and recovery capabilities of software agents representing functional machines in an assembly workshop. The overall goal is to increase the efficiency of the assembly workshop. This is achieved in two different ways, as described in the following.

The scenario from the production automation domain (Fig. 1) consists of several different software agents each being responsible for the machine it represents. Such an agent may be:

- a **pallet agent** (PA) representing the transportation of a production part and knowing the next machine to be reached by the real pallet,
- a **crossing agent** (CA) routing pallets towards the right direction according to a routing table,
- a **conveyor belt agent** (CBA) transporting pallets, with optionally speed control, from one crossing agent to another,
- a **machine agent** (MA) controlling robots of a docking station for e.g., painting or assembling product parts,
- a **strategy agent** (SA) which, based on the current usage rate of the production system, knows where to delegate pallets, so that by taking some business requirements, like order situation, into consideration, a product is created in an efficient way, or
- a **facility agent** (FA) which specifies the point in time when machines have to be turned off for inspection.

Fig. 1 shows a software simulator for a production system, in the concrete case for an assembly workshop. Such manufacturing systems are very complex and distributed. The usage of a digital simulator instead of a miniature hardware model has a lot of advantages like, low operating costs, the easy reconfiguration and parallel testing.

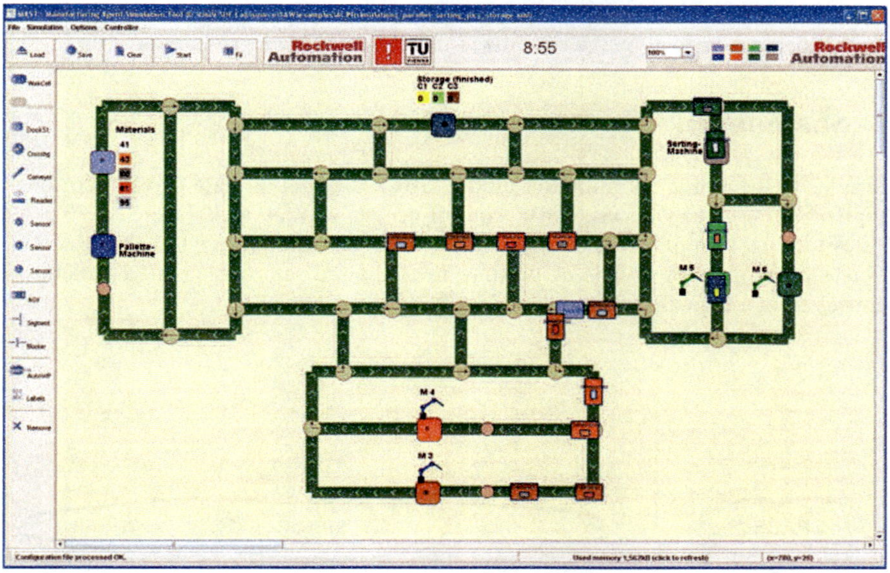

Fig. 1 View of a simulated Production Automation System[1] [55-57]

Multi-agent system (MAS) [58] is an accepted paradigm in safety-critical systems, like the production automation. A major challenge in production automation is the need to become more flexible. The requirement is to react quickly to changing business and market needs by efficiently switching to new production strategies and thus supporting the production of new market relevant products. However, the overall behavior of the many elements in a production automation system with distributed control can get hard to predict as these elements may interact in complex ways (e.g., timing of fault-tolerant transport system and machines) [59]. Therefore, an issue in this context refers to the implementation of agents with reduced complexity of their implementation by e.g., minimizing the communication effort to be managed by the agent.

An approach towards fast reactions may be the prioritization [60, 61] of pallets. Some special parts of the product with higher priority have to be favored by the agents rather than pallets with lower priority. This approach may help to a) produce a small number of products quickly, or b) to phase out products as soon as possible in order to free resources for brand new products to be assembled. Therefore, the aspect of priority has to be considered between all neighboring CAs and all CBAs connecting them. In the described scenario a CA has to check first, whether there is a pallet with high priority on one of the transporting conveyor belts. If this is the case that particular CBA may speed up its transportation speed as well as the CA may force the other conveyor belts to stop. This may happen by e.g. either not handling any pallets coming from them and so forcing those CBAs to stop, or by requesting the other CBAs to halt. So, the high priority pallet is

[1] Thanks to Rockwell Automation for the provision of the simulator.

routed earlier than the other pallets, and it overtakes other pallets which may occupy machines needed by the prioritized pallet based on its production tree.

4 Space-Based Computing

Similar to the Linda coordination model, SBC[2] is mainly a data-driven coordination model, but can be used in a control-driven way as well (see Sect. 0). As shown in Fig. 2, application components running on different physical nodes coordinate each other by means of writing, reading, and removing shared structured entries from a logically central space entity.

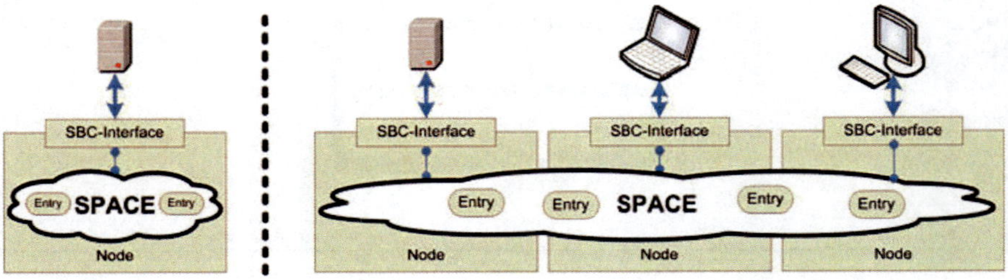

Fig. 2 High-level view of the Space-Based Computing Paradigm

An implementation of the SBC paradigm can be deployed on a physical central server, or on several multiple nodes. In the latter case, internal mechanisms have to make sure, that the shared data structures on the participating nodes are synchronized by taking into account use case specific requirements.

The following paragraphs summarize the XVSM[2] reference architecture based on the latest MozartSpaces[2] implementation in detail. Fig. 3 illustrates a general overview of the XVSM reference architecture divided into an application part (left side) and a space part (right side).

Container-Engine: As in Linda, in SBC application components coordinate each other by means of placing and retrieving data into/from a shared "space". In XVSM data is stored in so called containers that can be interpreted as a bag containing data entries. In XVSM multiple containers may exist at the same time and the number of containers defines the XVSM space. The responsibility of the container-engine layer is the creation and destruction of containers. In its basic form a container is similar to a tuple space - a collection of entries. The main difference to a tuple space is that a container

- extends the original Linda API with a *destroy* method
- introduces so called coordinators enabling a structuring of the space
- may be bounded to a maximum number of entries

[2] SBC has been realized in the eXtensible Virtual Shared Memory (XVSM) reference architecture which has been implemented among others in Java (mozartspaces.org) and .Net (xcoordination.com)

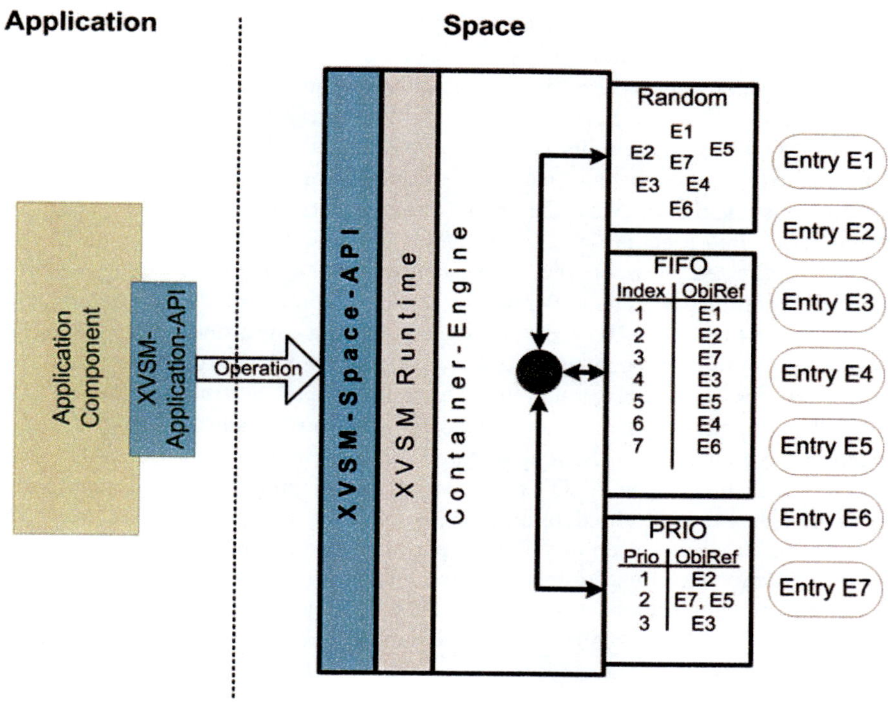

Fig. 3 XVSM architecture with a container hosting a random-, a FIFO-, and a PRIO coordinator structuring 7 entries [55]

Container-API: As in Linda (*out, in, rd*), a container's interface provides a simple API for *reading, taking*, and *writing* entries, but extends the original Linda API with a *destroy* operation. Similar to a take operation, a *destroy* operation removes an entry from the container. Although a *destroy* operation could be mapped onto a *take* operation where the result is omitted, it is still necessary to induct this kind of operation that does not return an operation value. The reason is that this way a lot of data traffic is avoided since the removed data does not need to be transferred back to the initiator of the operation.

The *destroy* operation is also helpful especially in the case of bulk operations [62]. Containers support bulk operations, so that it is possible to insert multiple entries into a container resp. to retrieve/remove multiple entries out of it within one operation.

While Linda makes an explicit distinction between blocking (*rd, in*) and non-blocking (*rdp, inp*) primitives, XVSM primitives are restricted to the mentioned four basic operations. Whether an operation blocks depends on the coordination policy a coordinator represents.

Coordinator: A container possesses one or multiple coordinators. Coordinators implement and are the programmable part of a container. They are responsible for managing certain views on the entries in the container. The aim of a coordinator is to represent a coordination policy. Each coordinator has its own internal data structures which help it perform its task. If the business coordination context and requirements are known beforehand, the coordinator can be implemented in an efficient way with respect to its policy. A coordination policy is represented in the implementation of each coordinator. This implies that the semantics of two coordinators may be the same, but they may be implemented in different ways; each of them taking into account different business specific requirements. A coordinator has an optimized view on the stored entries by taking into account scenario specific coordination requirements. Fig. 3 shows three exemplary coordinators (Random, FIFO, and PRIO Coordinator) referencing seven entries (E1-E7). The Random Coordinator contains all existing entries in the container and returns/removes an arbitrary entry in case of *read/take, destroy* operations. The FIFO Coordinator imitates a queue. It stores in the lowest index the entry that has been in the container for the longest time and in the highest index the entry that has been added last. The PRIO Coordinator groups references only to specific entries according to a priority defined by the software developer.

In general, whenever an operation is performed on a container, the parameters of the operation are collected in a so called selector. Every coordinator has its specific selector which can be interpreted as the coordinator's logical interface for the performed operation. Comparing the relation between selector and coordinator with OOP concepts, the selector is the interface and the coordinator the actual implementation of that interface. In case of *read*, *take*, and *destroy* operations the selector contains parameters (like a counter for the exact number of entries to be retrieved) for querying the view on the managed entries. In case of a *write* operation parameters influencing the coordinator in updating its view are required.

In case a container hosts several coordinators, operations may define multiple selectors as well. The number of specified selectors depends on the business coordination requirements and is not bound to the number of coordinators in the container. If more than one selector is used in querying operations, the outcome of the execution of the first selector will be used as input to the second and so on [63, 64]. The sequence of selectors in read operations is non-commutative *AND* concatenated (i.e. filter style). This means that it makes a crucial difference if 10 entries are selected from a FIFO Coordinator and then a template matching is performed or if the template matching is done first and then the FIFO Coordinator tries to return ten entries.

Before explaining how operations are executed two classes of coordinators have to be introduced. The software developer may declare a coordinator at the time of its creation to be either obligatory or optional. An obligatory coordinator must be called for every write operation on the container, so that a coordinator always has a complete view of all entries. An optional coordinator, however, only manages entries if it is explicitly addressed in the write operation, while other entries in the container remain invisible. The FIFO Coordinator can be used as an obligatory one since it does not need any additional parameters.

In the following the execution of the operations in the container-engine is explained in general. The given explanation does not consider the semantics of XVSM operation transactions or operation timeout. Those aspects are described in [65] in detail.

- **Write:** the *write* operation is executed on all optional coordinators for which parameters have been specified. Afterwards, the *write* operation is executed on all remaining obligatory coordinators even if the operation cannot provide parameters for those coordinators. When a *write* operation has to be blocked depends on the semantics of the coordinator. A semantic may be that an operation has to block if for instance, a Key Coordinator already has a key in its view that the *write* operation of a new entry uses too.
- **Read:** the container-engine iterates over the specified selectors of the operation and queries the corresponding coordinators. In case multiple selectors are specified the result set of the first queried coordinator is the set the next coordinator has to use to execute its query. A *read* operation has to be blocked in case the query cannot be satisfied.
- **Take:** the operation is executed the same way as a *read* operation whereas the result set of the last coordinator defines the set of entries which have to be removed from the container. Therefore, before returning the result set to the initiator of the operation the container-engine asks all coordinators which store a reference on the entries of that result set to remove the entry from their views. Similar to a *read* operation, a *take* operation has to be blocked in case the query cannot be satisfied.
- **Destroy:** the operation is executed like a *take* operation without returning the final result set to the initiator of the operation. Similar, a *destroy* operation has to be blocked in case the query cannot be satisfied.

The **XVSM Runtime** is a layer that is responsible for executing the basic operations by concurrent runtime threads. Operations executed in the container-engine are called requests in the XVSM Runtime layer. Beside the operation itself requests contain context specific meta-information (e.g., timeout, location of the receiver of the request result). Analogously to Linda primitives that block if a tuple does not match a specific template, the XVSM Runtime is also responsible for managing the blocking semantics of operations. The difference to the Linda coordination model is that software developers can alter the semantics of the initiated request. This is achieved by so called aspects (see below) that are treated by the Runtime as well.

Containers are Internet addressable using an URI of the addressing scheme "*xvsm://namespace/ContainerName*", like "*xvsm://host.mydomain.com:1234 /CName*". Every XVSM Runtime hosts several different transportation profiles responsible for accepting requests and sending responses over the physical network. Transportation profiles implement mechanisms for transporting data between nodes. The protocol type "*xvsm*" makes the usage of transportation profiles transparent to the application component. This means that as a transportation medium for accessing that particular container one of the transportation profiles is used without impact on the application component. The application component may specify the properties of transportation (e.g., reliability).

Aspects: The XVSM Runtime layer realizes aspect-oriented Programming (AOP) [66] by registering so called aspects [67] at different points, i.e. before the operation accesses the container-engine or when the operation returns from the container-engine. Aspects are executed on the node where the container is located and are triggered by operations either on a specific container (i.e. container aspect) or on operations related to the entire set of containers (i.e. space aspect). The join points of AOP are called interception points (IPoints). Interception points on container operations are referred to as local IPoints, whereas interception points on space operations are called global IPoints. IPoints are located before or after the execution of an operation, indicating two categories: pre and post. Local pre- and post-IPoints exist for read, take, destroy, write, local aspect appending, and local aspect removing. The following global pre- and post-IPoints exist: transaction creation, transaction commit, transaction rollback, container creation, container destruction, aspect add, and aspect delete. In case multiple aspects are installed on the same container, they are executed in the order they were added. Adding and removing aspects can be performed at any time during runtime.

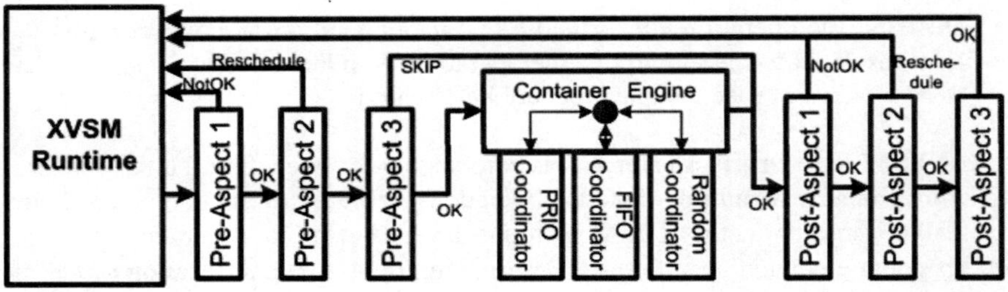

Fig. 4 Data- and control-flow in a container with three installed pre- and post-aspects [67]

Fig. 4 shows a container with three local pre and three post aspects and their various return values. The XVSM Runtime layer accepts incoming requests and passes them immediately to the first pre-aspect of the targeted container. The request passed to and analyzed by the aspect contains the parameters of the operation, like entries, transaction, selectors, operation timeout, and the aspect's context. The called aspect contains functionality that can either verify or log the current operation, or initiate external operations to other containers or third-party services. Aspects can be used to realize security (authorization and authentication) [68], the implementation of highly customizable notification mechanisms (see below), or the manipulation of already stored incoming or outgoing entries.

The central part of a container is the implementation of the container's business logic, i.e. the storage of the entries and the management of coordinators. A request is successful if it passed all pre-aspects, the container-engine, and all post-aspects without any errors. However, an aspect may return several values by which the execution of the request can be manipulated. The following return values are supported:

- OK: The execution of the current aspect has been finished and the execution of the next aspect or of the operation on the container proceeds.
- NotOK: The execution of the request is stopped and the transaction is rolled back. This can be used by e.g. a security aspect denying an operation if the user does not have adequate access rights.
- SKIP: This return value is only supported for pre-aspects and triggers the execution of the first post-aspect. This means that neither any other pre-aspects nor the operation on the container is executed.
- Reschedule: The execution of the request is stopped and will be rescheduled at a later time. This can be used to delay the execution of a request until an external event occurs.

Depending on the result of the last post-aspect the result of the request is either returned to the initiator of the request, or the request is rolled back.

The **XVSM-Application API** extends the XVSM-Space API with a notify method. It is a programming language specific implementation which communicates with the XVSM-Space API. The exchange of requests between the two APIs is performed in an asynchronous way. Fig. 5 shows the general structure of processing a notification in XVSM. In contrast to a specific notification mechanism in e.g., JavaSpaces, the introduced notification approach is flexible, thus can be adapted to business specific needs. In the example there is a container "X" and an application component 3 that wants to be notified whenever container X is accessed. When that application component invokes the notify method, XVSM Runtime registers an aspect (e.g., a so called notification aspect) on container X and creates a so called notification container. The notification aspect intercepts the processing of the operation on container X and writes data into the notification container. When the operation is intercepted (pre or post) information (e.g., a copy of the executed operation or use case specific information about the operation) is written into the notification container, depending on the scenario. A notification container is an "ordinary" container that is therefore capable of hosting additional pre and post-aspects for e.g., aggregation of entries.

Beside the notification aspect and the notification container, XVSM Runtime performs *take* operations on the notification container and specifies a virtual answer container where the result for that *take* operation has to be placed. The virtual answer container is addressed like an ordinary container but is bound to a call back method of the application component specified at the time of creating the notification. Therefore, whenever an entry is written into the virtual answer container the application component receives that entry. This way an application is notified about events on container X.

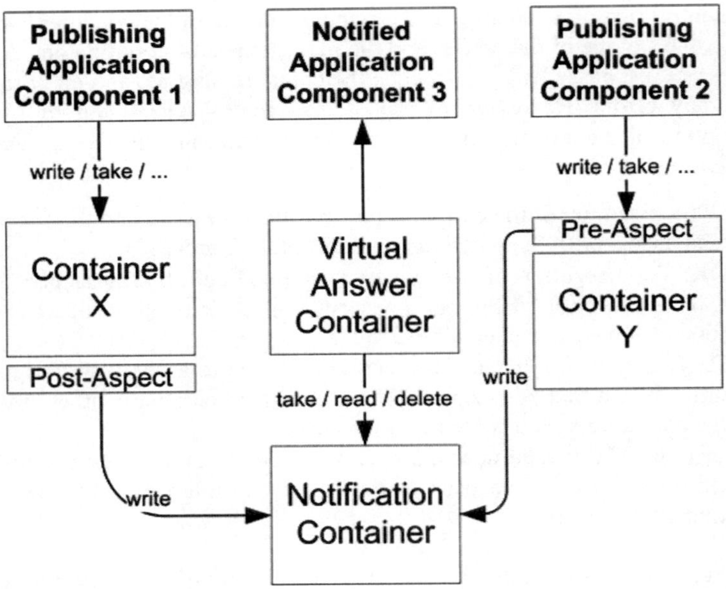

Fig. 5 General structure of an XVSM Notification [69]

As it can be seen, the introduced notification mechanism builds on already described XVSM architectural concepts. This allows software developers to create domain and application specific notification mechanisms which exactly meet given requirements. In Fig. 5 application component 2 wants to be notified in case entries were written. Since that component is not always online, notifications are temporarily and transparently stored in container Y [26].

The described mechanism shows several points where tuning of notification is possible. For instance, the notification aspect can be placed either before or after the execution of the operation on the container. If the aspect is installed as a pre-aspect, the application component cannot be sure whether the operation was really successfully executed on the container or had to be aborted due to errors. Furthermore, if the operation has to be blocked the application component is notified every time that operation comes to execution. The notification aspect can be registered for any XVSM operation (*read, write* ...) and therefore a notification cannot only be created when an entry is written. It is also possible to create notifications which notify a user when entries are *read, taken* or *deleted*. Furthermore, Fig. 5 does not define where the shown containers are placed physically. It is possible that the containers are on the same node or on different ones. The latter one enables the creation of durable subscriptions [26] by placing the notification container on a node which is always reachable. The notification events are collected in the notification container whether or not the client is reachable. When the subscribing application component is online again, the XVSM Runtime fetches the notifications from the notification container which contains new entries written and optionally aggregated during its absence and pushes them via the specified call back method to the application component.

5 Evaluation

A major challenge in production automation is the need to be flexible in order to support a fast and efficient reaction to changing business and market needs. An approach towards fast reactions may be the prioritization of pallets. As mentioned before, some special parts of the product with higher priority have to be favored by the agents to pallets with lower priority. Simplified, the scenario can be summarized as the following: entries have to be ordered by means of the sequence of writing and grouped according to the priority of the entry written. Then, the task is to remove the entry first written from the non-empty group with the highest priority. Additionally, a conveyor belt has only a limited amount of space available depending on the length of the conveyor. In the following the proposed SBC based architecture is compared with architectures based on JMS messaging middleware or the Linda tuple space.

5.1 Java Message Service

For communication between agents in the production automation system, JMS [4] queues are appropriate. With respect to the described statement, Fig. 6 depicts how queues would realize the coordination problem with three different priority categories whereas 1 is the highest priority. In contrast, Fig. 7 shows the realization with an XVSM space container containing a PRIO-FIFO Coordinator. The PRIO-FIFO Coordinator stores messages in a FIFO order grouped according to their priority. Additionally, both figures show the sequence to write an entry and to take the next entry with the highest priority from the FIFO perspective.

In case of queues there are two possible implementations. In the first variant there is one queue for each priority. In the second variant a single queue hosts all messages (i.e. entries) whereas parameters in the message header define its priority for which so called selectors allow querying.

In the first solution, when an agent (Agent A1) wants to place an entry into a queue it looks up its priority. Based on the entry's priority the send operation (operations 1, 2, or 3) of the proper queue is executed. This implies that the application component has to manage three different queue connections. However, before placing the entry into the queue the agent has to retrieve its size. If the number of stored messages is greater than the maximum of permitted ones, then the sender has to look for alternative routing paths. On the receiver side, the agent (Agent A2) has two options of how to receive an entry (operations 4, 5, or 6). Either it polls queues starting with the queue with the highest priority, or it is notified by JMS in case an entry has been written into one of the queues. If it polls, then the agent accesses the queue with the highest priority (Q-Priority 1, operation 4) first. If it is empty then it accesses the queue with the second highest priority (operation 5), and so on. Once a queue has been found that is not empty it removes the entry from the queue and processes it. If the agent is notified then messages are pushed to the subscribed agents. However, in this case the concepts of a queue have to be changed from QueueSession and QueueReceiver to e.g., TopicSubscriber and MessageConsumer triggering an update of the agent's implementation logic. The difference between the two approaches is mainly concerned with the question of who controls an agent. If the agent is notified then it has to process the pushed entry immediately. If the agent polls a queue it can act more autonomously since it can specify when to access a queue and according to which strategy (e.g., configuration of polling rate).

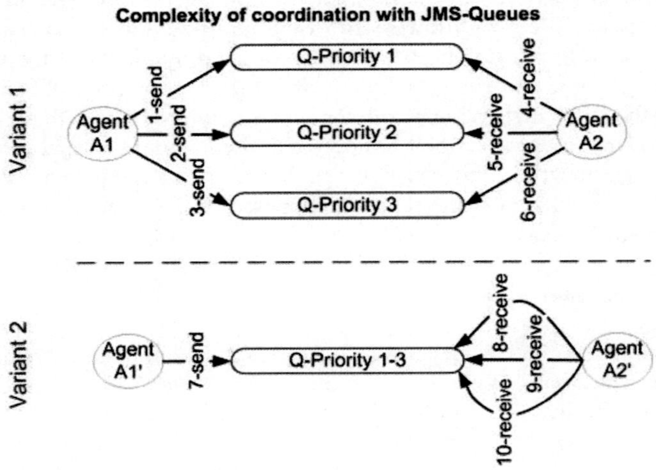

Fig. 6 Prioritized JMS queues

In the second implementation, agents (Agent A1' and A2') access a single queue. The difference to the first implementation is the usage of selectors specifying the priority of entries to be accessed. This means that instead of three different connections to a queue, three different selectors have to be used appropriately.

In the proposed SBC architecture (see Fig. 7), the usage of a "PRIO-FIFO" coordinator allows the software developer to specify the coordination policy transparent to the agents. A write operation needs a priority parameter and the entry. How entries are stored in the coordinator is up to the software developer and of no concern to the application component (coordination category). Since the coordination policy is represented in the coordinator the agent's take operation already reflects its semantics regarding priority restrictions. This means that the take operation does not need any parameters as the coordinator already knows that the entry with the highest possible priority has to be returned.

The migration from a take operation to a notification of written entries does not imply any change of concepts. The application component just executes a notify operation where it specifies the callback method. As described in the previous section, aspects make sure that consuming notifications are pushed to the application component. In contrast to the three queues, aspects can also help sort notifications according to the concurrently written entries' priorities before delivering them to the application component.

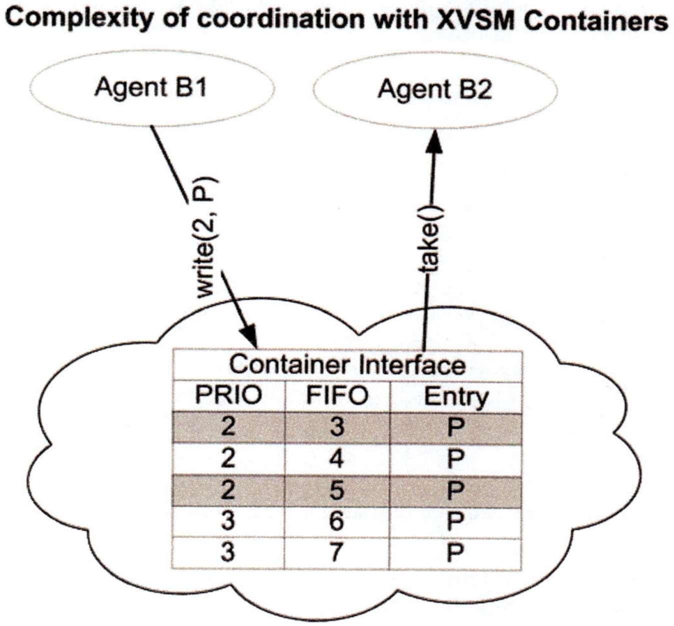

Fig. 7 Container with PRIO-FIFO coordinator (P..payload)

5.2 Linda Tuple Space

Fig. 8 depicts how the Linda tuple space approach would realize the coordination problem. Additionally, the diagrams show the sequence to write an entry and to take the next entry with the highest priority from the FIFO perspective.

For the implementation of a queue in Linda two additional tuples have to be placed into the tuple space. One tuple that represents the first index (i.e. beginning) of the queue (in-token) and one that represents the last index (i.e. end) of the queue (out-token). Therefore, each tuple in the space has to follow a specific structure. Either it is an index tuple containing information about its index type (in-token or out-token), the priority of the queue representing, and the actual value of the index, or it is a message type consisting of its type (i.e. message) and its index in the queue. Whenever a tuple is placed into the queue the last index tuple has to be taken out, the new tuple and an updated index tuple (i.e. index is increased by one) written into the space. Whenever the first tuple needs to be read, the first index tuple has to be found, its index read, and according to this information the tuple retrieved. Whenever the first tuple needs to be taken out, the first index tuple has to be found, its index read, the message based on this index taken out the space, and an updated index tuple (i.e. index is increased by one) written into the space. If no message can be retrieved then it implies that the current queue is empty. Therefore, the process has to be repeated until a message has been found with a lower priority.

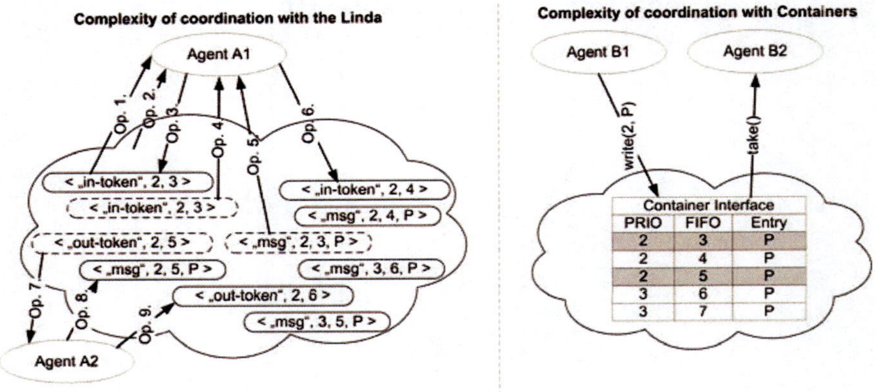

Fig. 8 Prioritized queue realized with the traditional Linda approach

Listing 1 shows how to retrieve an entry based on Fig. 8 as an example setting for stored entries in queues. It can be seen, that while the XVSM approach (see Fig. 7) needs a single API operation to write or to retrieve an entry from the space, the Linda tuple space approach requires three API operations: one to remove the index tuple, one to remove/write the message, and one to write back the index tuple. This is because the realization of a prioritized queue requires the agent taking over a part of the coordination problem.

Listing 1. Retrieving a FIFO sorted entry with Linda

Nr.	Operation
1.	`//retrieve index of first message with highest priority 1` `index = in("in-token", 1, ?int)`
2.	`//retrieve message from index with highest priority 1` `message = inp("msg", 1, index, ?P )`
3.	`// write back retrieved index tuple` `out("in-token", 1, index)`
4.	`//retrieve index of first message with new priority 2` `index = in("in-token", 2, ?int)`
5.	`//retrieve message from index with new priority 2` `message = inp("msg", 2, index, ?P )`
6.	`//write back new index tuple of new priority 2` `out("in-token", 2, index+1)`

Measured times required to retrieve the next entry, with highest priority, from a prioritized queue are shown in Table 1. A benchmark has been set up, which compares the performance of a JavaSpaces (as a Linda tuple space implementation), and a PRIO-FIFO coordinator. The benchmark demonstrates that a PRIO-FIFO coordinator is both able to retrieve entries faster than a coordinator with Linda pattern matching techniques and behaves retrieves entries in a constant access time, as expected from a FIFO queue.

Table 1 Time in ms to retrieve a single entry using different coordinators

Entries	Linda	PRIO-FIFO
10000	5,24	0,20
20000	15,15	0,20
30000	47,93	0,21
40000	58,66	0,20
50000	70,10	0,21

In order to run the benchmark the container was first filled with a specific amount of entries (10000, 20000, 30000, 40000 and 50000 entries). After that a take operation was issued, and the time needed to get the entry measured. The results of the benchmarks clearly show that the PRIO-FIFO coordinator is always the fastest. The results also show that the PRIO-FIFO coordinator offers constant access time, thus perfectly representing the coordination requirements within a single operation call. The benchmarks were run three times on a single node using an Intel Core2Duo T9500 with 4GB RAM to calculate the average access time.

6 Discussion

Besides XVSM the LuCe coordination framework, described in Sect. 0, offers the possibility to enrich the semantics of coordination operations. XVSM achieves this property by means of changeable coordinators. LuCe relies on the usage of so called reactions. Such reactions are hidden from the application and triggered transparently to the application whenever an entry is written or read. Reactions are also changeable and capable of simulating any kind of coordination policies.

However, LuCe just maps a single logical operation onto one or more system operations. This means that one operation in the application is mapped onto several Linda operations in the system. Therefore, the complexity of coordination policies has been just moved from the application (and consequently from the application developer) to the coordination middleware, thus to the software developer of that platform. In contrast, XVSM also moves the complexity of coordination from the application to the coordination middleware, but allows the usage of language specific primitives (i.e. the semantics of a FIFO coordination can be mapped on e.g., a java.list) which shift complexity further away from the software developer to the compiler of that language.

For example if LuCe had to support coordination models with ordering requirements, every tuple in the space had to be additionally wrapped into a tuple managed by reactions. This extra tuple stores meta-information, like the position in the queue, of each written tuple. Consequently, every incoming operation has to be adapted according to the new structure of the tuples, which decreases performance. In the MozartSpaces implementation of XVSM Java specific functions/libraries are used to organize entries in a queue resulting in a single and efficient operation.

Based on the fact that reactions in LuCe are implemented by means of Linda primitives, they cannot access other resources but the tuple space. Aspects in XVSM are written in higher-level languages and allow therefore the integration of other technologies, like web services or databases, into the coordination process. Furthermore, reactions cannot execute blocking operations. The limitation may arise due to the missing separation between reactions responsible for coordination and reactions responsible for e.g., tuple aggregation. Reactions must be non-blocking since strategies for synchronizations of reactions had to be implemented which would significantly decrease the performance of the platform.

Discussing similarities and differences of XVSM and control-driven coordination models like JMS then it can concluded that the FIFO coordination model represents the characteristics and behavior of messaging. However, in JMS the used interface for representing the FIFO coordination model is almost strongly coupled to underlying queuing technologies. This implies that in case of JMS the coordination of processes is not only limited to FIFO capabilities but also to the predefined middleware technology. On the other hand, XVMS's interface specifies only the way of coordination. Its interface is therefore capable of abstracting heterogeneous middleware technologies [70]. It allows injecting aspects e.g., used to coordinate services provision of a group rather than only of a single receiver. Additionally, aspects help manage different integration strategies depending on the used middleware technology. Adding the possibility to intercept communication methods in the XVSM platform minimizes the complexity of implementation. Compared to traditional integration solution XVSM abstracts any kind of middleware technologies. While in traditional solutions specific connectors between each used combination of different middleware technologies need to be implemented, the XVSM requires only the binding to the interface of the middleware adapter only. Although the approach of a common interface is not sophisticated, the benefit of it is a common interface with different transmission semantics. The semantic of the method, e.g. reliable or secure communication, depends on the capability of the middleware that is represented by that interface.

7 Conclusion

Today's software systems can be seen as complex systems in the sense that they usually interact with other software, systems, devices, sensors and people over distributed, heterogeneous, decentralized and interdependent environments while operated more often in dynamic and frequently unpredictable circumstances. Therefore, software developers have to deal with issues like heterogeneity and varying size of components, variety of protocols for interaction with internal and external components. Those software systems typically consist of mainly distributed application components representing higher-level business goals and a middleware technology usually representing an architectural style and abstracting the complexity concerns related to network and distribution.

The message-passing paradigm is a common concept allowing application components to interact with each other. But even asynchronous message-oriented middleware technologies are not suitable for complex coordination requirements since the processing and state of coordination have to be handled explicitly by the application component, thus increasing its complexity. Data-driven frameworks, like tuple spaces, support the coordination of application components, but have a limited number of coordination policies. Therefore, with respect to more complex coordination requirements application components still need to implement coordination functionality that is not directly supported by the coordination framework. Control-driven coordination models suit best in scenarios with point-to-point or 1:N communication requirements. Data-driven coordination models on the other hand are effective when several processes need to be synchronized to reach a common goal. The evaluation of the Simulation of Assembly Workshop (SAW) project shows that Space-Based Computing (SBC) is capable of representing both coordination models. The paradigm allows software developers to build applications being suitable for both coordination models and to switch between the models requiring small changes (regarding operation parameters) in the implementation of coordinating processes.

In the SBC paradigm coordination requirements are reflected in so called coordinators which explicitly distinguish between coordination data and payload. The evaluation of benchmark results shows that this distinction improves the efficiency of coordination significantly. This is due to the fact that a coordinator can be implemented efficiently by taking into account scenario specific context and coordination requirements.

With respect to complexity management the provided SBC concept of coordinators in containers moves the complexity of coordination requirement away from application components to a central point in the SBC coordination framework. The complexity of a coordination issue is concentrated at one point enabling a clear separation between business logic and coordination logic again. Process models comparing the number of processing steps needed to realize a coordination requirement show that by moving the complexity into the coordinator coordination requirements can be reduced to a single operation call on a container. Additionally, since coordination inherently consists of communication, aspects of communication can be abstracted as well by reducing the number of operations to a minimum.

Remaining future work refers to research topics such as the improvement of evaluation strategies for complexity measurement, investigation of scenarios with high-frequently changing conditions both of infrastructure and application requirements and capabilities, and wide-scale benchmarks of the proposed reference architecture with respect to scalability. Additionally, the proposed SBC paradigm will be further investigated in several research projects. In the research project

SecureSpace [68] the main issue is to develop a software platform for the secure communication and collaboration of autonomous participants across enterprise boundaries in the Internet and to prove its usability by means of industrial applications from the security domain. Moreover, in the research project AgiLog [71] the aspect of mobility is investigated in the context of SBC. Industrial scenarios from the logistics domain are used to evaluate the strengths and limitations of SBC with respect to development, configuration, and deployment of distributed applications running on mobile, embedded devices.

Acknowledgments. The work is funded by the Austrian Government under the program FIT-IT (Forschung, Innovation und Technologie für Informationstechnologien), project 825750 Secure Space - A Secure Space for Collaborative Security Services.

References

[1] Solomon, S., Shir, E.: Complexity; a science at 30. Europhysics News 34(2), 54–57 (2003)
[2] Cilliers, P.: Complexity and Postmodernism: Understanding Complex Systems. Routledge, London (1998)
[3] Broy, M.: The 'Grand Challenge' in Informatics: Engineering Software-Intensive Systems. Computer 39(10), 72–80 (2006)
[4] Monson-Haefel, R., Chappell, D.: Java Message Service, p. 220. O'Reilly & Associates, Inc., Sebastopol (2000)
[5] Hohpe, G., Woolf, B.: Enterprise Integration Patterns: Designing, Building, and Deploying Messaging Solutions. Addison-Wesley Longman Publishing Co., Inc., Boston (2003)
[6] Chappell, D.: Enterprise Service Bus. O'Reilly Media, Inc., Sebastopol (2004)
[7] Karhinen, A., Kuusela, J., Tallgren, T.: An architectural style decoupling coordination, computation and data. In: Proceedings of Third IEEE International Conference on Engineering of Complex Computer Systems (1997)
[8] Kühn, E., Mordinyi, R., Lang, M., Selimovic, A.: Towards Zero-Delay Recovery of Agents in Production Automation Systems. In: IEEE/WIC/ACM International Conference on Web Intelligence and Intelligent Agent Technology (IAT 2009), vol. 2, pp. 307–310 (2009)
[9] Auprasert, B., Limpiyakorn, Y.: Structuring Cognitive Information for Software Complexity Measurement. In: Proceedings of the 2009 WRI World Congress on Computer Science and Information Engineering, vol. 07. IEEE Computer Society, Los Alamitos (2009)
[10] McDermid, J.A.: Complexity: Concept, Causes and Control. In: 6th IEEE international Conference on Complex Computer Systems. IEEE Computer Society, Los Alamitos (2000)
[11] Gelernter, D.: Generative communication in Linda. ACM Trans. Program. Lang. Syst. 7(1), 80–112 (1985)
[12] Papadopoulos, G.A., Arbab, F.: Coordination Models and Languages. In: Advances in Computers (1998)

[13] Lehman, T.J., Cozzi, A., Xiong, Y., Gottschalk, J., Vasudevan, V., Landis, S., Davis, P., Khavar, B., Bowman, P.: Hitting the distributed computing sweet spot with TSpaces. Comput. Netw. 35(4), 457–472 (2001)

[14] Mordinyi, R.: Managing Complex and Dynamic Software Systems with Space-Based Computing. Phd Thesis, Vienna University of Technology (2010)

[15] Malone, T.W.: What is coordination theory? MIT Sloan School of Management, Cambridge (1988)

[16] Malone, T.W., Crowston, K.: What is coordination theory and how can it help design cooperative work systems? In: CSCW 1990: Proceedings of the 1990 ACM Conference on Computer-Supported Cooperative Work. ACM, New York (1990)

[17] Malone, T.W., Crowston, K.: The interdisciplinary study of coordination. ACM Comput. Surv. 26(1), 87–119 (1994)

[18] Weigand, H., van der Poll, F., de Moor, A.: Coordination through Communication. In: Proc. of the 8th International Working Conference on the Language-Action Perspective on Communication Modelling (LAP 2003), pp. 1–2 (2003)

[19] Ciancarini, P.: Coordination models and languages as software integrators. ACM Comput. Surv. 28(2), 300–302 (1996)

[20] Ciancarini, P., Jensen, K., Yankelevich, D.: On the operational semantics of a coordination language. In: Object-Based Models and Languages for Concurrent Systems, pp. 77–106 (1995)

[21] Zavattaro, G.: Coordination Models and Languages: Semantics and Expressiveness. Phd Thesis, Department of Computer Science, University of Bologna (2000)

[22] Sancese, S., Ciancarini, P., Messina, A.: Message Passing vs. Tuple Space Coordination in an Aerodynamics Application. In: Malyshkin, V.E. (ed.) PaCT 1999. LNCS, vol. 1662. Springer, Heidelberg (1999)

[23] Franklin, S.: Coordination without Communication. Inst. For Intelligent Systems, Univ. of Memphis (2008)

[24] Tanenbaum, A.S., Steen, M.v.: Distributed Systems: Principles and Paradigms, 2nd edn. Prentice-Hall, Inc., Englewood Cliffs (2006)

[25] Triantafillou, P., Aekaterinidis, I.: Content-based publish-subscribe over structured P2P networks. In: International Conference on Distributed Event-Based Systems (2004)

[26] Eugster, P.T., Felber, P.A., Guerraoui, R., Kermarrec, A.M.: The many faces of publish/subscribe. ACM Comput. Surv. 35(2), 114–131 (2003)

[27] Cugola, G., Di Nitto, E., Fuggetta, A.: The JEDI Event-Based Infrastructure and Its Application to the Development of the OPSS WFMS. IEEE Trans. Softw. Eng. 27(9), 827–850 (2001)

[28] Huang, Y., Garcia-Molina, H.: Publish/subscribe in a mobile environment. Wirel. Netw. 10(6), 643–652 (2004)

[29] Web services business process execution language version 2.0. OASIS Committee Specification 2007 (2007), `http://docs.oasis-open.org/wsbpel/2.0/wsbpel-v2.0.html`

[30] Arbab, F., Herman, I., Spilling, P.: Manifold: Concepts and Implementation. In: Proceedings of the Second Joint International Conference on Vector and Parallel Processing: Parallel Processing. Springer, Heidelberg (1992)

[31] Cruz, J.C., Ducasse, S.: A Group Based Approach for Coordinating Active Objects. In: Ciancarini, P., Wolf, A.L. (eds.) COORDINATION 1999. LNCS, vol. 1594, pp. 355–370. Springer, Heidelberg (1999)

[32] Jayadev, M.: Computation Orchestration - A basis for wide-area computing. In: Engineering Theories of Software Intensive Systems, pp. 285–330 (2005)
[33] Gelernter, D., Carriero, N.: Coordination languages and their significance. Commun. ACM 35(2), 96 (1992)
[34] Wells, G.C.: Coordination Languages: Back to the Future with Linda. In: Proceedings of the Second International Workshop on Coordination and Adaption Techniques for Software Entities (WCAT 2005), pp. 87–98 (2005)
[35] Zloof, M.M.: Query-by-example: the invocation and definition of tables and forms. In: Proceedings of the 1st International Conference on Very Large Data Bases. ACM, Framingham (1975)
[36] van der Goot, R., Schaeffer, J., Wilson, G.V.: Safer Tuple Spaces. In: Garlan, D., Le Métayer, D. (eds.) COORDINATION 1997. LNCS, vol. 1282. Springer, Heidelberg (1997)
[37] Freeman, E., Arnold, K., Hupfer, S.: JavaSpaces Principles, Patterns, and Practice. Addison-Wesley Longman Ltd., Essex (1999)
[38] Murphy, A.L., Picco, G.P., Roman, G.C.: LIME: A coordination model and middleware supporting mobility of hosts and agents. ACM Trans. Softw. Eng. Methodol. 15(3), 279–328 (2006)
[39] Picco, G.P., Murphy, A.L., Roman, G.C.: LIME: Linda meets mobility. In: ICSE 1999: Proceedings of the 21st International Conference on Software Engineering. IEEE Computer Society Press, Los Alamitos (1999)
[40] Cabri, G., Leonardi, L., Zambonelli, F.: MARS: a programmable coordination architecture for mobile agents. IEEE Internet Computing 4(4), 26–35 (2000)
[41] Cabri, G., Leonardi, L., Zambonelli, F.: Mobile Agent Coordination for Distributed Network Management. Journal of Network and Systems Management 9(4), 435–456 (2001)
[42] Cremonini, M., Omicini, A., Zambonelli, F.: Coordination and Access Control in Open Distributed Agent Systems: The TuCSoN Approach (2000)
[43] Omicini, A., Ricci, A.: MAS Organization within a Coordination Infrastructure: Experiments in TuCSoN (2004)
[44] Omicini, A., Zambonelli, F.: Coordination for Internet Application Development. Autonomous Agents and Multi-Agent Systems 2(3), 251–269 (1999)
[45] Wyckoff, P., McLaughry, S.W., Lehman, T.J., Ford, D.A.: T spaces. IBM Systems Journal 37(3), 454–474 (1998)
[46] Lehman, T.J., McLaughry, S.W., Wycko, P.: T-Spaces: The Next Wave. In: Hawaii Intl. Conf. on System Sciences (HICSS-32) (1999)
[47] Tolksdorf, R., Glaubitz, D.: Coordinating Web-Based Systems with Documents in XMLSpaces. In: CoopIS '01: Proceedings of the 9th International Conference on Cooperative Information Systems. Springer, London (2001)
[48] Tolksdorf, R., Liebsch, F., Nguyen, D.M.: XMLSpaces.NET: An Extensible Tuplespace as XML Middleware. Report B 03-08, Free University Berlin (2003), `ftp://ftp.inf.fu-berlin.de/pub/reports/tr-b-0308.pdf`; Open Research Questions in SOA 5-25 and Loose Coupling in Service Oriented Architectures (2004)
[49] Wells, G., Chalmers, A., Clayton, P.: Extending the matching facilities of linda. In: Arbab, F., Talcott, C. (eds.) COORDINATION 2002. LNCS, vol. 2315, p. 380. Springer, Heidelberg (2002)
[50] Wells, G.C.: A Programmable Matching Engine for Application Develoment in Linda. Phd Thesis, University of Bristol (2001)

[51] Wells, G.C.: New and improved: Linda in Java. Sci. Comput. Program. 59(1-2), 82–96 (2006)
[52] Denti, E., Natali, A., Omicini, A.: Programmable Coordination Media. In: Garlan, D., Le Métayer, D. (eds.) COORDINATION 1997. LNCS, vol. 1282. Springer, Heidelberg (1997)
[53] Denti, E., Omicini, A.: An architecture for tuple-based coordination of multi-agent systems. Softw. Pract. Exper. 29(12), 1103–1121 (1999)
[54] Denti, E., Omicini, A., Toschi, V.: Coordination Technology for the Development of Multi-Agent Systems on the Web. In: Proceedings of the 6th AI*IA Congress of the Italian Association for Artificial Intelligence (AI*IA 1999), pp. 29–38 (1999)
[55] Kühn, E., Mordinyi, R., Keszthelyi, L., Schreiber, C.: Introducing the concept of customizable structured spaces for agent coordination in the production automation domain. In: Proceedings of the 8th International Conference on Autonomous Agents and Multiagent Systems (AAMAS 2009), International Foundation for Autonomous Agents and Multiagent Systems, Richland (2009)
[56] Vrba, P.: MAST: manufacturing agent simulation tool (2003)
[57] Vrba, R., Marik, V., Merdan, M.: Physical Deployment of Agent-based Industrial Control Solutions: MAST Story. In: IEEE International Conference on Distributed Human-Machine Systems (2008)
[58] Wooldridge, M.: An Introduction to MultiAgent Systems. John Wiley & Sons, Inc., New York (2009)
[59] Lüder, A., Peschke, J., Sauter, T., Deter, S., Diep, D.: Distributed intelligence for plant automation based on multi-agent systems: the PABADIS approach. Production Planning and Control 15, 201–212 (2004)
[60] Kemppainen, K.: Priority scheduling revisited - dominant rules, open protocols and integrated order management. Phd Thesis, Acta Universitatis oeconomicae Helsingiensis. A. (2005)
[61] Rajendran, C., Holthaus, O.: A comparative study of dispatching rules in dynamic flowshops and jobshops. European Journal of Operational Research 116(1), 156–170 (1999)
[62] Hirmer, S., Kaiser, H., Merzky, A., Hutanu, A., Allen, G.: Generic support for bulk operations in grid applications. In: Proceedings of the 4th International Workshop on Middleware for Grid Computing. ACM, Melbourne (2006)
[63] Kühn, E., Mordinyi, R., Schreiber, C.: An Extensible Space-based Coordination Approach for Modeling Complex Patterns in Large Systems. In: 3rd International Symposium on Leveraging Applications of Formal Methods, Verification and Validation, Special Track on Formal Methods for Analysing and Verifying Very Large Systems (2008)
[64] Craß, S., Kühn, E., Salzert, G.: Algebraic foundation of a data model for an extensible space-based collaboration protocol. In: Proceedings of the 2009 International Database Engineering & Applications Symposium (IDEAS 2009). ACM, New York (2009)
[65] Crass, S.: A Formal Model of the Extensible Virtual Shared Memory (XVSM) and its Implementation in Haskell. Institute of Computer Languages, Vienna University of Technology (2010)
[66] Kiczales, G., Lamping, J., Mendhekar, A., Maeda, C., Lopes, C., Loingtier, J.-M., Irwin, J.: Aspect-oriented programming (1997)

[67] Kühn, E., Mordinyi, R., Keszthelyi, L., Schreiber, C., Bessler, S., Tomic, S.: Aspect-oriented Space Containers for Efficient Publish/Subscribe Scenarios in Intelligent Transportation Systems. In: Proceedings of the 11th International Symposium on Distributed Objects, Middleware, and Applications, DOA 2009 (2009)

[68] Secure Space - A Secure Space for Collaborative Security Services (2010), `http://tinyurl.com/34ymays` (cited)

[69] Kühn, E., Mordinyi, R., Schreiber, C.: Configurable Notifications for Event-based Systems, Vienna University of Technology (2008), TechRep. at `http://tinyurl.com/oht888`

[70] Mordinyi, R., Moser, T., Kuhn, E., Biffl, S., Mikula, A.: Foundations for a Model-Driven Integration of Business Services in a Safety-Critical Application Domain. In: Euromicro Conference on Software Engineering and Advanced Applications, pp. 267–274 (2009)

[71] Agile-Logistics. Komplexitätsreduzierende Middleware - Technologien für Agile Logistik (2010), `http://tinyurl.com/2u8qovc`

Glossary

AOP	Aspect-oriented Programming
API	Application Programming Interface
ESB	Enterprise Service Bus
JMS	Java Message Service
MAS	Multi-agent System
SAW	Simulation of an Assembly Workshop
SBC	Space-based Computing
XVSM	eXtensible Virtual Shared Memory

Chapter 2
Ant Colony Optimization and Data Mining

Ioannis Michelakos, Nikolaos Mallios,
Elpiniki Papageorgiou, and Michael Vassilakopoulos

Abstract. The Ant Colony Optimization (ACO) technique was inspired by the ants' behavior throughout their exploration for food. In nature, ants wander randomly, seeking for food. After succeeding, they return to their nest. During their move, they lay down pheromone that forms an evaporating chemical path. Other ants that locate this trail, follow it and reinforce it, since they also lay down pheromone. As a result, shorter paths to food have more pheromone and are more likely to be followed. ACO algorithms are probabilistic techniques for solving computational problems that are based in finding as good as possible paths through graphs by imitating the ants' search for food. The use of such techniques has been very successful for several problems. Besides, Data Mining (DM), a discipline that consists of techniques for discovering previously unknown, valid patterns and relationships in large data sets, has emerged as an important technology with numerous practical applications, due to wide availability of a vast amount of data. The collaborative use of ACO and DM (the use of ACO algorithms for DM tasks) is a very promising direction. In this chapter, we review ACO, DM, Classification and Clustering (two of the most popular DM tasks) and focus on the use of ACO for Classification and Clustering. Moreover, we briefly present related applications and examples and outline possible future trends of this promising collaborative use of techniques.

1 Introduction

The use of various optimization techniques has evolved over the years and a variety of methods have been proposed in order to approach the optimal solution, or a set of approximate solutions to a range of problems in specific areas.

Ioannis Michelakos · Michael Vassilakopoulos
2-4 Papasiopoulou st., 35100 Lamia, Greece
Dept. of Computer Science & Biomedical Informatics, University of Central Greece
e-mail: {imichelakos,mvasilako}@ucg.gr

Nikolaos Mallios · Elpiniki Papageorgiou
3rd Km Old National Road Lamia-Athens, 35100 Lamia, Greece
Dept. of Informatics and Computer Technology, Technological Educational Institute of Lamia, Greece
e-mail: {nmallios,epapageorgiou}@teilam.gr

N. Bessis, F. Xhafa (Eds.): Next Generation Data Technologies for CCI, SCI 352, pp. 31–60.
springerlink.com

Social insects like ants, perform a series of tasks as a group rather than atomically. Such behavior illustrates a high rate of swarm intelligence and classifies ants as collaborative agents. The Ant Colony Optimization (ACO) technique was introduced in the early 1990's by Marc Dorigo in his PhD Thesis [11] and was mainly inspired by the ants' behavior throughout their exploration for food. The introduction of the ACO technique [11] was followed by a number of research efforts that aimed at exploiting the behavior of ants' throughout their exploration for food in scientific problems. Computational models which apply the swarm behavior in various application areas such as finding the optimal routes (e.g. TSP problem) [13], solving hard combinatorial optimization problems (e.g. MAX-MIN Ant System) [51], biomedical data processing and classification [4], even character recognition [45] and many others have been presented.

Moreover, Data Mining (DM), a discipline that consists of techniques for discovering previously unknown [31], valid patterns and relationships in large data sets, has been acknowledged as a key research field and has emerged as an important technology with numerous practical applications, due to the wide availability of a vast amount of data. Large-scale organizations apply various DM techniques on their data, to extract useful information and patterns [24].

The objective of this survey chapter is to briefly present these two emerging technologies and outline the various ways that these technologies could be combined. The enabling technology which is derived from the collaborative use of ACO and DM (that has been rather recently proposed, for example in [41,42]) leads to improved algorithms and techniques with numerous usages in real problems and can be employed in next generation applications.

This chapter is organized as follows: Primarily, a short review of background material and state-of-the-art research on ACO, DM and their collaborative use is presented, in order to give the reader an overview of the area. Afterwards, in the next two sections, both technologies (DM and ACO) are outlined, focusing on the aspects that led to the collaboration of DM techniques with the ACO technique. Emphasis will be given in the two main ways in which ACO and DM are combined: data classification methods based on ACO [19,57] and ACO for data clustering [6,25] which are thoroughly presented in section 5. Finally, a description of a number of applications and examples where the collaborative use of ACO and DM contributes in various research areas e.g. Health, Marketing, Finance, Molecular Biology is given in section 6. Conclusions and possible future trends of research in this area follow.

2 State-of-the-Art

This section presents, in brief, background material and state-of-the-art research on ACO, DM and their collaborative use. Selected methods are presented in more detail, in the rest of the chapter.

2.1 Ant Colony Optimization State-of-the-Art

The original idea for ACO comes from observing the search of ants for food. Ants individually have limited cognitive abilities, but collectively are able to find the shortest path between a food source and their nest. In nature, ants wander randomly, seeking for food. After succeeding, they return to their nest. During their move, they lay down pheromone that forms an evaporating chemical path. Other ants that locate this trail, follow it and reinforce it, since they also lay down pheromone. As a result, shorter paths to food have more pheromone and are more likely to be followed. Thus, this positive feedback eventually leads all the ants following a single path. ACO algorithms are probabilistic techniques for solving computational problems that are based in finding as good as possible paths through graphs by imitating the ants' search for food [12,36].

ACO algorithms are inspired by the pheromone trail laying and the following behavior of some ant species, a behavior that was shown to allow real ant colonies to find shortest paths between their colony and food sources. Considering many aspects of the real ants behavior, mostly their indirect communication through pheromone trails, ACO has attracted a large number of researchers, e.g. [10,12]. During the first few years of ACO research, the focus was mainly on algorithmic advancements, trying to make ACO algorithms competitive with established metaheuristic techniques. Currently, the majority of the contributions concern, on one hand successful applications of ACO algorithms to a variety of challenging problems, while on the other hand algorithmic developments and theoretical studies for difficult optimization problems.

In the evolving area of bioinformatics a number of interesting contributions have been made. Among those, in [38] a novel ACO algorithm for the problem of predicting protein functions using the Gene Ontology (GO) structure was presented. Moreover, in [35] ACO was applied to the well-known bioinformatics problem of aligning several protein sequences. Furthermore, a number of papers, e.g. [48,49], have appeared concerning the two dimensional hydrophobic-polar (2D HP) protein folding problem. This problem is one of the most prominent problems in computational biology and with the aim of an appropriate ACO algorithm is successfully addressed.

A number of ACO algorithms has also been applied in industry in order to optimize every day's industrial problems. An indicative work is the one by Corry and Kozan [9], that tries to generate solid and better solutions in optimizing the trade-off between material handling and rearrangement costs under certain environments. Another interesting approach includes the use of ACO in scheduling cars along a line, while at the same time, satisfying capacity constraints. This car sequencing problem was described in [50] with the aim of using two different pheromone structures for the algorithm.

Additionally, a number of approaches concerning dynamic (respectively, stochastic) problems have been presented. A few, very recently proposed, ACO algorithms are presented here for the Traveling Salesman Problem (TSP) . In [29] Lopez and Blum dealt with the TSP problem with time window, which arises often in logistics. In their attempt a hybrid Beam-ACO algorithm (ACO and beam search combined) is proposed in order to minimize the travel-cost. Moreover a generalized TSP algorithm (GTSP) was presented in [59], extending the classical TSP problem. The algorithm introduces a mutation process and a local searching technique which turn to be effective.

Borkar and Das [2] introduced an ACO variant that is closer to real ants' behavior than most state-of-the-art ACO algorithms. Their algorithm uses no external supervision and the pheromone update mechanism is based only on differential path length.

Neumann, Sudholt, and Witt [37] presented a rigorous runtime analysis for several variants of ACO algorithms. Their work addresses the question of how long it takes until the algorithm finds an optimal solution for a specific problem.

Furthermore, a review on the current status of Multiple Objective Ant Colony Optimization was addressed in [1]. An extended taxonomy of ACO approaches to multiple objective optimization problems was proposed and many existing approaches were reviewed and described using this taxonomy. This taxonomy offers guidelines for the development and use of Multiple Objective Ant Colony Optimization algorithms.

2.2 Data Mining State-of-the-Art Elements

DM is the process of analyzing data in order to discover useful, possibly unexpected, patterns in data [16]. Two of the most important techniques of DM are classification and clustering. A classification model carries out the task of assigning a class label to an unknown input object after it has been trained with several examples from a given training data set. Clustering on the other hand is the partitioning of a set of input data into subsets (named clusters) so that data in the same subset have something in common. In this subsection, we briefly refer to some of the numerous research efforts in the area of DM. A textbook, like [16], is a more detailed recommended informative source on DM.

DM's contribution to scientific community is indisputable. As DM is becoming more popular, it is gaining wide acceptance in a large number of fields such as healthcare, biomedicine, stock market, fraud detection, telecommunication, text and web mining and others [20,52]. In biomedical research, DM research in DNA analysis has led to the discovery of genetic causes for many diseases and disabilities as well as approaches for disease diagnosis, prevention and treatment [22,46]. Additionally, DM for business continues to expand, as e-commerce and marketing becomes mainstream parts of the retail industry.

An approach proposed by Kargupta et al. describes the Collective Data Mining (CDM) approach, which provides a better approach to vertically partitioned datasets [21].

The design of DM languages, the development of effective and efficient data mining methods and systems, the construction of interactive and integrated data mining environments, and the applications of data mining to solve large-scale application problems, are important challenges for both data mining researchers and data mining system and application developers.

2.3 ACO and DM State-of-the-Art

An interesting research area for ACO is the combination with DM methods for classification and clustering decision making tasks. Modeling classification and clustering as graph search problems allows the use of ACO for finding optimal solutions to these DM tasks. Until today, ACO has been combined with DM methods for classification and clustering in a limited number of studies. A brief description of a number of papers is presented here. The reader is encouraged to read the rest of this chapter for more extensive descriptions and examples of the combination of ACO and DM algorithms.

In [19] Jin et al. proposed a classification rule mining algorithm which was based-on ACO. A number of improvements were implemented to intensify classification accuracy and simplicity of the rules. With these improvements, the overall performance of the algorithm is improved and classification predictive accuracy is enhanced.

Wang and Feng [57] proposed an improved ACO for rule mining classification which is called ACO-Miner. The purpose of ACO-Miner is to give efficient classification rules with accuracy and a simpler rule list based on Ant-Miner. Another interesting approach was proposed in [53] by Thangavel and Jaganathan. In this work, an enhanced ACO algorithm, called TACO-Miner, that has as its main purpose to provide classification rules with a simpler rule list and higher predictive accuracy was presented.

Parpinelli et al. [42] proposed the ACO algorithm for discovering classification rules with the Ant-Miner algorithm. Otero et al. [39] presented an extension to Ant-Miner, named cAnt-Miner (Ant-Miner coping with continuous attributes), which incorporates an entropy-based discretization method in order to cope with continuous attributes during the rule construction process. The same research group [40] introduced the cAnt-Miner2 for mining classification rules. The cAnt-Miner2 is a more flexible representation of continuous attributes' intervals and deposits pheromone on edges instead of vertices of the construction graph. Recently, Michelakos et al. [32] presented a hybrid algorithm for medical data mining, combining the cAnt-Miner2 and the mRMR feature selection algorithms.

In addition to the above research efforts, data clustering techniques have also been combined with ACO methods to discover the optimal solution to a number of problems. The classical clustering methods can be improved when these are combined with the concepts of ACO. More specific, the Ant K-Means algorithm modified the familiar K-means clustering algorithm by the probability of locating the objects in a cluster with the use of pheromone, while the rule of this update is obeying the Total Within Cluster Variance [25].

Tsai et al. [56] proposed a new ACO algorithm with a different favorable strategy, namely ACODF (Ant Colony Optimization with differently favorable strategy) by utilizing the main aspects of the classical Ant System. This algorithm exploits the well-known tournament selection strategy to choose the desired path for the clustering problem.

Chelokar et al. presented an ACO technique for clustering, using a matrix of pheromone values as a kind of adaptive memory, which directs other ants towards the optimal clustering solution [6]. Recently, Tiwari et al. [55] proposed two new techniques which slightly improve the general ACO algorithm for Data Clustering. The first technique avoids stagnation by initializing the pheromone values every 50 iterations, and the second technique, again initializes the pheromone values when there is no change on the path after 10 iterations.

3 Ant Colony Optimization

ACO was inspired by the observation of the behavior of real ants. Ant colonies consist of individuals ants with simple behavior, not capable to solve complex problems. However, at the collective level, these societies are capable of solving complex tasks, such as constructing optimal nest structure, or finding the shortest path to food source. Building of chains of ants [26], or formation of drops of ants [54] have been observed.

As it was briefly outlined in Section 2, when ants walking to a food source (Figure 1, state 1) from their nest following a way x, they deposit on the ground a chemical substance called pheromone. The pheromone deposited on the ground forms a pheromone trail y (Figure 1, state 1) which allows the ants to find food sources that have been previously identified by other ants and by following the path with the greatest amount of pheromone laid upon it. Pheromone trails evaporate if more ants do not come along to reinforce their strength. The ants that find the shortest route to the food will arrive back at the nest quicker than the others and will have laid more pheromone along this shortest path (Figure 1, state 2). Therefore, when new ants seek to travel to the food source, since they are guided by the amount of pheromone on the path, they will take the shortest route. It has been observed that eventually all foraging ants converge on the shortest path to a food source (Figure 1, state 3).

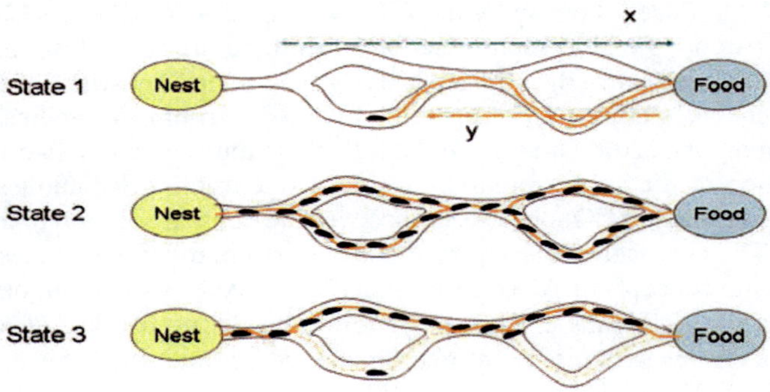

Fig. 1 Food finding procedure followed by ants.

The French entomologist Pierre-Paul Grasse used the term stigmergy [15] to describe this particular type of indirect communication in which "the workers are stimulated by the performance they have achieved." The term is derived from the Greek words stigma (mark, sign) and ergon (work, action), and captures the notion that an agent's actions leave signs in the environment, signs that other agents sense and that determine and incite their subsequent actions. Researchers investigated experimentally this pheromone laying and following behavior to better understand it.

The first ACO algorithm was published by Marco Dorigo under the name of Ant System (AS) [12]. The algorithm was initially applied on the Travelling Salesman Problem (TSP), where a salesperson wants to find the shortest possible trip through a set of cities on his/her tour of duty, visiting each and every city once and only once. The problem can be viewed as a weighted graph containing a set of nodes N representing the cities the salesperson has to visit. The cities are connected by set of edges E and the goal is to find a minimal-length closed tour of the graph.

In AS, m ants ($m \leq n$, where n is the number of the cities) build solutions to the TSP by moving on the problem graph from one city to another until they complete a tour.

For each ant, the transition from city i to city j at iteration t of the algorithm depends on:

1. Whether or not the city has been visited or not. A memory is maintained for each ant to hold the set of cities already visited in the tour which, in turn can be utilized to gather information about the cities that are to be visited when it is in the city i.
2. The inverse of the distance from city i to city j, $n_{ij} = 1/d_{ij}$, (d_{ij} expresses the distance from city i to city j, or in other words the weight of the edge from city i to city j) that is called visibility. Visibility is based on the local information and represents the heuristic desirability of choosing city j when in city i.
3. The amount of pheromone trail $\tau_{ij}(t)$ on the edge connecting city i to city j. The pheromone is updated as the ant moves from one city to another and represents the learned desirability of choosing city j when in city i. This update is performed as follows:

$$\tau_{ij} \leftarrow (1-\rho)\cdot\tau_{ij} + \sum_{k=1}^{m} \Delta\tau_{ij}^{k} \tag{1}$$

where ρ in (0,1] is the evaporation rate of pheromone, and $\Delta\tau_{ij}^{k} = \frac{1}{L_k}$ is the quantity of pheromone laid on edge (i,j) by the k-th ant in the case of the k-th ant used the edge (i,j) in its tour, otherwise this quantity equals to 0 (where L_k is its tour length).

The probability that the k-th ant will choose the city j as its next travel point is defined by a probability function. This function applied for ant k currently at city i during iteration t is of the form:

$$P_{ij}^{k}(t) = \frac{[\tau_{ij}(t)]^{a}\cdot[n_{ij}]^{\beta}}{\sum\limits_{k\in A_k}[\tau_{ik}(t)]^{a}\cdot[n_{ik}]^{\beta}} \tag{2}$$

In this expression the set A_k is the currently valid neighborhood for this ant, i.e. the set of cities not yet visited. This probability function is a combination of two components: the first is the strength of the pheromone trail and the second is a distance decay factor. If α=0 then the pheromone component has no impact and the probabilistic assignment is based on whichever city is closest, whilst if β=0 assignment is simply based on pheromone trail strength, which has been found to lead to stagnation of the solutions giving sub-optimal tours.

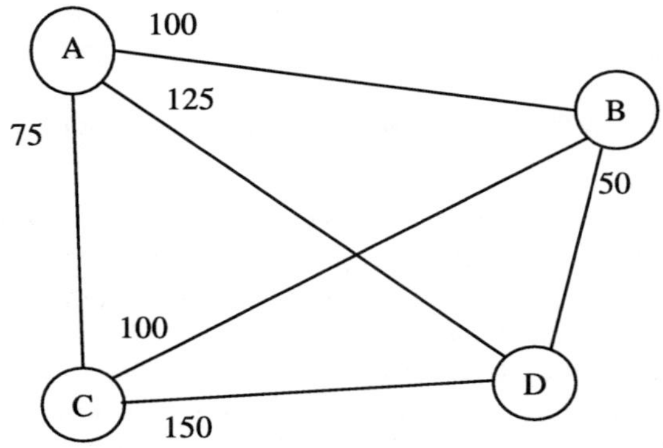

Fig. 2 A four city TSP problem

For example, consider the weighted graph for 4 cities as shown in Figure 2. The distance between cities is denoted along the edge connecting two cities. The first ant starts from city A and has to choose probabilistically one of the three remaining cities to visit, as shown in Figure 2. It does so according to the transition rule given in Equation (2). In the equation, α is set to 1, β is set to 2 and ρ is chosen to be equal to 0.1 [13]. The initial pheromone is set to be equal to 1. The probability the ant will choose the city B is:

$$P^1_{AB}(1) = \frac{(1/100)^2}{(1/100)^2 + (1/125)^2 + (1/75)^2} = 0.293$$

Similarly the probabilities for choosing cities C and D are:

$$P^1_{AD}(1) = 0.187$$

$$P^1_{AC}(1) = 0.520$$

Subsequently the ant chooses to visit city C as its next station. Continuing the iteration, the ant completes the tour by visiting the city B and then the city D. After completing the tour, the ant lays pheromone along the path of the tour. The amount of pheromone added is equal to the inverse of the total length of the tour.

$$\Delta\tau^1_{AC}(1) = \frac{1}{75 + 100 + 50 + 125} = 0.0029$$

The new pheromone levels are calculated using the Equation (1).

$$\tau^1_{AC}(1) = (1-0.1)\cdot 1 + 0.1\cdot(0.0029) = 0.90029$$

and,

$$\tau^1_{AB}(1) = (1-0.1)\cdot 1 + 0.1\cdot 0 = 0.9$$

Pheromone is updated in the same way to the other edges in the path. Finally the pheromone is decreased along all the edges in order to simulate the pheromone decay. You can see all the new values of pheromone level in the Figure 3. The next ants will start from the remaining cities (second ant will start from city B etc.) and will follow the same procedure in order to complete their tour. The pheromone updates will be done as earlier. The algorithm continues to find the shortest path until a certain number of solution constructions, fixed at the beginning of the algorithm is met. This number is also the terminating condition of the algorithm.

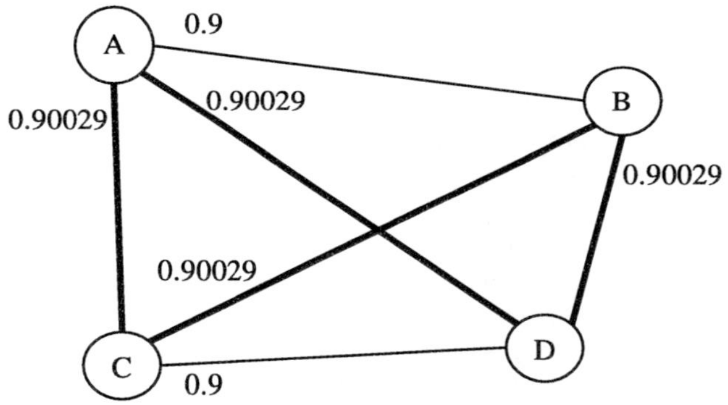

Fig. 3 Pheromone values for the graph in Figure 2 after the first ant finishes a tour.

Summing up, we could say that in order to design a new ant algorithm for a complex combinatorial problem, the problem can be modeled as a search of artificial ants for a best path through a graph. This graph consists of nodes and edges, where nodes represent the basic elements of a solution to the problem and each node is associated with an edge which measures the quality of a partial solution.

An ACO algorithm should have the following basic characteristics:

- an appropriate *problem representation* is required that allows the artificial ants to incrementally build a solution using a probabilistic transition rule. In AS for example the artificial ants build their solution for the TSP by moving on the problem graph from one city to another until they complete a closed tour;
- a *local heuristic* provides guidance to an ant in choosing the next node for the path it is building. This heuristic is problem dependant and for AS it is the inverse of the distance between two cities;

- a probabilistic *transition rule* which determines which node an artificial ant should visit next. The transition rule is dependent on the heuristic value and the pheromone level associated with an edge joining two nodes;
- a *constraint satisfaction* method that forces the construction of feasible rules and in the case of AS [12], an ant must visit each city once and only once during its solution construction;
- a *fitness function* which determines the quality of the solution built by an artificial ant. For the AS algorithm the ant that produces a closed tour of minimal length has the greatest quality, and finally;
- a *pheromone update rule* which specifies how the modification of the pheromone trail laid along the edges of the graph will happen. The pheromone levels are an essential part of the transition rule mentioned above.

Since the first publishing of the Ant System algorithm by Dorigo several versions of the ACO strategy have been proposed, but they all follow the same basic ideas:

- search performed by a population of ants
- incremental construction of solutions
- probabilistic choice of solution components based on stigmergic information
- no direct communication between the ants

Some of the most popular variations of the ACO algorithms other than the Ant System [12] are the Elitist Ant System [58], the Max-Min Ant System (MMAS) [51], the Rank-based Ant System (ASrank) [44] and the Continuous Orthogonal Ant Colony (COAC) system [17].

An ant colony system simulates the behavior of real-world ant colonies since artificial ants have preference for trails with larger amounts of pheromone, shorter paths have a stronger increment in pheromone and ants communicate indirectly with other ants in order to find the shortest path. On the other hand, Parpinelli et al. (2001) [41], showed that there are also some differences between real ants and artificial ants, such as that artificial ants have memory, they are completely blind and time is discrete.

4 Data Mining

The most significant reason which guided DM as a key research and practical area in Information Technology is the wide availability of a vast amount of data. Such data, combined with the availability of a variety of database clusters and other storage facilities, could be utilized to extract valuable pieces of information [16], which in turn could be used in a majority of industrial and scientific areas (e.g. Health, Finance, Marketing etc.).

DM has attracted, throughout the last two decades, a lot of attention and a great number of tools, techniques and algorithms, have been applied in unprocessed data, in order to discover new association rules, predict the outcome of an event, or describe, in convenient ways – e.g. patterns, unsolved problems.

DM is nowadays widely acknowledged as part of the overall Knowledge Discovery process (KDD) [31]. More specifically as stated in [31] the whole KDD process consists of three main phases, the phase of data pre-processing, the phase of data processing (DM) and the phase of data post-processing. DM process, depending on the task performed, may use two data types, namely labeled and unlabeled data. The first type of data contains a class attribute for each data item and mainly appears in training data sets used for classification, whereas in the second type of data no information exists about the attribute class and mainly appears in data sets to be clustered. DM that uses labeled data is characterized as supervised learning, contrary to DM performed upon unlabeled data which is characterized as unsupervised learning. In the remaining part of this section, a brief description of the two main techniques in the DM process, classification and clustering, is given.

4.1 Classification

Classification is a common task of the DM emerging field. With classification, data is arranged into predefined groups with certain characteristics [16]. For example you may use classification to predict whether a given patient is normal, or suffers from breast cancer.

Classification uses labeled data for creating its model. Each data object of the training data set has been allocated to exactly one class, which is described by a specific attribute, the class label attribute. The classification data model that is derived by considering this allocation can be in turn used to classify new data items (without the class label attribute) and more generally, to extract useful pieces of information, valid patterns, predict future trends, etc..

There exist numerous techniques for data classification. For more information the reader is encouraged to study [16,31]. The main most used techniques are briefly outlined here:

- *Decision trees:* A classical tree-structure flowchart, where starting from the root node of the tree, progression is made to the internal nodes, which represent a test on the value of one, or more data attributes. A decision is obtained, when a node representing no test is reached.
- *Association Rules:* A set of rules having type «if Condition then Prediction» where the Condition could be a conjunction of terms and the derived Prediction could be a possible solution that satisfies the Condition.
- *K-Nearest neighbors algorithms:* The training samples are portrayed by dimensional numeric attributes and with the use of the Euclidean distance between two samples, the K samples which are closest to the unknown sample are identified, and the most common class among them is identified.
- *Artificial Neural Networks:* A composite modeling technique based upon the model of a human neuron. The system made consists of simple parallel-functioning interconnected units (artificial neurons) that form a network called a neural network). The operations carried out by these units conclude to the prediction of one, or more events.

4.2 Clustering

Clustering, on the contrary, is an unsupervised learning technique, as it is performed upon unlabelled data and primarily depicts a method where objects of similar characteristics are grouped together to form clusters. Clustering mainly aims in forming the amount of unmanaged data to manageable piles, by discovering homogeneous groups. Clustering has numerous applications. For example, by using past buying records, clustering can be used for determining groups of customers with similar behavior, for marketing purposes. A basic discrimination of clustering techniques is presented below:

- *Hierarchical Clustering*: Basic type of the clustering methods, where a tree of classes is build, called a dendrogram. The fundamental idea for the tree is to start with each object in a cluster of its own and merge the closest pair of clusters, ending up in one cluster, enclosing everything.
- *Non-hierarchical Clustering*: In this type of clustering technique, classes which are not subclasses of each other are built. The fundamental technique representing non-hierarchical clustering is the k-means algorithm. The k-means algorithm uses the concept of a centroid, the median point in a group of points. Briefly, values of k points as the initial centroids are chosen, then an assignment for every object to the nearest to the centroid cluster is made, a recalculation for the centroids of the k clusters is performed and finally the last two steps are repeated until the centroids remain unaffected.

Several algorithms have been proposed in order to perform clustering techniques upon data. The selection of the appropriate algorithm to be used depends mainly on the type of data which is offered, as well as, on the particular purpose or the application that DM is applied to [16]. In [16] several clustering techniques are exhaustively explained and paradigms of the techniques are outlined, therefore the reader is encouraged to study further the Cluster Analysis chapter of [16], in order to acquire supplementary details.

5 Ant Colony Optimization and Data Mining Techniques

5.1 Data Classification and Ant Colony Optimization

The basic elements of the solution to the classification rule induction problem are the attribute terms. ACO algorithms used for classification aim to discover knowledge expressed in the form of IF-THEN classification rules: IF (conditions) THEN (class), where conditions follow the form ($term_1$) AND ($term_2$) AND ... AND ($term_n$). The class to be predicted by the classification rule is represented by the THEN part corresponding to the rule's consequent and the IF part corresponds to the rule's antecedent. An instance that satisfies the IF part will be assigned the class predicted by the rule. Each term in the rule is a triple (attribute, operator, value), such as <smoke=no>. The value is a value that belongs to the domain of the attribute. For example a simple rule for a weather dataset (Table 1) containing four predicting attributes namely outlook with three possible values {sunny, overcast, rainy}, temperature with three possible values {hot, mild, cool}, humidity with two possible values {high, normal} and windy with two possible values {true, false} concerning the problem of playing outside, or not (attribute play with two possible values {play, don't play}) could be IF <humidity = normal> THEN <play>.

An attribute term, $term_{ij}$, is in the form $A_i = V_{ij}$, where A_i is the i-th attribute and V_{ij} is the j-th value of the domain of A (e.g. humidity is the third attribute and normal is its' second possible value in the above example) . Terms of a predicting attribute and class attribute are called predicting terms and class terms, respectively (e.g. <humidity = normal> is a predicting term and <play=yes> is a class term in the above example). The process of construction of a rule is to search for a combination of predicting terms in the rule antecedent that best identifies a class term. Therefore, in the graph of a classification rule induction problem, the nodes represent attribute terms (e.g. <humidity=normal>) and edges model the quality of the attribute terms. An artificial ant then constructs a rule by visiting a set of possible nodes in the graph and forms a path that ends at a class term node (e.g. <play=yes>). A complete path is a constructed rule. The quality of the path is assessed by a global fitness function. The quality of a node is evaluated by a heuristic value and a pheromone level value associated with the node. These values provide a guide to the ant for which node should be visited next.

Table 1 Weather dataset subset of 10 instances [47].

Outlook	Temperature	Humidity	Windy	Play
Sunny	Hot	High	false	Don't Play
Sunny	Hot	High	true	Don't Play
Overcast	Hot	High	false	Play
Rain	Mild	High	false	Play
Rain	Cool	Normal	false	Play
Rain	Cool	Normal	true	Don't Play
Overcast	Cool	Normal	true	Play
Sunny	Mild	High	false	Don't Play
Sunny	Cool	Normal	false	Play
Rain	Mild	Normal	false	Play
Sunny	Mild	Normal	true	Play
Overcast	Mild	High	true	Play
Overcast	Hot	Normal	false	Play
Rain	Mild	High	true	Don't Play

Parpinelli et al. (2002) [42] proposed the ACO algorithm for discovering classification rules with the Ant-Miner algorithm. Starting from a training dataset, Ant-Miner generates a set of ordered rules through iteratively finding a "best" rule that covers a subset of the training data, adds the "best" rule to the induced rule list, and then removes the examples covered by the rule (e.g. the rule <humidity=normal> then <play=yes> covers six examples of the dataset given in Table 1), until a stop criterion is reached.

In Ant-Miner, an artificial ant follows three procedures to induce a rule from a current training dataset. Rule construction, rule pruning and pheromone updating. The artificial ant starts from an empty rule (no attribute terms in rule antecedent), and selects one attribute term, at a time, adding to its current partial rule based on the local problem-dependent heuristic value and the pheromone level associated with the term. Terms with higher heuristic value and pheromone level are preferred, and terms whose attributes are already present in the current rule antecedent are not considered. Two constraint rules must be satisfied when the ant selects a term. The first one is that two terms that belong to the same attribute must not appear in a rule and the second one is that a rule must cover at least a predefined minimum number of examples. In order to satisfy the first restriction, artificial ants must "remember" which terms are contained in the current partial rule. The second restriction helps to avoid over-fitting and improves the generality of a rule and should be satisfied both in rule construction and in the rule pruning process.

The construction stops when adding a term would make the rule coverage (the number of examples the rule covers) smaller than a user-specified threshold, or until all attributes have been used. The local heuristic function applied in Ant-Miner is an entropy measure of individual terms and is defined by:

$$H(C \mid A_i = V_{ij}) = -\sum_{c=1}^{k} (P(c \mid A_i = V_{ij}) \cdot \log_2 P(c \mid A_i = V_{ij})) \qquad (3)$$

where:

- C is the class attribute and k is the number of class values,
- A_i is the i-th attribute and V_{ij} is the j-th attribute value of the i-th attribute,
- $P(c \mid A_i = V_{jj})$ is the probability of observing class c conditional on observing $A_i = V_{jj}$.

For example, the entropy of the term "outlook = rain" in the training data in Table 1 using the Equation (3) is:

$$H(Play \mid outlook = rain) = -\frac{3}{5} \cdot \log_2(\frac{3}{5}) - \frac{2}{5} \cdot \log_2(\frac{2}{5}) = 0.97$$

The higher the entropy value of a term, the more uniformly distributed the classes are and, so, the smaller the probability that the current ant chooses this term to add to its partial rule. However, the ant prefers to choose a term with higher heuristic value. It, therefore, requires a proper normalization of the entropy values, which is handled by a normalized heuristic function:

$$n_{ij} = \frac{\log_2 k - H(C \mid A_i = V_{ij})}{\sum_{i=1}^{a} x_i \cdot \sum_{j=1}^{b_i} (\log_2 k - H(C \mid A_i = V_{ij}))} \tag{4}$$

where:

- a is the total number of attributes,
- x_i is set to 1 if the attribute A_i is not yet selected; otherwise, it is set to 0,
- b_i is the number of domain values of the i-th attribute.

For example, the heuristic value for the term "outlook = rain" in the training data in Table 1 using the Equation (4) is:

$$\eta_{(outlook=rain)} = \frac{1-0.97}{(1-0.97)+(1-0)+(1-0.97)+(1-0.906)+(1-0.811)+(1-0.984)-(1-0.65)+(1-0.811)} = 0.0158$$

The Ant-Miner [41] uses the transition rule given in Equation (5). Given an attribute-value pair, the transition rule gives the probability of adding the attribute value pair to the rule. The probability is calculated for all of the attribute- value pairs, and the one with the highest probability is added to the rule.

$$P_{ij} = \frac{n_{ij} \cdot \tau_{ij}(t)}{\sum_{i=1}^{a} x_i \sum_{j=1}^{b_i} (n_{ij} \cdot \tau_{ij}(t))} \tag{5}$$

where:

P_{ij} is the probability that $term_{ij}$ is selected for addition to the current partial rule antecedent with a range [0,1], η_{ij} is the heuristic value associated with $term_{ij}$,

- $\tau_{ij}(t)$ is the amount of pheromone associated with a $term_{ij}$ at iteration t,
- a is the total number of attributes,
- b_i is the number of domain values of the i-th attribute,
- x_i is set to 1 if the attribute A_i is not yet selected; otherwise, it is set to 0,

Once the artificial ant stops building a rule, the majority class among the examples covered by the rule antecedent is then assigned to the rule consequent.

After constructing the rule, the artificial ant performs the rule pruning procedure. The purpose of rule pruning is to increase the quality and comprehensibility of the rule built by simplifying the rule antecedent. This is done by iteratively removing one term at a time from the rule antecedent while the quality of the rule is improved. The quality of a rule, denoted by Q, is defined by the following formula:

$$Q = \frac{TP}{TP + FN} \cdot \frac{TN}{FP + TN} \tag{6}$$

- *TP* (true positive) is the number of examples covered by the rule that belong to the class predicted by the rule,
- *FP* (false positive) is the number of examples covered by the rule that belong to a class different from the class predicted by the rule,
- *FN* (false negative) is the number of examples that are not covered by the rule, but belong to the class predicted by the rule,
- *TN* (true negative) is the number of examples that are not covered by the rule and that do not belong to the class predicted by the rule.

For example, the quality of a rule, IF <outlook = sunny> AND <humidity = high> THEN <don't play>, of the training data in Table 1 is:

$$Q = \frac{TP}{TP + FN} \cdot \frac{TN}{FP + TN} = \frac{3}{3+0} \cdot \frac{3}{0+3} = 1$$

This fitness function evaluates the accuracy of a rule without considering rule simplicity. The accuracy consists of both accuracy among positive examples (called sensitivity) and accuracy among negative examples (called specificity). The range of Q values is in [0, 1]. In each iteration of rule pruning, every term in turn is temporarily removed from the rule, a new rule consequent is assigned and the quality of the rule is reconsidered. At the end of the iteration, only the term whose removal improves the rule quality most is actually left out. The rule pruning process stops when the removal of any term does not improve the rule quality or the rule has just one term. Once rule pruning is done, the artificial ant increases the pheromone level of a term in the rule antecedent according to the rule quality given by the following formula:

$$\tau_{ij}(t+1) = \tau_{ij}(t) + \tau_{ij}(t) \cdot Q \tag{7}$$

where:

- $\tau_{ij}(t)$ is the pheromone level of the $term_{ij}$ at iteration t,
- Q is the quality of the constructed rule.
- *i,j* belong to the constructed rule

For example, if the ant adds the rule IF <outlook = sunny> AND <humidity = high> THEN <don't play>, from the training data in Table 1, then the pheromone value at these nodes is:

$$\tau_{ij}(2) = \tau_{outlook=sunny}(1) + \tau_{outlook=sunny}(1) \cdot 1 = 2$$

In our example the initial pheromone level, in favor of simplicity, is equal to 1 but the actual initial pheromone level for each term is given by the type:

$$\tau_{ij}(t=0) = \frac{1}{\sum_{i=1}^{a} b_i}$$

where:

- α is the total number of attributes and
- b_i is the number of domain values of the i-th attribute.

The ant then normalizes the pheromone levels of all terms (each pheromone level is divided by the sum of all pheromone levels) which reinforces the pheromone levels of the terms occurring in the rule antecedent and decreases the pheromone levels of other terms that are not selected in the rule.

These procedures (rule construction, rule pruning and pheromone updating) by which an artificial ant induces a rule, are repeated until every artificial ant (number of ants is a user-defined parameter) has generated a rule, or the current rule has been generated by previous (maxRulesConvergence - 1) ants. The maxRulesConvergence is a user-defined parameter for testing the convergence of ants, which simulates the convergence of real ants to the shortest path between a food source and their nest.

The best rule among the rules generated by all ants is added to the induced rule set. The training dataset is appropriately updated by removing all the examples covered by the best rule. Ant-Miner uses this updated training dataset to induce a new rule that will be added to the rule set through the process described above. Different training datasets are different problems, similar to different food sources that real ants tackle, and, so, the pheromone level of terms needs to be re-initiated. In the end, Ant-Miner stops when the number of examples in the training dataset is smaller than a user-defined threshold (MaxUncoveredCases).

A significant difference between Ant-Miner and other ACO algorithms is the size of the population of ants required between two pheromone updates. Ant-Miner works with a population of a single ant. This ant constructs a rule and updates pheromone levels according to the quality of the rule. Other ACO algorithms normally require a group of artificial ants to work together, such that each ant finds a solution and the pheromone is updated according to the best solution among the solutions found.

Ant-Miner employs an ACO approach providing a mechanism for conducting a global search which is more effective than those provided by traditional covering algorithms. Analogous to the application of a genetic algorithm to classification rule induction, Ant-Miner copes better with attribute interaction than greedy rule induction algorithms do.

Ant-Miner, however, has some limitations. One of the limitations is that Ant-Miner supports only nominal (categorical, or discrete) attributes where the only valid relational operator is "=" and in a preprocessing step continuous attributes need to be discretized using other techniques, such as the C4.5- Disc discretization method [28].

Following the main aspects of Ant-Miner, a number of ACO variations were proposed. They involve different pruning and pheromone update procedures, new rule quality measures and heuristic functions, for discovering fuzzy classification rules, rules for multi-label classification problems and handling of continuous attributes. A typical example of an Ant-Miner variation able to cope with continuous attributes is the cAnt-Miner2 algorithm [40].

5.2 Data Clustering and Ant Colony Optimization

The basic model for data clustering techniques based on ideas coming from Ant Colonies was firstly introduced by Deneubourg et al. (1990) [10]. The main idea behind their method comprises the basic activities of an ant colony to gather items in order to form piles e.g. cluster dead bodies and sort them discriminating among different kind of items. The model proposed is a continuous model, where ants are represented as simple agents, which randomly move into a two-dimensional (square) grid, with a number of limitations in order to pile their corpses. Items distributed within such an environment could be picked-up with a probability $P_p = (\frac{a_1}{a_1 + f})^2$ or dropped-down with a probability $P_d = (\frac{f}{a_2 + f})^2$.

In each iteration step an ant explores its neighborhood and computes the above probabilities. Parameters a_1 and a_2 are threshold constants and their values are compared to the value of function f that denotes a high probability of picking up or dropping down an item. For example, if a_1 is much higher than f, then P_p converges to 1, thus making the probability of an ant to pick-up an item quite high. Function f is a function that encapsulates the notion of the average distance of elements [10].

This procedure is influenced by a number of parameters, within the agents' local neighborhood which are set empirically and may produce more clusters than the optimal number. Moreover, in the basic model the absence of pheromone could be critical in a number of cases. For that reason many improvements to this algorithm have been proposed. The main extension of Deneubourg's model was introduced by Lumer and Faieta (1994) [30] who use a continuous similarity function and define the idea of distance, or dissimilarity d between objects in the space of object attributes. This technique has been called *Ant Colony Clustering* and a variety of modifications have been proposed, which modify existing parameters, or introduce the notion of pheromone in the algorithm in order to reduce the large amount of computational time, or improve convergence of the algorithm [3,34].

A novel approach presented by Tsai et al. (2004) [56] is not only based on ideas coming from Ant Colonies, but utilizes the classical Ant System and proposes a new ACO algorithm with a different favorable strategy (Ant Colony Optimization with differently favorable strategy - ACODF). This algorithm initially uses favorable ants in order to provide a solid solution for the clustering problem, and then it uses simulated annealing in order to decrease possible paths and finally exploits the well-known tournament selection strategy to choose the desired path.

The basic steps of the algorithm are summarized in the following. In the initialization phase n data points are chosen and m ants are assigned to m nodes (n represents the number of nodes and m the number of ants). Then, a computation is performed concerning the number of nodes that ants should visit (initially for the first time and later randomly for each ant in arbitrary directions). Afterwards, a random selection of a number of trails is performed and with the aid of a selection mechanism (Tournament Selection in this case) the algorithm finds the pheromone trail with high quantity. In the next step this pheromone quantity of every trail is updated and an iteration of the above steps is executed, until all trails of pheromone quantity reach a stable state. In the last step, clustering is performed using the value of pheromone quantity.

Moreover, the results obtained with ACODF algorithm [56] were compared with two other well-known approaches for data clustering, Fast Self-Organizing Map (FSOM) combining K-means (aka FSOM+K-means) and Genetic K-means algorithm (GKA). The comparison showed that ACODF algorithm performs better in terms of time cost when the data sets used are data sets of 300 and 579 samples and the clustering methods used are both non-spherical and spherical. Additionally, ACODF produces a smaller number of errors (better clustering results) than the two other algorithms.

Other approaches include the improvement of classical clustering methods when these are combined with the concepts of ACO. The major paradigm of such an approach is presented in [25] where the *Ant K-Means* algorithm is introduced, which modifies the familiar K-means clustering algorithm by the probability of locating the objects in a cluster with the use of pheromone, while the rule of this update is according to the Total Within Cluster Variance (TWCV). The main disadvantage of techniques based on Ant K-means algorithm and its variations is that the number of the clusters and the corresponding centroids should be known in advance and are generated with the aim of the Ant System-based Clustering Algorithm (ASCA) which was also developed by the authors.

This algorithm consists of four sub-procedures (*divide, agglomerate_obj, agglomerate* and *remove*) and calculates the TWCV. The main algorithm introduced modifies the well-known K-means algorithm in the way the location of objects in a cluster is calculated and the probability used is modified by the pheromone (updating pheromone according to TWCV). The first step of AK (Ant K-means) algorithm is the initialization phase, where all the parameters including the number of clusters and its centroid are initialized. In the second step, equal amount of pheromone is laid on each path, and then each ant chooses the centroid with probability *P*,

$$P_{ij}^{k} = \frac{\tau_{ij}^{\alpha}\eta_{ij}^{\beta}}{\sum_{c}^{nc}\tau_{ic}^{a}\eta_{ic}^{\beta}},$$

where i is the start point, j is the end point which the ant k chooses eventually to move-in, c is the centroid and nc is the overall number of the centroids. The next step, is the update of pheromone by

$$\tau_{ij} \leftarrow \tau_{ij} + \frac{Q}{TWCV},$$

where Q is a constant as described in [25]. Afterwards, a calculation of the object $O_{center}(T_k)$ which is the center of all objects in T, where k=1,2,3,..*nc* is performed and a recalculation of TWCV is performed, if necessary. Parameter T describes the set which includes all used objects (maximal number is n). If TWCV is changed, probability P is recalculated in the third step. The final step is to run the procedure *Perturbation* in order to leap from the local minimal solution and if the number of iterations is accomplished the algorithm is stopped, otherwise P is recalculated.

The solution proposed in [25] is analytically compared with two other methods (Self Organizing Maps + K-means and Genetic K-means algorithms) via data sets which are generated by the Monte Carlo simulation. Moreover, another comparison is performed upon real case data (suggestions formulated for the price reduction in plasma TV's).

The list of clustering approaches using ACO incorporates one more approach, which is described in [6]. In this approach a matrix of pheromone values is used as a kind of adaptive memory, which directs other ants towards the optimal clustering solution. The algorithm outlines a methodology used to discover an optimal clustering technique in order to assign N objects to one of the K clusters. The pheromone matrix used is of size N *x* K, thus each object is associated with K pheromone concentrations. The matrix is updated during each iteration depending on the solutions produced. Initially, each ant starts with an empty solution string S and in order for a solution to be constructed, the agent utilizes the pheromone trail information to assign each element that resides in S to an appropriate cluster label.

During the first iteration of the algorithm each element of the matrix is initialized to the same values. As the algorithm carries on, the pheromone matrix is updated accordingly, depending upon the solutions produced. At each iteration, the agents or software ants produce trial solutions using pheromone trails in order to obtain the optimal or near-optimal partitioning of the given N objects into K clusters (groups). Subsequent to the generation of the trial solutions, a further improvement of the solutions proposed is achieved, by performing a local search. The pheromone probability is used in order to choose among the various clusters and is given by

$$p_{ij} = \frac{\tau_{ij}}{\sum_{\kappa=1}^{K} \tau_{ik}},$$

where p_{ij} is the normalized pheromone probability for element i that belongs to cluster j, and j=1,2…K.

Recently, Tiwari et al. (2010) [55] proposed two new techniques which slightly improve the general ACO algorithm for Data Clustering. In the generalized model, each agent initially begins with an empty solution string S and a pheromone matrix τ which maintains the ant's position in a specific cluster and is initialized to a small value τ_0. As the algorithm proceeds, the agent uses the pheromone trail information obtained during each iteration, in order to update the pheromone matrix τ and extend the solutions produced which show the probability of an ant that belongs to a specific cluster. Later on, a local search is performed, which re-organizes the pheromone matrix depending on the quality of the solutions produced. In order to optimize the solutions produced, the objective function, which is defined as the sum of squared Euclidian distances between each object and the center of belonging cluster, should be minimized. After a number of iterations is performed, the solution which has the lowest function value is chosen as the optimal solution. The first proposed technique in order to avoid stagnation, initializes the pheromone values every 50 iterations, and the second technique, again initializes the pheromone values when there is no change on the path after 10 iterations. This solution describes the optimal partitioning of objects of a given dataset into several groups [55].

6 Applications and Examples

Bursa and Lhotska (2007) in their work [4] describe the way clustering techniques use ant colonies. They used the ACO_DTree method [5] (a method based on the MAX-MIN Ant System algorithm [51]) together with Particle Swarm Optimization as a local search method. Their study examined two types of biological signals. Electrocardiograms (ECG) and Electroencephalogram (EEG). Electrocardiograms (ECG) an electrical recording of heart activity is one of the most important diagnostics techniques used in patients. Its processing consists of seven stages: signal pre-processing, signal transfer and/or storage, digital signal processing and feature extraction, clustering of the similar data, signal classification and expert validation. From the ECG signal, eight features have been automatically extracted [8] and two classes have been used (normal cardiac action and abnormal cardiac action) for the above mentioned study. Electroencephalogram (EEG) is an electrical recording of brain activity which is used in order to classify stages of sleep. The EEG recordings used contain eight EEG channels, Electrooculogram (EOG), Electromyogram (EMG), Respiratory channel (PNG) and Electrocardiogram (ECG). All these recordings have been classified by a medical expert into four classes (wake, quiet sleep, active sleep, movement artifact).

In the first stage of the ACO_DTree method, a population of random solutions is generated. In each iteration of the algorithm, the population is enhanced with new solutions driven by pheromone levels. The new updated population is evaluated and only a specified number of solutions is preserved. Pheromone updating is made by depositing a certain amount of pheromone balanced to the quality of best individuals and afterwards by pheromone evaporation. The matrix used for the pheromone updating procedure conforms to a full graph where nodes represent feature indexes and edges contain pheromone representing transition from one feature to another. Only the best solutions deposit an amount of pheromone, determined by the quality of the solution, into this matrix. This process is iterated up to maximum level of the tree and finally the trees are optimized using a local search technique (the Particle Swarm Optimization method which is a population approach inspired by the behavior of animals with swarm intelligence).

Data from the MIT-BIH database [14] with more than 80.000 for ECG and about 450.000 for EEG records were used for this study. The hybrid combination of DM algorithms for data partitioning and data classification with ACO allows better convergence leading to increased robustness and clearer structure with better clinical use [4].

Another interesting application presented by Kuo et al. (2007) [24] concerns a framework which integrates both the clustering analysis and association rules mining to discover the useful rules in the database through an ACO system. The first component of the proposed method is the clustering analysis and the second one is the association rules mining. The first stage employs the ant system-based clustering algorithm (ASCA) and ant K-means (AK) [25] to cluster the database, while the ant colony system-based association rules mining algorithm is applied to discover the useful rules for each group. The main reason for clustering the database is that this can dramatically decrease the mining time. In order to assess the proposed method, a database being provided by the National Health Insurance Plan of Taiwan Government is applied.

After encoding, clustering analysis was done with the two-stage clustering algorithm, which includes ASCA and AK. The application of the algorithm generated three clusters. These clusters were chosen for DM with the ACS-based association rule mining algorithm.

The main target of the application was to develop a decision support system about patient treatments that is able to extract important relationships or association rules between diseases in order to provide an alternative way to help diagnose the diseases and to specify treatments for them. Such a system could help the physicians pay more attention on important groups of patients and find out the hidden relation in these groups easier.

The computational results showed that the proposed method not only can extract the useful rules faster, but also can provide more precise rules for the medical doctors and let the researchers pay more attention on some important patient groups and find out the hidden relation in the groups easier.

Kumar and Rao (2009) [23] proposed a use of DM algorithms for the extraction of knowledge from a large set of flow shop schedules. In the first section of their work they describe the ACO algorithm used and the method to generate a population of the optimal sequences. The second section of their work deals with mining the solutions given by the ACO algorithm in order to extract from them decision rules. These rules are based on several attributes like processing time, position in the job, remaining time of the job or machine loading. Finally they used a Decision Tree, (See5 classifier –a commercial version of C4.5 [47]) in order to find their affection order of operation on all machines.

Finally, another interesting application was proposed by Phokharatkul et al. (2005) [45]. They presented a system of handwritten Thai character recognition, which is based on the Ant-miner algorithm. The system uses zoning for each Thai character to determine each character and three attributes of each character in each zone are extracted. These attributes are Head zone, End point, and Feature code and are used by the Ant-miner algorithm in order to classify 112 Thai characters (76 alphabet characters and 36 special symbols).

Thai characters are composed of lines, curves, circles and zigzags. The head is normally a starting point of writing a Thai language character. It is one of the distinctive features of Thai characters and it is defined as a circle or a closed loop in a character [7]. The end point is the point that has only one point connected to it [7] and finally the feature code is defined by the maximum number of points that the referent lines pass in its zone [7]. The data used in this application were collected from 100 persons where each person made 3 copies of a sheet with handwritten characters providing a total data set of 33600 characters.

On the first step of the model, each handwritten character is converted to bitmap by a scanner into a two-color bitmap file. On the next step, an algorithm [7] is used to convert each bitmap into a character image that is only 1 pixel in width. Afterward, each character is normalized to 128x128 pixels and segmented into 12, 9 and 15 zones with the same width and height for feature Head, Endpoint, and Feature code. In this Feature extraction step the features of each character are extracted and saved to a file. In the next step, the Ant-Miner algorithm is used for training the recognition system and finally, the data of 11200 samples are used in order to classify the characters into the next five groups (lower, middle and low, middle, middle and upper and upper characters) [45]. Finally data of each group are classified by the Ant-miner algorithm and the induced rule list is used as the recognition engine. The experimental results shown that the system can recognize 97% of the training set.

7 Conclusions

The audience of this chapter includes researchers, instructors, senior students and graduates of higher education, who are interested in next generation data technologies that handle (possibly distributed) data in a collaborative manner. More specifically, in this chapter, we reviewed a technique which is based on simple agents that collaborate in order to solve a problem. This technique was inspired from the physical behavior of real ants and the way they behave in order to solve problems, like finding food or sorting broods. This technique, named ACO, and its collaborative use with two DM techniques, classification and clustering, which are the most widely used tasks in DM, have been outlined. The chapter has focused in making a review of work on the use of ACO for classification and clustering purposes. The enabling technology which is derived from the collaborative use of ACO and DM leads to improved algorithms and techniques with numerous usages, as presented in Section 6 by providing contemporary real-world examples of various application areas e.g. Health, Marketing, Finance, Molecular Biology.

8 Future Trends

The heuristic function, the pheromone updating strategy and the pruning procedure used in an ACO algorithm are among the basic components of an ACO algorithm. These parts of the algorithm influence its performance and their fine tuning, or correct choice could lead to better accuracy. Several papers in the literature propose this tuning as a worthy target, e.g. [19,40]. We believe that, such a tuning, taking into account the respective real application areas, is also important for collaborative ACO-DM algorithms.

Since recently, ACO algorithms were not able to cope with continuous variables and a pre-processing step of discretization was mandatory. Otero et al. in a recent work [40] introduced a new promising algorithm able to cope with such variables, having the necessary discretization procedure embedded on the main algorithm procedure. The encapsulation of a discretization method in the rule construction process of the ACO algorithm used for classification showed that better results can be achieved. As future research direction, it would be interesting to investigate the performance of different discretization methods in the rule construction process.

Besides their main components, ACO algorithms have a number of system parameters that influence their performance and/or accuracy [18,27]. Detailed experimentation is needed to determine the effects of these parameters, and develop an understanding of methods that set parameters appropriately for particular problems. Michelakos et al. [33] recently studied various system parameter settings of the cAnt-Miner2 algorithm [40]. Further experiments, to study the influence of system parameters on the performance of ACO-DM algorithms for particular problem areas have been planned.

Another main issue that has emerged from collaborative ACO-DM algorithms is their computational cost [41]. This cost is extremely high when the search space (number of predicting attributes) is large. Most of the techniques presented in this chapter are dealing with a rather small amount of data (residing in main memory) and mainly with a single dimension. An interesting research direction could be the adaption of such techniques for applying on large amount of data, which (inevitably) reside on disk, in transactional Databases, Data Warehouses, or specialized disk based data structures and / or have more than one dimension. Apart from accuracy of the result, the I/O and CPU performance of such techniques could be studied.

Moreover the application of collaborative ACO-DM techniques on distributed data, resulting from possibly heterogeneous sources, like data streams, requires appropriate data collection and processing methods that aim at high accuracy and/or performance. This is also considered a challenging issue.

New possibilities might result, regarding the improvement of accuracy and/or performance, by the introduction of hybrid ACO techniques and their application for DM tasks. In a recent study [32], a hybrid algorithm for data classification was presented, combining the cAnt-Miner2 [40] and the mRMR feature selection [43] algorithms. The resulting algorithm was very promising and was experimentally compared to the (non hybrid) cAnt-Miner2 algorithm, using public medical data sets.

Another issue that is worth researching is the appropriate (for ACO use) modeling of other DM tasks, or modeling of different approaches to classification and/or clustering [19], since such a modeling is necessary in order to apply ACO to a problem (see Section 3). Since the accuracy of DM techniques is problem and data dependent, the application of ACO-DM techniques to diverse problem areas (related to current, or future applications) and their thorough comparison with other (state-of -the-art) DM techniques would be interesting. In general, a thorough comparison which will encompass a significant number of DM techniques already proposed, including ACO-DM ones, would be very useful and informative.

Finally, increasing attention could be given to even more challenging problems, involving dynamic data (temporal and/or spatio-temporal data) and their constraints. Dynamic problems are characterized by the fact that the search space changes in the course of time. Hence, the conditions of the search, the definition of the problem instance and, thus, the quality of the solutions already found may change while searching. It is crucial in such situations that the algorithm is able to adjust its search direction and follow the changes of the problem being solved, exhibiting (a kind of) self-adaptation.

9 Key Terms

Ant Colony Optimization (ACO): The ant colony optimization algorithm is a probabilistic technique for solving computational problems aiming at finding an optimal path in a graph, based on the behavior of ants seeking a path between their colony and a source of food.

Agent: an autonomous entity which observes and acts upon an environment and directs its activity towards achieving goals

Ant Colony: An ant colony is an underground lair where ants live. Ants are social insects that form colonies which range in size from a few up to millions of individuals.

Attributes: An attribute is frequently and generally a property of a property and can be considered metadata of an object, element, or file. A specification that defines a property.

Categorical (or Nominal) Attributes / Values: A categorical attribute has values that function as labels rather than as numbers. For example, a categorical attribute for gender might use the value 1 for male and 2 for female.

Continuous Attributes / Values: A continuous attribute has real numeric values such as 1, 2, 6.28, or -7. Examples of continuous attributes are blood pressure, height, weight, age.

Classification: Classification is the assignment of a class label to an input object. The term refers to either of the task, the problem of, and the result of such an assignment.

Classification Rule: IF-THEN classification rules are in the form: IF (conditions) THEN (class), where conditions follow the form ($term_1$) AND ($term_2$) AND ... AND ($term_n$).

Clustering: Clustering or cluster analysis is the assignment of a set of observations into subsets (called clusters) so that observations in the same cluster are similar in some sense.

Data Mining: Data mining is the process of analyzing data in order to discover of useful, possibly unexpected patterns in data.

Graph: Graph is a mathematical structure used to model pair wise relations between objects from a certain collection. A "graph" in this context refers to a collection of vertices or 'nodes' and a collection of edges that connect pairs of vertices. A graph may be undirected, meaning that there is no distinction between the two vertices associated with each edge, or its edges may be directed from one vertex to another.

Learning (Supervised): Supervised learning is a machine learning technique for deducing a function from training data. The task of the supervised learner is to predict the value of the function for any valid input object after having seen a number of training examples. One form of supervised learning is classification.

Learning (Unsupervised): In machine learning, unsupervised learning is a class of problems in which one seeks to determine how the data are organized. One form of unsupervised learning is clustering.

Optimization: Optimization refers to choosing the best possible element from some set of available alternatives.

Pheromone: A pheromone is a chemical substance that triggers a social response in members of the same species. Ants use pheromone in order to communicate indirectly.

Swarm Intelligence: Swarm intelligence describes the collective behavior of decentralized, self-organized systems, natural or artificial. These systems are typically made up of a population of simple agents interacting locally with one another and with their environment leading to the emergence of "intelligent" global behavior, unknown to the individual agents.

Stigmergy: Stigmergy is a mechanism of indirect coordination between agents. It is derived from the greek words stigma (mark, sign) and ergon (work, action), and captures the notion that an agent's actions leave signs in the environment, signs that it and other agents sense and that determine and incite their subsequent actions.

Acknowledgments. The first author acknowledges the financial support of "Heraclitus II" program (Operational Program "Life Long Learning"), funded by the European Union and the Greek State.

References

1. Angus, D., Woodward, C.: Multiple objective ant colony optimization. Swarm Intelligence 3(1), 69–85 (2009)
2. Borkar, V.S., Das, D.: A novel ACO algorithm for optimization via reinforcement and initial bias. Swarm Intelligence 3(1), 3–34 (2009)

3. Boryczka, U.: Finding Groups in Data: Cluster Analysis with Ants. Applied Soft Computing 9(1), 61–70 (2009)
4. Bursa, M., Lhotska, L.: Ant Colony Cooperative Strategy in Electrocardiogram and Electroencephalogram Data Clustering. In: Nature Inspired Cooperative Strategies for Optimization (NICSO 2007), pp. 323–333 (2007)
5. Bursa, M., Lhotska, L., Macas, M.: Hybridized swarm metaheuristics for evolutionary random forest generation. In: 7th International Conference on Hybrid Intelligent Systems (HIS 2007), pp. 150–155 (2007)
6. Chelokar, P.S., Jayaraman, V.K., Kulkarni, B.D.: An Ant Colony Approach for Clustering. Analytica Chimica Acta 509(2), 187–195 (2004)
7. Choruengwiwat, P.: Thai handwritten character recognition using extraction of distinctive features. Master's Thesis, Department of Electrical Engineering, Chulalongkorn University, Thailand (1998)
8. Chudacek, V., Lhotska, L.: Unsupervised creation of heart beats classes from long-term ECG monitoring. In: 18th International Conference of European Association for Signal Processing (EURASIP) Biosignals, pp. 199–201 (2006)
9. Corry, P., Kozan, E.: Ant colony optimisation for machine layout problems. Computational Optimization and Applications 28(3), 287–310 (2004)
10. Deneubourg, J.L., Goss, S., Franks, N., Sendova-Franks, A., Detrain, C., Chrétien, L.: The Dynamics of Collective Sorting: Robot-like Ants and Ant-like Robots. In: From Animals to Animats, 1st International Conference on Simulation of Adaptive Behaviour, pp. 356–363 (1990)
11. Dorigo, M.: Optimization, Learning and Natural Algorithms. PhD thesis, Politecnico di Milano, Italie (1992)
12. Dorigo, M., Maniezzo, V., Colorni, A.: Ant system: optimization by a colony of cooperating agents. IEEE Transactions on Systems, Man, and Cybernetics B 26(1), 29–41 (1996)
13. Dorigo, M., Gambardelaa, L.M.: Ant Colony System:A cooperative Learning Approach to Travelling Salesman Problem. IEEE Trans. Evol. Comp. 1, 53–66 (1997)
14. Goldberger, A.L., Amaral, L.A.N., Glass, L., Hausdorff, J.M., Ivanov, P., Mark, R., Mietus, J., Moody, G., Peng, C., Stanley, H.: PhysioBank, PhysioToolkit, and PhysioNet: Components of a New Research Resource for Complex Physiologic Signals. Circulation 101(23), 215–220 (2000)
15. Grasse, P.: La reconstruction du nid et les coordinations inter-individuelles chez bellicositermes natalensis et cubitermes sp. La theorie de la stigmergie: Essai d'interperation des termites constructeurs. Insectes Sociaux 6, 41–81 (1959)
16. Han, J., Kamber, M.: Data Mining: Concepts and Techniques, 2nd edn. Morgan Kaufmann Publishers, San Francisco (2006)
17. Hu, X., Zhang, J., Li, Y.: Orthogonal methods based ant colony search for solving continuous optimization problems. Journal of Computer Science and Technology 23(1), 2–18 (2008)
18. Jiang, W., Xu, Y., Xu, Y.: A novel data mining method based on ant colony algorithm. In: Li, X., Wang, S., Dong, Z.Y. (eds.) ADMA 2005. LNCS (LNAI), vol. 3584, pp. 284–291. Springer, Heidelberg (2005)
19. Jin, P., Zhu, Y., Hu, K., Li, S.: Classification Rule Mining Based on Ant Colony Optimization Algorithm. In: International Conference on Intelligent Computing (ICIC 2006). LNCIST, vol. 344, pp. 654–663. Springer, Heidelberg (2006)
20. Kantardzic, M., Zurada, J. (eds.): Next Generation of Data-Mining Applications. Wiley-IEEE Press, Chichester (2005)

21. Kargupta, H., et al.: Collective Data Mining. In: Karhgupta, Chan (eds.) Advances in Distributed Data Mining. MIT Press, Cambridge (2000)
22. Masaomi, K.: Application of Data Mining Techniques to the Data Analyses to Ensure Safety of Medicine Usage. In: Ponce, J., Karahoca, A. (eds.) Data Mining and Knowledge Discovery in Real Life Applications. I-Tech Education and Publishing (2009)
23. Kumar, S., Rao, C.: Application of ant colony, genetic algorithm and data mining-based techniques for scheduling. Robotics and Computer-Integrated Manufacturing 25, 901–908 (2009)
24. Kuo, R.J., Lin, S.Y., Shih, C.W.: Mining association rules through integration of clustering analysis and ant colony system for health insurance database in Taiwan. Expert Systems with Applications 33, 794–808 (2007)
25. Kuo, R.J., Wang, H.S., Hu, T.L., Chou, S.H.: Application of ant K-means on clustering analysis. Computers & Mathematics with Applications 50, 1709–1724 (2005)
26. Lioni, A., Sauwens, C., Theraulaz, G., Deneubourg, J.L.: Chain formation in Oecophylla longinoda. Journal of Insect Behavior 14, 679–696 (2001)
27. Liu, B., Abbass, H.A., McKay, B.: Classification rule discovery with ant colony optimization. IEEE Computational Intelligence Bulletin 3(1), 31–35 (2004)
28. Liu, H., Hussain, F., Tan, C.L., Dash, M.: Discretization: An enabling technique. Data Mining and Knowledge Discovery 6, 393–423 (2002)
29. Lopez-Ibanez, M., Blum, C.: Beam-ACO for the traveling salesman problem with time windows. Computers & Operations Research 37(9), 1570–1583 (2010)
30. Lumer, E.D., Faieta, B.: Diversity and Adaptation in Populations of Clustering Ants, From Animals to Animats. In: 3rd International Conference on the Simulation of Adaptive Behaviour, pp. 501–508 (1994)
31. Mhamdi, F., Elloumi, M.: A new survey on knowledge discovery and data mining. In: 2nd IEEE Int. Conf. on Research Challenges in Information Science, pp. 427–432 (2008)
32. Michelakos, I., Papageorgiou, E., Vasilakopoulos, M.: A Hybrid Classification Algorithm evaluated on Medical Data. In: 1st International Workshop on Cooperative Knowledge Discovery & Data Mining / 19th IEEE International Workshops on Enabling Technologies: Infrastructures for Collaborative Enterprises (CKDD / WETICE), pp. 98–103 (2010)
33. Michelakos, I., Papageorgiou, E., Vasilakopoulos, M.: A Study of cAnt-Miner2 Parameters Using Medical Data Sets. In: 1st International Workshop on Cooperative Knowledge Discovery & Data Mining / 19th IEEE International Workshops on Enabling Technologies: Infrastructures for Collaborative Enterprises (CKDD / WETICE), pp. 119–121 (2010)
34. Monmarche, N., Slimane, M., Venturini, G.: On improving clustering in numerical database with artificial ants. In: Floreano, D., Mondada, F. (eds.) ECAL 1999. LNCS (LNAI), vol. 1674, pp. 626–635. Springer, Heidelberg (1999)
35. Moss, J.D., Johnson, C.G.: An ant colony algorithm for multiple sequence alignment in bioinformatics. In: Pearson, D.W., Steele, N.C., Albrecht, R.F. (eds.) Artificial Neural Networks and Genetic Algorithms, pp. 182–186. Springer, Heidelberg (2003)
36. Mullen, R.J., Monekosso, D., Barman, S., Remagnino, P.: A review of ant algorithms. Expert Systems with Applications 36, 9608–9617 (2009)
37. Neumann, F., Sudholt, D., Witt, C.: Comparing variants of MMAS ACO algorithms on pseudo-boolean functions. In: Stützle, T., Birattari, M., Hoos, H.H. (eds.) SLS 2007. LNCS, vol. 4638, pp. 61–75. Springer, Heidelberg (2007)

38. Otero, F.E.B., Freitas, A.A., Johnson, C.G.: A Hierarchical Classification Ant Colony Algorithm for Predicting Gene Ontology Terms. In: Pizzuti, C., Ritchie, M.D., Giacobini, M. (eds.) EvoBIO 2009. LNCS, vol. 5483, pp. 339–357. Springer, Heidelberg (2009)
39. Otero, F.E.B., Freitas, A.A., Johnson, C.G.: cAnt-Miner: an ant colony classification algorithm to cope with continuous attributes. In: Dorigo, M., Birattari, M., Blum, C., Clerc, M., Stützle, T., Winfield, A.F.T. (eds.) ANTS 2008. LNCS, vol. 5217, pp. 48–59. Springer, Heidelberg (2008)
40. Otero, F.E.B., Freitas, A.A., Johnson, C.G.: Handling continuous attributes in ant colony classification algorithms. In: IEEE Symposium on Computational Intelligence in Data Mining (CIDM), pp. 225–231 (2009)
41. Parpinelli, R.S., Lopes, H.S., Freitas, A.A.: An ant colony based system for data mining: applications to medical data. In: Genetic and Evolutionary Computation Conference (GECC0 2001), pp. 791–797 (2001)
42. Parpinelli, R.S., Lopes, H.S., Freitas, A.A.: Data mining with an ant colony optimization algorithm. IEEE Transactions on Evolutionary Computation 6, 321–332 (2002)
43. Peng, H., Long, F., Ding, C.: Feature selection based on mutual information: criteria of max-dependency, max-relevance, and min-redundancy. IEEE Transactions on Pattern Analysis and Machine Intelligence 27, 1226–1238 (2005)
44. Pérez-Delgado, M.: Rank-Based Ant System to Solve the Undirected Rural Postman Problem. In: Distributed Computing, Artificial Intelligence, Bioinformatics, Soft Computing, and Ambient Assisted Living, pp. 507–514 (2009)
45. Phokharatkul, P., Sankhuangaw, K., Somkuarnpanit, S., Phaiboon, S., Kimpan, C.: Off-Line Hand Written Thai Character Recognition using Ant-Miner Algorithm. Transactions on ENFORMATIKA on Systems Sciences and Engineering 8, 276–281 (2005)
46. Prather, J.C., Lobach, D.F., Goodwin, L.K., Hales, J.W., Hage, M.L., Hammond, W.E.: Medical Data Mining: Knowledge Discovery in a Clinical Data Warehouse. In: Annual Conference of the American Medical Informatics Association, pp. 101–105 (1997)
47. Quinlan, J.R.: C4.5: Programs for Machine Learning. Morgan Kaufmann, San Francisco (1993)
48. Shmygelska, A., Aguirre-Hernández, R., Hoos, H.H.: An ant colony optimization algorithm for the 2D HP protein folding problem. In: Dorigo, M., Di Caro, G.A., Sampels, M. (eds.) ANTS 2002. LNCS, vol. 2463, pp. 40–52. Springer, Heidelberg (2002)
49. Shmygelska, A., Hoos, H.H.: An ant colony optimisation algorithm for the 2D and 3D hydrophobic polar protein folding problem. BioMed Central Bioinformatics 6(30) (2005)
50. Solnon, C.: Combining two pheromone structures for solving the car sequencing problem with Ant Colony Optimization. European Journal of Operational Research 191(3), 1043–1055 (2008)
51. Stützle, T., Hoos, H.H.: MAX MIN Ant System. Future Generation Computer Systems 16, 889–914 (2000)
52. Taniar, D. (ed.): Research and Trends in Data Mining Technologies and Applications. Idea Group Publishing, USA (2007)
53. Thangavel, K., Jaganathan, P.: Rule Mining Algorithm with a New Ant Colony Optimization Algorithm. In: International Conference on Computational Intelligence and Multimedia Applications, pp. 135–140 (2007)

54. Theraulaz, G., Bonabeau, E., Sauwens, C., Deneubourg, J.L., Lioni, A., Libert, F., Passera, L., Solé, R.: Model of droplet dynamics in the Argentine ant Linepithema humile (Mayr). Bulletin of Mathematical Biology 63, 1079–1093 (2001)
55. Tiwari, R., Husain, M., Gupta, S., Srivastava, A.: Improving ant colony optimization algorithm for data clustering. In: International Conference and Workshop on Emerging Trends in Technology, pp. 529–534 (2010)
56. Tsai, C.F., Tsai, C.W., Wu, H.C., Yang, T.: ACODF: a novel data clustering approach for data mining in large databases. Journal of Systems and Software 73(1), 133–145 (2004)
57. Wang, Z., Feng, B.: Classification rule mining with an improved ant colony algorithm. In: Webb, G.I., Yu, X. (eds.) AI 2004. LNCS (LNAI), vol. 3339, pp. 357–367. Springer, Heidelberg (2004)
58. White, T., Kaegi, S., Oda, T.: Revisiting elitism in ant colony optimization. In: Cantú-Paz, E., Foster, J.A., Deb, K., Davis, L., Roy, R., O'Reilly, U.-M., Beyer, H.-G., Kendall, G., Wilson, S.W., Harman, M., Wegener, J., Dasgupta, D., Potter, M.A., Schultz, A., Dowsland, K.A., Jonoska, N., Miller, J., Standish, R.K. (eds.) GECCO 2003. LNCS, vol. 2723, pp. 122–133. Springer, Heidelberg (2003)
59. Yang, J., Shi, X., Marchese, M., Liang, Y.: An ant colony optimization method for generalized TSP problem. Progress in Natural Science 18(11), 1417–1422 (2008)

Chapter 3
OpenSEA: A Framework for Semantic Interoperation between Enterprises

Shaun Bridges, Jeffrey Schiffel, and Simon Polovina

Abstract. The modus-operandi for information systems is shifting. Agility and adaptability will be the kingmakers in the decentralising enterprise architecture where on-premise and cloud systems have to be combined seamlessly. At the same time the wealth of data available to organisations needs to be understood and interpreted so as to provide information and inferences needed to generate the knowledge that drives competitive advantage. This chapter offers a high-level introduction to OpenSEA, a framework that combines the open semantics of TOGAF with the open syntax of ISO 24707:2007 Common Logic to provide an Open Semantic Enterprise Architecture. Because of its open nature it is free to adopt and extend, yet retains a root commonality to ensure all participating agents can agree on a common understanding without ambiguity, regardless of the underlying ontology or logic system used.

1 Introduction

A new frontier in enterprise architecture is being explored. This new frontier is the realisation of a distributed enterprise architecture. Like all frontiers it brings the possibilities of explosive growth and exploitation but also brings the undeniable fact that for every winner there will be a number of losers. Frontier dynamics were

Shaun Bridges
Open-SEA.org
e-mail: Shaun.Bridges@Open-SEA.org

Jeffrey Schiffel
The Boeing Company – Wichita Division
e-mail: jeffrey.a.schiffel@boeing.com

Simon Polovina
Sheffield Hallam University
e-mail: S.Polovina@shu.ac.uk

N. Bessis, F. Xhafa (Eds.): Next Generation Data Technologies for CCI, SCI 352, pp. 61–86.
springerlink.com

described by Pascale (2000), who likens organisations to organisms that need to adapt to survive or thrive. Pascale stated that evolutionary pressures are high where the surrounding environment shifts or opportunities to grow are offered in new environments. "A fish takes for granted the water in which it swims; when it learns about the land it is usually too late" (Pascale 2000, p.25).

So what can enterprises do to take advantage of this decentralised model? Organisations need to be able to respond rapidly to new offerings and to find and consume data, processes and services from other organisations. Such an undertaking requires protocols and common understanding. A number of new ontologies and definitions are being created to bring common semantics to the cloud architecture including a unified ontology of cloud computing (Youseff et. al., 2008) and a whitepaper on the taxonomy of cloud computing from the Cloud Computing Use-Case Discussion Group (2010).

It could be argued that this new architecture builds on Service Oriented Architecture since the web services that provided the foundation for SOA are now complemented by the XaaS services providing Infrastructure as a Service (e.g., Amazon Web Services), Platform as a Service (e.g., Force.com) and Software as a Service (NetSuite, SalesForce, Business By Design). The International Research Forum of 2008 explored the issues facing enterprises as they look to exploit this newly evolved Service Oriented Architecture and identified that service discovery is inefficient and that web services are too granular to be of value and need to be extended to provide functionality. They also described an 'integration debt,' where services are created in their own domains and data models. "The stack must rest on a firm architectural foundation and share a common language" (p. 57).

At the same time enterprises are reliant on data from within their own boundaries and data from the larger market place. Consuming and interpreting this wealth of data, i.e., drawing information from it, is central to an enterprises operation. Dashboards offer CEOs snapshots of every conceivable metric. The time relevancy of the data has created In Memory Databases, such as that used by SAP's High Performance Analytics Appliance (HANA) to provide up to the second accuracy in data. But without a reliable understanding of what the data means it is of little value. Beyond information and data, knowledge is the driving force that creates competitive advantage. Taking data, extracting its meaning, and then using rules and inferences to derive new knowledge will allow an enterprise to predict where it needs to be rather than responding to the current situation.

Boisot and Canals (2007, p. 39) described knowledge agents as being able to use models and information to act on the prevailing environment. If an enterprise can be confident of the models that drive its knowledge agents, and of the value of the information fed to the agent, then it can be reasonably confident that it is acting in the best way to exploit its environment and explore new opportunities.

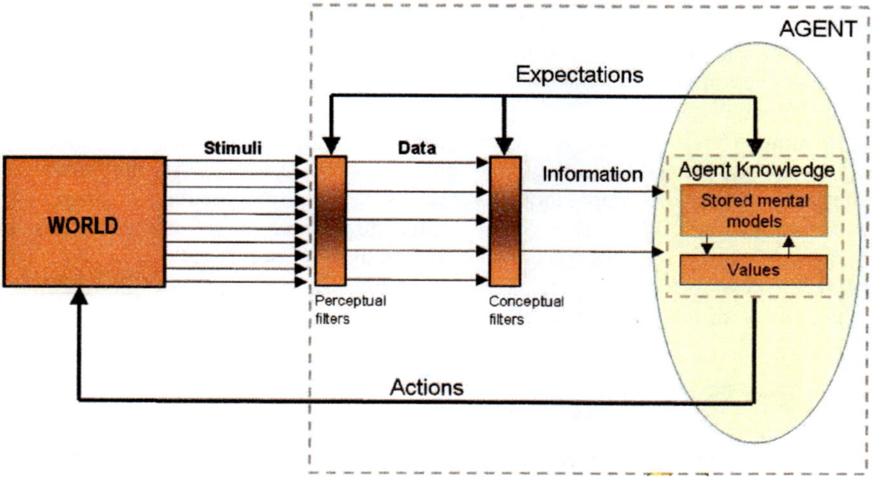

Fig. 1 Boisot's and Canal's Knowledge Agent; Actioning Filtered Stimuli to Effect the Environment.

Logic systems must be able to accurately filter out data from noise and combine these units of datum to draw the information. To achieve this the data may need to be annotated in such a way that the system can understand what the data 'means,' or information needs to be traded between different knowledge agents in a format that be used by both agents. Unfortunately, the wealth of different logic systems, and the lack of integration between them, means that different systems have to operate in isolation. As Sowa (2009) put it, "The proliferation of incompatible semantic systems is a scandal that is strangling the growth of the entire field" (p.119).

The International Research Forum also identified the need for semantic systems to be part of the future of Service Oriented Architecture including actions such as moving service discovery away from key words and domain specific ontologies, semantically enriching services (going beyond web services to include the XaaS offerings) and describing services in a holistic manner. This move towards adding meaning to data and combining different applications is a significant step towards Web 3.0. Google CEO Schmidt described Web 3.0 as "applications that are pieced together" (MacManus, 2009). Cap Gemini CTO Mulholland referred to the Web 3.0 Conference on his blog of July 2009 (Mulholland, 2009), noting the apparent shift in emphasis from machines to users as consumers of Web 3.0:

The goal of Web 3.0 is to reorganize information so users can capture what things are and how they are related. This seemingly simple concept will have a profound effect at every level of information consumption, from the individual end user to the enterprise. Web 3.0 technologies make the organization of information radically more fluid and allow for new types of analysis based on things like text semantics, machine learning, and what we call serendipity — the stumbling upon insights based on just having better organized and connected information.

In summary, if data is to be annotated in a meaningful way it should be expressed in a format that is equally consumable by machines and humans alike. The links and relationships should allow discrete snippets of data or sources of information (including those from unstructured data such as web sites or documents), to become part of a web of information, processes and services where inferences can be drawn, new knowledge gleaned and services provided and consumed at all levels of the XaaS stack.

1.1 OpenSEA

We have noted that SOAs need to evolve to include the XaaS landscape. Integration and interoperation are key to this new SOA and ideally it should be capable of including semantic systems within this integration. OpenSEA proposes that any approach needs to be based on open standards in order that enterprises are not divided by proprietary lock-in. The approach needs to be abstract and capable of extension and specialisation to allow different domains to be able to adapt it to their needs. And it needs to combine a common language of business with a common syntax for logic. In the next sections the case will be made for using The Open Group Architecture Framework as the abstract business language, and using ISO24707:2007 Common Logic as the abstract syntax.

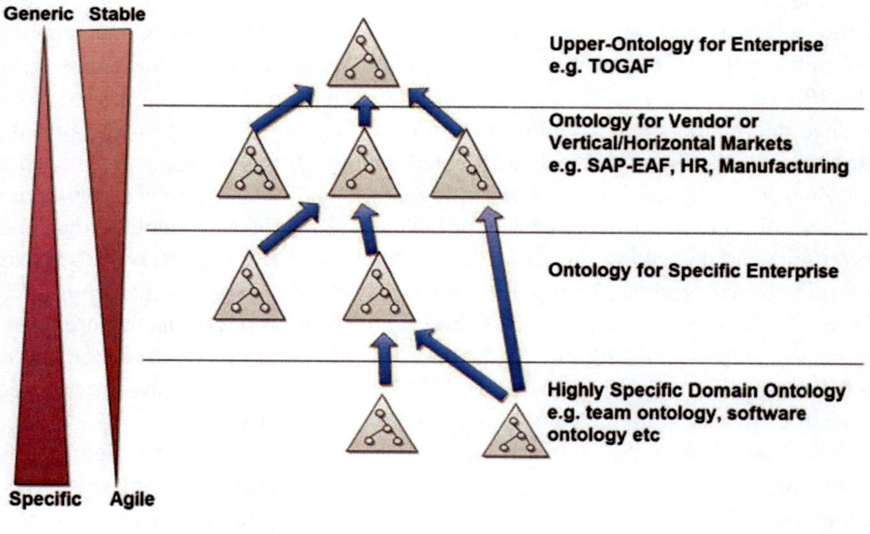

Fig. 2 Specialisation of The Upper Ontology in OpenSEA (from Bridges, 2010).

Figure 2 shows how the upper ontology created by TOGAF is generic yet stable and can be extended and specialised by vendors, markets or enterprises which can in turn be specialised. The different ontologies show how different domains only take the terms that are of use to them and may take them from one or more domains. By retaining a chain of generalisation/specialisation a common generalisation can be found between any two terms/definitions in different domains.

1.2 TOGAF – The Upper Ontology for OpenSEA

TOGAF9 is an open and freely licensed framework that is vendor neutral and sector neutral. It was developed by over 300 Architecture Forum members and companies from highly respected IT customers and vendors. It provides a framework for enterprises to extend and specialize, and provides a set of commonly used terms and definitions in the process. Some are formally represented as specific concepts and relations, others as textual descriptions. For example, TOGAF defined 'Enterprise' as "any collection of organizations that has a common set of goals. For example, an enterprise could be a government agency, a whole corporation, a division of a corporation, a single department, or a chain of geographically distant organizations linked together by common ownership" (The Open Group, 2009, p. 5) This abstract definition fits within the broad remit of providing an upper ontology that is loose (i.e., general) enough to be meaningful to entire markets, corporations, public bodies, and societies or groups working within these bodies. The Open Group also defined the purpose of Enterprise Architecture in such a way as to meet the requirements of 'enterprises' exchanging information openly and meaningfully "to optimize across the enterprise the often fragmented legacy of processes (both manual and automated) into an integrated environment that is responsive to change and supportive of the delivery of the business strategy" (p. 6). Enterprise architecture should therefore provide the platform for innovation and interoperation within and between units of operation of all sizes. It is not limited to information systems. As Tolido (2009) (Vice President of Cap Gemini Netherlands) pointed out Oracle OpenWorld 2009 focused heavily on innovation and being able to reuse existing resources effectively. Indeed, innovation, efficiency, collaboration and cooperation may all be born out of necessity in the cooler economic climate of the early 21st Century.

Because OpenSEA proposes to be a 'framework' for interoperation across and within enterprises, it is important to examine how The Open Group (2009) defined an 'Architecture Framework':

> a foundational structure, or set of structures, which can be used for developing a broad range of different architectures. It should describe a method for designing a target state of the enterprise in terms of a set of building blocks, and for showing how the building blocks fit together. It should contain a set of tools and provide a common vocabulary. It should also include a list of recommended standards and compliant products that can be used to implement the building blocks (p. 7).

The key points to note here are the common vocabulary and building blocks, which are fundamental components for a semantic market place or semantic interoperation between different enterprises. Note again the convenience of redefining the 'enterprise' to go beyond the individual corporations to the concept of a group of corporations trading together within the defined boundary of a market-place. Finally, one of the underlying aims of the TOGAF9 framework is to allow for enterprises to communicate without boundaries. Again, at its heart is the aim to breakdown silos and barriers between discrete units to promote communication and interplay. Information systems operating within architectures that have been guided by TOGAF should therefore experience 'boundaryless' information flow. 'Boundaryless Information Flow' is a trademark of The Open Group, and represents "access to integrated information to support business process improvements," representing a desired state of an enterprise's infrastructure specific to the business needs of the organization" (The Open Group, (2009, p. 27).

To summarise, The Open Group Architecture Framework has, at its core, many of the implicit semantics required for the integration of disparate and distinct domains. It provides the broad terms and definitions aimed to provide 'Boundaryless Information Flow' without specifying any prerequisites or restrictions based on size or market.

1.3 ISO 24707:2007 – The Meta-ontology for OpenSEA

Uschold (2003) identified the major evolutionary paths of the Web as finding resources in an ever growing pool, redefining the Web for human and machine consumption, changing the Web from a pool of resources to a pool of services and semantically enriching those resources. Semantic systems should provide the capability to recognise, represent and react to the meaning of data in the context of the goals of the user (Sowa, 2009, p. 33). Types of semantic systems include deductive databases, expert systems, knowledge based systems and the Semantic Web and its associated applications. However, they can be built on any one of a number of logical languages or formats and interoperation between different systems using different semantic structures or different ontologies can be difficult or impossible. Two legacy systems can be brought to interoperate better than two new, semantically-enabled systems that use different ontologies (Sowa 2009).

As systems start to interact with other systems and corporations look to operate seamlessly with other organisations in an 'extended enterprise' (Kuhlin & Thielmann, 2005) this problem becomes global in scale. Common Logic (referred to as CL throughout the Chapter) proposes a standardized approach to develop interoperation between systems using different formalisms and representations. The CL standard outlines its aims as, "The intent is that the content of any system using first-order logic can be represented in this International Standard. The purpose is to facilitate interchange of first-order logic-based information between systems" (ISO/IEC 24707, 2007). It provides a standard for a logical framework for the exchange of data and information across networks, including open networks such as the Internet.

CL dialects must be compliant with the semantics of First Order Logic but CL does not impose any formal syntax, rather it provides an abstract syntax and thereby allows for the reliable translation between languages. The three dialects that currently support CL are;

- CGIF – Provides a serialised representation for conceptual graphs.
- CLIF – A syntax similar to the Knowledge Interchange Format which has become the de facto standard notation for many applications of logic
- XCL – An XML notation for CL that is the intended interchange language for communicating CL across networks.

1.3.1 Common Logic and RDF

ISO/IEC24707:2007, section 5.1.2c, states that "The syntax should relate to existing conventions; in particular, it should be capable of rendering any content expressible in RDF, RDFS, or OWL." This has been demonstrated by Pat Hayes in the following example (Hayes, 2006).

```
<owl:Class rdf:id="#ChildOfUSCitizenPost1955">
 <owl:intersectionOf rdf:parseType="Collection">
  <owl:Restriction>
   <owl:onProperty rdf:resource="#parentOf" />
   <owl:allValuesFrom>
    <owl:Restriction>
     <owl:onProperty rdf:resource="#isCitizenOf" />
     <owl:hasValue rdf:resource="#USA" />
    </owl:Restriction>
   <owl:Restriction>
    <owl:onProperty rdf:resource="#dateOfBirth" />
    <owl:allvaluesFrom rdf:resource="#YearsSince1955" />
   </owl:Restriction>
 </owl:intersectionOf>
</owl:Class>
```

Maps to

```
(= ChildOfUSCitizenPost1955
   (And (AllAre parentOf (MustBe isCitizenOf USA))
     (AllAre dateOfBirth YearsSince1955) )
```

A further possibility, however, was provided by Hayes (2009) in his keynote speech to a recent International Semantic Web Conference 2009, in which he showed that RDF is almost Peircian in notation, and how RDF Redux theme could become a fully expressive CL dialect. Doing so would allow integration with other CL compliant dialects and greatly simplify the semantic stack (or 'layer cake'). Figure 3 shows how RDF simplifies the semantic stack of tools and protocols.

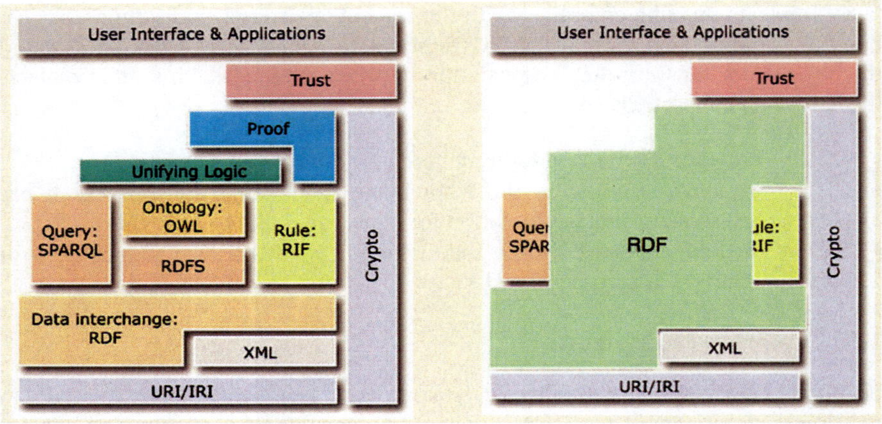

Fig. 3 Hayes' Vision of RDF within Common Logic (Hayes, 2009).

1.4 OpenSEA – Developing an Upper Ontology with CL

By expressing the terms and definitions used by TOGAF in a CL compliant format, and allowing these definitions to be extended and specialized freely, OpenSEA aims to create the abstract framework required for different domains to exchange information without requiring a rigid ontology nor a specific system or language.

In this chapter we use Conceptual Graphs which can be expressed in a CL dialect CGIF. Graphical notation is used as the primary representation to express how full logic can be portrayed in a format that is readily consumed by human agents yet can also be expressed in a compact linear notation that can be converted to any other CL compliant dialect without loss of meaning. The graphs were developed using CharGer (sourceforge.net/projects/charger), Amine (sourceforge.net/projects/amine-platform) and CoGui (www.lirmm.fr/cogui/).

1.5 Interlinked Domains

Key to the framework is the need for a chain of generalization or specialization in order for two disparate domains using different ontologies to come to an common understanding. one can imagine the chain as one of simple 'IS-A' relationships where all concepts and relations ultimately link back to a definition in the upper ontology as provided by TOGAF. The approach is similar to the idea of the Domain Naming System which relies on lookups and, at its heart, has the '.' domain above all others to provide the link between top level domains.

John Sowa (via email private correspondance) has provided the following examples of linear expression for this generalisation/specialisation concept, as shown below.

```
CLIF:
(forall ((R1 MonadicRelation) (R2 MonadicRelation) (x) (y))
 (if (and (GeneralizationOf R1 R2) (R2 x y)) (R1 x y)))

CGIF:
[MonadicRelation @every *R1] [MonadicRelation @every *R2]
[Entity: @every *x] [Entity: @every *y]
[If (GeneralizationOf ?R1 ?R2) (#?R2 ?x ?y) [Then (#?R1 ?x ?y)]]
```

This says that for all monadic relations R1 and R2 and any x and y, if R1 is a generalization of R2 and R2(x,y), then R1(x,y). Once the GeneralisationOf statement is made then the type hierarchy can be listed as a simple collection of assertions:

```
CLIF;
 (and (GeneralizationOf Architect Business_Analyst)
  (GeneralizationOf Architect Information_Analyst)
  (GeneralizationOf Information_Analyst Data_Analyst)
  (GeneralizationOf Information_Analyst Technical_Analyst))
```

For example, in TOGAF the relation 'Performs Task In' is formalized as a canon (common usage) as relating an 'Actor' to a 'Role'. Suppose a health-specific domain specializes the term 'Actor' to cover the concepts of 'Doctor' and 'Patient' whilst a Sales and Distribution domain specializes the same term to concepts such as 'Salesperson', 'Lead' and 'Customer'. If the domains were required to interact, for example a pharmacy needed to purchase drugs, the two domains could agree on the fact that a 'Doctor' and a 'Salesman' share some commonality in that they both perform tasks in their respective roles.

This example can be expressed as follows:

```
TOGAF:
[Actor: @every *t]
 (PerformsTaskIn ?t [Role])
HealthCare
(GeneralizationOf Agent Doctor)
(GeneralizationOf Role Healthcare)
Sales
(GeneralizationOf Agent Salesman)
(GeneralizationOf Role Sales)
```

that we can translate to the CLIF form:

```
CLIF:
[Doctor: @every *t]
 (PerformsTaskIn ?t [Healthcare])
And
[Salesman: @every *t]
 (PerformsTaskIn ?t [Sales])
```

Any logic or inferences that can be made in the TOGAF domain would be equally expressed in both the sub domains and any knowledge farms that can make those generalised rules has the capacity to gather new information from data collected from disparate domains.

It is worth noting that unlike DNS, OpenSEA relies on the fact that the specialization and generalization is not a simple hierarchy but allows for concepts to be specialisations of one or more 'master' concepts. The classic example is ANGEL (Sowa, 1984, p. 408) where ANGEL is a specialisation of both ANIMATE and MOBILE-ENTITY. This example also includes the specification that ANGEL < ¬PHYSOBJ ie an angel is not a physical object, the IS-NOT specification being a powerful tool for future developments of OpenSEA.This notion of specialisation extends from concepts to instances. For example, Doctor Bob Smith is a keen motorsport fan. Within the health domain a patient is unlikely to be interested in which team Dr Smith follows so the information is superfluous. Yet a domain specializing in providing executive travel to Formula 1 events may be very interested in the fact that the instance 'Bob Smith' IS-A 'Doctor' and Bob Smith 'IS-A' 'McLaren fan' if they have an inference engine that could deduce that 'if a fan performs tasks within certain roles then the fan has disposable income'.

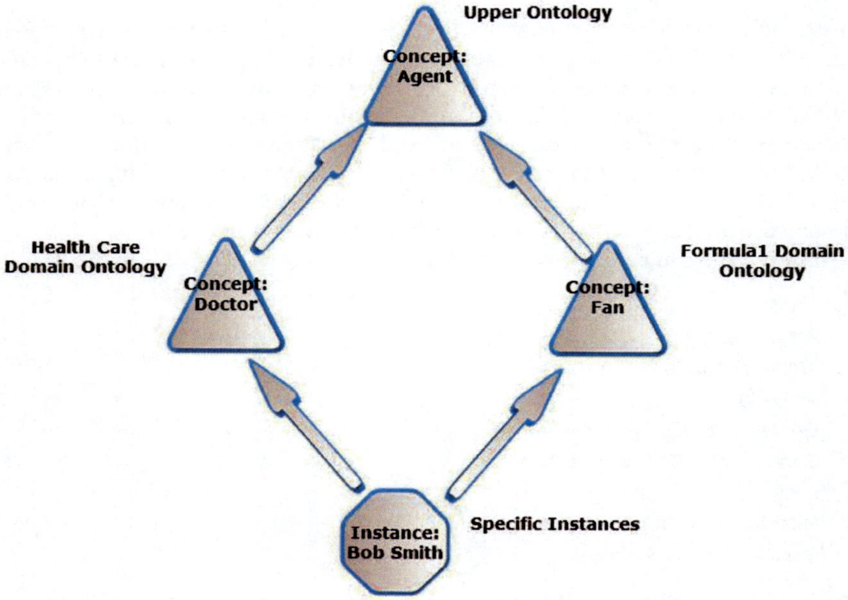

Fig. 4 Simple example of An Instance Being A Specialisation of Different Concepts

2 Developing an Upper Ontology from TOGAF

This section draws on the research of (Bridges, 2010; Bridges & Polovina, 2010). One of the problems faced was the task of viewing terms abstractly, even if they are as well defined as those in the TOGAF9 material. Bridges observed the problem lay with disengaging from the precepts that any individual holds and the assumptions that shape how an individual perceives the world.

2.1 Sowa's Conceptual Catalogue

Bridges referred extensively to the conceptual catalogue created by Sowa (1984, pp. 405-424) (herein referred to as SCC) which provides a number of 'canons' (i.e., common meanings) for a broad range of terms. However, the real value of the catalogue is in what has been canonised as well as how it has been described. It is possible to see how a few high level concepts and relations can provide a broad range of conceptual structures and move towards an unbiased and abstracted definition (or canon) for terms that are so common they are hard coded into the mind of the individual without any clear and logical structure.

By reusing the SCC OpenSEA adheres to the principle advocated by Berners Lee and Kagal (2008) to not reinvent the wheel and to use existing ontologies wherever possible. It is also hoped that using something as established as the SCC would provide some base commonality with other ontologies that have also been influenced by it.

The broad concepts of many enterprise architectures (What, Where, When, How, Why and Who) provide a useful approach to determining upper level concepts as high level contexts. This is mirrored to some extent in the SCC. For example, the context of 'What' could map to Sowa's ENTITY concept and specialisations of this concept would all have a common general meaning, whether they are Data Entities or Abstract Objects.

Similarly, a high level concept of 'how' could provide the specialisations of Process, Business Function, Business Service, Information System Service etc (again, examples taken from the TOGAF9 definitions).

This approach is shown in Figure 5 as part of an initial concept type hierarchy.

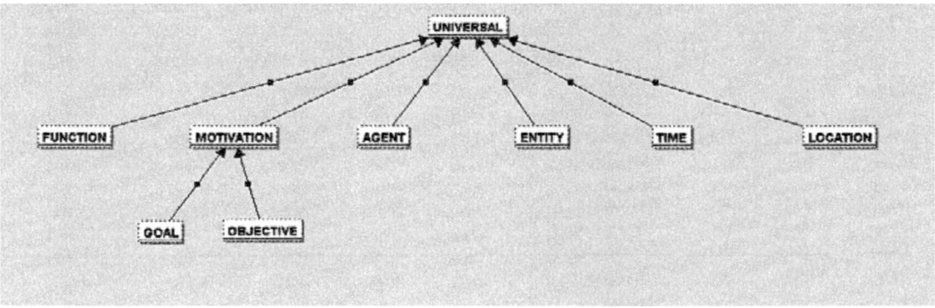

Fig. 5 An Initial Type Hierarchy for TOGAF Terms (Bridges, 2010)

TOGAF describes Data Entity as, "an encapsulation of data that is recognized by a business domain expert as a thing. Logical data entities can be tied to applications, repositories, and services and may be structured according to implementation considerations" (The Open Group, 2009).

In other words, it too is covered by Sowa's canon of 'physical objects as well as extractions'. If we continue to focus on the elements of an enterprise, resources and agents can be seen as specialisations of ENTITY within the SCC. Events are also present within the SCC. Events could be seen as being part of a process involving different states (another SCC concept) and it is easy to see how 'who', 'how' and 'when' could be linked to a process.By including the concept of 'why' the business goals and objectives from TOGAF9 are brought within the Architecture framework.

Figure 6 builds on the basic concepts of Figure 5 by adding the EVENT concept. It also adds STATE as this is an integral part to a process that should be involved with changing the state of an entity.

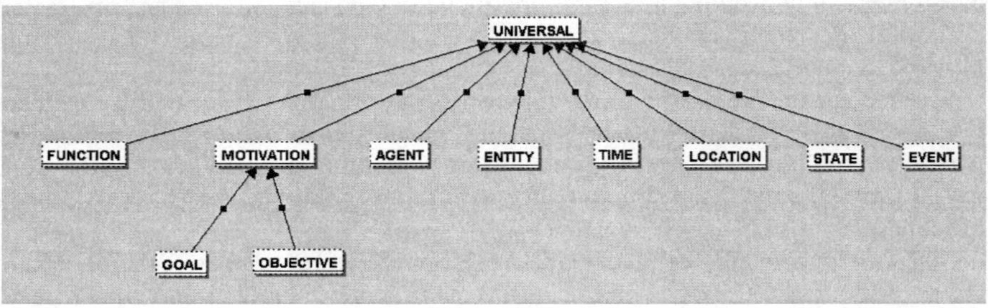

Fig. 6 Development of the Type Hierarchy.

Within the SCC Sowa described 'CHARACTERISTIC' (a specialisation of the concept ATTRIBUTE) as being 'essential in nature'. By this canon the object attributes shown in Table 1 can also be seen as characteristics, a confusion that needs to be addressed in the ontology if the SCC is to be used to any extent.

Table 1 TOGAF Attributes for all Metamodel Objects

Metamodel Attribute	Description
ID	Unique identifier for the architecture object.
Name	Brief name of the architecture object.
Description	Textual description of the architecture object.
Category	User-definable categorization taxonomy for each metamodel object.
Source	Location from where the information was collected.
Owner	Owner of the architecture object

Attributes that are essential could be linked to their associated objects by a 'chrc' relation to denote the fact that they are essential in describing individuals of those concepts and could provide a very useful means of defining individuals within a global market place. For example, all metamodel objects within an enterprise must have the attributes of ID, Name, Description, Category, Source and Owner (Table 1). These same attributes could be used to provide the information needed to help identify resources, services, processes etc within an

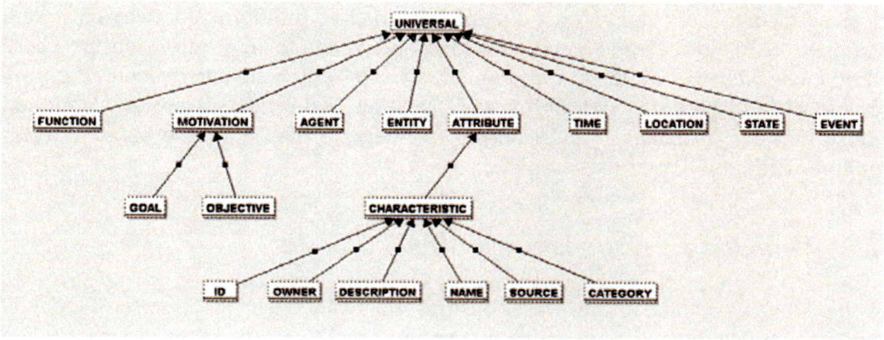

Fig. 7a Common Attributes Within The Type Hierarchy.

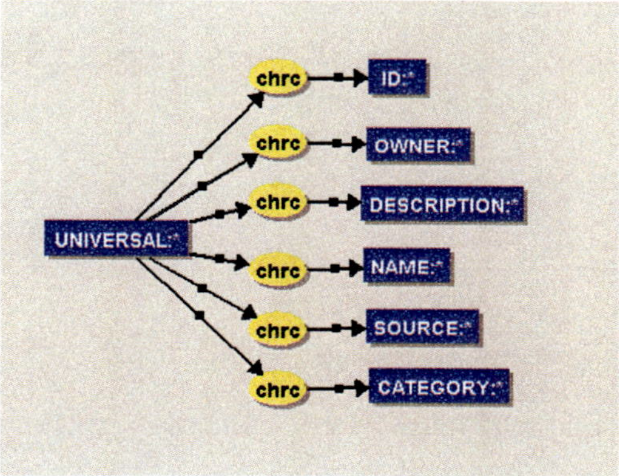

Fig. 7b A Conceptual Graph Representation for all Objects and Their Minimal Metadata.

open market and provide the commonality to assist with interoperation. This approach the expansion of Sowa's type hierarchy (Figures 7a and 7b).

In CGIF, Figure 7b would be represented as

```
[Universal: @every *t]
 (chrc ?t [Category])
 (chrc ?t [Description])
 (chrc ?t [ID])
 (chrc ?t [Name])
 (chrc ?t [Owner])
 (chrc ?t [Source])
```

In the framework the ID, Category, Source and Owner are all represented using URLs, and the name and description by a human-readable text string. The attributes are extended with 'Definition', which provides for a CL based definition of how the object is defined by other objects. 'Category' is used within OpenSEA as the means of embedding the well recognized 'IS-A' relationship..

Knowledge bases can harvest this information and build links between different domains using the shared generalisations and generate new information. These 'knowledge farms' could be the pioneers of the new, distributed architecture by acting as both brokers (identifying what services and processes are available and those that are in demand) and offering to integrate the enterprise business rules engine within the larger market place.

2.2 *Modelling Relationships within TOGAF9*

Table 2 shows some of the relationships that can occur within the TOGAF9 metamodel. Relationships can also be captured in CG as shown in Figures 8a and 8b, in this example the relationship 'Resolves' is shown with a signature consisting of an Actor and an Event, as is the relationship that an Actor 'generates' an Event.

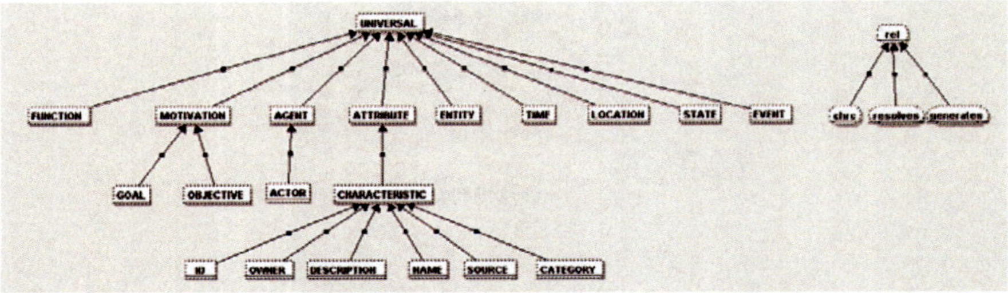

Fig. 8a Capturing Relationships within the Type Hierarchy.

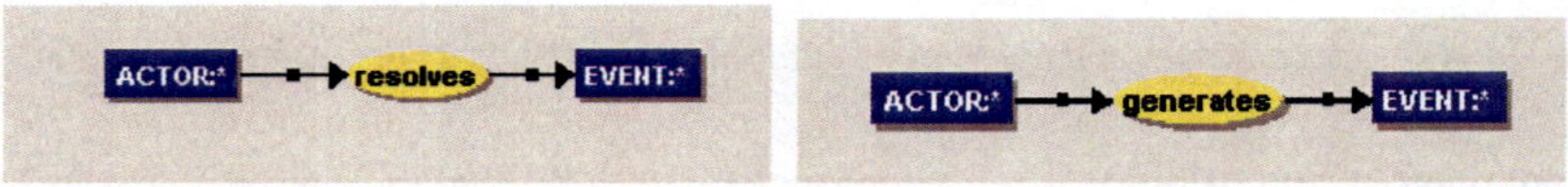

Fig. 8b Graphical Representations of Simple Canons.

2.3 *Nested Graphs*

Zachman and Sowa (1992) showed that concepts could be seen as nested graphs so that, for example, 'how' something is achieved is represented using symbols relevant to the agent at that level. In Figure 9 Zachman and Sowa illustrated the difference between how agents operating at the Enterprise Level would view 'what' in terms of an entity where as those operating at the level of Information System analysis would regard 'what' in terms of data yet the two contexts are related by a relation ('NAME'). Similarly, business process experts would consider the business process when considering 'how' something is done but this has to be mapped to the system analyst's contextual view in terms of the functions called within the system. In this case the two are connected by the 'MODL' relation.

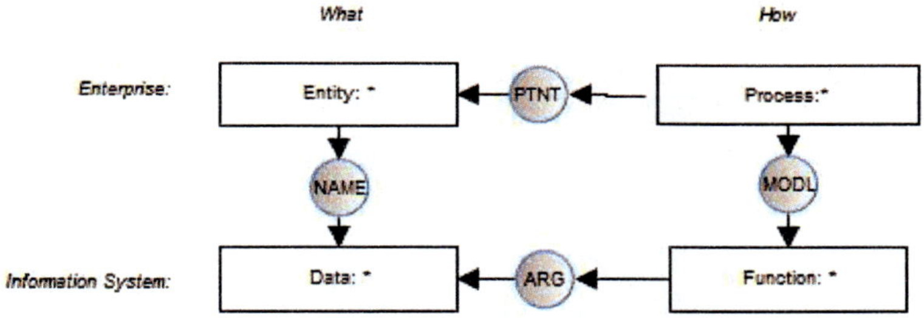

Fig. 9 Inter-related Contexts (from Zachman and Sowa, 1992, p. 610).

As already identified, TOGAF9 models the architectures of Business, Information (Architecture and Data) and Technology. Each tier of the TOGAF model could be considered to be a context in a similar fashion to Zachman's and Sowa's formalisation of the ISA . Within the Business Architecture a Business Service ('what') could be a unit that is 'owned and governed by' an Organisational Unit ('who') which in turn 'operates in' a location ('where'). From a System Analyst perspective the same Business Service 'provides' or 'consumes' a Data Entity ('what') which 'resides within' a Logical Data Component (which could be argued to be an abstract object, i.e. another entity) which is, in turn 'realised by' a Physical Data Component (another entity). Figure 10 shows the type hierarchy and relations being developed for this purpose.

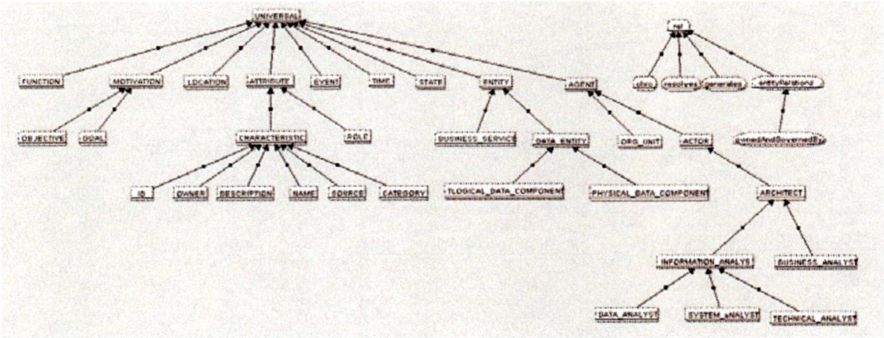

Fig. 10 TOGAF Objects and Relations Captured in the Type Hierarchy.

2.4 Contextualisation of Information

Figure 11 shows a nested graph within the concept 'Business Analyst' and illustrates the contextual meaning of a Business Service to a Business Analyst (i.e. an agent working at the layer of business architecture). The types and relations are highlighted in Figure 12 which focuses on the agents and relationships within the type hierarchy.

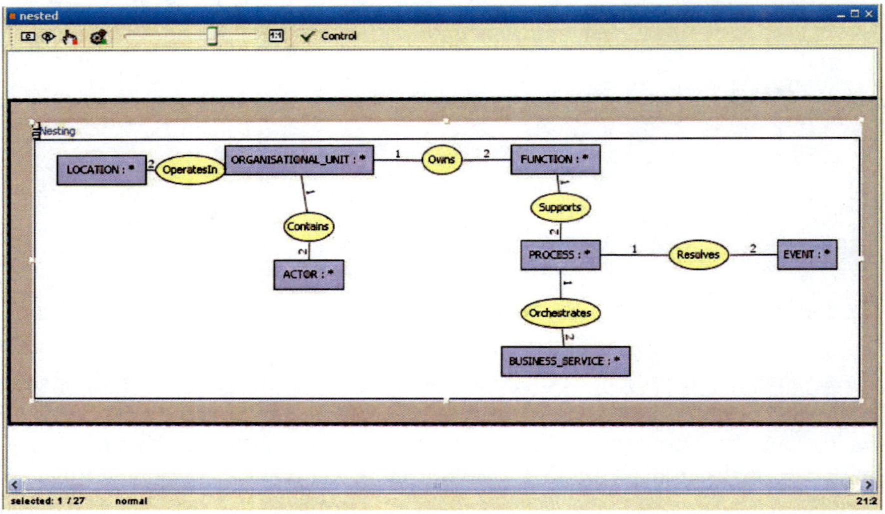

Fig. 11 'Business Service' Modelled Within the Perception of a Business Analyst (Bridges, 2010).

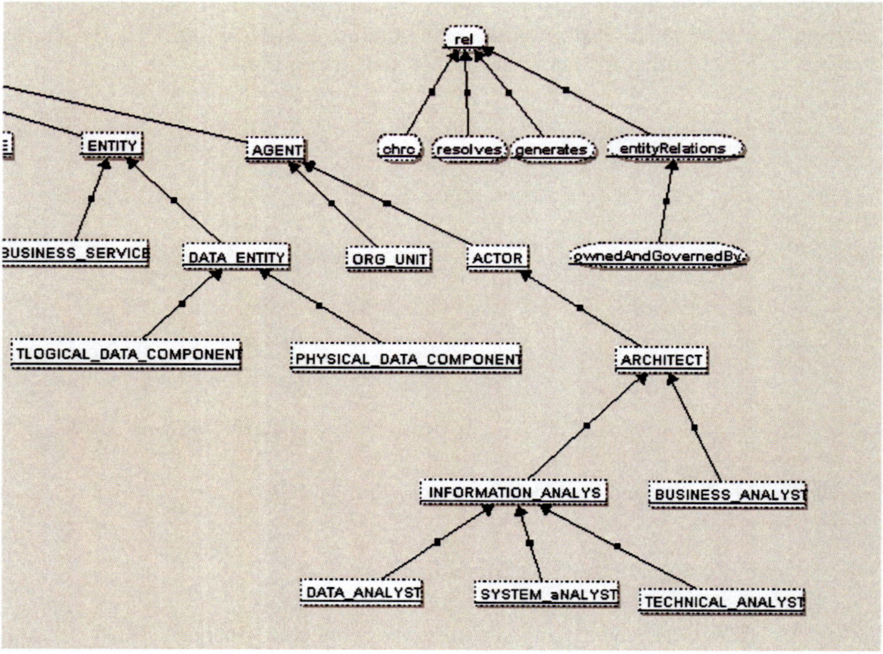

Fig. 12 Focus on the Relations and Actors In the Type Hierarchy.

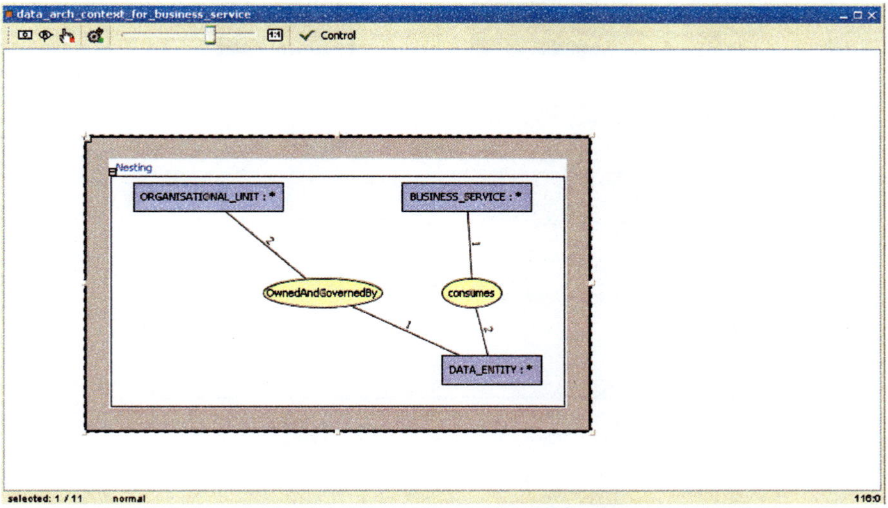

Fig. 13 'Business Service' as Perceived by a Data Analyst (Bridges, 2010).

In Figure 13 the concept of a business service is nested within the context of a Data Analyst to show what a Business Service means to an agent operating within the Data Architecture layer of an enterprise.

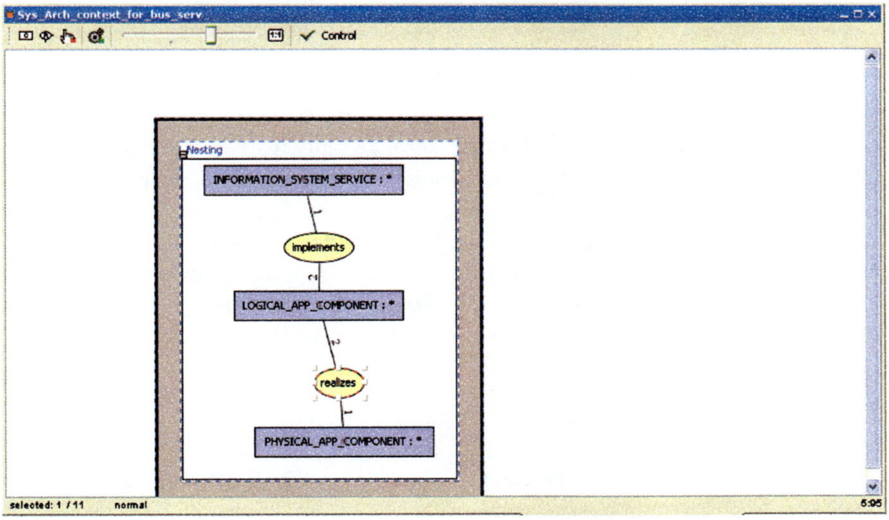

Fig. 14 'Business Service' As Perceived by a Systems Analyst (Bridges, 2010).

In Figure 14 the context of a business service is shown as per a Systems Analyst. The Business Service is shown as the more specialised 'Information System' inferring that it is a fully automated business service (a fact that would need representing in the type hierarchy).

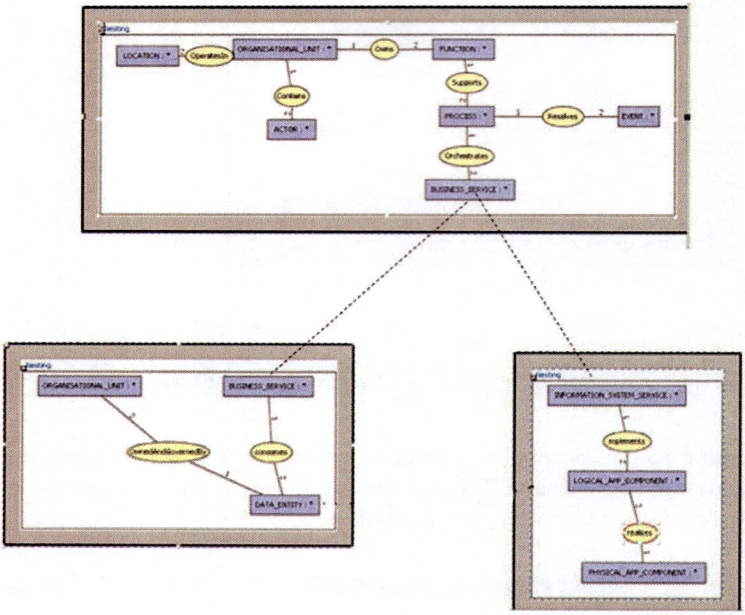

Fig. 15 Different Perceptions Modelled With Co-Referents (Bridges, 2010).

Figure 15 shows the three different nested graphs representing the different contextual perceptions of what a business service means interlinked by co-referents (the dotted lines). The co-referents allow the three agents to interconnect and for changes in one tier to be integrated within the other tiers as a single version of the truth. Relationships between different nested graphs could be used to connect different contexts and so this simple example shows how the different interpretations for what a concept means to different agents can be interlinked.

2.5 *OpenSEA and the Cloud*

The TOGAF semantics may be used to describe the XaaS offerings of the cloud by compartmentalizing the services between the external face and inner workings. Consumers of the services are interested in different aspects of the service to those that run it, as shown in Figure 16.

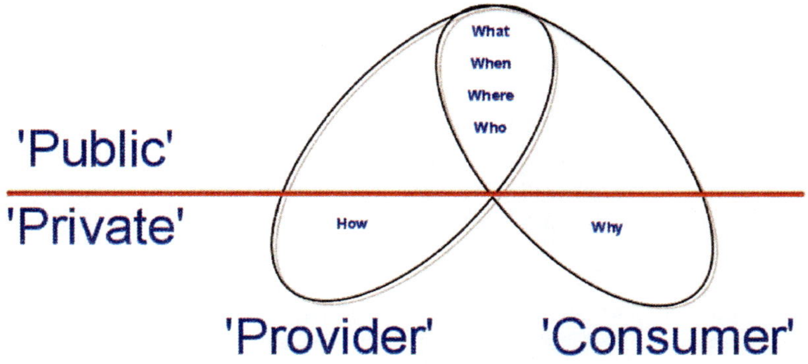

Fig. 16 The Public/Private Differentiation of XaaS (Bridges, 2010).

Figure 16 shows that the provider is 'privately' interested in 'how' the service is provided whilst the consumer has little concern of 'how' the service is realised. Within TOGAF this would relate to the lower tiers and the contexts relevant to software engineers, hardware engineers, etc. TOGAF semantics used at this level could include 'Physical Technology Component' 'is hosted in' a 'Location' and 'Realizes' a 'Physical Application Component'. Similarly, a 'Service' 'is Realized through' a 'Logical Application Component' and is 'Implemented on' a 'Logical Technology Component'

The consumer has their own individual objectives and goals related to the 'why' a service should be used and these may vary from consumer to consumer and is of little or no interest to the provider. TOGAF terminology relevant to the consumer but not the provider could be a 'Goal' 'is realized through' an 'Objective' or 'Addresses' a 'Driver'.

The terms used will be determined by the service provided, for example Hardware as a Service is more interested in the physical aspects of the infrastructure such as RAM, storage and processing power, infrastructure may include all this plus virtual machines, operating systems, software and scalability. Platform as a Service offers the chance to build and deliver web applications but may not have any reference to the underlying hardware or infrastructure.

The *consumption* of a service, however is dependent on 'what', 'when', 'where' and 'who' and consumers and providers could advertise their need/provision of the service in these terms for brokers to 'matchmake' or agree contracts. TOGAF terms that would be used in agreeing the contract could include a 'Service' 'Provides' or 'Consumes' a 'Data Entity' and 'Provides Governed Interface Access' to a 'Function' whilst 'Service Quality' 'Applies to' a 'Contract' and 'is Tracked Against' a 'Measure'.

This shift in emphasis from service provision to transaction fulfilment could be referred to as a step away from *Service* Oriented Architecture to *Transaction* Oriented Architecture where the provision and consumption of services are perceived as part of a whole. After all, a Service means nothing without a consumer; it is defined by consumption.

2.6 OpenSEA and Web3.0

In the introduction reference was made to the advantages inherent in users and consumers being able to capture what things are and the relationships between them (Mulholland 2009). If web sites were annotated with either XCL or the less verbose CLIF or CGIF strings (using XML tags to identify the strings) the user could use simple client tools to express the inter-relations graphically and visualize how the web-site sits within the greater web. A web cache could become a powerful information set, moving beyond a series or URL's to a web of knowledge and guided reference. Similarly, search engines could gather XCL or other CL dialect from any Web sites or data sets that support OpenSEA to create powerful semantic searches, broker services or generate new knowledge through rules engines.

2.7 OpenSEA and the Software Engineer

Trapp (2009) outlined how Semantic Web technologies can can add some elements of Knowledge Management to enterprise software. By introducing the expert's knowledge to the data and functions within a software the entire system (by which we mean the human agents and the software) can become 'intelligent'. This can be incorporated within the software through metadata (e.g. XCL definitions of business objects or processes), reasoning and visualisation amongst others.

Trapp refered to the 'Design Time Type Information' Open Source project available under Apache. DTTI combines a base ontology with REST web services that expose RDF data about objects within SAP systems. He suggests some of the benefits include formalising the architecture, using reasoning to detect direct and indirect dependencies and forbidden dependencies. All this is possible in the OpenSEA framework.

2.7.1 Example

Within TOGAF9 (2009) a Data Entity has the following interactions with its environment:

Table 2 Concepts and Relations Referring to 'Data Entity'

Source Object	Target Object	Relationship
Data Entity	Logical Application Component	Is processed by
Data Entity	Logical Data Component	Resides within
Data Entity	Service	Is accessed and updated through
Data Entity	Data Entity	Decomposes
Data Entity	Data Entity	Relates to

Added to the Type Hierarchy, we now have the following.

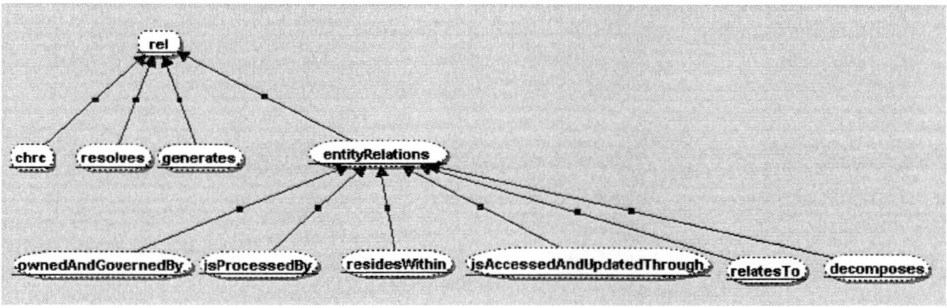

Fig. 14 Relations for Data Entity captured within a Type Hierarchy

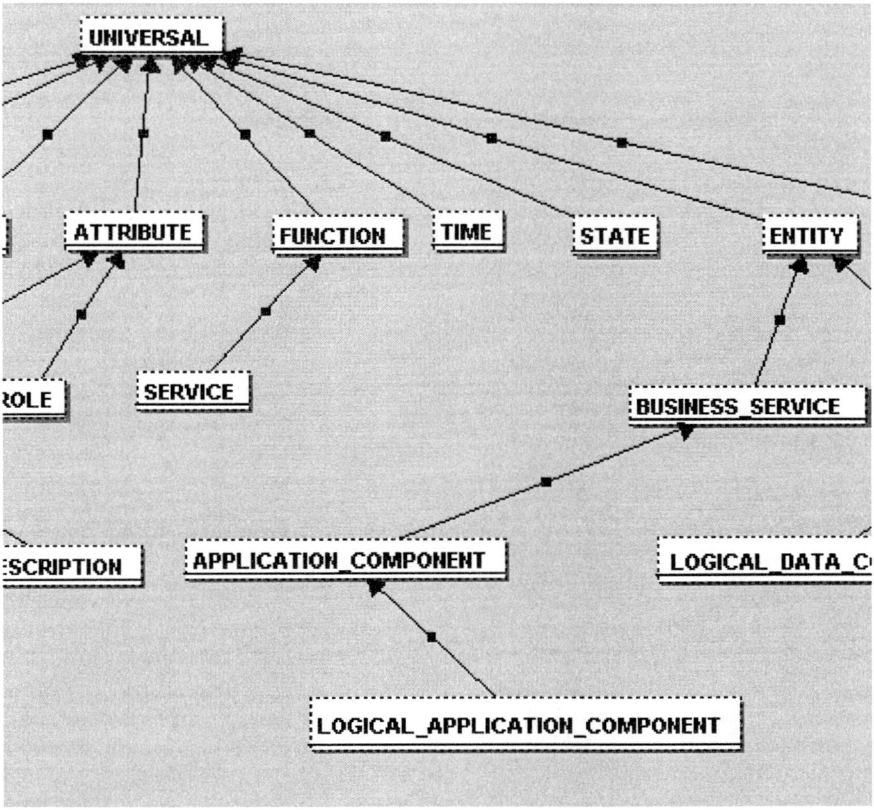

Fig. 15 Specialisations of the Concept 'ENTITY'

Data Entities are stored within the knowledge base of all participating domains as:

```
CGIF:
[DEFINITION: "[DATA_ENTITY:*x1] [SERVICE:*x2] (isAccessedAndServicedThrough ?x1 ?x2) "] [NAME: Data Entity]
[CATEGORY: OpenSEA.org/Universal] [SOURCE: "http://www.opengroup.org/architecture/togaf9-doc/arch/index2.html"]
[ID: "OpenSEA.org/DATA_ENTITY"] [OWNER: OpenSEA] [DESCRIPTION: "AN ENCAPSULATION OF DATA..."]
[DATA_ENTITY: *x1]
(chrc ?x1 OpenSEA) (chrc ?x1 Universal) (chrc ?x1 "http://www.opengroup.org/architecture/togaf9-
doc/arch/index2.html") (chrc ?x1 "AN ENCAPSULATION OF DATA...") (chrc ?x1 "OpenSEA.org/DATA_ENTITY") (chrc
?x1 Data Entity) (chrc ?x1 ?x1] [SERVICE:*x2] (isAccessedAndServicedThrough ?x1 ?x2) ")
```

In the example, the generic DATA_ENTITY contains its own DNA containing examples of how it can be used, (in the DEFINITION), what it's a specialisation of (CATEGORY), a plain text description, the URL at which it is defined and so on. Any specialisations of the generic DATA_ENTITY refers to this ID as its category, thereby retaining the links that make up the web of references.

3 Further Investigations

3.1 Integration of OpenSEA and GoodRelations

GoodRelations (Hepp, 2008) provides a standard vocabulary for expressing services and products that are offered on web sites. OpenSEA should not be seen as an alternative to an established ontology as the aims are similar but the ontology of GoodRelations could be investigated as a specialization of TOGAF. Furthermore, Hayes (2009) suggestion that RDF has the capacity to become CL compliant means the RDF expression of GoodRelations could be expressed in XCL or as a CLIF string within XML brackets. This could introduce the potential for users of GoodRelations to integrate with other domains of knowledge through OpenSEA and human consumers of web-sites could use simple clients to visualize how the information on the site relates to the greater Web.

3.2 Knowledge, Inference and Information Generation through OpenSEA

OpenSEA can, by definition, be used as part of an information generating rules engine through the application of rules (knowledge) on the information available to infer new, un-tapped information or making decisions on new courses of action.

Sowa has reported that SBVR is capable of full CL syntax and as such it could be included as part of the OpenSEA framework by providing the meanings used in Business Rules. The overlap between SBVR and Controlled Natural Language could act as a bridge between human and machine agents by bringing the two representations closer together. For example, (Baisley et. al., 2005),

Below is a description of the semantic formulation of the rule above expressed in terms of the SBVR Logical Formulations of Semantics Vocabulary. It is easily seen that SBVR is not meant to provide a concise formal language, but rather, to provide for detailed communication about meaning. The description is verbose, when separated into simple sentences. But it captures the full structure of the rule as a collection of facts about it.

The rule is meant by an obligation claim.
That obligation claim embeds a logical negation.
The negand of the logical negation is an existential quantification.
The existential quantification introduces a first variable.
The first variable ranges over the concept 'barred driver'.
The existential quantification scopes over a second existential quantification.
That second existential quantification introduces a second variable.
The second variable ranges over the concept 'rental'.
The second existential quantification scopes over an atomic formulation.
The atomic formulation is based on the fact type 'rental has driver'.
The atomic formulation has a role binding.
The role binding is of the fact type role 'rental' of the fact type.
The role binding binds to the second variable.
The atomic formulation has a second role binding.
The second role binding is of the fact type role 'driver' of the fact type.
The second role binding binds to the first variable.

Note that designations like 'rental' and 'driver' are used above to refer to concepts. The semantic formulations involve the concepts themselves, so identifying the concept 'driver' by another designation (such as from another language) does not change the formulation.

4 Conclusion

OpenSEA is a framework for unified modelling tools for enterprise architecture. Zachman and Sowa (1992) provided examples of how Conceptual Graphs could be used to model Zachman's Information System Architecture. The same ideas are contained within the OpenSEA framework as it seeks to formalise architectures that are aligned with TOGAF using CL compliant dialects which, of course, includes CGIF which has a graphical representation.

By formalising the links between the horizontal components (in the ISA these include Enterprise Model, System Model, Technology Model and Component Level, similar to the tiers of TOGAF9) and vertical components (What, How, Where etc) of the ISA Zachman and Sowa identified that each unit could be represented with respect to the other units and each unit could be represented using graphical CGs which had the power to embed full predicate calculus within an easily accessible form.

It is worth noting some quotes taken from Sowa's and Zachman's paper. The quotes continue to be relevant today and hold particular significance when viewed in light of the findings of the International Research Form 2008:

> Dramatic improvements in the price-performance of information technology and the escalation of the rate of change show no signs of abatement. In the words of Alvin Toffler, "Knowledge is change . . . , and accelerating knowledge, fuelling the great engine of technology, means accelerating change." Gone are the days of computers for simple calculations. We are only now beginning to see the enormous complexity of integrating information technology into the very fabric of our enterprises. *Soon, the enterprise of the information age will find itself immobilized if it does not have the ability to tap the information resources within and without its boundaries* [italics ours] (Zachman & Sowa, 1992, p.613)..
>
> an enterprise will form into a free market structure if the nature of the transaction between two organization units is simple, well defined, and universally understood. In this case, the organization (or person) with work to assign would survey all possible workers to find one who is acceptable in terms of availability and cost. This method is much like a stock buyer who scans the pool of stockbrokers to find one who will execute a buy within an agreeable time and for a reasonable fee (Zachman & Sowa 1992, p. 596).
>
> Tools exist for both users and developers. Tools such as online help are there specifically for users, and attempt to use the language of the user. Many different tools exist for different types of developers, but they suffer from the lack of a common language that is required to bring the system together. It is difficult, if not impossible, in the current state of the tools market to have one tool interoperate with another tool (The Open Group, 2009, p.418).

It is worthwhile noting that if the nature of the dependency between cells could be understood and stored in the repository along with the cell models, it would constitute a very powerful capability for understanding the total impact of a change to any one of the models, if not a capability for managing the actual assimilation of the changes (Zachman & Sowa, 1992, p.603).

4.1 www.Open-SEA.org

The domain Open-SEA.org has been created to facilitate open discussion and development of the ideas generated in this chapter. Please email shaun.bridges@Open-SEA.org for more information.

References

Baisley, D.E., Hall, J., Chapin, D.: Semantic Formulations in SBVR. Paper presented at W3C Workshop on Rule Languages for Interoperability, Washington, D.C (April 27-28, 2005), `http://www.w3.org/2004/12/rules-ws/paper/67/` (retrieved December 15, 2010)

Berners-Lee, T., Kagal, L.: The fractal nature of the Semantic Web. AI Magazine 29(3), 29 (2008)

Boisot, M., Canals, A.: Data, information, and knowledge: Have we got it right? In: Boisot, M., MacMillan, I., Han, K.S. (eds.) Explorations in Information Space: Knowledge, Agents, and Organization, pp. 15–47. Oxford University Press, Oxford (2007)

Bridges, S.: The Extent and Appropriateness of Semantic Enterprise Interoperability with TOGAF9 and ISO Common Logic. Unpublished Dissertation, Sheffield Hallam University, Sheffield, UK (2010)

Bridges, S., Polovina, S.: An OpenSEA framework using ISO24707 common logic. In: Xhafa, F., Stavros, D., Santi, C., Ajith, A. (eds.) Proceedings of 2nd International Conference on Intelligent Networking and Collaborative Systems, Thessaloniki, Greece, November 24-26, pp. 335–336. IEEE Computer Society, Los Alamitos (2010)

Cloud Computing Use Case Discussion Group, Cloud Computing Use Cases, Version 4.0 (2010), http://opencloudmanifesto.org/Cloud_Computing_Use_Cases_Whitepaper-4_0.pdf (retrieved December 20, 2010)

Hayes, P.: IKL presentation for Ontolog (2006), http://www.slideshare.net/PatHayes/ikl-presentation-for-ontolog (accessed October 18, 2009) (retrieved December 21, 2010)

Hayes, P.: BLOGIC or Now What's in a Link? (November 24, 2009), http://videolectures.net/iswc09_hayes_blogic/ (retrieved December 22, 2010)

Hepp, M.: GoodRelations: An ontology for describing products and services offers on the Web. In: Gangemi, A., Euzenat, J. (eds.) EKAW 2008. LNCS (LNAI), vol. 5268, pp. 332–347. Springer, Heidelberg (2008)

Heuser, L., Alsdorf, C., Woods, D.: International Research Forum 2008, 1st edn. Evolved Technologies Press, New York (2009)

ISO/IEC 42010, Recommended practice for architectural description of software-intensive systems (2007), http://www.iso.org/iso/catalogue_detail.htm?csnumber=45991 (retrieved October 16, 2009)

ISO/IEC 24707, Common Logic (CL): A framework for a family of logic-based languages (2007), http://standards.iso.org/ittf/PubliclyAvailableStandards/index.html (retrieved November 1, 2009)

Kuhlin, B., Thielmann, H.: The Practical Real-Time Enterprise. Springer, Heidelberg (2005), http://www.springerlink.com/index/10.1007/b138980 (retrieved October 16, 2009)

MacManus, R.: Eric Schmidt Defines Web 3.0 (2009), http://www.readwriteweb.com/archives/eric_schmidt_defines_web_30.php (retrieved October 20, 2009)

Mulholland, A.: Time to return to the Semantic Web again! | CTO Blog | Cap Gemini | Consulting, Technology, Outsourcing (2009), http://www.capgemini.com/ctoblog/2009/07/time_to_return_to_the_semantic.php (retrieved October 21, 2009)

Muholland, A.: Genuine progress on clouds, September 6 (2010), http://www.capgemini.com/ctoblog/uncategorized/genuine_progress_on_clouds-php/ (retrieved December 12, 2010)

Pascale, R.T.: Surfing the Edge of Chaos: The Laws of Nature and the New Laws of Business. Crown Business, New York (2000)

Sowa, J.: Controlled natural languages for semantic systems - A roadmap of directions to explore (2009), http://www.jfsowa.com/talks/cnl4ss.pdf (retrieved October 21, 2009)

Sowa, J.F.: Conceptual Structures: Information Processing in Mind and Machine. Addison-Wesley, Reading (1984)

The Open Group. TOGAF Version 9, Van Haren Publishing, Zaltbommel (2009)

Tolido, R.: Oracle OpenWorld: Innovation, 2009 style. I CTO Blog I Capgemini I Consulting, Technology, Outsourcing (2009), http://www.capgemini.com/ctoblog/2009/10/oracle_openworld_innovation_20.php (retrieved October 21, 2009)

Trapp, T.: ABAP Software Ontologies (2009), http://www.sdn.sap.com/irj/scn/weblogs?blog=/pub/wlg/15917%3Fpage%3Dlast%26x-maxdepth%3D0 (retrieved September 30, 2009)

Uschold, M.: Where are the semantics in the Semantic web? AI Magazine 24(3), 25–36 (2003)

Youseff, L., Butrico, M., Da Silva, D.: Toward a unified ontology of cloud computing. Paper presented at Grid Computing Environments Workshop, Austin, TX, (November 16, 2008), http://freedomhui.com/wp-content/uploads/2010/03/CloudOntology.pdf (retrieved December 12, 2010)

Zachman, J., Sowa, J.: Extending and formalizing the framework for information systems architecture. IBM System Journal 31(3), 590–617 (1992)

Acronyms

CG (or **CGs**)**:** Conceptual Graph, Conceptual Graphs
CL: Common Logic
ISO: International Standards Organisation (International Organization for Standards)
OpenSEA: Open Semantic Enterprise Architecture
TOGAF: The Open Group Enterprise Architecture Framework

Glossary

Conceptual Graphs (CGs) are a formal way of knowledge representation. Originally used to represent the conceptual schemas used in database systems, CGs have been applied a wide range of topics in artificial intelligence, computational intelligence, computer science, cognitive science and enterprise architectures.

ISO 24707:2007 Common Logic (CL) is a framework for a family of logic languages, based on first-order logic, intended to facilitate the exchange and transmission of knowledge in computer-based systems. The standard includes specifications for three dialects, the Common Logic Interchange Format (CLIF), the Conceptual Graph Interchange Format (CGIF), and an XML-based notation for Common Logic (XCL). Many other logic-based languages could also be defined as subsets of CL by means of similar translations; among them are the Semantic Web's RDF and OWL.

OpenSEA is a framework that combines the open semantics of TOGAF with the open syntax of ISO 24707:2007 Common Logic to provide an Open Semantic Enterprise Architecture.

Chapter 4
Building Collective Tag Intelligence through Folksonomy Coordination

G. Varese and S. Castano

Abstract. In this chapter, we provide techniques for automatically classifying and coordinating tags extracted from one or more folksonomies, with the aim of building *collective tag intelligence* which can then be exploited to improve the conventional searching functionalities provided by tagging systems. Collective tag intelligence is organized in form of *tag equivalence clusters* with corresponding semantic, terminological, and linguistic relations. For building tag collective intelligence, we define i) normalization techniques to identify equivalence clusters of tags and extract the relations holding between them and ii) similarity techniques to match tags on the basis of available collective tag intelligence. Finally, we describe the evaluation of the proposed techniques over real datasets extracted from del.icio.us and Flickr folksonomies and a real application example of exploiting the collective tag intelligence for similarity-based resource retrieval.

1 Introduction

In the recent years, tagging systems have acquired a great popularity. Tagging systems allow users to annotate web resources (e.g., text documents, images, videos, web pages) by associating them a set of *tags*. Tags are terms arbitrarily chosen by users for their capability to describe the content of web resources. The popularity of tagging systems is mainly due to their ease of use. In fact, users can easily classify web resources without having any technical knowledge and without being constrained by specific conventions. A further important feature of tagging systems is that tags can be shared across users, in a way that classification of web resources can be exploited by others leading to *social tagging* and *folksonomies* [26]. Even if through tagging systems annotation and classification activities

G. Varese · S. Castano
Università degli Studi di Milano, Dipartimento di Informatica e Comunicazione
Via Comelico 39, 20135, Milano (MI), Italy
e-mail: {varese,castano}@dico.unimi.it

N. Bessis and F. Xhafa (Eds): Next Generation Data Technologies for CCI, SCI 352, pp. 87–112.
springerlink.com

shifted from an individual level to a collective level, further issues need to be addressed in order to exploit the collective knowledge emerging from social tagging in a more effective way. In fact, the complete freedom of choosing any term for the annotation of web resources inevitably leads to the generation of messy sets of tags. A lot of research is currently focused on trying to organize folksonomies [3, 22] and associate with them a certain degree of semantics [19, 25, 7], in order to enable semantic resource search. Some typical problems that need to be faced include the capability of dealing with tag heterogeneity, typographical errors, acronyms, abbreviations, compound words, slangs, or even nonsense words. Moreover, linguistic, terminological, and semantic relations between tags in folksonomies need also to be identified and represented with the aim of providing more effective search capabilities.

In this chapter, we go in this direction and we propose an approach to extract *collective tag intelligence* from one or more folksonomies, which is conceptualized in form of *tag equivalence clusters*. The proposed approach relies on i) *normalization techniques* to group tags into equivalence clusters and to discover semantic, terminological, and linguistic relations between them and ii) *similarity techniques* for tag matching and coordination on the basis of available collective tag intelligence. Proposed techniques have been evaluated over two real tag datasets extracted from del.icio.us and Flickr folksonomies. Using collective tag intelligence and similarity techniques improves search results, in that, given a target keyword, a set of relevant resources is retrieved larger than the one returned by existing folksonomies and related search functionalities.

This chapter is organized as follows. In Section 2, we discuss related work. In Section 3, we introduce collective tag intelligence and normalization techniques for its extraction from a set of tag assignments. In Section 4, we describe the similarity techniques for tag matching based on collective tag intelligence. In Section 5, we discuss the evaluation and application of the proposed techniques with reference to real datasets composed of tag assignments extracted from del.icio.us and Flickr systems. Finally, concluding remarks are given.

2 From Folksonomies to Collective Tag Intelligence

Before going into detail of the various approaches proposed in the literature, we want to briefly recall the meaning of the terms *folksonomy*, *taxonomy*, and *ontology*.

Folksonomy. A folksonomy is a collection of free text labels assigned by users to heterogeneous resources (e.g., images, documents, web pages) as the result of a collaborative annotation process [9]. The annotation process does not generally impose any kind of restriction on tag choice/definition. As a consequence, terms in a folksonomy are freely chosen by the users, without complying with structure or semantic constraints for their specification and organization.

Taxonomy. A taxonomy is a collection of terms of a controlled vocabulary organized into a hierarchical structure according to a generalization/specialization relationship by which a parent term has a more general meaning than a corresponding child term. In a taxonomy, an inheritance relationship can hold between parent and child terms, by which if properties, behaviour, and/or constraints are specified for the parent, they are automatically specified also for the child, which, in turn, can add one or more of them.

Ontology. An ontology is a vocabulary of terms which denote concepts representing a set of individuals [1]. Terms in an ontology are formally interpreted according to a well-defined semantics and they are organized according to semantic relations, constraints, and rules, including also the hierarchical relation typical of taxonomies.

Folksonomies cannot offer the expressivity and formality of ontologies and are affected by inconsistency/redundancy problems due to subjective definition of tags. However, they are widely used for web resource annotation in real systems due to their ease of use and management. Research work is currently focused on developing solutions for improving folksonomy organization borrowing some formal and semantic properties from ontologies and taxonomies, in order to obtain more structured tag organizations with a certain degree of semantics.

In the following, we discuss state of the art work by distinguishing two main categories of approaches dealing with social tagging system management and by framing the contributions of this chapter with respect to them.

2.1 Tag Classification Approaches

These approaches are focused on extracting collective tag intelligence in the form of taxonomies or ontologies from folksonomies using some kind of tag classification technique. For example, the approaches presented in [15] and [17] rely on the use of the WordNet lexical dictionary [20] to detect correct relations between tags. In [25], authors propose a methodology to build an ontology starting from a set of tags by exploiting information harvested from WordNet, Google, Wikipedia and other similar knowledge repositories available in the Web. This way, it is possible to automatically detect terminological relations between tags like synonymy or hyponymy to be used for tag classification. Schmitz [23] proposes a probability model to build an ontology from tags extracted from Flickr. Subsumption relations between tags are mined on the basis of the conditional probability between pairs of tags, by considering the number of resources containing each tag and the number of users who used each tag. An alternative approach for building a taxonomy starting from a set of tag assignments is presented in [2]. In this work, a parent-child or a sibling relation between each tag and its most frequently co-occurring tag is established. The choice about the kind of relation to consider is taken with the help of WordNet.

Mika [19] provides a model of semantic-social networks for extracting lightweight ontologies from del.icio.us, which exploits co-occurrence information for clustering tags over relevant concepts. Heymann and Garcia-Molina [11] propose a method for building a hierarchical taxonomy according to a defined measure of tag centrality in the tag graph. A similar approach is presented in [8], where authors distinguish between subjective tags, which reflect user's ideas about resources (e.g., "cool", "funny"), and objective tags, which are related to the resources themselves (e.g., "tutorial", "webtechnology"). Thus, the tag taxonomy is created by taking into account only the objective tags.

A different approach to deal with folksonomy mapping into ontologies is presented in [7]. In this work, authors propose to build an RDF description of a generic folksonomy, where the ontology concepts represent the elements of the folksonomy itself, rather than general concepts.

2.2 *Similarity-Based Search Approaches*

Several contributions deal with the issue of defining similarity-based techniques for social annotations with the goal of improving web resource search and retrieval. A survey of similarity measures for collaborative tagging systems is provided in [18]. Cattuto et al. [6] propose a method for creating networks of similar web resources. In particular, similarity between resources is determined by analyzing the tags used for their annotation, their respective TF-IDF value, and their intersection. The TF-IDF value (Term Frequency - Inverse Document Frequency) [21] is a measure which is used in information retrieval to evaluate the importance of a word for a specific document in a collection of documents. The importance increases proportionally to the number of times a word appears in the document but is offset by the frequency of the word in the whole collection. Applied to social tagging, the TF-IDF value can be used to evaluate the importance of a tag for a specific resource, by counting the number of times the tag has been used to annotate such resource and the number of times the tag has been globally used to annotate other resources. In [26], authors propose an application, called DBin, where networks of similar users are created in order to collaboratively build RDFS ontologies over domains of interest starting from the del.icio.us tags. Similarity techniques exploiting the co-occurrence between tags are described in [3] for tag clustering and in [24] to provide meaningful suggestions during the tagging phase of photos in Flickr. A formal model to enhance the information retrieval functionalities of folksonomies is provided in [14, 13]. In particular, Hotho et al. [14] propose a method for converting a folksonomy into an undirected weighted network, used for computing a modified PageRank algorithm called FolkRank for ranking query results. In [13], authors propose to use FolkRank in order to identify the relevance of each resource, user and tag, with respect to a specific target resource, user and tag. In [12], authors study the impact that social tagging can have in the traditional web search, analyzing tags in del.icio.us, with respect to the web pages they are associated with.

2.3 Contributions of the Chapter

With respect to the works reported in the previous sections, the main contributions of our work can be summarized as follows.

- **Automated identification of relations between tags.** In our approach, we consider different kinds of relations (i.e., semantic, linguistic, terminological) between tags and we provide normalization techniques for automatically discovering such relations. In doing this, we rely as much as possible on existing on-line dictionaries and lexical tools to make the approach general and applicable in different contexts. In particular, we propose techniques that take into account at the same time not only information coming from the on-line lexical system WordNet, but also co-occurrence and linguistic information carried by compound tags and abbreviations. This in order to automatically discover as many tag relations as possible, ranging from conventional terminological relations (already considered by several approaches described in previous sections) to semantic and linguistic relations. We want to stress that the capability to manage all different kinds of relations at the same time is a new contribution of our work.
- **Automated similarity evaluation between tags.** In this chapter we propose a family of similarity functions for tag matching in order to automatically identify tags which are similar to each other. These functions have been conceived to fully exploit information provided by all the different kinds of tag relations for flexibly ranking similar tags according to different characteristics (i.e., syntactic, semantic, linguistic, terminological similarity). Such similarity functions can be used separately or in combination to evaluate tag similarity and thus to enforce a more effective web resource search.

3 Normalization Techniques for Collective Tag Intelligence Extraction

In this section, we introduce a set of normalization techniques for extracting collective tag intelligence from an input collection of *tag assignments* belonging to one or more folksonomies. Each tag assignment *ta* is a triple of the form:

$$ta = < u, r, TS >$$

meaning that user *u* has annotated the web resource *r* with a set of tags $TS = \{t_1, ..., t_n\}$. The whole collection of tag assignments is denoted by TA.

In our approach, collective tag intelligence is defined as a set of tag equivalence clusters and relations between them, as shown in the conceptual schema of Figure 1.

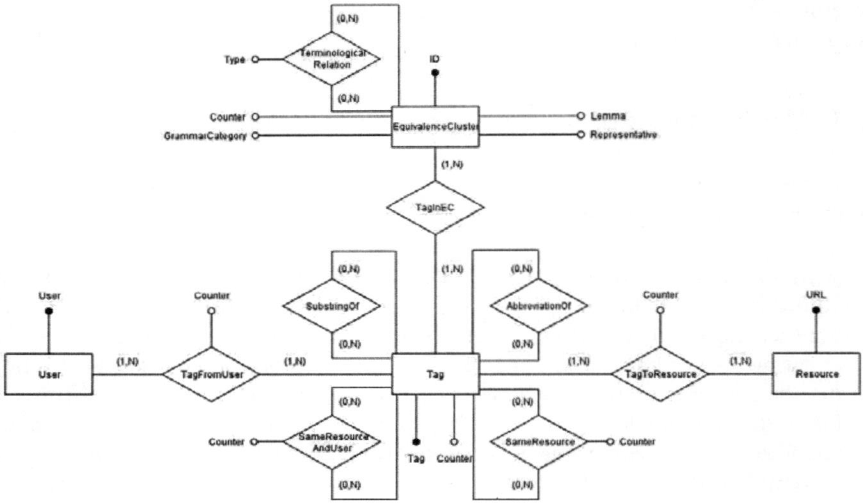

Fig. 1 Conceptual representation of collective tag intelligence

In the following, we first present the collective tag intelligence elements, and then we describe the normalization techniques to extract collective tag intelligence out of the input tag assignments.

3.1 Tag Equivalence Clusters

The *Tag* entity represents all the tags included in the input tag assignments. In the following, we will refer to the whole collection of such tags as *T*. Each tag $t \in T$ is associated with its frequency $f(t)$, corresponding to the number of occurrences of *t* in *T*. Thus, *T* is formally defined as a multiset of ordered pairs $\{(t_1, f(t_1)), \ldots, (t_n, f(t_n))\}$, where $t_1, \ldots, t_n$ are the distinct tags in *T*, and $f(t_1), \ldots, f(t_n)$ are their respective frequencies. Each tag is related to the resources it has been associated with in the corresponding tagging system (*TagToResource* relationship), and to the users who used it (*TagFromUser* relationship).

Tags having the same lemma are grouped together into *equivalence clusters*. This way, singular and plural forms of the same noun are included in the same equivalence cluster, as well as the different declined forms of the same verb. An equivalence cluster *ec* is defined as a 6-tuple of the form:

$$ec = \langle ID, lemma, ECS, grammarCategory, representative, counter \rangle$$

where:

- *ID* is the unique identifier of *ec*;
- *lemma* is the stem which characterizes all the tags included in *ec*;
- $ECS = \{t_1, \ldots, t_n\}$ is the set of tags in *ec*;

- *grammarCategory* ∈ *{common noun, proper noun, verb, adjective, adverb}* is the grammar category of *lemma*;
- *representative* is the representative tag of *ec*, namely the tag $t_i \in ECS$ having the highest frequency $f(t_i)$;
- *counter* is the sum of the frequency of all tags in *ec*.

Each tag can be included in one or more equivalence clusters, depending on the number of different lemmas and/or grammar categories it can be associated with. For example, the tag "playing" can be considered as a noun with lemma "playing", defined as the act of playing a musical instrument, or it can be considered as the gerund form of the verb "play", thus having the lemma "play". In such a case, the tag "playing" will be included in two different equivalence clusters: the one with lemma "playing" and grammar category "noun", and the one with lemma "play" and grammar category "verb".

In the following, we will refer to the whole set of equivalence clusters in the collective tag intelligence repository as *EC*.

In Section 3.5, we will describe our technique for tag equivalence cluster construction based on the use of WordNet.

Tags are then linked to each other through semantic, terminological, and linguistic relations, which are described in the following sections.

3.2 Semantic Relations

Semantic relations $SameResource \subseteq T \times T$ and $SameResourceAndUser \subseteq T \times T$ are defined between tags that have been used to annotate the same web resource. Thus, such kind of relations denote the *co-occurrence* between tags. The number of times that tags are used together (co-occur) in the input collection of tag assignments is also taken into account for tag matching purposes. In particular, the function $counterSR: T \times T \rightarrow \mathbb{N}$ is defined to count the number of times that a given pair of tags $(t_i, t_j) \in SameResource$ has been used to annotate the same web resource, even by different users, while the function $counterSRU: T \times T \rightarrow \mathbb{N}$ is defined to count the number of times that a given pair of tags $(t_i, t_j) \in SameResourceAndUser$ has been used to annotate the same web resource within the same tag assignment (i.e., by the same user). Formally, the value of *counterSR* and *counterSR* for a given pair of tags (t_i, t_j) are calculated as follows.

$$counterSR(t_i, t_j) = \sum (ta_h, ta_k) \in TA \times TA \text{ such that } t_i \in TS(ta_h), t_j \in TS(ta_k), r(ta_h) = r(ta_k)$$

$$counterSRU(t_i, t_j) = \sum ta_h \in TA \text{ such that } t_i \in TS(ta_h), t_j \in TS(ta_h)$$

Where *TA* is the set of all the input tag assignments, *TS(ta)* is the set of tags included in a given tag assignment *ta*, and *r(ta)* is the resource a given tag assignment *ta* is referred to.

As the order in which resources have been tagged does not matter, we have that $counterSR(t_i, t_j) = counterSR(t_j, t_i)$ and $counterSRU(t_i, t_j) = counterSRU(t_j, t_i)$.

We note that the *SameResourceAndUser* relation denotes a stronger semantic relation between t_i and t_j than the *SameResource* relation, in that tag pairs $(t_i, t_j) \in$ *SameResourceAndUser* are a subset of tag pairs $(t_h, t_k) \in$ *SameResource*. Moreover, for each pair of tags (t_i, t_j), we have that:

$$counterSRU(t_i, t_j) \leq counterSR(t_i, t_j) \leq \min\{f(t_i), f(t_j)\}$$

In order to capture the difference between the two kinds of semantic relations, consider the tag assignments of Table 1.

Table 1 Example of tag assignments

User (u)	Resource (r)	Tag Set (TS)
u_1	http://www.w3schools.com/	web, tutorials
u_2	http://www.w3schools.com/	html
u_3	http://www.htmldog.com/	web
u_3	http://www.webreference.com/	web, tutorials
u_4	http://www.htmldog.com/	html

For example, the user u_1 has used the tags "web" and "tutorials" to annotate the web page "http://www.w3schools.com/". The semantic relations (i.e., *SameResource* and *SameResourceAndUser*) that can be extracted from tag assignments in Table 1 are shown in Table 2 and Table 3, respectively.

Table 2 *SameResource* relations resulting from the example in Table 1

t_i	t_j	$counterSR(t_i, t_j)$
web	html	2
web	tutorials	2
html	tutorials	1

Table 3 *SameResourceAndUser* relations resulting from the example in Table 1

t_i	t_j	$counterSRU(t_i, t_j)$
web	html	0
web	tutorials	2
html	tutorials	0

For example, we have that *("web", "html")* ∈ *SameResource* and *counterSR("web", "html")* = 2, because both tags "web" and "html" have been used to annotate two different web pages (i.e., "http://www.w3schools.com/" and "http://www.htmldog.com/"). However, they have been used within different tag assignments (i.e., by different users), and thus *("web", "html")* ∉ *SameResourceAndUser* (i.e., *counterSRU("web", "html")* = 0). On the contrary, we have that *("web", "tutorials")* ∈ *SameResourceAndUser* and *counterSRU ("web", "tutorials")* = 2, because tags "web" and "tutorials" have been used together (i.e., within the same tag assignment) to annotate two different web pages (i.e., "http://www.w3schools.com/" and "http://www.webreference.com/").

3.3 Linguistic Relations

Linguistic relations *SubstringOf* $\subseteq T \times T$ and *AbbreviationOf* $\subseteq T \times T$ are defined between tags denoting different forms of the same expression (e.g., compound words, abbreviations). In particular, the *SubstringOf* relation links together pairs of tags (t_i, t_j) such that t_i is a substring of t_j, where both t_i and t_j belong to T. For example, when the tag "design&technology" is processed, supposing that the tags "design" and "technology" both belong to T, we set *("design", "design&technology")* ∈ *SubstringOf* and *("technology", "design&technology")* ∈ *SubstringOf*. The *AbbreviationOf* relation links together pairs of tags (t_i, t_j) such that t_i is an abbreviation of t_j, where both t_i and t_j belong to T. For example, when the tag "nyc" is processed, supposing that the tag "newyorkcity" belongs to T, we set *("nyc", "newyorkcity")* ∈ *AbbreviationOf*.

3.4 Terminological Relations

Terminological relations $TR \subseteq EC \times EC$ are defined between tag equivalence clusters. Terminological relations considered in our approach are the following:

- *SynonymOf (SYN)*: is a synonym of;
- *HypernymOf/HyponymOf (BT/NT)*: is more general than/is more specific than;
- *HolonymOf/MeronymOf (RT)*: includes/is a part of;
- *InstanceOf/HasInstance (IS)*: is an instance of/is the type of.

Thus, a terminological relation TR_h, with $TR_h \in \{SYN, BT, NT, RT, IS\}$, is defined between two equivalence clusters ec_i and ec_j if their respective lemmas verify the relation TR_h (i.e., $(ec_i, ec_j) \in TR_h$). For example, having an equivalence cluster ec_i with lemma "technology", and an equivalence cluster ec_j with lemma "engineering", a *SYN* terminological relation holds between ec_i and ec_j (i.e., $(ec_i, ec_j) \in$ *SYN*), because "technology" and "engineering" are synonyms. Terminological relations are automatically identified with the help of WordNet, as described in Section 3.5.

3.5 Normalization Techniques

In this section, we describe normalization techniques for extracting collective tag intelligence starting from an input collection of tag assignments. Each tag *t* belonging to the input collection is normalized according to the process shown in Figure 2.

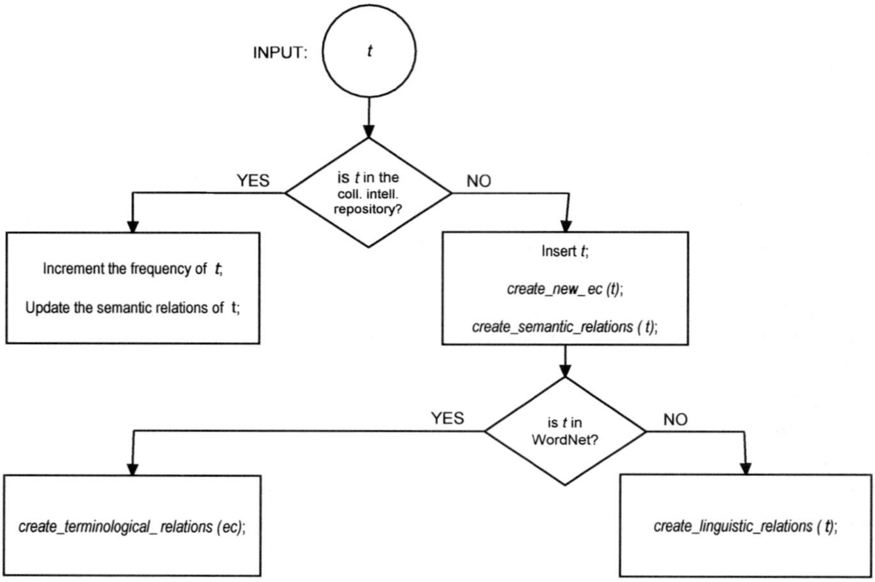

Fig. 2 Tag normalization process

If the input tag *t* is already stored in the collective tag intelligence repository, its frequency *f(t)* is incremented. The counters of the semantic relations *SameResource* and *SameResourceAndUser* (i.e., *countrerSR* and *countrerSRU*) between *t* and the other tags in the collective tag intelligence repository are also updated.

If *t* is a new tag, it is pre-processed in order to identify possible special characters (e.g., _, -, +, *) or numbers, and its equivalence clusters (one or more) are created using WordNet. In particular, the WordNet search is performed in three steps. In the first step, *t* is searched as it is. If no WordNet entry is found, special characters and numbers are discarded, if any, and a new search is launched. If no WordNet entry is found, special characters and/or upper case characters are exploited to tokenize *t*, and the tokenized version of *t* is searched again in WordNet.

The *create_new_ec* procedure is invoked to create the proper equivalence clusters for tag *t*, distinguishing two cases.

Case 1: t has at least one WordNet entry. In this case, the WordNet entry itself becomes the equivalence cluster lemma and its grammar category is taken from WordNet. The grammar categories considered in WordNet are: nouns, verbs, adjectives, and adverbs. Moreover, we distinguish among common nouns and proper nouns, on the basis of the terminological relations defined in WordNet for the lemma. In particular, if tag *t* is instance of something, its grammar category is set to "proper noun"; otherwise, the "common noun" category is defined. Terminological relations between the equivalence clusters of *t* and the other equivalence clusters in the collective tag intelligence repository are defined as well. The *create_terminological_relations* procedure defines the terminological relations *SYN*, *BT*, *NT*, *RT*, and *IS* between equivalence clusters, on the basis of the WordNet relations holding between their corresponding lemmas.

Case 2: t does not have a WordNet entry. A new equivalence cluster *ec* is created for *t*. *t* becomes the lemma of *ec* and then it is submitted to the *create_linguistic_relations* procedure.

In both *Case 1* and *Case 2*, the semantic relations between *t* and the other tags in the collective tag intelligence repository are created by calling the *create_semantic_relations* procedure.

3.5.1 Dealing with Compound Words and Abbreviations

For each input tag *t* without a WordNet entry, two different scenarios are possible.

- ***t* is a compound word.** A great amount of tags within folksonomies are compound words. Since white spaces are not allowed in a single tag, the components of a compound tag are sometimes separated by special/upper case characters but, most of the times, they are not separated at all. In this case, the *create_linguistic_relations* procedure is called in order to determine if *t* can be decomposed into component tags already included in the collective tag intelligence repository. For each component tag t_i recognized as substring of *t*, we set $(t_i, t) \in$ *SubstringOf*. *t* is tokenized until a valid WordNet entry is found for it, if any. If at least one WordNet entry is found for the tokenization of *t*, the lemma of *t* and its corresponding grammar category can be retrieved from WordNet, and new equivalence clusters are created for *t* following the procedure.
- ***t* is an abbreviation.** In this case, the *create_linguistic_relations* procedure is called in order to determine if *t* is an abbreviation of other tags in the collective tag intelligence repository. To this end, the abbreviations dictionary Abbreviations.com[1] is exploited. Given a tag *t* in input to the abbreviations dictionary, a set of possible extensions of *t* is returned. Thus, for each extension *e* of *t* which is stored in the collective tag intelligence repository, we set $(t, e) \in$ *AbbreviationOf*. If no such relations can be built, each extension *e* of *t* is tokenized, and for each single token e_i of *e* which is included in the collective tag intelligence repository, we set $(e_i, t) \in$ *AbbreviationOf*.

An example of compound words and abbreviations management is shown in Figure 3.

[1] http://www.abbreviations.com/

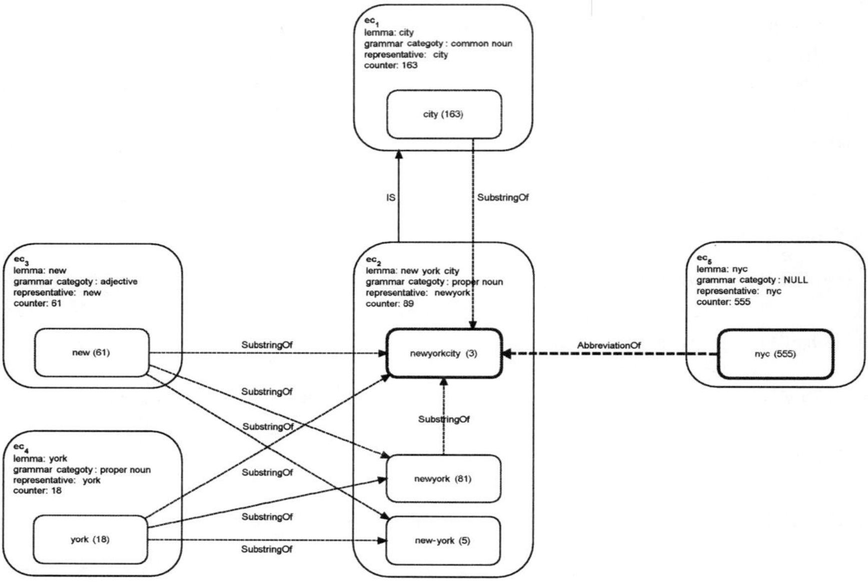

Fig. 3 Example of compound words and abbreviations normalization

Suppose to analyze the tag "newyorkcity", which is highlighted in Figure 3. It is a compound word, and a WordNet entry for it is not found. Suppose that the tags "new", "york", and "city" are tags already processed in the collective tag intelligence repository. When executing the *create_linguistic_relations* procedure, these tags are recognized to be substrings of "newyorkcity", and the corresponding *SubstringOf* relations are created. Thus, the word "newyorkcity" is tokenized into the string "new york city", for which a WordNet entry is found. So, its lemma (i.e., "new york city") and its corresponding grammar category (i.e., "proper noun") are retrieved from WordNet and are associated to the equivalence cluster created for it. Terminological relations between "new york city" equivalence cluster and other equivalence clusters in the collective tag intelligence repository are defined. For example, the *IS* relation is created between the "new york city" equivalence cluster and the "city" equivalence cluster, as "new york city" is defined as an instance of "city" in WordNet. Tags "newyork" and "new-york" are also placed in the equivalence cluster of "newyorkcity", because they are recognized as compound words as well. Now, suppose to analyze the tag "nyc", which is highlighted in Figure 3. Since it does not have a WordNet entry and it is not a substring of any other tag in the collective intelligent repository, it is searched in the abbreviations dictionary, where we discover that it is the abbreviation of "new york city". Thus, the corresponding *AbbreviationOf* relation is created between the tags "nyc" and "newyorkcity". An overall example of collective tag intelligence elements is shown in Figure 4.

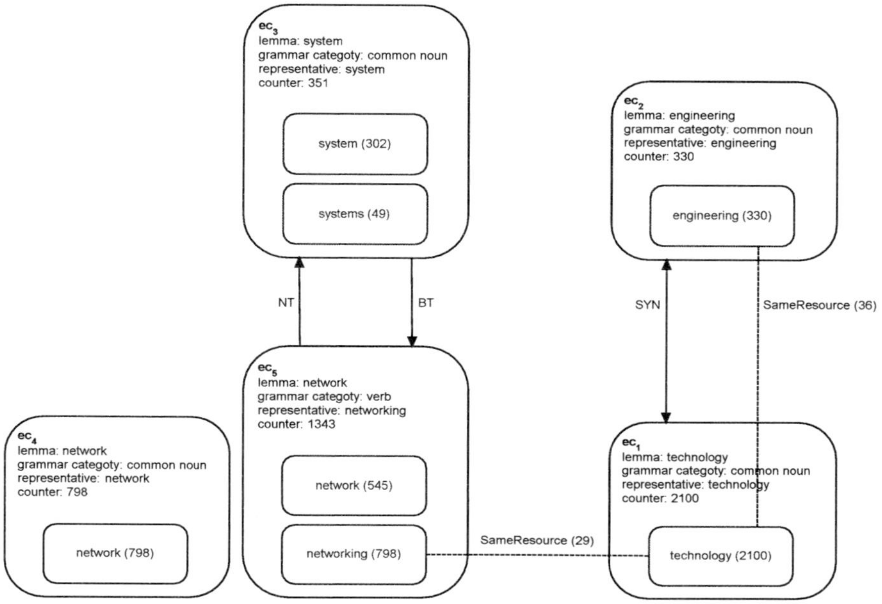

Fig. 4 Example of collective tag intelligence

4 Similarity-Based Techniques for Tag Matching

The generated collective tag intelligence and, in particular, the relations defined between tags and equivalence clusters can be exploited to match tags, in order to find out, given a target tag, the most similar tags to it. To this end, a set of *similarity functions* is defined. Such functions take into account knowledge deriving from equivalence clusters and from the different kinds of relations (i.e., semantic, terminological, linguistic) in the collective tag intelligence. We define a family of similarity functions each one devoted to capture a different kind of similarity, namely syntactic similarity, semantic similarity, terminological similarity, and linguistic similarity. The *syntactic similarity* is calculated by using conventional string matching functions, and it is mainly suited to recognize syntactic variations of the same term, including for instance typographical errors, or similar terms belonging to different grammar categories. The *semantic similarity* determines the level of matching on the basis of the co-occurrence between tags. The *terminological similarity* exploits the relations defined between equivalence clusters according to WordNet. Finally, the *linguistic similarity* takes into account knowledge about compound words/abbreviations and their related tags.

The different kinds of similarity functions are described in detail in the following sections.

4.1 Syntactic Similarity Function

The *syntactic similarity function* analyzes the syntactic similarity of a pair of tags (t_i, t_j). To calculate such similarity, we used the open source SimMetrics library[2], which provides the most popular string matching functions, such as the Levenshtein Distance, the Cosine Similarity, the Jaccard Similarity, the Jaro Distance, the Q-Gram Distance. Formally, the *syntactic similarity function* is defined as follows.

$$sim_{syntactic}(t_i, t_j) = getSimilarity(t_i, t_j)$$

Where *getSimilarity* is the specific string matching function used for calculating the syntactic similarity of the pair of tags (t_i, t_j). For the evaluation, we used as default the Levenshtein Distance, which has been selected because it works well in most situations occurring in the analyzed datasets. The Levenshtein Distance [24] of a given pair of strings (s_i, s_j) is calculated as the minimum number of edits (i.e., insertions, deletions, substitutions of single characters) needed to transform s_i into s_j. In the *getSimilarity* function, the Levenshtein Distance of a given pair of tags (t_i, t_j) is normalized with the length of the longer tag among t_i and t_j, as follows.

$$getSimilarity(t_i, t_j) = 1 - \frac{LevenshteinDistance(t_i, t_j)}{\max\{length(t_i), length(t_j)\}}$$

Where *LevenshteinDistance* is the function which calculates the Levenshtein Distance of the pair of tags (t_i, t_j), and $length(t_i)$ and $length(t_j)$ are the functions which calculate the length of t_i and t_j, respectively.

Example. The results of matching the tag "technology" against the del.icio.us and Flickr datasets using the syntactic similarity function are shown in Table 4.

Table 4 Syntactic similarity results

Keyword	Top-10 matching tags	Similarity value
technology	technoloogy	0.91
technology	technologie	0.82
technology	ethnology	0.81
technology	biotechnology	0.78
technology	webtechnology	0.78
technology	technologies	0.75
technology	terminology	0.73
technology	nanotechnology	0.71
technology	teknologi	0.69
technology	tech-blog	0.69
technology	technological	0.68

[2] http://www.dcs.shef.ac.uk/~sam/simmetrics.html

Using this kind of similarity, the resulting matching tags are syntactically similar to the target keyword. Thus, tags containing typographical errors (e.g., "technoloogy", "technologie"), or which are non-English words (e.g., "teknologi") are also returned as matching. These results can also be useful in that they are related to the tag "technology" as well. However, some of the results can be misleading (e.g., "ethnology"), as they have nothing to do with "technology", even if they are syntactically similar to it. This kind of situation can be avoided by applying more sophisticated similarity functions that exploit semantic knowledge to better discriminate.

4.2 Semantic Similarity Function

The *semantic similarity function* analyzes the semantic similarity of a pair of tags (t_i, t_j) considering the *SameResource* and *SameResourceAndUser* semantic relations defined between tags as well as their frequency.

In order to properly assess the impact of semantic relations in the semantic similarity computation, the counter associated with them is also considered. The idea is that the more frequently two tags t_i and t_j co-occur, the more they are likely to be similar. In particular, considering how the semantic relations are defined, the semantic similarity of two tags is directly proportional to the number of different resources both of them are associated with. In fact, the more such number is high, the more the semantic relation between t_i and t_j can be considered to be valid in general, and not only dependent from the specific content of the web resource at hand or from the user's choice in a certain situation. The *semantic similarity function* has been conceived to allow the choice of which semantic relation to consider, under the consideration that *SameResourceAndUser* relation is stricter than *SameResource* relation.

Moreover, the *semantic similarity function* also takes into account information coming from the frequency (i.e., the IDF value) of t_i and t_j within the collective tag intelligence repository. The rationale is that we want to avoid to give too high importance to co-occurring tags which are very frequent in the collection. In particular, the semantic similarity of two tags t_i and t_j is inversely proportional to their frequency, and thus directly proportional to their respective IDF values. In fact, the more t_i and t_j rarely appear in the tag collection, the more likely their co-occurrence denotes a semantic similarity between them.

In order to combine the information coming from tag co-occurrence and frequency, the *semantic similarity function* is defined as follows.

$$sim_{semantic}(t_i, t_j) = sim_{co-occurrence}(t_i, t_j) \cdot \left[\left(\frac{idf(t_i)}{\text{MAX IDF}} + \frac{idf(t_j)}{\text{MAX IDF}} \right) \Big/ 2 \right]$$

Where the $sim_{co\text{-}occurrence}$ function evaluates the similarity deriving from the co-occurrence of t_i and t_j, $idf(t_i)$ and $idf(t_j)$ are the functions which evaluate the IDF value of t_i and t_j, respectively, and MAX IDF is the IDF value of the tag having the smallest frequency within the collective tag intelligence repository. Both the $sim_{co\text{-}occurrence}$ and the *idf* functions are calculated by taking into account all the tags belonging to the equivalence clusters t_i and t_j belong to, because we want to consider all the tags having the same lemma of them, respectively.

The $sim_{co\text{-}occurrence}$ function is defined as follows.

$$sim_{co-occurrence}(t_i, t_j) = \frac{2 \cdot \sum_{t_h \in T(t_i)} \sum_{t_k \in T(t_j)} counterSemanticRelation(t_h, t_k)}{\sum_{t_h \in T(t_i)} f(t_h) + \sum_{t_k \in T(t_j)} f(t_k)}$$

Where $T(t_i)$ is the set of all the tags belonging to the equivalence clusters of t_i, $T(t_j)$ is the set of all the tags belonging to the equivalence clusters of t_j, and *counterSemanticRelation*$(t_h, t_k) \in \{counterSR(t_h, t_k), counterSRU(t_h, t_k)\}$. For each pair of tags (t_h, t_k), with $t_h \in T(t_i)$ and $t_k \in T(t_j)$, $sim_{co\text{-}occurrence}(t_i, t_j)$ normalizes the total number of co-occurrences of t_h and t_k against the total frequency of all tags in $T(t_i)$ and $T(t_j)$. Note that the normalization of the formula is guaranteed because, for each pair of tags (t_h, t_k), $counterSemanticRelation(t_h, t_k) \leq \min\{f(t_h), f(t_k)\}$.

The *idf* function is defined as follows.

$$idf(t_i) = -\log \frac{\sum_{t_h \in T(t_i)} f(t_h)}{\sum_{ec_k \in EC} counter(ec_k)}$$

Where $T(t_i)$ is the set of all the tags belonging to the equivalence clusters of t_i, *counter(ec)* is the counter of a given equivalence cluster *ec*, and *EC* is the set of all the equivalence clusters in the collective tag intelligence repository.

Example. The results of matching the tag "technology" against the del.icio.us and Flickr datasets using the semantic similarity function are shown in Table 5. In particular, the semantic similarity of each pair of tags is calculated considering the *SameResource* relations defined between them.

Table 5 Semantic similarity results

Keyword	Top-10 matching tags	Similarity value
technology	web	0.09
technology	computer	0.08
technology	geek	0.07
technology	internet	0.06
technology	software	0.05
technology	tech	0.05
technology	programming	0.04
technology	it	0.04
technology	news	0.04
technology	hardware	0.04

With this kind of similarity, matching tags are more semantically related to "technology" than those returned by syntactic similarity, even if their similarity value with it is quite low. This is due to the fact that all matching tags are very frequent in the collective tag intelligence repository, and thus both the $sim_{co\text{-}occurrence}$ and the *idf* functions produce a rather low value.

4.3 Terminological Similarity Function

The *terminological similarity function* analyzes the terminological similarity of a pair of tags (t_i, t_j) by exploiting the terminological relations (i.e., *SYN*, *BT*, *NT*, *RT*, *IS*) defined between the equivalence clusters tags belong to. The idea is to assess the similarity of two tags t_i and t_j on the basis of the kind of the terminological relations defined between the equivalence clusters t_i and t_j belong to. To this end, a weight w is defined for each kind of terminological relation to assess its strength in determining the level of similarity, with $w(SYN) \geq w(BT) \geq w(NT) \geq w(IS) \geq w(RT)$. Specific weights defined for terminological relations are:

- $w(SYN) = 1.0$
- $w(BT) = w(NT) = w(IS) = 0.8$
- $w(RT) = 0.6$

Weights for terminological relationships have been borrowed from our HMatch ontology matching system [4] where they have been defined after extensive experimentation on several ontology matching cases. We performed experimentations using them also on several tag matching cases and we have seen that they work well also for tag matching.

Formally, the *terminological similarity function* is defined as follows.

$$sim_{terminological}(t_i, t_j) = \begin{cases} 1.0 & \text{if } t_i \text{ and } t_j \text{ share an equivalence class} \\ \text{MAX}\{w(TR)\} & \forall (ec_h \in EC(t_i), ec_k \in EC(t_j)) \in TR, \text{otherwise} \end{cases}$$

Where $TR \in \{SYN, BT, NT, RT, IS\}$ is a terminological relation, $EC(t_i)$ is the set of equivalence clusters t_i belongs to, and $EC(t_j)$ is the set of equivalence clusters t_j belongs to. If t_i and t_j share at least an equivalence cluster, their terminological similarity is 1. Otherwise, it is calculated as the weight of the strongest terminological relation holding between the equivalence clusters of t_i and t_j.

Example. The results of matching the tag "technology" against the del.icio.us and Flickr datasets using the terminological similarity function are shown in Table 6.

Table 6 Terminological similarity results

Keyword	Top-10 matching tags	Similarity value
technology	technologies	1.0
technology	engineering	1.0
technology	application	0.8
technology	applications	0.8
technology	nanotechnology	0.8
technology	computer+science	0.8
technology	computerscience	0.8
technology	biotechnology	0.8
technology	it	0.8
technology	hightech	0.8

Using this kind of similarity in the similarity computation process provides as a result matching tags which are terminologically related with the target. In particular, the first result (i.e., "technologies") has the same lemma of "technology", and thus it belongs to the same equivalence cluster. The second result (i.e., "engineering") is a synonym of "technology". All remaining matching tags are either hypernyms (e.g., "application" , "applications") or hyponyms (e.g., "nanotechnology", "computer+science", "computerscience", "biotechnology", "it", "hightech") of "technology". The tag "computer+science" is a compound word which is recognized in WordNet after the pre-processing step, replacing the special character "+" with a space. The other compound words (e.g., "nanotechnology", "computerscience", "biotechnology", "hightech") are recognized in WordNet after their tokenization in the respective component substrings belonging to the collective tag intelligence repository.

4.4 Linguistic Similarity Function

The *linguistic similarity function* determines the linguistic similarity of a pair of tags (t_i, t_j) by exploiting the linguistic relations (i.e., *SubstringOf*, *AbbreviationOf*) defined between tags. The idea is to consider t_i and t_j similar if t_i is an abbreviation or a substring of t_j or, vice versa, t_i is an extension or a compound form of t_j.

Formally, the *linguistic similarity function* is defined as follows.

$$sim_{linguistic}(t_i, t_j) = \begin{cases} 0.8 & \text{if } ((t_i, t_j) \in AbbreviationOf \vee (t_j, t_i) \in AbbreviationOf) \\ 0.6 & \text{if } ((t_i, t_j) \in SubstringOf \vee (t_j, t_i) \in SubstringOf) \end{cases}$$

The *linguistic similarity function* checks if at least a *SubstringOf* or *AbbreviationOf* relation exists between t_i and t_j, and returns a corresponding similarity value. If no linguistic relations exist between t_i and t_j, their linguistic similarity is set to zero. Otherwise, a constant value is returned depending on the kind of linguistic relation holding between t_i and t_j. We set the similarity value for the *AbbreviationOf* relation higher than that of the *SubstringOf* relation to reflect a higher probability for t_i and t_j to be related terms in the former case. In fact, the *SubstringOf* relation can sometimes be misleading, as short tags can be included in many other tags, even if no real semantic connection exists between them.

Example. The results of matching the tag "technology" against the del.icio.us and Flickr datasets using the linguistic similarity function are shown in Table 7.

Table 7 Linguistic similarity results

Keyword	Top-10 matching tags	Similarity value
technology	tech	0.8
technology	tec	0.8
technology	it	0.8
technology	informationtechnology	0.6
technology	design&technology	0.6
technology	music_technology	0.6
technology	nanotechnology	0.6
technology	computer-technology	0.6
technology	computersandtechnology	0.6
technology	science_and_technology	0.6
technology	emerging-technology	0.6

The application of this kind of similarity in the similarity computation process provides a set of matching tags which are compound or abbreviated forms of the keyword.

5 Evaluation and Application to a Real Scenario

The proposed approach has been evaluated and applied to two real datasets (i.e., the PINTS Experiments Data Sets[3] [10]), containing tags crawled during 2006 and 2007 from two different tagging systems, namely del.icio.us[4] and Flickr[5]. Both such datasets consist in a collection of tag assignments. In particular, the del.icio.us dataset contains 634736 tags, 213428 resources, and 6234 users, while the Flickr dataset contains 1389350 tags, 380001 resources, and 16235 users. Starting from the input tag assignments, the corresponding collective tag intelligence is built, by applying the presented normalization techniques.

5.1 Evaluation Issues

In this section, we present the results obtained by applying the proposed approach to the del.icio.us and Flickr datasets. Some of the main features of the considered datasets are reported in Table 8.

---

3 http://www.uni-koblenz-landau.de/koblenz/fb4/AGStaab/Research/DataSets/PINTSExperimentsDataSets/index_html
4 http://del.icio.us
5 http://www.flickr.com

Table 8 Datasets analysis

	del.icio.us	Flickr
Total number of tags	634736	1389350
Number of distinct tags	38309	80041
Average frequency of each tag	17	17
Number of resources	213428	380001
Average number of tags for each resource	3	4
Number of users	6234	16235
Average number of tags for each user	102	86

The evaluation results of the WordNet-based tag pre-processing are shown in Table 9.

Table 9 Pre-processing evaluation

	del.icio.us	Flickr
Number of tags having a WordNet entry	441936	956218
Percentage of tags having a WordNet entry	70 %	69 %
Number of distinct tags having a WordNet entry	13153	29893
Percentage of distinct tags having a WordNet entry	34 %	37 %
Percentage of tags recognized without pre-processing	82 %	94 %
Percentage of tags recognized by removing special characters	15 %	5 %
Percentage of tags recognized by replacing special characters	3 %	1 %

We note that the percentage of tags having a WordNet entry in the two datasets is quite high, but leaves out a good number of tags in both datasets. Tags coming from the Flickr folksonomy contain less special characters than the ones coming from del.icio.us, and they are directly recognized without any pre-processing action.

Once input tags have been analyzed and classified, the proper equivalence clusters are created, and the results are reported in Table 10.

Also in this case, we discovered that the proportion of equivalence clusters having a WordNet lemma, the average number of tags for each equivalence cluster, and the percentage distribution of the equivalence clusters lemmas in the different grammar categories are similar in both datasets.

Table 10 Equivalence clusters evaluation

	del.icio.us	Flickr
Number of equivalence clusters	37240	73821
Number of equivalence clusters having a WordNet lemma	12084	23673
Percentage of equivalence clusters having a WordNet lemma	32 %	32 %
Average number of tags for each equivalence cluster	25	28
Percentage of common noun lemmas	59 %	58 %
Percentage of proper noun lemmas	10 %	14 %
Percentage of verb lemmas	16 %	14 %
Percentage of adjective lemmas	13 %	13 %
Percentage of adverb lemmas	2 %	1 %

Finally, Table 11 shows some of the evaluation results for tags without a WordNet entry.

Table 11 Compound words and abbreviations management evaluation

	del.icio.us	Flickr
Number of non-recognized tags (before the analysis)	27329	59011
Number of compound words recognized in WordNet	2173	8863
Percentage of compound words recognized in WordNet	8 %	15 %
Number of tags having substrings	21389	47955
Percentage of tags having substrings	78 %	81 %
Number of abbreviations recognized	1594	2351
Percentage of abbreviations recognized	6 %	4 %

The percentage of compound words recognized in WordNet is greater in the Flickr folksonomy. In particular, figures in the table refer to the quantity of tags for which a corresponding WordNet entry has been found after the tokenization procedure. Thus, the higher percentage of recognized compound words in the Flickr dataset with respect to the del.icio.us one is probably due to its larger dimension. In fact, the higher is the number of tags, the higher is the probability of finding the component substrings of a compound word. However, the percentage of tags having substrings within the del.icio.us and the Flickr datasets is quite similar. An interpretation of this can be that the Flickr dataset contains more compound words than the del.icio.us one.

The percentage of the tags not normalized by our techniques, which is about the 6% in both datasets, includes special kinds of typographical errors, non-English words, or nonsense words. By manually analyzing such unmanaged tags, we discovered that most of them are proper nouns (referred for example to business trademarks, or technology products). This kind of tags could be managed by adding a special-purpose dictionary where the relevant terms and their variations are stored. In our evaluation, we worked with on-line dictionaries only, to provide a more generally applicable approach.

5.2 *An Example of Similarity-Based Resource Retrieval*

In this section, we consider a real scenario composed of about 600000 web resources of the PINTS datasets related to del.icio.us and Flickr systems. Our goal is to show how the collective tag intelligence can be exploited for *similarity-based resource retrieval*. In particular, we will show that the availability of the collective tag intelligence gives an improved support for retrieving web resources of interest with respect to the traditional search functionalities provided by tagging systems. For example, a user searching for the tag "technology" within a tagging system (e.g., del.icio.us, Flickr) can only find web resources which have been tagged with the word "technology" itself. By contrast, by exploiting the collective tag intelligence, additional web resources can be retrieved, namely those tagged with "technologies", "web", "engineering", "tech", "informationtechnology".

A real example of similarity-based resource retrieval executed over the collective tag intelligence obtained by analyzing the del.icio.us and Flickr folksonomies and their underlying resources is shown in Figure 5.

The user specifies a target keyword (i.e., a tag) t, chooses one or more similarity functions to be used for tag matching, and a value for the threshold k, to be used for selection of the top-k matching tags. In Figure 5, the user specifies the tag "technology" as target keyword, and he chooses the semantic similarity function with a threshold $k = 10$. For each tag t_i in the collective tag intelligence, a similarity value $sim(t, t_i)$ is calculated, and the top-k matching tags are returned by the matching process. The top-10 matching tags obtained by matching "technology" against the collective tag intelligence using the semantic similarity function are shown in Table 5. Finally, all the web resources which have been annotated with t or with one of the top-k tags are returned to the user. Thus, in the example of Figure 5, all web resources r_1, r_2, r_3, and r_4 are returned to the user. On the contrary, by using searching functionalities provided by del.icio.us only for example, only the web resource r_2 can be retrieved and returned to the user, as it is the only one which have been annotated with the keyword "technology" itself and is stored into the del.icio.us system. Moreover, the capability of building collective tag intelligence over multiple tagging systems enables the coordinated access and retrieval to their underlying resources. In our example, by exploiting the collective tag intelligence built out of del.icio.us and Flickr, both web pages (coming from del.icio.us) and images (coming from Flickr) related to "technology" can be returned at the same time to the user.

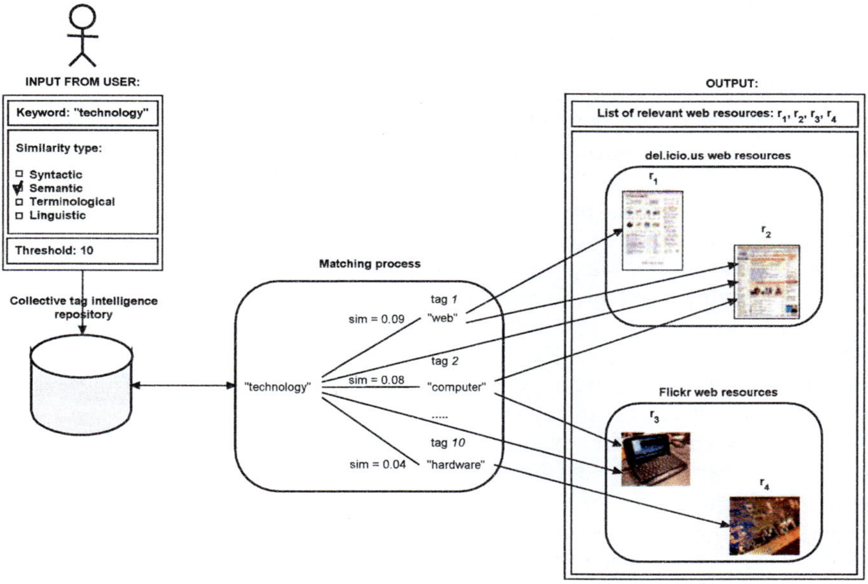

Fig. 5 Example of similarity-based resource retrieval

We have developed a Java prototype for supporting similarity-based resource retrieval based on collective tag intelligence. The collective tag intelligence repository is implemented as a relational database organized according to the ER schema shown in Figure 1. In the prototype, the different similarity functions (i.e., syntactic, semantic, terminological, linguistic) can also be combined to provide a comprehensive similarity value for two tags as follows:

$$
\begin{aligned}
sim(t_i, t_j) = \; & w_{syn} \cdot sim_{syntactic}(t_i, t_j) \; + \\
& w_{sem} \cdot sim_{semantic}(t_i, t_j) \; + \\
& w_{ter} \cdot sim_{terminological}(t_i, t_j) \; + \\
& w_{lin} \cdot sim_{linguistic}(t_i, t_j)
\end{aligned}
$$

where w_{syn}, w_{sem}, w_{ter}, and w_{lin} are weights assigned to the syntactic similarity, the semantic similarity, the terminological similarity, and the linguistic similarity, respectively, with $w_{syn} + w_{sem} + w_{ter} + w_{lin} = 1$.

The weight associated with each kind of similarity can be set by the user according to the specific need. In particular, the different kinds of similarity can be analyzed in an independent or combined way. If the user decides to consider only one kind of similarity, the weight associated with the corresponding similarity function is set to 1, and remaining weights to zero. If the user decides to combine $n \in \{1, 2, 3, 4\}$ different kinds of similarity, the weight associated with each corresponding similarity function is set to $1/n$, and remaining weights to zero. Evaluation results described in this chapter have been produced by considering a single similarity function at a time.

6 Conclusions

In this chapter, we presented normalization and similarity techniques to extract collective tag intelligence and perform similarity-based resource retrieval. Collective tag intelligence has been defined in form of tag equivalence clusters with semantic, terminological, and linguistic relations between them. We described how the collective tag intelligence can be extracted starting from an input set of tag assignments, and how the relations between tags and equivalence clusters can be automatically identified by relying on conventional, on-line dictionaries like WordNet and Abbreviations.com. We presented an application example showing how the collective tag intelligence built from a significant dataset of del.icio.us and Flickr systems can be exploited to provide enhanced similarity-based search functionalities of underlying web resources. Evaluation results obtained by applying the proposed techniques over datasets extracted from two of the most popular tagging systems, namely del.icio.us and Flickr, have been discussed.

Goals of the future work regards the capability to automatically identify semantic relations between the different term-components of compound words and to manage tags that remain unmanaged in the current approach (e.g., non-English words, proper nouns of trademarks and products). In particular, the semantic relations between the term-components of compound words can be identified by manually analyzing the set of recurrent composition patterns of compound tags within folksonomies, and by defining heuristics to automatically identify the most important component of each compound tag. Tags which are non-English words or proper nouns of trademarks and products can be managed by exploiting different external sources, such as multi-language dictionaries and/or web-based encyclopedias like Wikipedia or special-purpose dictionaries. The collective tag intelligence repository can also be extended by including web resources extracted from other sources than folksonomies, such as for example social networks, blogging systems, or ontologies. Some preliminary results in this direction are presented in [5].

References

1. Baader, F., Calvanese, D., McGuinness, D.L., Nardi, D., Patel-Schneider, P.F.: The Description Logic Handbook: Theory, Implementation, and Applications. Cambridge University Press, Cambridge (2003)
2. Barla, M., Bielikovà, M.: On Deriving Tagsonomies: Keyword Relations Coming from Crowd. In: Proceedings of the 18th International Conference on Computational Collective Tag Intelligence (2009)
3. Begelman, G., Keller, P., Smadja, F.: Automated Tag Clustering: Improving Search and Exploration in the Tag Space. In: Proceedings of the Collaborative Web Tagging Workshop co-located with the 15th International World Wide Web Conference (2006)
4. Castano, S., Ferrara, A., Montanelli, S.: Matching Ontologies in Open Networked Systems. Techniques and Applications Journal on Data Semantics (2006)
5. Castano, S., Ferrara, A., Montanelli, S., Varese, G.: Matching Micro-Data. In: Proceedings of the 18th Italian Symposium on Advanced Database Systems (2010)
6. Cattuto, C., Baldassarri, A., Servedio, V.D.P., Loreto, V.: Emergent Community Structure in Social Tagging Systems. Advances in Complex Systems (2008)

7. Echarte, F., Astrain, J., Córdoba, A., Villadangos, J.: Ontology of Folksonomy: A New Modeling Method. In: Proceedings of the Semantic Authoring, Annotation and Knowledge Markup (2007)
8. Eda, T., Yoshikawa, M., Uchiyama, T., Uchiyama, T.: The Effectiveness of Latent Semantic Analysis for Building Up a Bottom-Up Taxonomy from Folksonomy Tags. In: Proceedings of the 18th International World Wide Web Conference (2009)
9. Golder, S.A., Huberman, B.A.: The Structure of Collaborative Tagging Systems. Journal of Information Science 32(2), 198–208 (2005)
10. Görlitz, O., Sizov, S., Staab, S.: PINTS: Peer-to-Peer Infrastructure for Tagging Systems. In: Proceedings of the 7th International Workshop on Peer-to-Peer Systems (2008)
11. Heymann, P., Garcia-Molina, H.: Collaborative Creation of Communal Hierarchical Taxonomies in Social Tagging Systems. Technical Report, Stanford University, Computer Science Department (2006)
12. Heymann, P., Koutrika, G., Garcia-Molina, H.: Can Social Bookmarking Improve Web Search? In: Proceedings of the International Conference on Web Search and Data Mining (2008)
13. Hotho, A., Jäschke, R., Schmitz, C., Stumme, G.: FolkRank: A Ranking Algorithm for Folksonomies. In: Proceedings of the Workshop on Information Retrieval of the Special Interest Group on Information Retrieval (2006)
14. Hotho, A., Jäschke, R., Schmitz, C., Stumme, G.: Information Retrieval in Folksonomies: Search and Ranking. In: Sure, Y., Domingue, J. (eds.) ESWC 2006. LNCS, vol. 4011, pp. 411–426. Springer, Heidelberg (2006)
15. Laniado, D., Eynard, D., Colombetti, M.: Using WordNet to Turn a Folksonomy into a Hierarchy of Concepts. In: Proceedings of the 4th Italian Semantic Web Workshop (2007)
16. Levenshtein, V.I.: Binary Codes Capable of Correcting Deletions, Insertions, and Reversals. Soviet Physics Doklady 10(8), 707–710 (1966)
17. Lin, H., Davis, J., Zhou, Y.: An integrated approach to extracting ontological structures from folksonomies. In: Aroyo, L., Traverso, P., Ciravegna, F., Cimiano, P., Heath, T., Hyvönen, E., Mizoguchi, R., Oren, E., Sabou, M., Simperl, E. (eds.) ESWC 2009. LNCS, vol. 5554, pp. 654–668. Springer, Heidelberg (2009)
18. Markines, B., Cattuto, C., Menczer, F., Benz, D., Hotho, A., Stumme, G.: Evaluating Similarity Measures for Emergent Semantics of Social Tagging. In: Proceedings of the 18th International World Wide Web Conference (2009)
19. Mika, P.: Ontologies are Us: A Unified Model of Social Networks and Semantics. Web Semantics: Science, Services and Agents on the World Wide Web 5, 5–15 (2007)
20. Miller, G.: WordNet: A Lexical Database for English. Communications of the ACM 38, 39–41 (1995)
21. Salton, G., Buckley, C.: Term-Weighting Approaches in Automatic Text Retrieval. Information Processing & Management (1988)
22. Schmitz, C., Grahl, M., Hotho, A., Stumme, G., Cattuto, C., Baldassarri, A., Loreto, V., Servedio, V.D.P.: Network Properties of Folksonomies. AI Communications 20, 245–262 (2007)
23. Schmitz, P.: Inducing Ontology from Flickr Tags. In: Proceedings of the Collaborative Web Tagging Workshop co-located with the 15th International World Wide Web Conference (2006)
24. Sigurbjörnsson, B., Van Zwol, R.: Flickr Tag Recommendation Based on Collective Knowledge. In: Proceedings of the 17th International World Wide Web Conference (2008)

25. Specia, L., Motta, E.: Integrating folksonomies with the semantic web. In: Franconi, E., Kifer, M., May, W. (eds.) ESWC 2007. LNCS, vol. 4519, pp. 624–639. Springer, Heidelberg (2007)
26. Tummarello, G., Morbidoni, C.: Collaboratively Building Structured Knowledge with DBin: from del.icio.us tags to an "RDFS Folksonomy". In: Proceedings of the Social and Collaborative Construction of Structured Knowledge Workshop co-located with the 16th International World Wide Web Conference (2007)

Glossary of Terms and Acronyms

C

Collective tag intelligence: collection of tags, organized in form of tag equivalence clusters, with their corresponding semantic, terminological, and linguistic relations.

F

Folksonomy: collection of free text labels assigned by users to heterogeneous resources (e.g., images, documents, web pages) as the result of a collaborative annotation process.

L

Linguistic relations: relations defined between tags denoting different forms of the same expression (e.g., compound words, abbreviations).

N

Normalization techniques: set of techniques for extracting collective tag intelligence from an input collection of tag assignments.

S

Semantic relations: relations defined between tags that have been used to annotate the same web resource.

Similarity functions: set of functions measuring the different kinds of similarity between tags.

$sim_{linguistic}$: linguistic similarity function.

$sim_{semantic}$: semantic similarity function.

$sim_{syntactic}$: syntactic similarity function.

$sim_{terminological}$: terminological similarity function.

Similarity-based resource retrieval: method of retrieving web resources of interest based on the exploitation of the collective tag intelligence and the developed similarity techniques.

Social tagging: collaborative process of annotation of web resources.

T

Tag: free text label assigned by a user to a resource.

Tag assignment: assignment of a set of tags to a resource performed by a user.

Tag equivalence cluster: set of tags having the same lemma.

Terminological relations: relations between tag equivalence clusters borrowed from WordNet.

BT/NT: *HypernymOf/HyponymOf* relation.

IS: *InstanceOf/HasInstance* relation.

RT: *HolonymOf/MeronymOf* relation.

SYN: *SynonymOf* relation.

W

WordNet: lexical dictionary for the Engligh language.

Chapter 5
Trust-Based Techniques for Collective Intelligence in Social Search Systems

Pierpaolo Dondio and Luca Longo

Abstract. A key-issue for the effectiveness of collaborative decision support systems is the problem of the trustworthiness of the entities involved in the process. Trust has been always used by humans as a form of collective intelligence to support effective decision making process. Computational trust models are becoming now a popular technique across many applications such as cloud computing, p2p networks, wikis, e-commerce sites, social network. The chapter provides an overview of the current landscape of computational models of trust and reputation, and it presents an experimental study case in the domain of social search, where we show how trust techniques can be applied to enhance the quality of social search engine predictions.

1 Introduction

A key issue to the success of collaborative decision support systems, and indeed to any effective analysis of collaboratively generated content, is the reliability and trustworthiness of the entities involved. As user-generated content is no more regarded as a second-class source of information, but rather a complex mine of valuable insights, it is critical to develop techniques to effectively filter and discern good and reliable content. *The Wisdom of the Crowd* is not always sufficient to support good decisions, and many situations require the ability to spot the *Wisdom in the crowd*. One of the main challenges concerns how to effectively mine a large set of complex data affected by a great level of noise, represented by non-pertinent, untrustworthy or even malicious data. The proposed solution has to resist malicious attacks, spot low quality information and preserve privacy. Computational model

Pierpaolo Dondio
Department of Computer Science and Statistics, Trinity College Dublin
e-mail: `pdondio@cs.tcd.ie`

Luca Longo
Department of Computer Science and Statistics, Trinity College Dublin
e-mail: `longol@cs.tcd.ie`

N. Bessis, F. Xhafa (Eds.): Next Generation Data Technologies for CCI, SCI 352, pp. 113–135.
springerlink.com

of Trust and Reputation appear to be essential candidates to enhance and support an effective analysis of web activity. These mechanisms could help filter, interpret and rank web-users' behaviour to assign the relevance of web-search results and deliver the most reliable and adequate content. Similarly, they may be helpful in defining user-based anti-spam techniques, in supporting web-analytics applications that mine only trustworthy sites and users' activity, and helping users' segmentation and decisions support tools for online marketing. This chapter presents the current landscape of computational trust models, and describes how such techniques can be used to enhance the quality of collective computed intelligence.

Computational models of the human notion of trust have emerged in the last decade with the aim of predict and quantify the trustworthiness of digital entities in open and collaborative environments. The word Trust is used here to define a quantifiable prediction about user's expected ability to fulfill a task. When applied to computational intelligence, a trust computation helps predicting which peers will likely produce useful and reliable content for the community. A level of trust in our context is therefore a concept that overlaps competence, expertise and reliability. In particular, we present an experimental study case where we apply a trust function in a collaborative distributed domain. The domain chosen is Social Search, a fast-growing information retrieval paradigm where documents are ranked according to how the web-community is sharing and consuming them. Social search represents an ideal study case due to its collaborative, decentralized and large-scale nature. Our experimental study shows how trust techniques improve the quality of Social Search engines, confirming their central role in deploying effective collective intelligence in the age of Global Computing.

This chapter is organized as follows: Section 2 introduces the core concept of collective intelligence and distributed decision making, section 3 describes the current landscape of computational models of trust, and how trust models can be used as decision support tools. Section 4 introduces the concept of Social search, describing briefly the main trends and challenges of this paradigm. Section 5 describes a practical social search technology used in our study case along with the definition and implementation of a trust model for social search. Section 6 describes our experimental results and section 7 concludes the chapter underlining future directions.

2 State-of-the-Art: Distributed Decision-Making and Collaboration

Collaboration is a process where people interact with each other towards a common goal, by sharing their knowledge, learning and building consensus. This concept does not require a leadership figure and it can deliver good results if applied in decentralised distributed systems. The Internet is the most popular scenario where entities are widely distributed, without any kind of authority. The Web 2.0 is the evolution of the World Wide Web. This term refers to applications in which users can contribute independently, sharing information towards new collaborative architectures, creating worldwide network effects. The contribution is intended as a process where an entity, usually an individual, provides a judgement about another entity, either digital or human, by using specific graded relevance systems such as numbers, letters, descriptions. Wikipedia is a good example in which a good degree of collaboration can be achieved. Here humans collaborate towards the development of an open encyclopaedia on a distributed world wide scale, by creating and editing articles about a disparate range of fields. The fact that this online encyclopaedia is written by an open community of users around the world and the majority of its articles can be edited by anyone with access to the Internet underlines the intrinsic degree of collaboration; several studies suggest that its content can be as accurate as other encyclopaedias [42].

The collaboration applied to Web 2.0 applications supports a new kind of shared intelligence, named *Collective Intelligence*. Here users are able to generate their own content building up an infrastructure where contributions are not merely quantitative but also qualitative [43]. *Collective Intelligence* has been defined in several ways. A shared agreement suggests that it is a group/shared intelligence that emerges from the collaboration and competition of many entities, either human or digital. Collecting judjement from a large group of people allows drawing statistical conclusions about the group that no individual would have known by themselves. The resulting information structures can be seen as reflecting the collective knowledge of a community of users and can be used for different purposes. For instance, as in collaborative tagging systems such as Del.icio.us[1], where users assign tags to resources and Web-entities shared with other users, the emerged community's knowledge, due to users' interaction, can be used to construct folksonomy graphs, which can be efficiently partitioned to obtain a form of community or shared vocabulary [38].

Although techniques for Collective Intelligence existed before the advent of the Internet, the ability to capture and collect information from thousands or millions of people on the World Wide Web has accelerated the proposal of new practical technologies aimed to provide applicable intelligence in the decision-making process. *Social Search* may be considered one of these technologies, an application of *Collective Intelligence*. Here multiple entities' behaviour is taken into account in order to deliver a usable supporting tool for classifying and ranking web-resources, therefore predicting web-users' requirements.

3 Computational Trust

Trust and Reputation are two indisputably recognised relevant factors in human societies. Several studies have been carried out in several fields: psychology [20], sociology [4], economy [6] and philosophy [16] . Computational models of trust emerged in the last decade with the aim of exploiting the human notion of trust in open and decentralized environments. According to Luhmann [29], trust is adopted by humans to decrease the complexity of the society we are living by using delegation.Trust has emerged as a key element of decision-support solution helping agents in the selection of good and trustworthy collaborative partners, in the identification

[1] http://www.delicious.com

of reliable pieces of information or as part of soft-security applications. Several definitions of Trust have been proposed. As suggested by Gambetta, trust is a prediction (subjective probability) that the trustee entity will fulfil trustier' s expectations in the context of a specific task [13]. A typical computational trust solution follows the high-level architecture depicted in 1, modelled after the Secure trust engine [9]. In a typical distributed environment, an agent - the trustier - is acting in a domain where he needs to trust other agents or objects, whose ability and reliability are unknown. The trustier agent queries the trust system to gather more knowledge about the trustee agent and better ground its decision

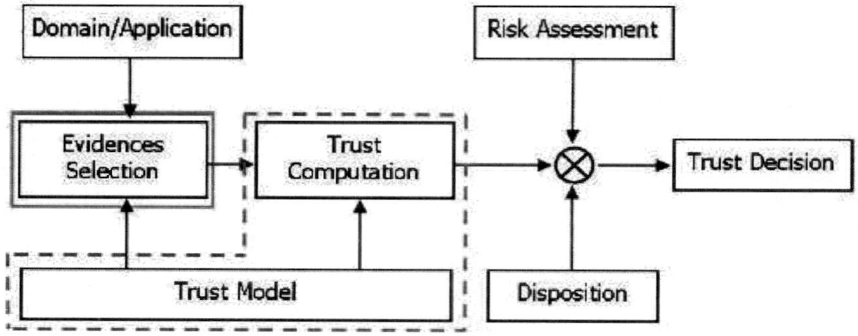

Fig. 1 A computational trust solution.

A trust-based decision in a specific domain is a multi-stage process. The first step is the identification and selection of the appropriate input data. These data are in general domain-specific and identified throught an analysis conducted over the application. We refer to this process as evidence selection and to the inputs used to compute trust as trust evidence. Evidence selection is driven by an underlying trust model that contains the notion of trust on which the entire system is centered. A trust model represents the intelligence used to justify which elements are selected as trust evidence, why some elements are selected and other discarded, and it informs the computation over the selected evidence. A trust model contains the definition of the notion of trust, its dynamics, how it evolves over time and with new evidences, and the mechanisms of trust used in the computation. After evidence selection, a trust computation is performed over evidence to produce trust values, the estimation of the trustworthiness of entities in a particular domain. A trust computation requires the formalization of a computable version of those mechanisms defined in the trust model. Examples of such mechanisms are the past-outcomes one, reputation and recommendation, but also temporal and social factors, similarity, categorization and so forth. For instance, a classical trust system uses two set of evidence: recommendations and past experience. Each of them is quantified separately and then aggregated into a final value. In this final aggregation stage, exogenous factors such as risk and trustier's disposition can also be considered. The output is presented as quantitative trust values and as a set of justifications. Fig 1 depicts the main component of the trust system described so far.

3.1 Computational Models of Trust

Current trust systems can be divided in the following macro-areas:

- Security-oriented approach
- Explicit-feedback systems
- Rule-based systems
- Probability-based systems, or past-outcomes, implicit learning systems
- Game Theoretical
- Cognitive models and computational trust models

With security oriented approach we intend a situation in which the focus is still the on the possession of a valid object, usually a credential, that allows an entity to access some resources and therefore to be trusted. Questionably, they are not trust system but security systems. Examples are PKI infrastructure, with third trusted party of decentralized as in a PGP scenario. A dedicated infrastructure, separated from the application, is in place to gather the required object and transfer among the peers community. The trust intelligence encoded in such systems is limited to the transitivity of trust, that means the fact that trust is propagated through a chain of trusted individuals. Transitivity is the mechanisms at the core of social networking applications [15], with the difference that what is propagated is social information, usually a level of acquaintance between two entities. Information sharing is also at the core of feedback systems, such as reputation or recommendation systems. In such systems users share recommendations in order to have a better idea of their peers. While reputation is a visible global value, expressing the consensus of a group [40], recommendation is an opinion privately shared. Advanced recommendation systems consider the level of trustworthiness of the recommender's peers, or better their ability to provide recommendations; these systems consider situational factors [18], the noise associated to the length of the chain [40], the consensus or conflicts among various sources [18]. In a rule-based system, trust is a collection of rules identified by domain experts that deliver the trust solutions. In the past-outcome paradigm, or direct experience, trust is computed using evidence that the trustier gathered directly from previous interactions in order to predict trustee's future behaviours. A clear definition of this computation, and the correlated notion of trust, is the one produced by the research group Trustcomp.org: 'Trust is a prediction of future behaviour based on past evidences'[2]. There are many different incarnations of the past outcome paradigm, but they all share a common basic scheme. The central notion is that a trust value is computed using the outcomes of all the pertinent past interactions. The value is updated when a new interaction occurs, proportionally to the outcome of this interaction. Examples are found in Quercia's model [37],

[2] Trustcomp online community, www.trustcomp.org

Wang p2p trust engine, the Secure project. In the Probability-based approach trust is represented and manipulated (predicted) as a probability distribution function that typically models the expected behaviour of a trustee. Advantages are a clear (but limited) semantic meaning and effective computational tools. In the use of beta-distribution and the Bayesian Inference, probability offers one of the most powerful tools for computing trust, where probability becomes not only a meaningful trust representation; it goes further, offering also mechanisms for updating and learning trust. The beta-distribution is a family of pdfs used to represent a distribution over binary outcomes. A beta-distribution is completely defined by two positive numbers. The two parameters define completely the expected value and the shape of the distribution. As an example, figure 2 presents a beta distribution with the value of $(1,1)$ on the left and $(8,2)$ on the right.

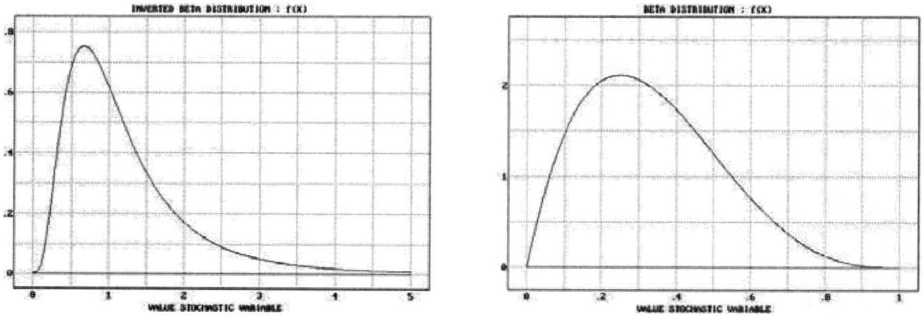

Fig. 2 Beta Distributions

This behaviour maps a representation of a trust value based on evidence. Usually, the two parameters a and b are the numbers of positive and negative evidence regarding the trustee, and the pdf distribution characteristics (expected value, variance) are used for trust values computation and uncertainty assessment. The method is used by Josang [18] or in [40], where r and s are respectively the good and bad evidence regarding a trustee, and $a = r + 1$ and $b = s + 1$ define the corresponding beta pdf. Referring to figure 1, when no information is available about an agent, $(r = s = 0)$, the beta distribution $(1,1)$ is uniform: no value is more likely than others and the uncertainty is at its maximum value. When, for instance, an agent holds 7 positive pieces of evidence and 1 negative, the corresponding beta distribution $(8,2)$ is distributed around the average value of 0.8 with a small variance. In the Game Theoretical approach, as described by Sierra and Sabater in [40], trust and reputation are the result of a pragmatic game with utility functions. This approach starts from the hypothesis that agents are rational entities that chose according to the utility attached to each action considering others' possible moves. Action could be predicted by recognizing an equilibrium to which all the agents are supposed to tend in order to maximize their collective utility. The Game Theoretical approach in trust can also be encoded in the design of the application. In this case, the application is designed so that trust is encoded in the equilibrium of the repeated game the agents are playing. Thus, for rational players trustworthy behavior is enforced.

Finally, a formal notion of trust has been formalized in many rich trust models, notably cognitive models of trust. These models present an articulated and composite notion of trust, and their aim is to define trust as a computational concept. Marsh's first model of trust is still a benchmark. His work gives many insights on the notion of trust as it emerges from social science and it provides a formalization of key aspects, such as basic trust, disposition, reciprocity, situational trust, the concept of a cooperation threshold to start an interaction. A cognitive model of trust defines the mental processes, mechanisms and dynamics of trust. These models stress the nature of trust as a complex structure of beliefs and goals, implying that the trustier must have a "theory of the mind" of the trustee [8]. Trust becomes a function of the degree of these beliefs. Cognitive models present a rich notion of trust, and reject the reduction of trust to a probability-based computation, that is seen as a simple and limited approach, as described by Castelfranchi and Falcone in [8]. Dondio [11] proposes a model of trust/reputation based on defeasible reasoning and knowledge engineering. This model considers the action of evaluating entity's trustworthiness an argumentation process. The form of such argumentation is represented by a defeasible reasoning semantic. A knowledge-based model of trust, as it emerges from social science, provides the content of each argument involved in the trust computation. The model is applicable to a large series of Web 2.0 applications such as Wikis and Online Communities.

4 Social Search

The phenomenon of *Social Search* has been acquiring importance in the World Wide Web with the proliferation of large-scale collaborative digital environments. A social search engine is a type of web search technique that takes into account the interactions or contributions of end-users in order to enhance the relevance of web-search results. The main advantage of such a system is that the value of Web-pages is determined by considering the end-user's perspective, rather than merely the perspective of page authors. This approach takes many forms, from the simplest based on sharing bookmarks [14], to more sophisticated techniques that combine human intelligence with computational paradigms [7]. The recent *Social Search* approach contrasts with established algorithmic or machine-based approaches such as the one of the leading searching engine, Google, whose Page-Rank algorithm [34] relies on the link structure of the Web to find the most authoritative pages. A key challenge in designing *Social Search* systems is to automatically identify human values in the Web. In other words, instead of analysing web-links among web-pages, social search aims to analyse human behaviour. As a consequence, capturing and collecting humans' values is the first step towards the inference of the relevance of web-resources. As mentioned before, a Social Search engine ranks web-resources according to how users of a community consume and judge those resources in relation to some searching needs. A particular practical problem for any potential solution based on gathering end-users' behaviour on a web-page is that they tend to be resistant to explicit and invasive techniques and as a consequence it is not easy to generate strong recommendations. In contrast, implicit feedback techniques capture activity performed by users over Web-pages indirectly.

As suggested in [25] , there are two ways for providing judgements:

- explicitly: users can provide feedback using a specific metric, by using letters, numbers or complex structures. The most popular examples are eBay[3] and Amazon[4] where buyers and sellers can rate transactions using a given graded system;
- implicitly: implicit judgements are automatically inferred by analysing users' behaviour while performing a specific task. Their behaviour is captured by data-mining software that generates logs, row data that need to be analysed, filtered and aggregated in order to extract meaningful information. A web-proxy monitor is an example of logger: it is a piece of software embedded in a web-proxy server, a special computer that acts as an intermediary for requests from others computers seeking web-resources. This software can capture web-site requests, URLs, request time, IP addresses, all potential behavioural information. A lower-level logger is represented by browser-plugins or add-ons, special software able to capture events such as scrolling, reading time, bookmarking, cut-paste, form filling, saving pictures, generated by Web-browsers such as Internet Explorer or Mozilla Firefox. These browser-events are all considered relevant implicit sources of user preferences [21].

Independently from the solutions adopted, whether explicit or implicit, there is a key problem to take into consideration: the trustworthiness of those entities who provide judgments. If entities who provide recommendations are malicious or untrustworthy, the resulting quality of the rank of web-resources is negatively affected. Computational trust techniques can be successfully applied in the context of search to enhance the quality of social Search engines. Here a trust module may be integrated in order to filter data and to make an engine's predictions more accurate. The users' level of trust, for example, may be assessed by considering their expertise in gathering information within the Web, and their ability to fulfil a searching problem. In other words, trustworthy users are the ones able to find the most relevant information when they need it.

5 Computational Trust to Enhance Social Search Ranking: A Practical Study-Case

We have developed a prototype of a search engine based on user-activity containing dedicated algorithms to rank pages, identify search sessions, query boundaries and group similar queries. The Prototype can incorporate a Trust computation to rank each peer based on its activity and use this value to weight its contribution, giving more importance to the most trustworthy peers. The Prototype components are:

[3] http://www.ebay.com
[4] http://www.amazon.com

1. Prototype Plugin: a software component responsible for monitoring a user's activity, storing it locally in a structured file. The plugin captures all the major browser events and generates a well-structured XML string, easy to parse for different purposes. It contains the activity occurred in each window and each tab of the browser; it saves the URL and the title of the opened web-pages along with the start time, the finishing time and the focused time. It gathers the main events that may occur during an Internet session, with the related time-stamp such as bookmark, printing, submitting a form, saving as, cutting, pasting and so forth. The logger also triggers an event every n-seconds of inactivity (set to 10 seconds). Furthermore, the logger traces users' searching sessions. Each time a user submits a query to the Google search engine the logger stores the keywords used, and the ordered list of the search engine results for the query, along with the pages browsed in that search session, and these are identified by analysing the outgoing links from the search engine result page.
2. Prototype Engine: a software component, installed locally with the Prototype Plugin or remotely connected to the plugin, that is responsible for collecting the data generated by the plugin and processing them. The engine is composed by three procedures:

   - Session/Queries Identifier: this algorithm finds the boundaries of a search session, from the starting query to the last page of the last query of the session, the set of pages relating to each query of the session, it interconnects queries belonging to the same session.
   - Evidence Selection: this software component is responsible for interpreting the raw user data coming from the plugin and identifying activity patterns that will be used in the following rank computation.
   - Local Evidence-Based Computation: this algorithm processes locally the information extracted from the evidence selection components and it generates indicators regarding the pertinence of each page to the query it belongs to.

3. Reasoning Engine: this component is responsible for computing a rank for each page and each peer, and connects queries and search session together. It is composed of:

   - Trust Computation (peers): this algorithm processes users' activity and assigns to each peer a trust value that measures the peer's ability to perform and complete a search session.
   - Rank Computation: this algorithm performs the computation of the global indexRanking ranking for each page browsed in the context of the specific query and search session by processing the structured evidence (the arguments) identified by the Prototype Engine. The component takes the structured user data of each peer as an input and computes a global ranking for each page in the context of a specific query and session.
   - Model Definition: where users can edit their own models.

4. Prototype DB: this database contains information about rank pages, queries, search session; peers trust value in a structured way. Data are organized by queries interconnected to each other and by peers. It is composed of:
   - Query Clustering: this component aggregates, links and clusters queries and groups of documents.
5. Prototype Interface: the end-user interface is used by a peer to query the Prototype DB via the Results Manager component.

5.1 The Functioning

A peer connects to the Prototype Community. The Prototype Plugin continuously monitors users' browsing activity, saving it into a local Raw Activity Data (RAD). Every time a peer submits a query to a search engine, the Prototype Plugin saves information related to the set of keywords used and the list of documents proposed by the search engine as result of a search query. The local raw activity file is periodically analyzed by the Prototype Engine that extracts information about a peer's activity for each document. The information extracted is organized into a complex set of evidence that forms an argument against or in favour of the pertinence of the page browsed in the context of a search session. The Prototype Engine sends the processed data, the Structured Activity Data (SAD) to the Reasoning Engine. The Aggregator computes a rank for each page in the SAD by means of a reasoning process. Data about the global activity of the Prototype community is retrieved by the reasoning engine to perform its computation. Each ranked page is saved in the Prototype Database, following a query-based data organization. The information about page ranks, along with the arguments used, is saved into the Prototype DB. Data are organized by query or keywords. A cluster matching component periodically organizes the information contained in the Prototype DB by clustering and grouping queries. The component adds logical links to queries that are considered similar or relevant using a page- and activity-based clustering approach. When a peer starts a search session and he wants to exploit the Prototype Page Rank, the query is sent to the Prototype Database which retrieves the pertinent documents for the query.

5.2 Computational Trust to Enhance the Social Search Engine

In this section, we present an experimental case study that shows how computational trust can be used as a form of collective intelligence to enhance social search, chosen as an example of distributed collaborative application. The ability to search an increasingly large and noisy Web has become a non trivial expertise, involving cognitive and practical abilities, and a familiarity with the browser technology. It is reasonable to assume that web users will exhibit a different level of expertise, directly linked to their ability to correctly identify the most pertinent information.

Therefore, there are reliable and trustworthy users - providing pertinent pages and meaningful results - and untrustworthy ones - generating incorrect results or even noise. Therefore, in our context, trustworthy users capable of finding the most relevant information, that means expert search users .Three main factors are involved in the ability of users to deliver good search results:

1. Cognitive skills, that means, the ability to read quickly, scan information, analytic/global thinking abilities;
2. Search Experience, that means, the familiarity with searching and browsing technology;
3. Domain specific knowledge, that means the expertise and interests that users might have in specific topics.

In this experimental study we define a trust function (or expertise function) to indicate reliable searchers modelled around the second factor. The other two factors, complementary to the second factor, are well-studied and are not discussed or used in the definition of the trust function. Our discussion will be mainly descriptive, focusing on the main concept rather than the technical details. Another important limitation to mention is the fact that this study only considers navigational queries (queries pointing to a specific piece of information) and not to informational query (open-ended). The following picture shows the trust computation is integrated into the Prototype Social Search Engine.

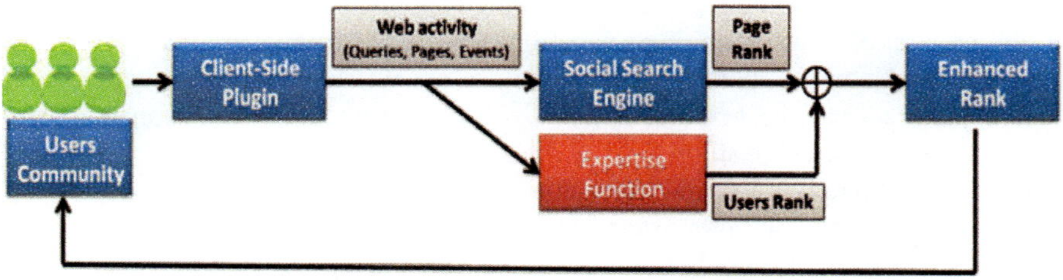

Fig. 3 Search Engine and the expertise Function

Our Prototype is used to gather data (client side plugin). We study how to enhance these search engines with an expertise function that assigns a level of searching expertise to each peer, and use this level to make engines' prediction more accurate. In order to keep more generality, we do not study the effect of the introduction of the trust function directly on our results, but rather we test the effect over a series of performance criteria commonly used by the majority of social search engine. Our trust function computes a level of expertise in the domain of searching based on the three concepts of browsing experience, competence and quality of past-performance. The hypothesis is that, by relying only on the super-users identified, the ability of a social search engine to spot relevant pieces of information in relation to a query will be enhanced. The enhancement is expected for two reasons: first, engines will generate a rank based on a smaller set of data; second, the rank will be based on a subset of information of better quality. We present a case-study performed over a set of 90 users, using their search logs collected for a period of 3 months and the results of lab-based search problems.

5.3 Leveraging Users' Features to Rank Pages

As discussed earlier, there are three main user features that certainly impact page ranking, namely:

1. Domain Expertise
2. Search Experience
3. Cognitive Style and Ability

Our trust function focuses on the second factor, although the other factors could also be embedded into a trust function, since they are complementary features needed for a complete exploitation of users' features. The first feature, Domain Expertise, considers the user's familiarity with the subject on which he or she is searching. The third feature, Cognitive Style, concerns the cognitive ability of the searcher and the type of intelligence. Searching effectiveness is influenced by the way users assimilate new information, users ability to read, memorize, scan documents, analyze text, images, colours. The Search Experience of users - the core of this chapter - can be leveraged to improve search engines ranking. Users who have a better understanding of the breadth of a search engine's capabilities have more ways to go about finding information. Knowing about Boolean operators, exact strings, filtering controls, and having proven strategies to exploit search, allows you a much richer toolset at your disposal. No quantitative study has been proposed to study the impact of search experience on web ranking. In [3] the authors focused on tasks such as classifying the users with regard to computer usage proficiency or making a detailed assessment of how long it took users to fill in fields of a form. The work provides hits and evidence for our analysis, since it was the IT familiarity with the browsing technology to be analyzed as we investigate. The information is yet not used to improve the ranking but rather to investigate the universe of web users and classifying patterns. According to a study performed by Hasting Research in the SEO world[5] the use of Boolean operators and other advanced query filters and connectors proved to narrow search results and improve the relevance of results. Some data samples suggest that these search skills are known to a relatively small percentage of users, but their effectiveness has been proved[6]. Anyway, the works clearly show that relevance is improved, and therefore the use of such advanced features could be used as a proxy to identify expert reliable users. The work performed by Augrihm [2] is described in the next section.

[5] Hastings Research databases of real-time searches and web server logs, 1995-2008.

[6] Hastings Research databases of real-time searches and web server logs, 1995-2008.

5.4 Performance Criteria

In this section we provide a set of criteria to test the impact of our expertise function on search ranking. The way implicit feedback algorithm processes user information suggests that ranking is obviously improved the more users find relevant information, and indeed by the way relevant information is found. Search engine works better when relevant information is found minimizing noise, represented by useless extra clicks, page browsed, queries reformulated and so forth. Even when a query is successful, situations in which the noise is minimized should be rewarded. The criteria we used to assess the quality of the information provided for a specific search problem are:

1. Percentage of peers that successfully completed the (navigational) query, since the more users reached the web page containing the information searched, the more chances the page had to be ranked high in the social engine rank;
2. Number of pages visited for successful queries. An ideal query is the one where only the page containing the answer is visited, since there is no ambiguity in ranking that page high in relation to the query. On the contrary, the more pages are visited the more noise is introduced into ranking algorithms, leading to a higher probability that the correct page will be mixed with useless ones;
3. Number of queries submitted. Analogous to the previous criteria, a limited number of queries refinements helps to better identify the search problem the user is looking for, that means the search context, important information to feed query classification and similarity/matching algorithms, reducing the probability of ambiguity in the search problem definition.
4. Time spent for a query. For navigational queries, short time on a query is a plausible evidence that the information has been found. An ideal navigational query is the one where non pertinent pages are discarded quickly. Moreover, in navigational query type, it is common that a relevant page is quickly analyzed and registers low activity if the information is clearly shown on the page.

The trust function

The trust function used to identify trustworthy users encompasses 3 factors: past performance, experience and competence

Past performance factor
Past-performance can be implemented by looking at the outcome of previous search sessions and computing a level of expertise for each individual proportional to the number of search session completed positively. The factor is easily implemented for navigational queries type. The query is successful if the user can actually find that information, probably duplicated in more than one web-site. In order to compute a level of past-performance for each of our users, we defined a set of search problems S_n for which the piece of information I_n that satisfies the problem is known a priori (Dublin solves the query 'capital of Ireland'). Users are asked to perform S_n and

explicitly write the answer. We could presume that a user is asked to perform some test queries the first time he joins the social search network, following a similar pattern to other reputation-based social networks, or that a human-assisted procedure will analyze some of his queries. A user gets a full score, say 1, for each query if he gets the right answer, otherwise the proximity to the web page containing In is considered. If the user actually browsed the page containing I_n, without identifying I_n, the query failed but its browsing activity contains at least the correct page and a non-null score < 1 is assigned to the user. The user at least had provided the relevant page. With the same reasoning if the user browsed a page P one-hop close to the page containing I_n - assuming that P is not a search engine result page, usually it is a page in the same domain - a smaller not null value is assigned to the user. Past performance were computed using a beta distribution $\beta(a+1,b+1)$ where a is the index of success while b is the index of failure. a and b after the query $t+1$ are computed from the values of a and b based on the first t queries:

$$a_{t+1} = \begin{cases} a_t + 1 & \text{if the query was satisfied} \\ a_t + 0.5 & \text{if the user visited a page 1-hop from the page with the required information} \\ a_t & \text{elsewhere} \end{cases}$$

$$b_{t+1} = \begin{cases} b_t + 1 \quad \text{iff } a_{t+1} = a_t \text{ (query failed)} \end{cases}$$

A value of past performance TPP for a specific user U is given by the average value of the beta distribution $\beta(a+1,b+1)$ related to user U.

Enhancing past performance

The past-performance mechanism can be enhanced by considering not only the outcome of the query, but also the level of difficulty of the query. To consider the difficulty of a query we could consider:

1. the percentage of success of that query P_{suc},
2. time spent for a query T_q
3. number of pages visited $PGtot$,
4. number of queries reformulation N_{qr} for the query.
5. hop-distance of the solution to the search engine page N^u_{hop}

Note the overlapping with our evaluation criteria: the most difficult queries are those more difficult to be properly ranked, where the information searched is hidden and/or far from search engine results. The hop distance for a query is computed as the average distance of hops for the page containing In using the logs of all the successful queries. The level of difficulty for a query is computed by ranking all the five above indicators and by aggregating them, giving more weight to the P_{suc} as it is the only explicit indicator of query difficulty.

$$Qdiff = f(P_{suc}, T_q, PG_{tot}, N_{qr}, N^u_{hop})$$

Experience factor
Our concept of experience is based on how much and for how long the user searched on the web. Experience has to do with the previous quantity of searches, rather than the quality of them, as it was for past-performance. Aligned to other studies, we presume that an experienced internet user has more chance to perform a correct query than an inexperienced one. Experience has been assessed in the following way. All the users were asked to install the plugin logger for a period of about 3 months. The data collected contains all users' browsing activity, including search sessions and queries executed. Using a special feature, we traced if a page belonged to a search session, i.e if there is a click-chain connecting the page and a search engine result page. Data are summarized in table 1, containing the average occurrences per user of all the browsing events monitored. A value of experience is computed considering the time each user spent on the web and the time spent searching. The time spent searching is approximated by the number of queries submitted and the total number of pages browsed in the context of a search session. Experience is therefore the aggregation of 3 ranked indicators:

- $Time_{web}$: total time spent on the web;
- N_q: total number of queries performed;
- P_s: number of pages linked to a search session.

$$Experience = f(N_q, Time_{web}, P_s)$$

Competence
Competence here is defined as the familiarity of users with search engine features and online search dynamics. While our concept of Experience is based purely on quantitative indicators, Competence analyzes the way and style a user performs in its web activity. Competence has not to be confused with the competence of a user in a particular topic. Here it is exclusively a metric of the competence of the user with the task of web searching. In order to compute competence we use the data collected in the 3-months monitoring of our user population (table 1). First, we select a set of events that can be plausibly linked to advanced searching skills, supported by the already cited studies [17][7]. The events E_x were:

1. find in page
2. use of Boolean/special connectors,
3. use of exact string (1-3 all proven to increase searching relevance),
4. use of multitab/multiwindows browsing (evidence of ability to perform parallel analysis of information [12] [17])
5. percentage of results skipped next or/and above, evidence of ability of users to discard non pertinence information in advance [19].

The idea is now to assign a level of competence to each browsing event, and compute the user competence as the weighted average of the events he performed. The

[7] Hastings Research databases of real-time searches and web server logs, 1995-2008.

competence weight of each event is the inverse of the probability that a user will perform that particular action, computed over the population's distribution of events (1). Therefore, less likely events have more impact on competence. The likelihood of a find in page (E1) was computed over the total number of pages browsed, the usage of Boolean connectors (E3) and exact string (E2) over the number of queries, the multi-tab and multi-window navigation (E4-14) over the total number of windows/tab opened. For instance, since 59 find in page have been recorded in 1254 pages, the probability that a user will perform that action is 0.047, and therefore the competence level of a find in page is 21.25. The formulas are shown below (E_{ix} = occurrences of events i, E_i^u occurrences of the event i for user u, N_p = number of pages, $N_q = numberofqueries$).

$$Competence_{E_1} = \left(\frac{E_{ix}}{N_p}\right)^{-1}$$

$$Competence_{E_{2,3}} = (\frac{E_{2,3}}{N_q})^{-1}$$

$$Competence_{E_{4-14}} = (\frac{E_{4-14}}{N_p})^{-1}$$

$$Competence_u(Competence_{E_i}, E_i)$$

Table 1 Browsing activity report

T	Browsing event	Avg. Occurrences per user
T	Total time (Hours)	2.14
N_p	Number of pages	4880
N_q	Query	539
P_s	pages belonging to a search query	1254
E_1	Find in page	59
E_2	Usage of Exact String	121
E_3	Logical Connectors or special command in the query	6.2
E_4	1 window (# of times user used only 1 window to browse)	4880
E_5	2 windows (# of times user used only 2 windows to browse)	291
E_6	3 windows (# of times user used only 3 windows to browse)	232.5
E_7	4 windows (# of times user used only 4 windows to browse)	57
E_8	5+ windows (# of times user used 5 or more windows to browse)	19
E_9	1 tab (# of times user used only 1 tab)	4880
E_{10}	2 tabs (# of times user used only 2 tabs)	1098
E_{11}	3 tabs (# of times user used only 3 tabs)	549.3
E_{12}	4 tabs (# of times user used only 4 tabs)	221.5
E_{13}	5 tabs (# of times user used only 5 tabs)	56
E_{14}	6+ tabs (# of times user used 6 or more tabs)	101

6 The Experimental Results

In this section we describe a first evaluation of the trust metric for social search. The controlled experiment was performed on a population of 93 users and focuses on navigational queries.

6.1 Navigational Queries

We defined a set of 20 navigational queries of different levels of difficulty that were executed by our population of users. For instance, the following are examples of queries of different levels of difficulty:

- Easy
  - Q1. Price of a ticket to enter Malahide Castle
  - Q5. Find if GB Shaw won a Nobel Prize in literature
- Medium
  - Q5. How much was the child benefit in Ireland in 2009?
  - Q9. How do you say 'when' in Latin?
- Hard
  - Q10. Number of 0-0 in the Premier League 2008-2009
  - Q15: When did the company Georgia Gulf double its price last summer in one day? On which news?

6.2 Enhancement of Past Performance

The past performance value is generated by the formula described in the previous section. Queries are ranked according to their degree of difficulty, quantified by Q_{diff} (section 5.4). Referring to the above set of 6 navigational queries, table 2 displays the value of Q_{diff} normalized from 0 to 1.

Table 2 Queries Difficulty Levels

Query ID	Q_{diff} normalised	P_{suc}
10	1	28%
15	0.95	44%
4	0.55	81%
9	0.5	88%
1	0.1	100%
5	0.05	100%

6.3 Baseline Results without Trust Function

We now compute our four performance criteria using all the population of 90 users. Table 3 shows the results for the top two most difficult queries (Q10 and Q15) and the others organized into 6 groups of 3 queries.

Table 3 Baseline results without trust

Query or Group	P_{suc}	N_p	N_q	T
10	28%	18.78	6.2	10′22″
15	44%	14.9	4.2	6′53″
G1	78%	13.77	3.4	6′05″
G2	78%	7.2	2.12	3′56″
G3	91%	4.56	2.6	2′23″
G4	91%	5.15	2.1	3′04″
G5	97.9%	4.89	1.78	2′03″
G6	100%	2.89	1.19	1′34″

6.4 Experiment 1 - with Past Performance

In this first experiment, we test the entire trust function, composed by enhanced past performance, competence and experience. The trust function used assigns more weight to the past-performance factor. User's U trust value SE_u is computed as:

$$SE_u = 2 \cdot T_{pp} + T_{comp} + T_{exp}$$

The 90 users were ranked according to their SE_u and normalized from 1 (the most trustworthy) to 0. In order to compute the past-performance value, the 20 queries were divided into train and test group of balanced difficulties, each containing 10 queries. The four performance criteria were now recomputed using a weighted average, using the trust value as the weight. Table 4 shows a comparison between the indicators for each query.

Results are encouraging and show the benefits of our expertise function in a social search context. The performance gain is evident for the most difficult queries. Q10 has now a success rate of 53%, from a baseline of 28%, 6 pages less visited on average, a slightly diminished number of query reformulation and a gain of 20% of time. This implies that a social search engine would have ranked pages in relation to that query better, with an increased likelihood to find the correct page at the top. In general, this trend is respected over the four criteria for all the queries groups with few exceptions, with an increasing significance when the difficulty of the query increases. The last column contains the results of a Z-test performed with a Z_{crit} level of 90%. We note how 4 out of 4 of the most difficult queries have a statistically

Table 4 Full expert function results

Query or Group	P_{suc}	N_p	N_q	T	Z-test
10	53%	12.45	4.95	8′03″	yes
15	69%	8.17	4.17	4′47″	yes
G1	86%	13.00	3.30	4′57″	yes
G2	87%	6.06	1.79	3′23″	yes
G3	93%	4.97	2.67	2′29″	no
G4	92%	4.52	2.00	2′47″	no
G5	99%	3.95	1.81	2′20″	no
G6	100%	2.89	1.05	1′20″	no

significant improvement, while there is no statistical difference in the low difficulty queries, that already show very high values for all the criteria.

6.5 Experiment 2 - No Past Performance

We wondered if the results would still remain valid by removing the past-performance factor from the expertise function. We noted that by removing this factor, by far the most used in trust models, the remaining computation can be easily implemented in an autonomic way, and therefore scalable. The results obtained are displayed in table 5. The four graphs in figure 4 display, for each criteria, the difference between baseline values and the full (gray line) and limited (dark grey line) trust function, as a percentage of the difference over the baseline values. Therefore, a positive value means that a gain is actually achieved in respect to the baseline function, while negative values mean how the function is actually under-performing the baseline value. Tables show how results are deteriorating, but there is still a significant gain for Q10 and Q15. Nevertheless, there is a significant gap between the two trust computations for the most difficult queries.

Table 5 Limited expert function results

Query or Group	P_{suc}	N_p	N_q	T
10	47%	14.65	5.95	9′12″
15	57.3%	9.17	4.2	5′21″
G1	84.2%	13.98	3.24	5′52″
G2	81.5%	7.04	1.82	3′25″
G3	92.4%	4.83	2.54	2′12″
G4	92.6%	4.78	2.10	2′56″
G5	97.9%	4.86	1.81	1′58″
G6	100%	2.73	1.10	1′28″

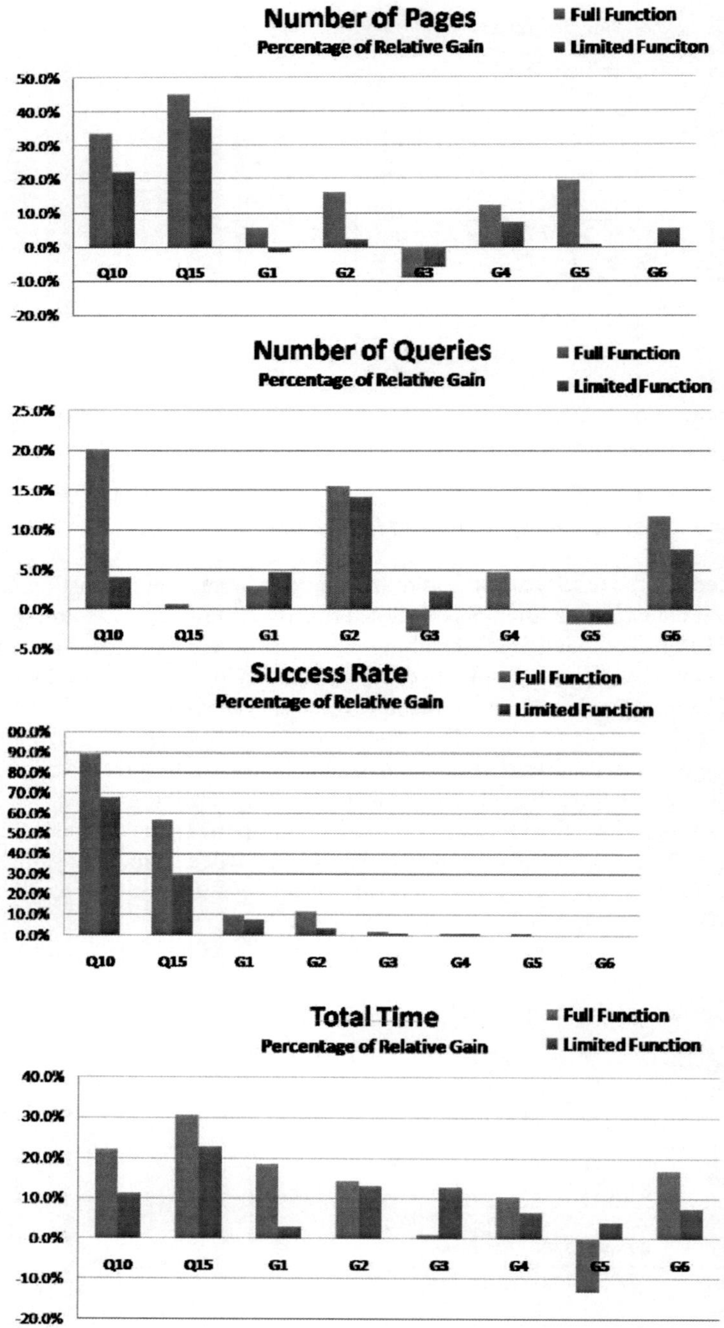

Fig. 4 Percentage of relative gain

7 Future Direction and Open Issues

With the proliferation of large-scale collaborative computing collaborative environments a new problem has emerged in the last few years: the reliability and trustworthiness of the entities involved. Trust is a form of collective intelligence used by humans as a powerful tool for the decision making process. This has enabled the proposal of computational trust models in the last decade aimed at filtering harmful entities. This chapter presented techniques to computationally assess the trustworthiness of entities involved in a collaborative context. An experimental study case was presented in the context of Social Search, where we developed and tested a trust function to spot reliable users. The experimental results show how the page ranking of the Social Search engine can be improved and the time spent to search info diminished, if enough information are collected from the community. The method proved to be more effective for difficult queries, where it is easier to get advantage of expert users. The study showed how trust techniques can improve the quality of Social Search engines, confirming their central role in deploying effective collective intelligence in the age of Global Computing.

Future work include a further expansion and enhancement of the proposed computational model by considering the effort spent by users in consuming activity over the World Wide Web [28]. This will be done in a lower-level of details, by considering the set of actions done by users while consuming the content of a particular web page, that means by considering activity such as clicking, scrolling, cut & paste, finding, bookmarking and so forth over the time dimension. This activity will be mapped to cognitive theories for cognitive effort [27] available in the fields of cognitive science, psychology, in order to further enhance the quality of predictions of a Social Search Engine by filtering not trustworthy entities based on their cognitive effort over the web.

References

1. Agichtein, E., Brill, E., Dumais, S.: Improving Web Search Ranking by Incorporating User Behavior Information. In: SIGIR 2006, Seattle, USA (2006)
2. Agichtein, E., Zheng, Z.: Identifying Best Bet Web Search Results by Mining Past User Behavior. In: KDD 2006, Philadelphia, Pennsylvaia, USA (2006)
3. Atterer, R., et al.: Knowing the User's Every Move - User Activity Tracking for Website Usability Evaluation and Implicit Interaction. In: WWW 2006, Edinburgh (May 23-26, 2006)
4. Buskens, V.: The Social Structure of Trust. Social Networks (20), 265–298 (1998)
5. Ball, E., Chadwick, D., Basden, A.: The Implementation of a System for evaluating Trust in a Pki Environment. In: Proceedings of Trust in the Network Economy, Evolaris (2003)
6. Celentani, M., Fudenberg, D., Levine, D.K., Psendorfer, W.: Maitaining a Reputation Against a Long-Lived Opponent. Econometria 64(3), 691–704 (1966)
7. Chi, E.H.: Information Seeking Can Be Social. Computer 42(3), 42–46 (2009)
8. Castelfranchi, C., Falcone, R.: Trust is much more than Web Probability. In: 32nd Hawaii Int. Conference (2000)
9. Cahill, V., et al.: Using Trust for Secure Collaboration in Uncertain Environments. IEEE Pervasive Computing Magazine 2(3), Special Issue (July-September 2003)

10. Dondio, P., Barrett, S., Weber, S., Seigneur, J.M.: Extracting Trust from Domain Analysis: a Study on Wikipedia. In: IEEE ATC, Wuhan, China (2006)
11. Dondio, P.: Trust as a Form of Defeasible Reasoning. Phd Thesis, Trinity College Dublin
12. Ford, N., et al.: Web Search Strategies and Human Individual Differences: Cognitive and Demographic Factors, Internet Attitudes, and Approaches. Journal of Am. Soc. Inf. Sci. Technol. 56, 7 (2005)
13. Gambetta, D.: Can we trust trust? . In: Trust: Making and Breaking Cooperative Relations, pp. 213–237 (2000)
14. Golder, S.A., Huberman, B.A.: Usage Patterns of Collaborative Tagging Systems. Journal of Information Science 32(2), 198–208 (2006)
15. Golbeck, J.: Trust Networks on the Semantic Web. University of Maryland, USA (2002)
16. Hume, D.: A Treatise of Human Nature. Clarendon Press, Oxford (1737) (1975)
17. Hlscher, C., Strube, G.: Web Search Behavior of Internet Experts and Newbies (2000)
18. Josang, A., Pope, S.: Semantic Constraints for Trust Transitivity. In: 2nd Conference on Conceptual Modelling (2005)
19. Joachims, T.: Optimizing Search Engines Using Clickthrough Data. In: The Proceedings of SIGKDD (2002)
20. Karlins, M., Abelson, H.I.: Persuasion, how Opinion and Attitudes are Changed. Crosby Lockwood & Son (1970)
21. Kelly, D., et al.: Reading Time, Scrolling and Interaction: Exploring Implicit Sources of User Preferences for Relevance Feedback During Interactive Information Retrieval. In: SIGIR 2001, New Orleans, USA (2001)
22. Abdi, H.: Kendall Rank Correlation. In: Salkind, N.J. (ed.) Encyclopaedia of Measurement and Statistics. Sage, Thousand Oaks (2007)
23. Kitajima, M., Blackmon, M.H., Polson, P.G.: Cognitive Architecture for Website Design and Usability evaluation: Comprehension and Information Scent in Performing by Exploration. HCI, Las Vegas (2005)
24. Kleinberg, J.: Authoritative Sources in a Hyperlinked Environment. Journal of the ACM 46(5), 604–632 (1999)
25. Longo, L., Barrett, S., Dondio, P.: Toward Social Search: from Explicit to Implicit Collaboration to Predict Users' Interests. In: WebIST 2009 (2009)
26. Longo, L., Dondio, P., Barrett, S.: Temporal Factors to Evaluate Trustworthiness of Virtual Identities. In: IEEE SECOVAL 2007, Third International Workshop on the Value of Security through Collaboration, SECURECOMM 2007, Nice, France (September 2007)
27. Longo, L., Barrett, S.: Cognitive Effort for Multi-Agent Systems. In: Yao, Y., Sun, R., Poggio, T., Liu, J., Zhong, N., Huang, J. (eds.) BI 2010. LNCS, vol. 6334, pp. 55–66. Springer, Heidelberg (2010)
28. Longo, L., Dondio, P., Barrett, S.: Information Foraging Theory as a Form Of Collective Intelligence for Social Search. In: 1st International Conference on Computational Collective Intelligence Semantic Web, Social Networks & Multiagent Systems, Wroclaw, Poland, (October 5-7, 2009)
29. Luhmann, N.: Familiarity, Confidence, Trust: Problems and Alternatives. In: Trust: Making and Breaking Cooperative Relations, pp. 213–237 (2000)
30. Marsh, S.: Formalizing Trust as Computational Concept. PhD, Stirling (1994)
31. Miller, C.S., Remington, R.W.: Modeling Information Navigation: implications for Information Architecture. In: HCI (2004)
32. Montaner, M., Lopez, B., De La Rosa, J.: Developing Trust in Recommender Agents. In: Proceedings of the First International Joint Conference on Autonomous Agents and Multiagent Systems (AAMAS 2002), Bologna, Italy, pp. 304–305 (2002)

33. Morita, M., Shinoda, Y.: Information Filtering Based on User Behavior analysis and Best Match Text Retrieval. In: 17th ACM SIGIR (1996)
34. Page, L., Brin, S., Motwani, R., Winograd, T.: The PageRank Citation Ranking: Bringing Order to the Web. Standford University, Standford (1999)
35. Pirolli, P.: Information Foraging Theory. Adaptive Interaction with Information. Oxford University Press, Oxford (2007)
36. Pirolli, P., Fu, W.: SNIF-ACT: A Model of Information Foraging on the World Wide Web. In: Brusilovsky, P., Corbett, A.T., de Rosis, F. (eds.) UM 2003. LNCS, vol. 2702. Springer, Heidelberg (2003)
37. Quercia, D.: STRUDEL: Supporting Trust in the Establishment of Peering Coalitions. In: ACM SAC 2006, pp. 1870–1874 (2006)
38. Robu, V., Halpin, H., Shepherd, H.: Emergence of Consensus and Shared Vocabularies in Collaborative Tagging Systems. ACM Transactions on the Web (TWeb) 3(4), article 14 (September 2009)
39. Stephens, D.W., Krebs, J.R.: Foraging Theory, Princeton, NJ (1986)
40. Sabater, J., Sierra, C.: REGRET: A reputation Model for Gregarious Societies. In: 4th Workshop on Fraud and Trust in Agent Societies, Montreal, Canada, pp. 61–69 (2001)
41. Velayathan, G., Yamada, S.: Behavior-based Web Page Evaluation. In: WWW 2007, Banff, Alberta, Canada, May 8-12 (2007)
42. Viégas, B.F., Wattenberg, M., Kushal, D.: Studying Cooperation and Conflict between Authors with History from Visualizations, MIT Media Lab. and IBM Research
43. Weiss, A.: The Power of Collective Intelligence. Collective Intelligence, 19–23 (2005)

Part II
Advanced Models and Practices

Chapter 6
Visualising Computational Intelligence through Converting Data into Formal Concepts

Simon Andrews, Constantinos Orphanides, and Simon Polovina

Abstract. Formal Concept Analysis (FCA) is an emerging data technology that complements collective intelligence such as that identified in the Semantic Web, by visualising the hidden meaning in disparate and distributed data. The chapter demonstrates the discovery of these novel semantics through a set of FCA open source software tools, *FcaBedrock* and *In-Close*, that were developed by the authors. These tools add computational intelligence by converting data into a Boolean form called a Formal Context, prepare this data for analysis by creating focused and manageable sub-contexts and then analyse the prepared data using a visualisation called a Concept Lattice. The Formal Concepts thus visualised highlight how data itself contains meaning, and how FCA tools thereby extract data's inherent semantics. The chapter describes how this will be further developed in a project called "Combining and Uniting Business Intelligence with Semantic Technologies" (CUBIST), to provide in-data-warehouse visual analytics for Resource Description Framework (RDF)-based triple stores.

Keywords: Formal Concept Analysis (FCA), Formal Context, Formal Concept, visualisation, Concept Lattice, data warehousing, in-warehouse analytics, objects and attributes, Galois connection, Semantic Web, RDF, distributed data, disparate data.

1 Introduction

As its core, the Semantic Web comprises of design principles, collaborative working groups and a variety of enabling technologies [28]. It includes formal specifications such as the *Resource Description Framework (RDF), Web Ontology*

Simon Andrews · Constantinos Orphanides · Simon Polovina
Conceptual Structures Research Group, Communication and Computing Research Centre
Faculty of Arts, Computing, Engineering and Sciences
Sheffield Hallam University, Sheffield, UK
e-mail: {s.andrews,c.orphanides,s.polovina}@shu.ac.uk

N. Bessis, F. Xhafa (Eds.): Next Generation Data Technologies for CCI, SCI 352, pp. 139–165.
springerlink.com

Language (OWL) and a variety of data interchange formats, such as *RDF/XML* and *N-Triples* [24]. These technologies provide a formal description of concepts, terms and relationships that capture and integrate meaning with distributed data within a given domain. New data technologies are emerging that can analyse, annotate and visualise such data and promote collective intelligence. In this vein, a data analysis method that has been rapidly developed during the past two decades for knowledge representation, information management and identifying conceptual structures in semantic data is *Formal Concept Analysis (FCA).*

2 Formal Concept Analysis

Formal Concept Analysis (FCA) was introduced in the 1990s by Rudolf Wille and Bernhard Ganter [12], building on applied lattice and order theory developed by Birkhoff and others in the 1930s. It was initially developed as a subsection of Applied Mathematics based on the mathematisation of concepts and concepts hierarchy, where a concept is constituted by its *extension*, comprising of all objects which belong to the concept, and its *intension*, comprising of all attributes (properties, meanings) which apply to all objects of the extension [30]. The set of objects and attributes, together with their relation to each other, form a *Formal Context*, which can be represented by a cross table [21].

Airlines	Latin America	Europe	Canada	Asia Pacific	Middle east	Africa	Mexico	Caribbean	USA
Air Canada	×	×	×	×	×		×	×	×
Air New Zealand		×		×					×
Nippon Airways		×		×					×
Ansett Australia				×					
Austrian Airlines		×	×	×	×	×			×

The cross-table above shows a formal context representing destinations for five airlines. The elements on the left side are formal objects; the elements at the top are formal attributes. If an object has a specific property (formal attribute), it is indicated by placing a cross in the corresponding cell of the table. An empty cell indicates that the corresponding object does not have the corresponding attribute. In the Airlines context above, Air Canada flies to Latin America (since the corresponding cell contains a cross) but does not fly to Africa (since the corresponding cell is empty). However, an empty cell might also mean that it is unknown whether the corresponding object has the corresponding attribute [31].

In mathematical terms, a Formal Context is defined as a triple $\mathbb{K} := (G,M,I)$, with G being a set of objects, M a set of attributes and I a relation defined between G and M. The relation I is understood to be a subset of the cross product between the sets it relates, so $I \subseteq G \times M$. If an object g has an attribute m, then $g \in G$ relates to m by I, so we write $(g,m) \in I$, or gIm. For a subset of objects $A \subseteq G$, a derivation operator $'$ is defined to obtain the set of attributes, common to the objects in A, as follows:

$$A' = \{m \in M \mid \forall g \in A : gIm\}$$

Similarly, for a subset of attributes $B \subseteq M$, the derivation operator $'$ is defined to obtain the set of objects, common to the attributes in B, as follows:

$$B' = \{g \in G \mid \forall m \in B : gIm\}$$

Now, a pair *(A, B)* is a *Formal Concept* in a given formal context (G,M,I) only if $A \subseteq G$, $B \subseteq M$, $A' = B$ and $B' = A$. The set A is the extent of the concept and the set B is the intent of the concept. A formal concept is, therefore, a closed set of object/attribute relations, in that its extension contains all objects that have the attributes in its intension, and the intension contains all attributes shared by the objects in its extension. In the Airlines example, it can be seen from the cross-table that Air Canada and Austrian Airlines fly to both USA and Europe. However, this does not constitute a formal concept because both airlines also fly to Asia Pacific, Canada and the Middle East. Adding these destinations completes (closes) the formal concept:

$$(\{Air\ Canada, Austrian\ Airlines\}, \{Europe, USA, Asia\ Pacific, Canada, Middle\ East\}).$$

Another central notion of FCA is a duality called a 'Galois connection', which is often observed between items that relate to each other in a given domain, such as objects and attributes. A Galois connection implies that "if one makes the sets of one type larger, they correspond to smaller sets of the other type, and vice versa" [21]. Using the formal concept above as an example, if Africa is added to the list of destinations, the set of airlines reduces to $\{Austrian\ Airlines\}$.

The Galois connections between the formal concepts of a formal context can be visualized in a *Concept Lattice* (Figure 1), which is an intuitive way of discovering hitherto undiscovered information in data and portraying the natural hierarchy of concepts that exist in a formal context.

A concept lattice consists of the set of concepts of a formal context and the subconcept-superconcept relation between the concepts [21]. The nodes in Figure 1 represent formal concepts. Formal objects are noted slightly below and formal attributes slightly above the nodes, which they label.

A concept lattice can provide valuable information when one knows how to read it. As an example, the node which is labeled with the formal attribute 'Asia Pacific' shall be referred to as *Concept A*. To retrieve the extension of Concept A (the objects which feature the attribute 'Asia Pacific'), one begins from the node where the attribute is labeled and traces all paths which lead down from the node. Any objects one meets along the way are the objects which have that particular attribute. Looking at the lattice in Figure 1, if one takes the attribute 'Asia Pacific' and traces all paths which lead down from the node, one will collect all the objects. Thus Concept A can be interpreted as 'All airlines fly to Asia Pacific'. Similarly, the node that is labelled with the formal object 'Air New Zealand' shall be referred to as *Concept B*. To retrieve the intension of Concept B (the attributes of 'Air New Zealand'), one begins by the node where the object is labeled and traces all paths which lead up from the node. Any attributes one meets along the way, are the attributes of that particular object. Looking at the lattice once again, if one takes the object 'Air New Zealand' and traces all paths which lead up from the node, one will collect the attributes USA, Europe, and Asia Pacific. This can be interpreted as 'The Air New Zealand airline flies to USA, Europe and Asia Pacific'. As a further example, the formal concept involving Air Canada and Austrian Airlines, from above, can be clearly seen in the concept lattice as the third node down from the top of the lattice.

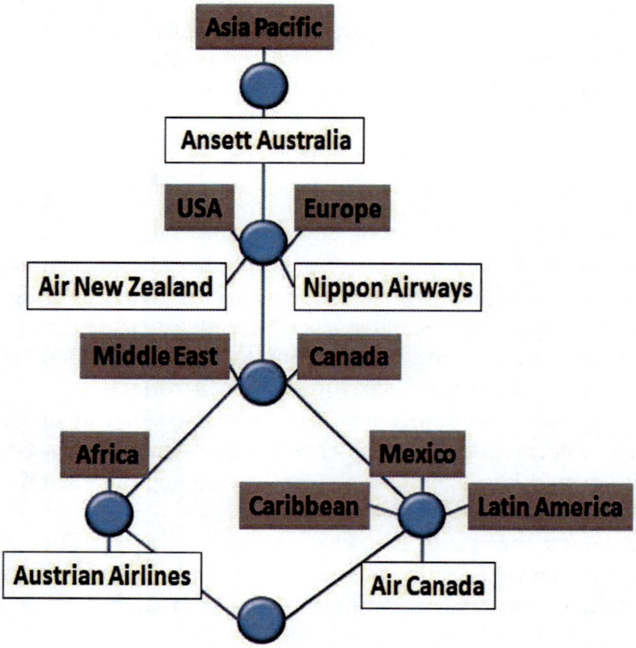

Fig. 1 A Lattice corresponding to the Airlines context.

Although the Airline context is a small example of FCA, visualising the formal context clearly shows that concept lattices provide richer information than by looking at the cross-table alone.

2.1 Formal Concept Analysis Scaling

In data terms, formal contexts are Boolean data. Of course data predominantly exists in non-Boolean form, so FCA introduces non-Boolean attributes via many-valued contexts. In FCA, *conceptual scaling* is used to transform many-valued contexts into single-valued contexts. Each non-Boolean attribute is given a *scale*, each scale being a context itself. In [31], an example is given of a table containing the sex and age of eight persons. In order to convert this many-valued context into a formal context, two scales (contexts) were created; the first scale represented their gender, containing 'm' for male and 'f' for female as possible values. The second scale represented their age, but instead of having the actual ages for each person, five formal attributes (the so called scale attributes) were produced; '<18', '<40', '<=65', '>65' and '>=80'. The first object is 21 years old, so the object has the '<40' and '<=65' formal attributes, as the object is both younger than 40 and younger than 65 (but not younger than 18, nor older than 65 or 80). The seventh object is 90 years old, so the object has the '>65' and '>=80' formal attributes, as the object is both older than 65 and 80 (but not younger than 18, 40 or 65). The two scales were then merged to form the complete formal context representing 'the age and gender of eight persons' (Figure 2). These notions are central to the process of formal context creation.

	sex	age
ADAM	m	21
BETTY	f	50
CHRIS	?	66
DORA	f	88
EVA	f	17
FRED	m	?
GEORGE	m	90
HARRY	m	60

→

	sex		age				
	m	f	<18	<40	<=65	>65	>=80
ADAM	X			X	X		
BETTY		X			X		
CHRIS						X	
DORA		X				X	X
EVA		X	X	X	X		
FRED	X						
GEORGE	X					X	X
HARRY	X				X		

Fig. 2 The transformation of Data into contexts (after [31]).

Figure 3 shows the age and gender context as a concept lattice, visualised in an open source FCA tool called *Concept Explorer* (ConExp) [32].

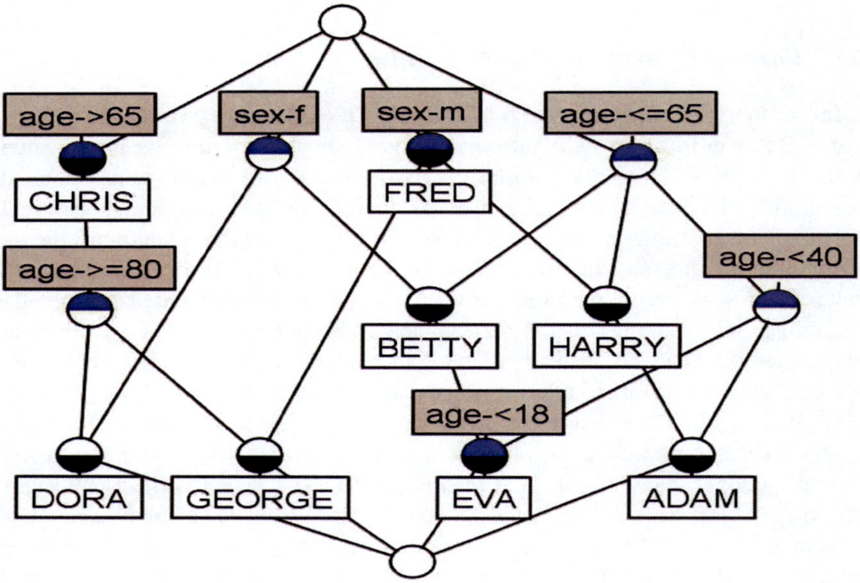

Fig. 3 The Concept Lattice for the 'age and gender' context (after [31]).

2.2 *Formal Context Formats*

To allow interoperability between FCA tools and applications, a number of file formats have been developed to represent formal contexts [23], the most popular being the *Burmeister* format, used for example by ConExp. As such, it is a sensible choice for developing new FCA tools. Outside of the FCA community, similar data analysis is carried out in the data mining domain [33]. A popular file format for data in this area is the *Frequent Itemset Mining Implementations* (FIMI) format [10]. This is therefore an appropriate choice of data file format to cater for if the interoperability of tools is to be extended beyond FCA. The two formats are briefly described below.

Burmeister (.cxt)

A Burmeister file begins with the letter 'B', to denote it is a Burmeister file (although the authors are unsure of its exact significance; it has merely become a convention to place this letter at the beginning of these files). It is followed by the number of objects, followed by the number of attributes and then lists the objects, followed by the attributes. It then stores the body of the formal context as a grid, using crosses for True values and dots for False values. File 1 is a cxt file for the ‘age and gender’ context and was used by ConExp to generate the concept lattice in Figure 3. In this figure the upper half-node when shaded means that the concept has its own attributes (i.e. they are not inherited). The lower half-node when shaded means that the concept has its own objects.

```
B

8
7

ADAM
BETTY
CHRIS
DORA
EVA
FRED
GEORGE
HARRY
sex-m
sex-f
age-<18
age-<40
age-<=65
age->65
age->=80
X..XX..
.X..X..
.....X.
.X...XX
.XXXX..
X......
X....XX
X...X..
```

File 1 `age-gender.cxt`, *Burmeister* context file.

FIMI (.dat)

The *Frequent Itemset Mining Implementations* (FIMI) [10] data format (.dat) is used in data mining, particularly in testing the efficiency of algorithms [13]. As opposed to Burmeister, a FIMI file only consists of rows of numbers; each row represents an object and each number represents a formal attribute. The ordering of the attributes is as one would expect from the formal context, taking the first column of the context to be attribute one and so on. File 2 represents the body of the ‘age and gender’ formal context in the FIMI format.

```
1 4 5
2 5
6
2 6 7
2 3 4 5
1
1 6 7
1 5
```

File 2 `age-gender.dat`, *FIMI* file.

3 Large Data Sets

Going beyond the simple example above, it has been shown that FCA can be usefully applied to large sets of data and that FCA has applications in data mining [19]. Rather than X-sized data it can handle XX-sized data. Programs, such as *In-Close* [1], exist which are capable of handling and processing large formal contexts. However, issues arise when trying to visualise the concept lattice. Formal contexts that have tens of attributes and thousands of objects can easily contain tens, if not hundreds of thousands of formal concepts, as work on parallel processing such volumes has identified [18]. The *Mushroom* data set [6], for example, has 23 attributes (properties of mushrooms) and 8124 objects (mushrooms). The formal context that results from the data set contains over 220,000 formal concepts. Lattice visualisation software does not exist that can compute lattices with such large numbers of nodes. Even if such tools existed, the results would be highly complex and unreadable, unless a sophisticated means of managing the lattice was employed.

Issues also arise when acquiring data for FCA; as described above, most tools for FCA require data to be in a formal context format. Clearly, the majority of typical data sets are not in this format, making accessibility to FCA difficult. To achieve interoperability with FCA tools and applications, data sets first need to be converted. In the process of conversion, different existing data set formats require different treatment, as do different data types. Creating the `age-gender.cxt` file was relatively convenient as the formal context already existed and was comparatively small. Converting large data sets into formal contexts and their corresponding cxt files would require additional effort to get them into a suitable form.

More problems arise when attributes and attribute values can be interpreted in different ways, having, as a result, the production of inconsistent conversions. For example, in choosing scales for continuous values or in dealing with missing values different approaches may be taken [2]. This can lead to apparently conflicting analyses of the same data and make the comparison of FCA tools and algorithms more difficult.

Another issue is the fact that disparate and distributed data do not, by definition, exist in a unified form. It is possible to regard a formal context as a unifying form, but existing data need to be converted into formal contexts, first, in order for FCA to be carried out.

As we shall explicate later on, data set conversion for FCA is conducted by *discretising* and *Booleanising* data; that is to take each many-valued attribute in a data set and convert it into as many Boolean attributes as it has values and by scaling continuous values using ranges. Although some issues of data discretisation and Booleanisation are understood (see FCA scaling, above), more work was required in this area to consider data interpretation, large scale data and automated conversion, and tools do not exist that facilitate these processes for FCA.

Thus, the task of converting data into formal contexts can be time consuming, is open to interpretation and usually requires a programming element to cope with large data sets. *FcaBedrock*, is a tool that has been developed to carry out this process [4], and is described later on.

4 Interpreting Data for FCA

Whereas FCA has used the term many-valued attribute to encompass such things as age and gender, such data may be distinguished by considering gender to be a *nominal* or *categorical* attribute [6], with the categories *male* and *female*, whilst age may be described as being a continuous attribute. Both are many-valued, but each needs to be treated differently in interpretation and conversion. The distinguishing characteristics of categorical and continuous are useful when considering these processes and how they might be automated. Furthermore, the process of discretising applies most clearly to continuous attributes, whereas the process of Booleanising is that of creating True/False attributes from many-valued ones, whether those values be categories or discretised values.

4.1 Data Discretisation

Data discretisation is defined as "a process of converting continuous data attribute values into a finite set of intervals with minimal loss of information" [17] and with simpler terminology as the process of scaling continuous values using ranges [4]. Data mining tasks often involve dealing with continuous attributes and this can decrease performance [17]. This can be resolved by producing discretised values of a continuous attribute and replacing it with the new values.

An example of data discretisation could be student grades for a module. Each student can have any grade value from 0 to 100 (including decimal values). Discretising the attribute *'grade'* can be accomplished by producing ranges and assigning each student grade to one of the produced ranges; any grades smaller than 40% could be categorized as *'FAIL'* and any grades equal or higher than 40% as *'PASS'*. This approach makes continuous attributes easier to handle and increases the performance of mining tasks.

Note that the notion of 'categories' may still be applied to a discretised continuous attribute, however in this instance referring to the ranges or boundary values used in the discretisation.

4.2 Data Booleanisation

Data Booleanisation is the process of taking each many-valued attribute in a data set and converting it into as many Boolean attributes as it has values [16]. This is also the approach used to convert many-valued attributes in FCA, as they are converted by creating a formal context attribute for each of the values [2].

The gender example above, is an instance of Booleanisation. Another example would be considering the attribute *eye-colour*, which might be converted into the formal attributes *eye-colour-blue*, *eye-colour-brown*, *eye-colour-green*, *eye-colour-grey*. Rarer colours could be captured by *eye-colour-other* and missing values by *eye-colour-missing*.

4.3 FcaBedrock

FcaBedrock is a tool for creating Formal contexts for FCA [4] by converting many-valued data into formal contexts (many-valued attributes into formal attributes). The tool is an open-source project at *Sourceforge* [5]. By using a process of *guided automation*, the tool obtains the metadata required for conversion, such as attribute names and attributes types, and converts corresponding data sets, using these metadata, into formal context files. The term guided automation refers to the fact that the process of conversion is automatic but the user first has to provide guidance as to how the data set should be interpreted. Further automation is provided in that FcaBedrock can be used to examine a data file to automatically determine certain types of metadata (see below) and thus create a basic interpretation and corresponding formal context without use guidance.

A separate file, called a Bedrock file, is used by FcaBedrock to store the metadata. This can be edited directly in or outside of the tool, used for subsequent conversions and acts as a record of how a data set was interpreted (Figure 4).

The tool currently takes many-column CSV and 3-column CSV data files as input, but a version is being developed that also takes RDF-S and OWL formats [20]. Large data sets are easily converted by FcaBedrock into the two popular FCA formats, Burmeister (.cxt) [22] and FIMI (.dat).

Figure 5 shows FcaBedrock with metadata for the *Mushroom* data set (a publicly available data set from the UCI Machine Learning Repository [6]). The following can be seen: the attribute numbers; the names of the attributes; whether they will be converted as part of a formal context; their type (in this case, all but one of the mushroom attributes are categorical); their categories and the actual category values as they appear in the data file (in this case, nominal letters are used in the mushroom data file to represent the various categories of each attribute). The names of the categories were obtained from a data description document accompanying the data file. It is often the case that data sets are accompanied with a description of the data and such descriptions are a key source of metadata.

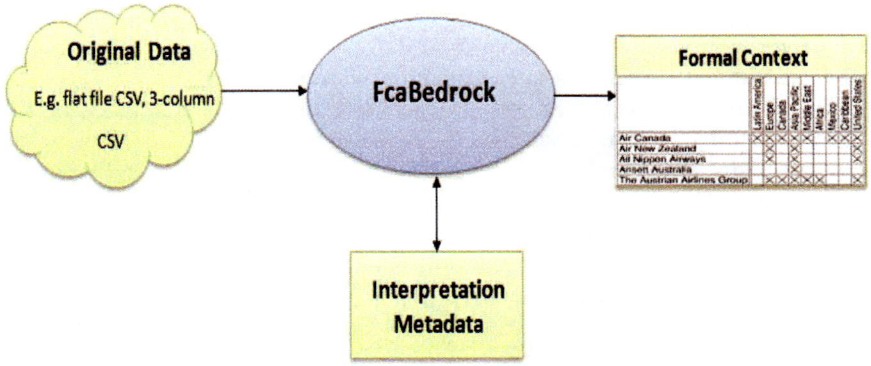

Fig. 4 FcaBedrock Process.

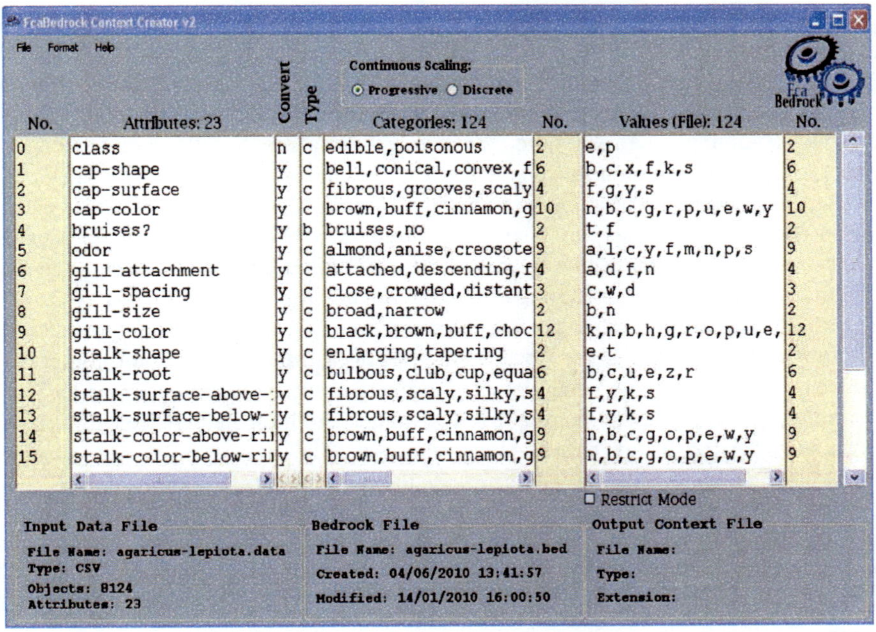

Fig. 5 Metadata of the *Mushroom* [6] data set in FcaBedrock.

4.3.1 Attribute Types

FcaBedrock uses three types of attribute: categorical, Boolean and continuous. These types are represented in FcaBedrock by the letters 'c', 'b' and 'o', respectively.

Categorical is the typical many-valued attribute and is converted by creating one formal attribute for each of the attribute categories. As an example, a categorical attribute *'height'* with categories *'short'*, *'normal'* and *'tall'* will be converted by FcaBedrock by creating three corresponding formal attributes and naming them by concatenating the attribute and category names; thus *'height-short'*, *'height-normal'* and *'height-tall'*.

Boolean attributes can be interpreted as a single formal attribute. In a data set, a Boolean attribute typically has two categories that represent True or False. For a Boolean attribute, FcaBedrock uses the first category value as the True value. However, both categorical and Boolean attribute types were dealt with keeping in mind freedom of interpretation. For example, an attribute *'Married?'* with values *'Yes'* and *'No'* can be interpreted as Boolean, where only *'Married?-Yes'* will appear in the formal context, or as categorical, where both *'Married?-Yes'* and *'Married?-No'* will appear in the formal context.

Continuous attributes are dealt with in FcaBedrock by either discretising the data using user defined ranges (e.g. *0-9*, *10-19*, *20-29*), or by progressive scaling (e.g. *<10*, *<20*, *<30*), which is the approach to convert continuous attributes in classical FCA [31] (see above).

4.3.2 Auto-detection of Metadata

FcaBedrock can auto-detect metadata, directly from the input file, if desired. It will initially assume that all attributes are categorical. It will add each new value it finds in a data-column to a corresponding list of category values. It will assume that each attribute is to be converted. It will set the category names to the category values. If FcaBedrock determines that there is a high proportion of different numerical values for an attribute, it will suggest that the attribute is continuous and ask the user if ranges are to be automatically calculated. If so, FcaBedrock will discretise the attribute using five, equal-sized ranges, basing the division on the highest and lowest values detected. If not, FcaBedrock will list the first 100 values and indicate that the addition of new categories for the attribute has been truncated. For non-numerical values (when the attribute being detected is free-text or some form of ID value, for example), FcaBedrock will list the first 100 values and indicate that the addition of new categories for the attribute has been truncated. The metadata obtained through auto-detection can then be edited to provide the required interpretation.

Figure 6 shows auto-detected metadata for the *Adult* data set (also from UCI [6]), which contains US census data of 32,561 adult citizens. The Adult data set is a typical CSV file of data and does not contain the attribute names, only columns of their values. Thus the attribute names are set by FcaBedrock as their attribute number. The user can usually obtain attribute names from the data description that accompanies the data file. Note that several of the attributes have been determined by FcaBedrock to be continuous and suitable discretisation has been automatically applied.

4.3.3 Attribute Exclusion and Restriction

FcaBedrock can be also used as a data preparation, or preprocessing tool, by allowing the exclusion of attributes from a conversion if they are not of particular interest in, or appropriate for, the analysis undertaken. However, these attribute and category

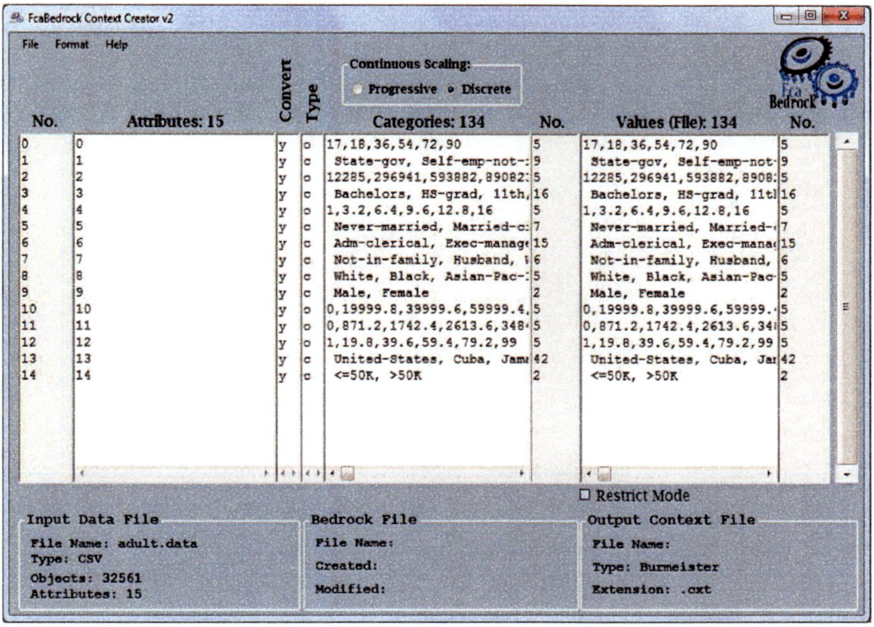

Fig. 6 Auto-detecting the metadata of the *Adult* data set [6] in FcaBedrock.

exclusions do not exclude objects from the formal context; all the rows (objects) of the data file will be included in the conversion.

On the other hand, sub-contexts can be created by restricting the conversion to user-specified attribute values. By specifying one or more category values of one or more attributes, the formal context will only contain objects with those values.

5 Concept and Lattice Generation

Lattice visualisation is made possible through several FCA tools, such as the *ToscanaJ* kit [8], a complete suit of tools for creating and using Conceptual Information Systems [7] and *ConExp* [32], a tool for analysing formal contexts, exploring dependencies between attributes, counting formal concepts and building concept lattices using contexts, in popular FCA formats, as input.

However, as mentioned earlier, the actual usefulness of the lattices that lattice visualisation software produce, rely on the numbers of formal attributes and formal concepts of the formal contexts given as input; formal contexts with a large number of formal concepts can result in performance issues and the production of unmanageable, unreadable lattices. These issues can be adressed by using FcaBedrock's data preparation features. By providing the ability to create sub-contexts, based on exclusion and restriction criteria set by the user, this means of focusing the analysis can result in smaller, more manageable lattices.

To illustrate an example of producing concept lattices from data sets by using sub-contexts, the *Mushroom* data set [6] will be used. The data set originally comprises of 8124 edible and poisonous mushrooms of the families *agaricus* and *lepiota*. It has 23 attributes describing properties such as edibility, stalk color and ring type (Figure 7).

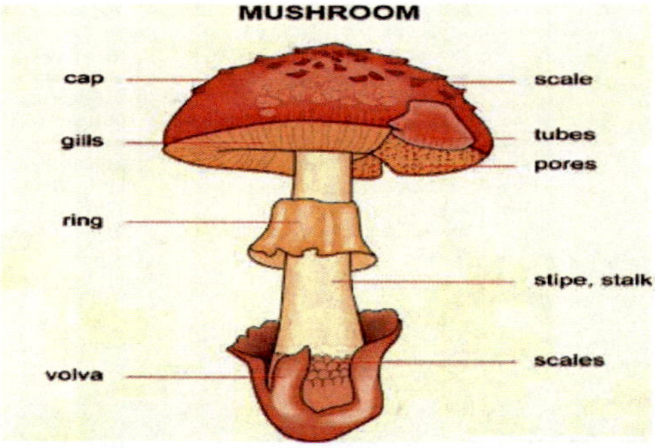

Fig. 7 Parts of a mushroom. (Source: http://www.infovisual.info)

Figure 8 shows an example of two mushrooms of the agaricus family; *Agaricus Arvensis* and *Agaricus Xanthodermus*. While they both resemble edible mushrooms on first sight, *Agaricus Xanthodermus* has poisonous properties. This is a good example of how mushroom identification can become confusing for non-experts who are unable to distinguish differences between similar-looking mushrooms.

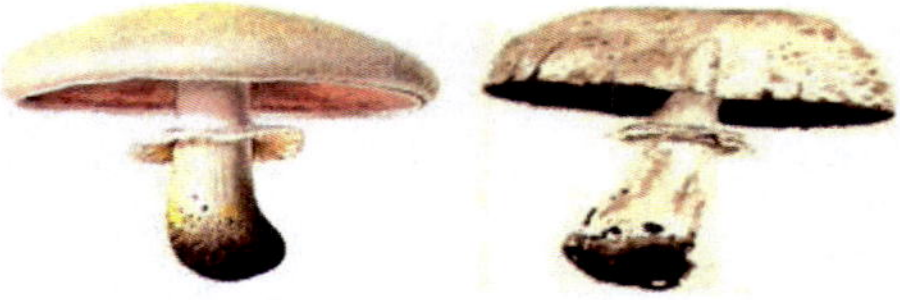

Fig. 8 *Agaricus Arvensis* (to the left) and *Agaricus Xanthodermus* (to the right). Source: [http://www.gracesmall.com/Tees/mushrooms.htm]

Within this context, Figure 9 shows how the restriction and exclusion capabilities of FcaBedrock can be applied to create a sub-context of the *Mushroom* data set and help determining whether a mushroom is edible or not based on some of it's properties. 20 attributes were excluded from the conversion, except from the *class* (poisonous and edible), *cap-color* and *habitat* attributes. Furthermore, the cap-color attribute was restricted to *red*, *pink*, *green*, *purple* and the habitat attribute was restricted to *woods*, *urban*, *meadows* and *grasses*. The restrictions set returned 996 objects and 19 formal attributes. Key to the analysis is the fact that this sub-context contains few enough formal concepts to make visualisation as a concept lattice practical.

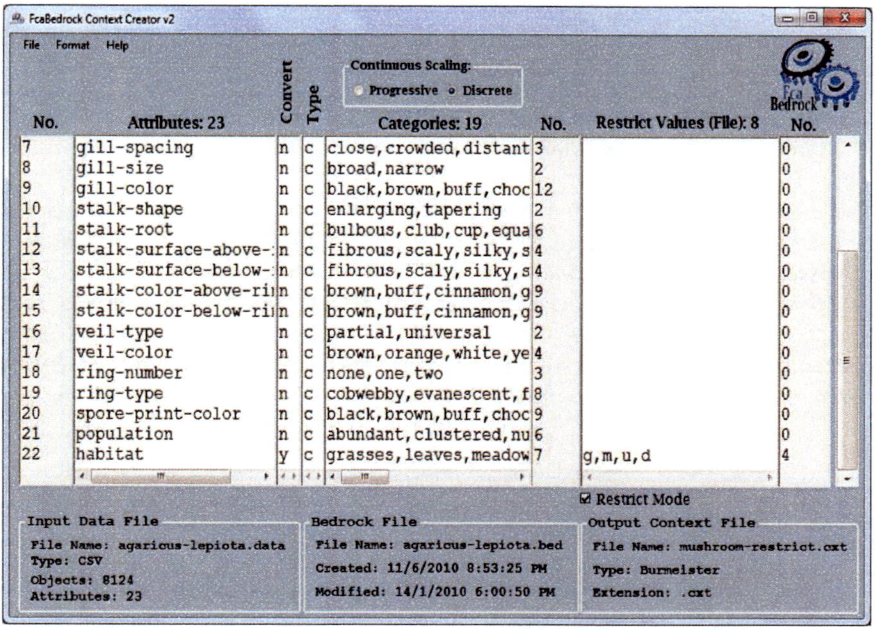

Fig. 9 Creating a *class-habitat-cap color* sub-context from the *Mushroom* data set [6].

Visualising the sub-context in ConExp (Figure 10) produced some interesting information, particularly as a feature of ConExp has been used to label each concept in the lattice with the object count and percentage of the overall number of objects. For example, in the sample, 98% of all the mushrooms live in woods, out of which 61% of them are safe to eat. The most dominant mushrooms are red-capped, of which 576 are edible and 300 are poisonous. This might indicate that mushroom cap colors are not the safest indicator for determining their edibility. On the other hand, pink-capped mushrooms can be found in woods and sometimes in grasses or meadows and all of them are poisonous. Green-capped and purple-capped are extremely rare, but if one is lucky enough to find some they are safe to eat.

As another illustration of using attribute exclusion, Figure 11 shows an example of a sub-context being created in FcaBedrock from two continuous attributes (*age* and *hours-per-week*) of the *Adult* US census data set [6], using progressive scaling. The corresponding concept lattice in ConExp is shown in Figure 12. The first scale represents the age of the objects, using the ranges *<20*, *<40*, *<60* and *all*. The second scale represents the hours worked per week, using the ranges *<20*, *<40*, *<60*, *<80* and *all*.

Various facts concerning the hours worked per week according to age of the population can be determined by reading the concept lattice. For example, a simple analysis is that 8% of the population work more than 60 hours per week. A more detailed analysis is that about a quarter of young adults (less than 20 years old) work less than 20 hours per week and about three-quarters work less than 40 hours per week. This can be compared to the population as a whole, where only 5% work less than 20 hours and only a quarter work less than 40 hours.

6 Dealing with Concept Simplification and Complexity

The examples in Figures 10 and 12 have shown how FcaBedrock's features can be used to produce smaller and easier to visualise lattices. However, this has been achieved by significantly reducing the size of the formal context by restricting the data conversion to objects matching certain criteria set by the user. An alternative approach, that involves all of the data, is to focus the analysis on large concepts. Small concepts might not include a significant, to the analysis, amount of objects/attributes and can add unnecessary complexity to the lattice. Filtering out these concepts from

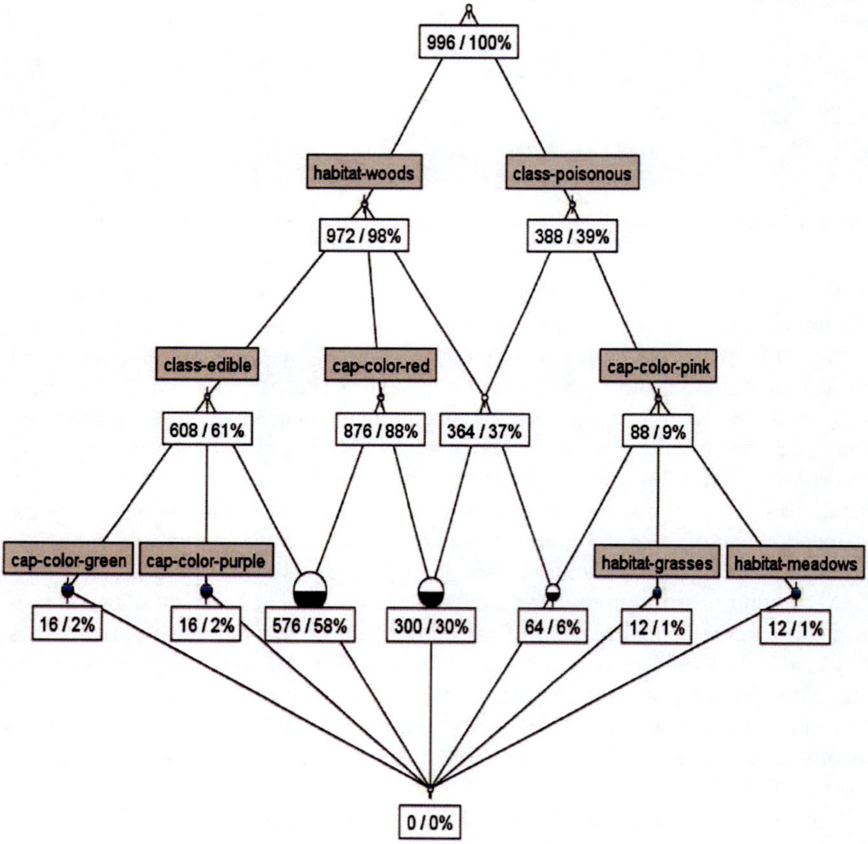

Fig. 10 Visualising the mushroom *class-habitat-cap color* sub-context, created by FcaBedrock, in ConExp.

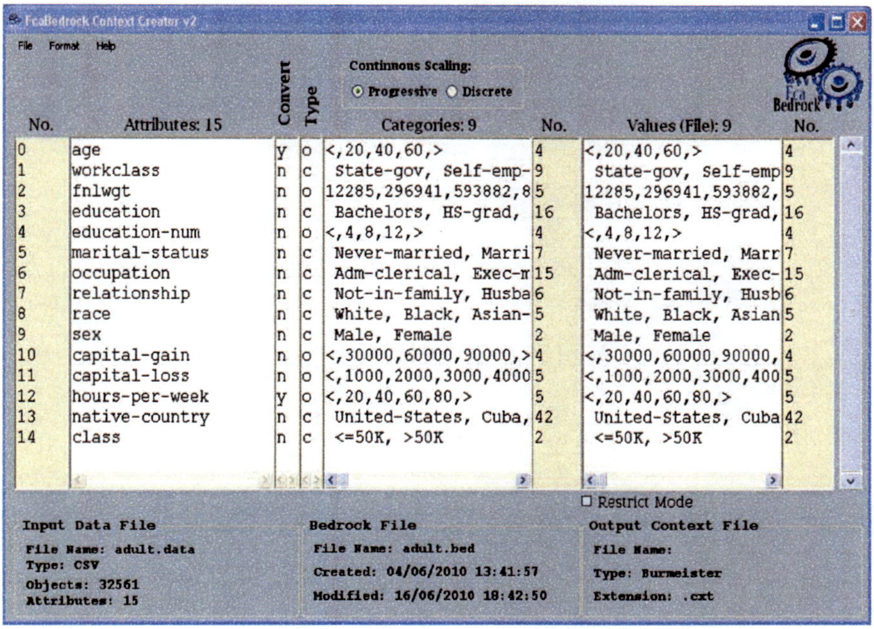

Fig. 11 Creating an *age-hours per week* sub-context from the *Adult* data set [6].

a context can thus allow the production of lattices which contain all the data, but can still produce useful and usable information. A tool that can filter out concepts from a context is *In-Close* [3]. In-Close accepts as input formal contexts in the Burmeister (.cxt) format and computes its concepts [1]. In-Close allows the user to exclude from the computation concepts with fewer than user-specified numbers of attributes and objects (the so-called *minimum support for intent* and *minimum support for extent*). After computing the concepts, In-Close outputs the same Burmeister file, but with only those concepts that have the minimum support set by the user (Figure 13). This is an approach similar to the *Iceberg Concept Lattices* of Stumme et al [27], although they do not consider minimum support for intent and do not produce a new context; rather their approach is to truncate the concept lattice by removing nodes with fewer objects than a specified number. They also leave the analysis of iceberg lattices as an open question for further research.

Using In-Close, the minimum support can be set high enough so that a relatively small number of concepts are computed. In this way, simplified contexts can be produced which result in readable concept lattices (Figure 14). Combining the exclusion and restriction capabilities of FcaBedrock with the minimum support features of In-Close adds further possibilities in the visualisation of computational intelligence (Figure 15).

In order to compare features of edible and poisonous mushrooms in the *Mushroom* data set, FcaBedrock was used to create two sub-contexts; one containing all

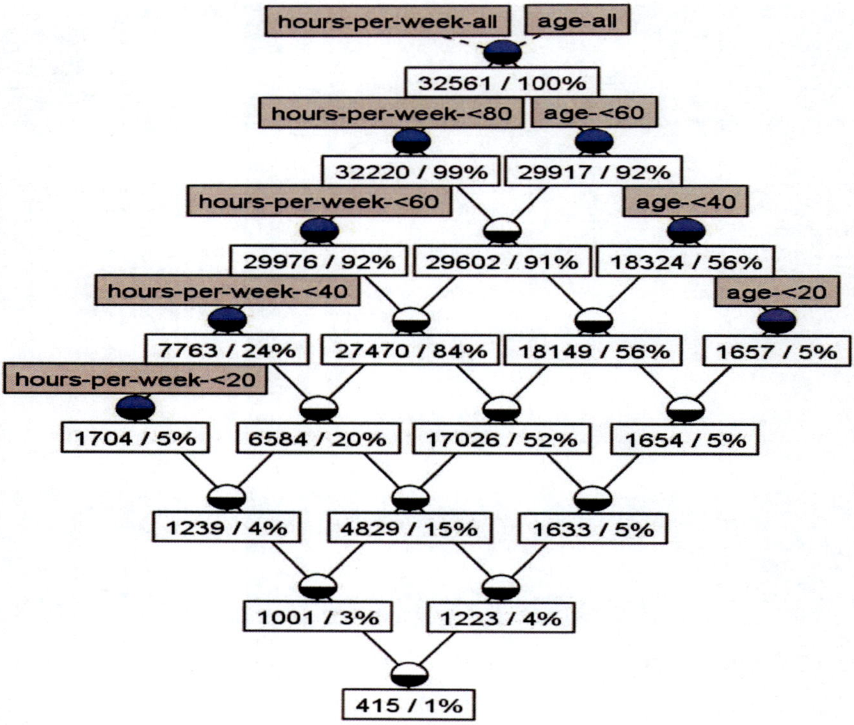

Fig. 12 Visualising the US census *age-hours per week* sub-context, created by FcaBedrock, in ConExp.

	a0	a1	a2	a3	a4	a5
o0	×	×	×			
o1	×	×	×	×		×
o2		×	×	×		
o3	×				×	

→

	a0	a1	a2	a3	a4	a5
o0	×	×	×			
o1	×	×	×	×		
o2		×	×	×		
o3						

Fig. 13 Simplifying a context using *In-Close*.

the edible mushrooms and one containing all the poisonous ones. This resulted in one sub-context containing 4208 edible mushrooms and one sub-context containing 3916 poisonous mushrooms. Each sub-context was then processed by In-Close to produce a manageable number of concepts, by setting appropriately large values for minimum support. Figure 16 is a command-line screen shot of In-Close being used to compute all 92,543 concepts in the edible mushroom sub-context (i.e. with no minimum support specified). When the minimum number of attributes was set to

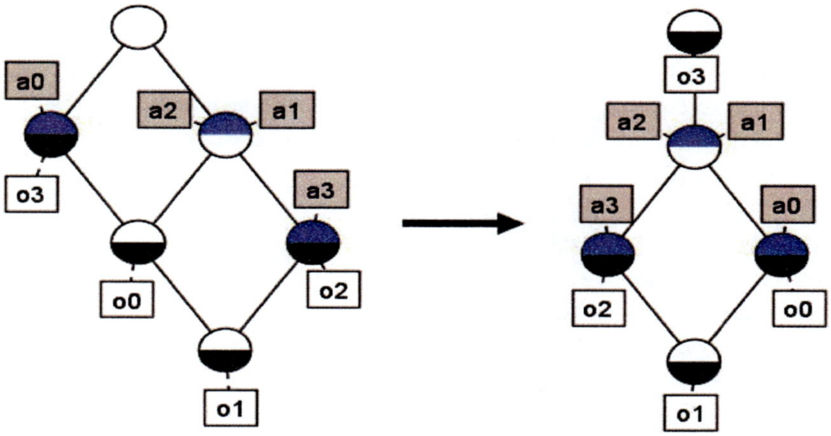

Fig. 14 Visualising the original (to the left) and simplified (to the right) contexts using *ConExp*.

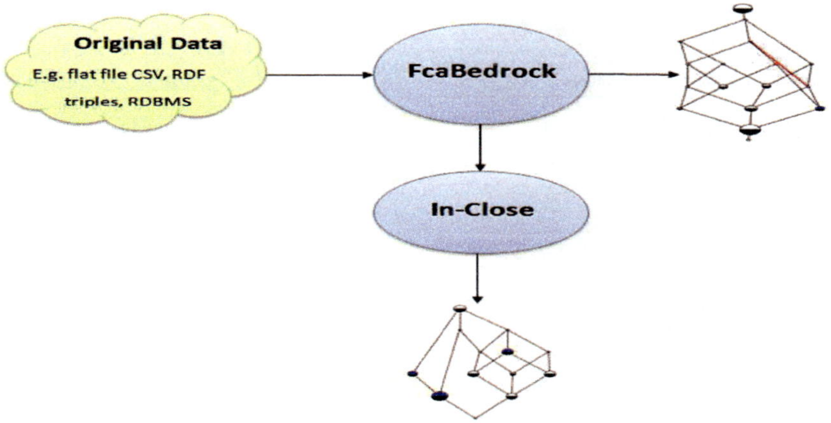

Fig. 15 Visualising formal contexts using FcaBedrock and In-Close.

10 and the minimum number of objects was set to 1500, In-Close computed only 9 formal concepts. For the poisonous mushroom sub-context, when the minimum number of attributes was set to 9 and the minimum number of objects was set to 1000, this resulted in 7 formal concepts. Although somewhat arbitrary, these sizes were arrived at to provide lattices with a similar small number of formal concepts. Figures 17 and 18 show the resulting concept lattices in ConExp.

In both cases, the resulting concepts involved most of the mushrooms in the corresponding 'pre-simplified' sub-context (2368 out of 4208 edible mushrooms and 3344 out of 3916 poisonous mushrooms). The attributes featured by both edible and poisonous mushrooms lattices (such as *veil-color-white*, *ring-number-one* and

```
*****     In-Close 2.0 Concept Miner     *****

Enter cxt file name including extension: edible.cxt
Enter minimum size of intent (no. attributes): 0
Enter minimum size of extent (no. objects): 0
Reading data...Done.
n: 126
m: 4208
x's: 94608
Mining concepts...Done.
Number of concepts: 92543
Output concepts to file? (y/n): n
Outputting context file...Done.
Hit <enter> to finish
```

Fig. 16 *In-Close* computing all Concepts in the edible mushroom context.

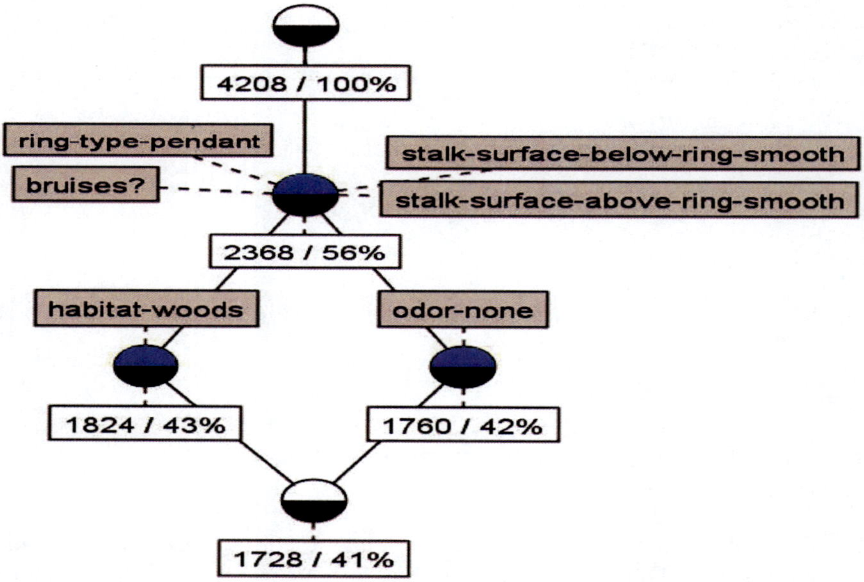

Fig. 17 Edible Mushroom Concept Lattice.

gill-attachment-free) were hidden, as the purpose was to highlight the difference between the sub-contexts, not the similarities. This also resulted in the formal concepts of the edible mushrooms reducing to 5. Hiding the common attributes that were commonly supported in both lattices gave a clearer overview between the differences of the two sub-contexts. For example, a smooth stalk would seem to indicate that a mushroom was safe to eat, where as those with silky stalks should be avoided. Those with a white spore print, a narrow gill and an evanescent ring should be avoided; better to try those with a pendant ring. Less surprising might be the fact that foul smelling mushrooms should be left alone; the mushrooms with no smell look like a safer choice. The fact that edible mushrooms have bruises looks like an interesting observation, perhaps indicating that edible mushrooms are more likely to show damage from foraging animals.

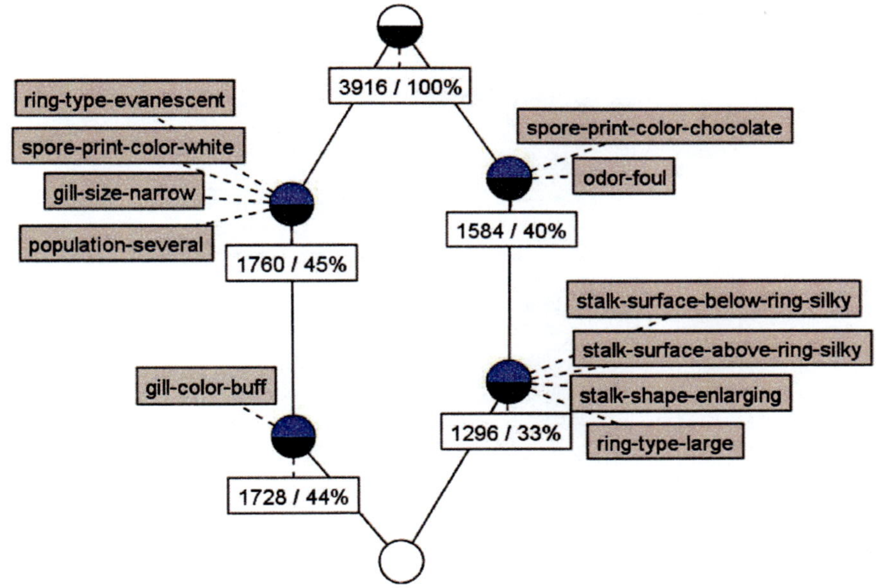

Fig. 18 Poisonous Mushroom Concept Lattice.

7 An Overall Process

An overall process for visualising computational intelligence through converting data into formal concepts is shown in Figure 19. It depicts the practical operation of the three open-source tools *FcaBedrock* (context creation), *In-Close* (context simplification) and *ConExp* (lattice visualisation). Start-to-finish analysis of a data set can be carried out in real time on a standard PC. Much of the process is automated, although if no Bedrock file exists for the data set being analysed, metadata not detectable by FcaBedrock from the data file must be entered manually (there is currently no 'magic' process by which this information can be automatically extracted from a free-text data description document). However, more metadata is available through auto-detection if 3-column, *object-attribute-attribute value*, data files are being analysed.

Managing concept complexity and the level at which there are few enough concepts to make a visualisation readable, is currently a non-scientific process; In-Close can quickly determine the number of concepts in a cxt file created by FcaBedrock but, if there are too many, an iterative trial-and-error process is required to reduce them to a practical number.

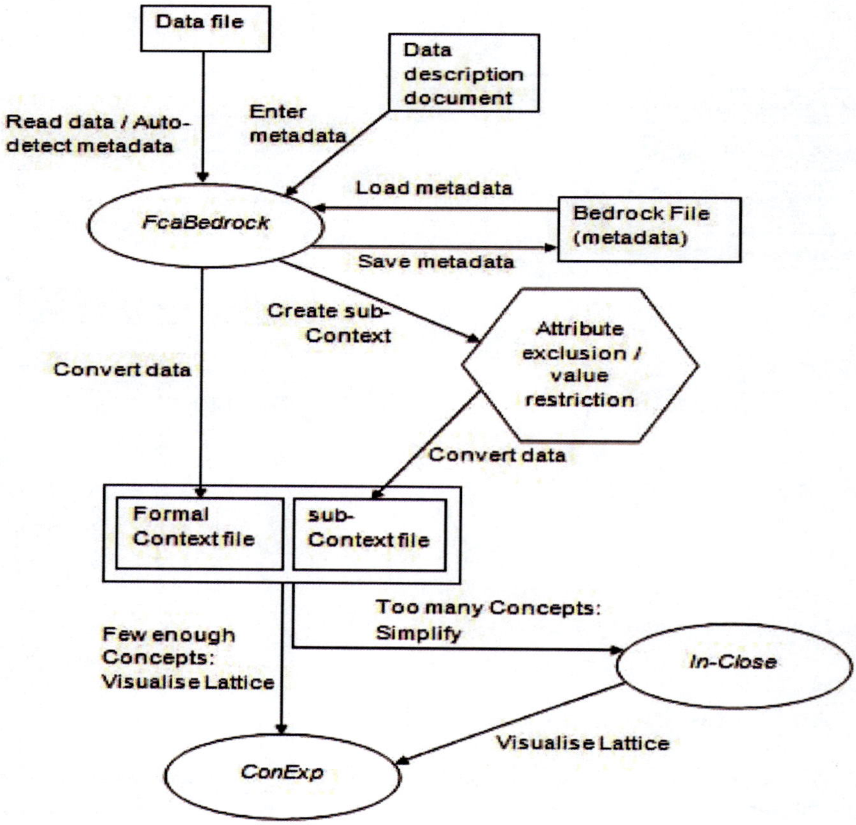

Fig. 19 Overall process incorporating the three open-source FCA tools

7.1 Performance Considerations

In-Close and similar high-performance programs have been shown to compute tens of thousands of formal concepts in seconds [1, 18]. Work on optimising these programs has brought these times down to fractions of a second, enabling real-time analysis of lager and larger data sets, as evidenced in this chapter. On a standard desktop PC, the computation of the 220,000 concepts in the complete Mushroom context took 0.5 seconds, for example, and the computation of the 80,000 concepts in the complete Adult context took 0.17 seconds. Computing concepts when a minimum support is specified is even faster. Nevertheless, applying this analysis to even larger data sets will require further improvements in performance, the primary directions of this work being in multiprocessing and parallel programming.

Moving towards creating formal contexts from data in triples form (be it 3-column CSV or RDF triples) will require enhancements to FcaBedrock. Currently, contexts from large traditional flat-file CSV data files can be created in less than a second. However, with the increased parsing required with data in triples form, this time increases to several seconds. The Adult data set, for example, requires over 500,000 triples to represent the 16 many-valued attributes of the 32,561 objects in the data, and FcaBedrock takes about 7 seconds to read the file. However, a current thrust of data warehousing using triple-stores is towards efficiency of querying that stems from the integer-indexing of triples [14]. This can be exploited by data conversion tools such as FcaBedrock.

Work in these areas, to increase the performance of FCA tools and thus increase the scope for visualising hitherto hidden information from larger and larger data sources, is going to be an important part of a new European Commission funded project called *CUBIST*.

8 CUBIST

The approach described in this chapter will form a core part of the "Combining and Uniting Business Intelligence with Semantic Technologies" (CUBIST) research project awarded under the European Union's 7th Framework Programme, 5th ICT call, topic 4.3: Intelligent Information Management; STREP Project No.: FP7 257403. CUBIST aims to develop an approach for Business Intelligence that augments Semantic Technologies with BI capabilities and provides conceptually relevant and user-friendly FCA-based visual analytics. CUBIST will find applications within the Semantic Web through its use of RDF. CUBIST aims to deliver high performance in-warehouse interactive visual analytics for information warehouses and triple stores.

8.1 Semantic Web, RDF and OWL

For the Semantic Web, FcaBedrock is being developed to accept RDF files as input [9, 20, 24]. RDF uses the same logic for both input types currently supported by

FcaBedrock; a normal form, where attributes are nested within their corresponding object, which is the same logic used for the flat-file CSV format and the *subject-predicate-object* logic, which is the same logic used for the 3-column CSV format. File 3 is an RDF/XML miniature version of the *Adult* data set [6], using only one object and five attributes: *age*, *education*, *employment*, *sex*, *us_citizen* and *class*. The file structure is straightforward and easy to follow; the example is referring to a 39 year old man who holds a Bachelors degree, works as a clerk, is a US citizen and earns less than 50K per year. A model of this file is depicted at Figure 20.

Functionality is also to be added for deriving data encoded in RDF vocabularies such as *Friend of a Friend (FOAF)* [11] and the Web Ontology Language *OWL* [15]. By using FcaBedrock as a semantic data preparation tool, FCA can find further applications in the Semantic Web and make knowledge representation, information management and visualizing conceptual structures among semantic data possible (Figure 21).

```
<?xml version="1.0"?>
<rdf:RDF
  xmlns:rdf="http://www.w3.org/1999/02/22-rdf-syntax-ns#"
  xmlns:ad="http://mini-adult.net/adult#">
  <rdf:Description
  rdf:about="http://mini-adult.net/adult/person1">
    <ad:age>39</ad:age>
    <ad:education>Bachelors</ad:education>
    <ad:employment>Clerical</ad:employment>
    <ad:sex>Male</ad:sex>
    <ad:us_citizen>Yes</ad:us_citizen>
    <ad:class>&lt;=50K</ad:class>
  </rdf:Description>
</rdf:RDF>
```

File 3 `mini-adult.xml`, *RDF/XML* file.

Number	Subject	Predicate	Object
1	http://mini-adult.net/adult/person1	http://mini-adult.net/adult#age	"39"
2	http://mini-adult.net/adult/person1	http://mini-adult.net/adult#education	"Bachelors"
3	http://mini-adult.net/adult/person1	http://mini-adult.net/adult#employment	"Clerical"
4	http://mini-adult.net/adult/person1	http://mini-adult.net/adult#sex	"Male"
5	http://mini-adult.net/adult/person1	http://mini-adult.net/adult#us_citizen	"Yes"
6	http://mini-adult.net/adult/person1	http://mini-adult.net/adult#class	"<=50K"

Fig. 20 Model of the `mini-adult.xml` RDF file (see above).

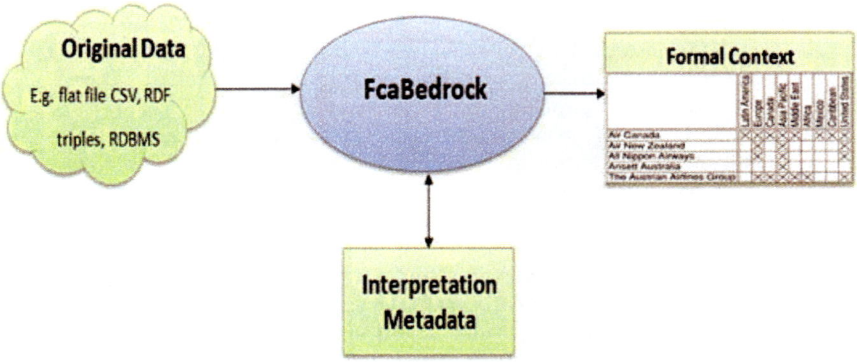

Fig. 21 Proposed Top-Level System Architecture of FcaBedrock.

9 Conclusion

Formal Concept Analysis (FCA) is an emerging data technology that complements collective computational intelligence such as that identified in the Semantic Web. Using the FCA open source software tools *FcaBedrock* and *In-Close* that were developed by the authors, and the open source lattice visualisation tool *ConExp*, disparate and distributed data can be visualised to discover its hitherto hidden meanings.

This chapter has also demonstrated how collective computational intelligence is facilitated by interoperation of data analysis tools; FcaBedrock produces files of a standard FCA format for ConExp and In-Close; In-Close also produces such files (simplified contexts) for ConExp.

This chapter has shown how FCA's visualisation can be applied to large-scale data sets, triples, and the Semantic Web's RDF and OWL. Key to this visualisation of data is that it is an inherent part of an intuitive and responsive interface, as initially demonstrated in this chapter. The CUBIST project will develop this enhancement, and provide further use cases demonstrating the integration of FCA with the Semantic Web. CUBIST is bringing together European data warehousing companies, universities with expertise in FCA and commercial use-case partners to develop powerful, insightful and intuitive RDF-based FCA Visual Analytics for BI. The disparate data will come from a range of structured and unstructured sources, providing rich and complex challenges for CUBIST to provide collective computational intelligence through the use of FCA as an emerging data technology. In the interim this chapter evidences that, by visualising this intelligence through converting data into formal concepts, the underlying knowledge that data depicts begins to be unlocked from the innumerable and increasing mass of it that is being recorded but not fully understood.

References

1. Andrews, S.: In-Close, A Fast Algorithm for Computing Formal Concepts (2009), http://sunsite.informatik.rwth-aachen.de/Publications/CEUR-WS/Vol-483/paper1.pdf
2. Andrews, S.: Data conversion and interoperability for FCA. In: Conceptual Structures Tools Interoperability Workshop, 17th International Conference on Conceptual Structures (ICCS 2009), Moscow (2009), http://www.kde.cs.uni-kassel.de/ws/cs-tiw2009/proceedings_final_15July.pdf
3. Andrews, S.: In-Close (2010), http://sourceforge.net/projects/inclose
4. Andrews, S., Orphanides, C.: FcaBedrock, a Formal Context Creator. In: Croitoru, M., Ferré, S., Lukose, D. (eds.) ICCS 2010. LNCS, vol. 6208, pp. 181–184. Springer, Heidelberg (2010)
5. Andrews, S., Orphanides, C.: FcaBedrock, a Formal Context Creator (2010), http://sourceforge.net/projects/fcabedrock
6. Asuncion, A., Newman, D.J.: UCI Machine Learning Repository. University of California, School of Information and Computer Science, Irvine (2007), http://www.ics.uci.edu/~mlearn/MLRepository.html
7. Becker, P., Correia, J.H.: The ToscanaJ Suite for Implementing Conceptual Information Systems. In: Ganter, B., Stumme, G., Wille, R. (eds.) Formal Concept Analysis. LNCS (LNAI), vol. 3626, pp. 324–348. Springer, Heidelberg (2005)
8. Becker, P., Correia, J.H.: ToscanaJ (2005), http://sourceforge.net/projects/toscanaj
9. Berners-Lee, T.: Why RDF model is different from the XML model (1998), http://www.w3.org/DesignIssues/RDF-XML
10. Frequent Itemset Mining Implementations Repository, http://fimi.cs.helsinki.fi
11. The Friend of a Friend (FOAF) project, http://www.foaf-project.org/
12. Ganter, B., Wille, R.: Formal Concept Analysis: Mathematical Foundations. Springer, Berlin (1998); Translated by C. Franzke
13. Goethals, B., Zaki, M.: Advances in Frequent Itemset Mining Implementations: Report on FIMI 2003. SIGKDD Explorations Newsletter 6(1), 109–117 (2004)
14. Harris, S., Gibbins, N.: 3store: Efficient bulk RDF storage. In: Proceedings of the 1st International Workshop on Practical and Scalable Semantic Web Systems (PSSS) 2003, pp. 1–15 (2003), http://km.aifb.kit.edu/ws/psss03/proceedings/harris-et-al.pdf
15. Horrocks, I., Patel-Schneider, P.F., Van Harmelen, F.: From SHIQ and RDF to OWL: the making of a Web Ontology Language. Web Semantics: Science, Services and Agents on the World Wide Web 1(1), 7–26 (2003), doi:dx.doi.org/10.1016/j.websem.2003.07.001
16. Imberman, S., Domanski, B.: Finding Association Rules from Quantitative Data using Data Booleanization (1999), http://citeseerx.ist.psu.edu/viewdoc/download?doi=10.1.1.14.4447&rep=rep1&type=pdf
17. Jin, R., Breitbart, Y., Muoh, C.: Data discretization unification. Knowledge and Information Systems 19(1), 1–29 (2009)
18. Krajca, P., Outrata, J., Vychodil, V.: Parallel Recursive Algorithm for FCA. In: Belohlavek, R., Kuznetsov, S.O. (eds.) Proceeding of the Sixth International Conference on Concept Lattices and their Applications, pp. 71–82. Palacky University, Olomouc (2008)

19. Kaytoue-Uberall, M., Duplesssis, S., Napoli, A.: Using Formal concept Analysis for the Extraction of Groups of Co-expressed Genes. In: Le Thi, H.A., Bouvry, P., Pham Dinh, T. (eds.) MCO 2008. CCIS, vol. 14, pp. 439–449. Springer, Heidelberg (2008)
20. Passin, T.B.: Explorer's Guide to the Semantic Web. Manning, Greenwich (2004)
21. Priss, U.: Formal Concept Analysis in Information Science. In: Cronin, B. (ed.) Annual Review of Information Science and Technology. ASIST, vol. 40 (2008)
22. Priss, U.: FcaStone - FCA File Format and Interoperability Software. In: Croitoru, M., Jaschkë, R., Rudolph, S. (eds.) Conceptual Structures and the Web, Proceedings of the Third Conceptual Structures and Tool Interoperability Workshop, pp. 33–43 (2008)
23. Priss, U.: FCA Software Interoperability. In: Belohlavek, R., Kuznetsov, S.O. (eds.) Proceeding of the Sixth International Conference on Concept Lattices and Their Applications, pp. 133–144 (2008)
24. Semantic Web. The Semantic Web (2010), `http://semanticweb.org/wiki/Main_Page`
25. Slezak, D., Wroblewski, J., Eastwood, V., Synak, P.: Brighthouse: an analytic data warehouse for ad-hoc queries. In: Proceedings of the VLDB Endowment, vol. 1(2), pp. 1337–1345. ACM Digital Library (2008)
26. SPARQL Query Language for RDF, `http://www.w3.org/TR/rdf-sparql-query/`
27. Stumme, G., Taouil, R., Bastide, Y., Lakhal, L.: Conceptual Clustering with Iceberg Concept Lattices. In: Proceedings of GI-Fachgruppentreffen Maschinelles Lernen 2001, Universitat Dortmund (2001)
28. World Wide Web Consortium. Design Issues (2010), `http://www.w3.org/DesignIssues/`
29. White, P.W., French, C.D.: Database system with methodology for storing a database table by vertically partitioning all columns of the table. US Patent 5,794,229, August 11 (1998)
30. Wille, R.: Formal Concept Analysis as Mathematical Theory of concepts. In: Ganter, B., Stumme, G., Wille, R. (eds.) Formal Concept Analysis: Foundations and Applications, pp. 1–6. Springer, Berlin (2005)
31. Wolff, K.E.: A First Course in Formal Concept Analysis (1993), `http://www.fbmn.h-da.de/home/wolff/Publikationen/A_First_Course_in_Formal_Concept_Analysis.pdf`
32. Yevtushenko, S.: ConExp. (2006), `http://sourceforge.net/projects/conexp`
33. Zaki, M.J., Hsiao, C.-J.: Efficient Algorithms for Mining Closed Itemsets and Their Lattice Structure. IEEE Transactions on Knowledge and Data Mining 17(4) (2005)

Chapter 7
Constructing Ensemble Classifiers from GEP-Induced Expression Trees

Joanna Jędrzejowicz and Piotr Jędrzejowicz

Abstract. The goal of the chapter is to construct high quality classifiers through applying collective computational techniques to the field of machine learning. Among the computational intelligence techniques one can distinguish a class referred to as the collective computational intelligence. The chapter proposes and reviews a family of ensemble classifiers constructed from expression trees. We propose to construct classifiers using collective computational intelligence paradigms at two levels. At the lower level the so-called weak classifiers are produced taking advantage of the benefits of co-operation between individuals evolved iteratively using gene expression programming and cellular evolutionary algorithms. At the upper level, cooperating individuals, which in our case are expression trees, are combined with a view to achieve better classification results through exploiting the collective intelligence property. Expression trees are induced using gene expression programming and cellular evolutionary algorithm. Ensemble classifiers are constructed from the weak classifiers obtained at the lower level of collaboration. To construct ensemble classifiers several standard techniques including majority voting, boosting and Dempster-Shafer theory of evidence, are used. To validate the approach a computational experiment has been carried-out using several well known datasets. The experiment aimed at comparison of the proposed classifiers performance with that of several widely used and popular classifiers with some of them also built through applying some collective computational intelligence tools. Experiment results confirm that next

Joanna Jędrzejowicz
Institute of Informatics, Gdańsk University,
Wita Stwosza 57, 80-952 Gdańsk, Poland
e-mail: jj@inf.ug.edu.pl

Piotr Jędrzejowicz
Department of Information Systems, Gdynia Maritime University,
Morska 83, 81-225 Gdynia, Poland
e-mail: pj@am.gdynia.pl

N. Bessis, F. Xhafa (Eds.): Next Generation Data Technologies for CCI, SCI 352, pp. 167–193.
springerlink.com

generation collective computational intelligence techniques like gene expression programming and cellular evolutionary algorithms, when applied to the field of machine learning, can offer an advantage that can be attributed to their collaborative and synergetic features.

Keywords: computational intelligence; collective computational intelligence; classification algorithms; gene expression programming; cellular evolutionary algorithms; ensemble classifiers.

1 Introduction

According to Engelbrecht [12] computational intelligence (CI) involves the study of adaptive mechanisms to enable or facilitate intelligent behavior in complex and changing environments. Among the CI paradigms Engelbrecht [12] mentions artificial neural networks, evolutionary computation swarm intelligence, artificial immune systems and fuzzy systems. Similar view is expressed by Fulcher [17] for whom the term CI means the use of artificial neural networks, evolutionary and/or fuzzy techniques, and more especially hybrids or synergistic combination/ensembles of these complementary approaches (as well as occasionally incorporating rule-based and/or statistical ones). Among the CI techniques one can distinguish a class referred to as the collective computational intelligence (CCI). In CCI individuals within the group interact to solve global objective by exchanging locally available information. Individuals, called also agents, within the group communicate with each other – either directly or indirectly, by acting on their local environment.

The goal of the chapter is to construct high quality classifiers through applying CCI techniques to the field of machine learning. More specifically, we propose to construct classifiers using CCI paradigms at two levels. At the lower level the so-called weak classifiers are produced taking advantage of the benefits of cooperation between individuals evolved iteratively using gene expression programming and cellular evolutionary algorithms. At the upper level cooperating individuals, which in our case are expression trees, are combined with a view to achieve better classification results through exploiting the collective intelligence property.

Appropriately combining information sources to form a more effective output than any of the individual sources is an effective approach used in a variety of areas. In machine learning there are several situations motivating combining multiple learners. For example, Benett [5] argues that "it may not be possible to train using all the data because data privacy and security concerns prevent sharing the data. However, a classifier can be trained over different data subsets and the predictions they issue may be shared. In other cases, the computation burden of the base classifier may motivate classifier combination. When a classifier with a nonlinear training or prediction cost is used, computational gains can be realized by partitioning the data and applying an instance of the classifier to each subset. In other situations, combining classifiers can be seen as a way of extending the hypothesis space or relaxing the bias of the original base classifier". Among methods based on classifier combination one can mention cascade generalization [18], stacking [41], and boosting [33], [4]. In this chapter we propose an ensemble method integrating the gene expression programming with the AdaBoost algorithm. Ensemble methods are methods that first solve a classification or regression problem by creating multiple learners each attempting to solve the task independently, then use a procedure specified by the particular ensemble method for selecting or weighing the individual learners. Apart from the already mentioned stacking and boosting, other example combination methods include composite classifiers [8], voting pool of classifiers [3], combination of multiple classifiers [20] and [27], classifier ensembles [9], [26], and many others. An excellent review of the classifier combination methods can be found in [26].

As Polikar [31] observes there are two types of classifier combination: classifier selection and classifier fusion. In classifier selection, each classifier is trained to become an expert in some local area of the feature space. In classifier fusion all classifiers are trained over entire feature space. In this case, the classifier combination process involves merging individual (weaker) classifiers to obtain a single (stronger) expert of superior performance. The ensemble method proposed in this chapter belongs to the classifier fusion category.

Although theoretical results [10] indicate there is no a priori choice of algorithm which will perform best over all problems, experience has shown that some algorithms can dominate large classes of problems. However, even when an algorithm outperforms another algorithm across a problem set, combining the algorithms can lead to better results than either alone. There are many concrete situations where weak classifiers can help improve the performance of a strong classifier [5].

In this chapter we focus on ensemble classifiers constructed from expression trees and combined using several different methods. The approach involves two stages where the collective computational intelligence techniques are applied. The first stage is inducing a set of expression trees using gene expression programming (GEP) or its extensions. The second is combining the trees into an ensemble classifier. The chapter is organized in sections. Section 2 proposes two procedures to induce expression trees using gene expression programming approach. In Section 3 several ensemble classifiers constructed from expression trees are discussed. In Section 4 results of the computational experiment are presented. Final section contains conclusions and ideas for future research.

2 Using Gene Expression Programming to Induce Classifiers

Gene expression programming introduced by Ferreira [14] is an automatic programming approach. In GEP computer programs are represented as linear character strings of fixed-length called chromosomes which, in the subsequent fitness evaluation, can be expressed as expression trees of different sizes and shapes. The approach has flexibility and power to explore the entire search space, which comes from the separation of genotype and phenotype. As it has been observed by Ferreira [15] GEP can be used to design decision trees, with the advantage that all the decisions concerning the growth of the tree are made by the algorithm itself without any human input, that is, the growth of the tree is totally determined and refined by evolution. Ferreira in her paper [15] proposed two different algorithms to grow the trees. The first one induces decision trees with nominal attributes, and the second one was developed for handling numeric attributes but, in fact, can handle all kinds of attributes.

The ability of GEP to generate decision trees makes it a natural tool for solving classification problems. Ferreira [15] showed several example applications of GEP including classification. The approach was based on the tree induction algorithm proposed by the author. Weinert and Lopes [39] apply GEP to the data mining task of classification by inducing rules. The authors proposed a new method for rule encoding and genetic operators that preserve rule integrity. They also implemented a system, named GEPCLASS which allows for the automatic discovery of flexible rules, better fitted to data. Duan with co-authors [10] claimed to improve efficiency of GEP used as a classification tool. Their contribution includes proposing new strategies for generating the classification threshold dynamically and designing a new approach called Distance Guided Evolution Algorithm. Zeng with co-authors [43] proposed a novel Immune Gene Expression Programming as a tool for rule mining. Another approach to evolving classification rules with gene expression programming was proposed in [44]. A different example of GEP application to classification problems was proposed by Li and co-authors [29]. They proposed a new representation scheme based on prefix notation which brings some advantages as compared with the traditional approach. Wang and co-authors [38] proposed a GEP decision tree system. The system can construct decision tree for classification without prior knowledge about the distribution of data. Karakasis and Stafylopatis [25] proposed a hybrid evolutionary technique for data mining tasks, which combines the Clonal Selection Principle with Gene Expression Programming. The authors claim that their approach outperforms GEP in terms of convergence rate and computational efficiency.

In this chapter classification of data with numeric and categorical attributes is considered. Gene expression programming is used to induce expression trees which correspond to rules. The data that satisfy the rule are classified as the first class and those for which the rule does not work - as the second class.

As usual when applying GEP methodology, the algorithm uses a population of chromosomes, selects them according to fitness and introduces genetic variation using several genetic operators. Each chromosome is composed of a single gene divided into two parts as in the original head-tail method [14]. The size of the head (h) is determined by the user with the suggested size not less than the number of attributes in the dataset. The size of the tail (t) is computed as $t = h(n-1)+1$ where n is the largest arity found in the function set. In the computational experiments the functions are: logical AND, OR and NOT. Thus $n = 2$ and the size of the chromosome is $h + t = 2h + 1$. The terminal set contains triples $(op, attrib, const)$ where op is one of relational operators $<$, $\leq$, $>$, $\geq$, $=$, $\neq$, $attrib$ is the attribute number, and finally $const$ is a value belonging to the domain of the attribute $attrib$. As usual in GEP, the tail part of a gene always contains terminals and head can have both, terminals and functions. Observe that in this model each chromosome is syntactically correct and corresponds to a valid expression. Each attribute can appear once, many times or not at all. This allows to define flexible characteristics like for example ($attribute1 > 0.57$) AND ($attribute1 < 0.80$). On the other hand, it can also introduce inconsistencies like for example ($attribute1 > 0.57$) AND ($attribute1 < 0.40$). This does not cause problems since a decision subtree corresponding to such a subexpression would evaluate it to $false$. Besides, when studying the structure of the best classifiers in our experiments the above inconsistencies did not appear.

The expression of this kind of genes is done in exactly the same manner as in all GEP systems. In Fig. 1 an example of a gene and a corresponding expression tree is given. The start position (position 0) in the chromosome corresponds to the root of the expression tree (OR, in the example). Then, below each function branches are attached and there are as many of them as the arity of the function - 2 in our case. The following symbols in the chromosome are attached to the branches on a given level. The process is complete when each branch is completed with a terminal. The number of symbols from the chromosome to form the expression tree is denoted as the termination point. For the discussed example, the termination point is 4 therefore further symbols are not meaningful and are denoted by $\cdots$ in Fig.1. Subsequent symbols in the gene are separated by dots and in the tree the terminals are drawn as ovals. The rule corresponding to the expression tree from Fig. 1 is:

IF ($attribute1 > 0.57$) OR NOT ($attribute10 \leq 0.16$) THEN Class1.
In fact, the above rule was taken as an example from the solution to the Sonar dataset from [2].

The algorithm for learning the best classifier using GEP works as follows. In the initial step the minimal and maximal value of each attribute is calculated and a random population of chromosomes is generated. For each chromosome the symbols in the head part are randomly selected from the set of functions AND, OR, NOT and the set of terminals of type $(op, attrib, const)$, where

chromosome)

0	1	2	3	4
OR	$(>, 1, 0.57)$	NOT	$(\leq, 10, 0.16)$	$\cdots$

expression tree)

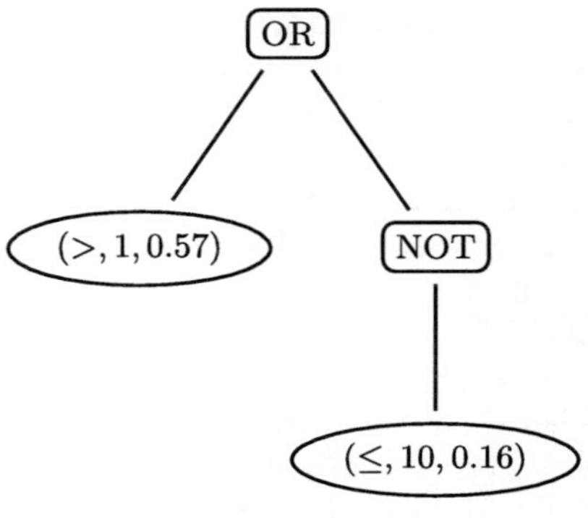

Fig. 1 One chromosome and a corresponding expression tree

the value of *const* is in the range of *attrib*. The symbols in the tail part are all terminals. To introduce variation in the population the following genetic operators are used:

- mutation,
- transposition of insertion sequence elements (IS transposition),
- root transposition (RIS transposition),
- one-point recombination,
- two-point recombination.

Mutation can occur anywhere in the chromosome. We consider one-point mutation which means that with a probability, called mutation rate, one symbol in a chromosome is changed. In case of a functional symbol it is replaced by another randomly selected function, otherwise for $g = (op, attrib, const)$ a random relational operator op', an attribute $attrib'$ and a constant $const'$ in the range of $attrib'$ are selected. Note that mutation can change the respective expression tree since a function of one argument may be mutated into a function of two arguments, or vice versa.

Transposition stands for moving part of a chromosome to another location. Here we consider two kinds of transposable elements. In case of transposition of insertion sequence (IS) three values are randomly chosen: a position in the chromosome (start of IS), the length of the sequence and the target site in the **head** - a bond between two positions. For example consider the chromosome C with head=6, defined below. The termination point is 7.

0	1	2	3	4	5	6	7	8	9
OR	AND	AND	$(>, 1, 0)$	$(=, 2, 05)$	$(>, 1, 2)$	$(<, 1, 10)$	$(>, 3, 0)$	$(<, 3, 5)$	$\cdots$

Suppose that IS is defined as: start position=6, length=2, target=0. Then a cut is made in the bond defined by the target site (in the example between symbol 0 and 1), and the insertion sequence (the symbols from positions 6 and 7) is copied into the site of the insertion. The sequence downstream from the copied IS element loses, at the end of the head, as many symbols as the length of the transposon. The resulting chromosome is shown below:

0	1	2	3	4	5	6	7	8	9
OR	$(<, 1, 10)$	$(>, 3, 0)$	AND	AND	$(>, 1, 0)$	$(<, 1, 10)$	$(>, 3, 0)$	$(<, 3, 5)$	$\cdots$

Observe that since the target site is in the head, the newly created individual is always syntactically correct though it can reshape the tree quite dramatically as in the above case. The termination point is 3 for the new chromosome.

In case of root transposition, a position in the head is randomly selected, the first function following this position is chosen - it is the start of the RIS element. If no function is found then no change is performed. The length of the insertion sequence is chosen. The insertion sequence is copied at the root position and at the same time the last symbols of the head (as many as RIS length) are deleted. For the chromosome C defined before and RIS sequence starting with the function AND at the position 2, of length 2 the resulting chromosome is defined as:

0	1	2	3	4	5	6	7	8	9
AND	$(>, 1, 0)$	OR	AND	AND	$(>, 1, 0)$	$(<, 1, 10)$	$(>, 3, 0)$	$(<, 3, 5)$	$\cdots$

Again the change has quite an effect since the termination point is now 9.

For both kinds of recombination two parent chromosomes P_1, P_2 are randomly chosen and two new child chromosomes C_1, C_2 are formed. In case of one-point recombination one position is randomly generated and both parent chromosomes are split by this position into two parts. Child chromosomes C_1 (respectively, C_2) is formed as containing the first part from P_1 (respectively, P_2) and the second part from P_2 (and P_1). In two-point recombination two positions are randomly chosen and the symbols between recombination positions are exchanged between two parent chromosomes forming two new child chromosomes. Observe that again, in both cases, the newly formed chromosomes are syntactically correct no matter whether the recombination positions were taken from the head or tail.

In this chapter we propose two approaches to learning decision trees - GEP learning and cellular GEP learning. In GEP learning C is the set of categorical classes which are denoted $1, \ldots, |C|$. We assume that the learning algorithm is provided with the learning instances $LD = \{< d, c > \mid d \in D, c \in C\} \subset D \times C$, where D is the space of attribute vectors $d = (w_1^d, \ldots, w_n^d)$ with w_i^d being symbolic or numeric values. The learning algorithm is used to find the best possible approximation $\bar{f}$ of the unknown function f such that $f(d) = c$. Then $\bar{f}$ can be used to find the class $\bar{c} = \bar{f}(\bar{d})$ for any $\bar{d} \in D - LD|D$. The set of learning instances LD consists of two subsets $LD = TD \cup TS$ that is TD-training set and TS - testing set.

The learning stage is class-specific:

- the fitness function is class dependent: for class cl and gene g the value of $fitness^{cl}(g)$ is defined as the difference between the number of rows from class cl for which g is 'true', and the number of rows from classes $\neq cl$ for which g is 'true',
- learning takes place separately for each class $cl \in C$ and the first step of each algorithm is data preprocessing: the training data $TD \subset LD$ is divided into $|C|$ sets $TD_1, \ldots, TD_{|C|}$ containing separately data from different classes.

In the process of learning the following parameters were used: P-the number of supporting genes (usually much smaller than the initial size of the population), NG - number of iterations. Algorithm INC (incremental learning) makes use of the procedure GEP class oriented to obtain one classifier $Cl = (Cl^1, \ldots, Cl^{|C|})$ where $Cl^i = (g_1^i, \ldots, g_{j_i}^i)$ for $i = 1, \ldots, |C|$ is the support for class i containing $j_i \leq P$ genes.

Algorithm 1. GEP class-oriented

```
Require: class cl, training data, integer NG
create random genes of the initial population
for i = 1 to NG do
    express genes as expression trees,
    calculate fitness of each gene with respect to cl,
    keep best gene
    select genes
    mutation
    IStransposition
    RIStransposition
    one-point recombination
    two-point recombination
end for
calculate fitness and keep the best gene
return the best gene for class cl
```

Traditionally, GEP algorithms work on a single population of genes. However within the cellular GEP benefits of structuring the population by defining neighborhood are explored. In cellular GEP learning individuals are arranged on a torus-like grid of dimension $xmax \times ymax$. Each point of the grid has a neighborhood that overlaps the neighborhood of nearby individuals; all the neighborhoods have the same size and identical shape. The boundary individuals of the grid are connected to the individuals located in the opposite borders in the same row/column, depending on the case. This results in toroidal

Algorithm 2. INC (incremental - learning)

Require: training data divided into $|C|$ subsets $TD_1, \ldots, TD_{|C|}$, integer P

```
for i = 1 to |C| do
    while TD_i ≠ ∅ and (population size does not exceed P) do
        call procedure GEP to find one best gene g for class i
        add g to the population
        delete from data set TD_i those data for which g evaluates to 'true'
    end while
end for
return classifier Cl
```

Algorithm 3. Testing

Require: classifier Cl, testing data TS
Ensure: qc quality of the majority vote classifier

```
qc ← 0
for all (x, y) ∈ TS do
    if Cl classifies x as y then
        qc ← qc + 1
    end if
end for
qc ← qc/|TS|
return qc
```

grid and all the individuals have exactly the same number of neighbors. In the experiments the L5, or NEWS neighborhood - 4 nearest neighbors in a given axial (north, east, west, south) direction neighborhood is applied. Details of the cellular evolutionary algorithms can be found in [1].

3 Ensemble Classifiers

In the chapter we consider three techniques for constructing ensemble classifiers – AdaBoost (see [31]), the majority voting and Dempster rule of combination through applying triplet mass functions (introduced in [6]).

3.1 AdaBoost-Based Ensemble Classifier with GEP Learning (AB-GEP)

The general idea is to create an ensemble of classifiers by resampling the training dataset and creating a classifier for each sample. Freund and Schapire [16]

Algorithm 4. cGEP learning

```
Require: class cl, training data, integer NG, integer noInd
create the grid xmax × ymax with a random population Pop
for i = 1 to NG do
   express genes as expression trees,
   calculate fitness of each gene,
   keep best gene
   for all g ∈ Pop do
      nghbrs←CalculateNeighourhood(g)
      offspring1←One-pointRecomb(g, nghbrs)
      offspring2←Two-pointRecomb(g, nghbrs)
      gNew ← the better fitted of two offsprings
      mutation(gNew)
      IStransposition(gNew)
      RIStransposition(gNew)
      Replacement(position(g), AuxiliaryPop, gNew)
   end for
   Pop ← AuxiliaryPop
end for
return noInd best genes from Pop
```

suggested a refinement of a boosting algorithm called AdaBoost. The idea of Algorithm 5 is that in each iteration t the distribution weights of those instances that were correctly classified are reduced by a factor β_t and the weights of the misclassified instances stay unchanged. After the normalization the weights of instances misclassified are raised and they add up to 1/2, and the weights of the correctly classified instances are lowered and they also add up to 1/2. What is more, since it is required that the weak classifier has an error less than 1/2, it is guaranteed to correctly classify at least one previously misclassified instance. In the ensemble decision those classifiers which produced small error and β_t is close to zero, have a large voting role since $1/\beta_t$ and logarithm of $1/\beta_t$ are large.

3.2 AdaBoost-Based Ensemble Classifiers with Cellular GEP Learning (AB-cGEP)

The approach is similar to the one described in the Subsection 3.1 except that weak classifiers are induced using cGEP (Algorithm 4). To be more precise, AB-cGEP is a variant of Algorithm 5 where in the incremental learning (line 4) Algorithm 1 is replaced by Algorithm 4.

Algorithm 5. AB-GEP

Require: training data TD of size N, test dataset TS, integer T, integer $M \leq N$ - size of the selected dataset
Ensure: qc quality of the AdaBoost classifier.
initialize the distribution $D_1(i) = \frac{1}{N}, i = 1, \ldots, N$
for $t = 1$ to TT **do**
  for the current distribution D_t select a training dataset $S_t \subset TD$ of size M,
  call Algorithm INC for the dataset S_t, receive the classifier C_t
  using the majority voting for C_t calculate the error $\epsilon_t = \sum_{C_t(\mathbf{x}_i) \neq y_i} D_t(i)$
  if $\epsilon_t > 0.5$ **then**
    abort
  else
    $\beta_t = \epsilon_t / (1 - \epsilon_t)$
  end if
  { update the distribution}
  for $i = 1$ to N **do**
    if $C_t(\mathbf{x}_i) = y_i$ **then**
      $D_t(i) \leftarrow D_t(i) \times \beta_t$
    end if
    normalize the distribution $D_{t+1}(i) = D_t(i)/Z_t, \; Z_t = \sum_i D_t(i)$
  end for
end for
{test the ensemble classifier $C_1, C_2, \ldots, C_T$ in the test dataset TS}
$qc \leftarrow 0$
for all $(\mathbf{x}, \mathbf{y}) \in$ **TS do**
  $V_i = \sum_{C_t(\mathbf{x})=i} \log(1/\beta_t), \; i = 1, \ldots, |C|$
  $c \leftarrow arg\max_{1 \leq j \leq |C|} V_j$
  if $c = y$ **then**
    $qc \leftarrow qc + 1$
  end if
end for
$qc \leftarrow qc/|TS|$
return qc

3.3 Majority-Voting-Based Ensemble Classifier with GEP Learning (MV-GEP and MVC-GEP)

We propose two variants of the majority-voting-based Ensemble Classifier with GEP learning: simple majority voting (MV) and majority voting with clustering (MVC). Given a new instance d the decision profile is calculated:

$$DP(d) = \begin{pmatrix} g_1^1(d) & \ldots & g_P^1(d) \\ \ldots & \ldots & \ldots \\ g_1^i(d) & \ldots & g_P^i(d) \\ \ldots & \ldots & \ldots \\ g_1^{|C|}(d) & \ldots & g_P^{|C|}(d) \end{pmatrix}$$

Algorithm 6. MV (majority voting - learning)

```
Require: training data TD divided into |C| subsets, integer P,two parameters:
    supp_cl (support in class), supp_ncl (support outside class)
for i = 1 to |C| do
  repeat
    call procedure GEP to find gene g
    g is 'true' for at least supp_cl % of rows from class i
    g is 'true' for not more than supp_ncl % of rows from classes other than i
  until population for class i contains at least P elements
end for
return population of P genes
```

where the row $(g_1^i(d), \ldots, g_P^i(d))$ is the support for class i. And $g_j^i(d) = 1$ if the expression tree for gene g_j^i is 'true' for instance d, otherwise $g_j^i(d) = 0$. Observe that after an ideal learning, if i is the right class for data d then in matrix $DP(d)$ all the elements in row i are 1 and 0 everywhere else.

To assign a class label j to instance d, the entire decision profile $DP(d)$ is considered and

$$j = arg \max_{1 \leq j \leq |C|} (\sum_{k=1}^{P} g_k^j(d))$$

that is, the majority voting is performed among genes which evaluate to 'true'. For MVC algorithm the k-means algorithm [19] is used for each class i to partition the training data set TD_i so that the resulting intercluster similarity is high but the intracluster similarity is low. Cluster similarity is measured in regard to the mean value of the objects in a cluster. Observe

Algorithm 7. MVC (cluster - learning)

```
Require: training data TD divided into |C| subsets, number of clusters k, integer
    P,
for i = 1 to |C| do
  use k-means algorithm to partition the data set TD_i into k clusters
  CL^i_1, ..., CL^i_k with centroids CNT^i_j
  for j = 1 to k do
    l ← 0
    repeat
      l ← l + 1
      call procedure GEP to find gene g best fitting the class i with data set
      CL^i_j
    until l = P
  end for
end for
return population of P genes, centroids CNT
```

that in this case the population is a three dimensional structure since for each class i the gene matrix contains $k \times P$ genes denoted g^i_j. The testing algorithm for MVC is performed as follows. For instance d and each class i, $1 \leq i \leq |C|$ the nearest cluster $C^{j_i}_i$ is found, i.e

$$j_i = arg \min_{1 \leq l \leq k} (dist(d, CNT^i_l))$$

Then for each class i, genes generated for cluster j_i make up for the decision profile:

$$DP(d) = \begin{pmatrix} g^{j_1}_1(d) & \dots & g^{j_1}_P(d) \\ \dots & \dots & \dots \\ g^{j_i}_1(d) & \dots & g^{j_i}_P(d) \\ \dots & \dots & \dots \\ g^{j_{|C|}}_1(d) & \dots & g^{j_{|C|}}_P(d) \end{pmatrix}$$

3.4 Majority-Voting-Based Ensemble Classifier with Cellular GEP Learning (MV-cGEP)

Learning takes place in two steps, shown as Algorithms 8 and Algorithm 9. Firstly, for each class (separately) the population of genes best fitting the class is found. During this step, in each generation the algorithm cGEP is applied and using the domination relation best individuals are copied to next generation. The objective of the second step, called meta learning, is to select subsets of informative genes from the population defined in the first step in order to obtain high classification accuracy. In the process of meta-learning genetic algorithms are applied in order to select a subset of genes obtained in the process of learning and resulting in the population of $|C| \times sizeBest$ genes. Now individuals are defined as matrices of type $MG = \{0,1\}^{|C| \times sizeBest}$ which correspond to the distribution of genes from the population *population_best*. For the matrix $mg \in MG$ the set $\{i : mg[cl,i] = 1\}$ picks up genes from

Algorithm 8. MV-cGEP first step

Require: training data with correct labels representing $|C|$ classes, integer $sizePop$, $sizeBest$
Ensure: population of $sizeBest$ genes for each class
1: **for all** $cl \in C$ **do**
2:     $population\_best^{cl} \leftarrow \phi$
3:     **repeat**
4:         call algorithm cGEP for class *cl* to generate *pop*
5:         add to $population\_best^{cl}$ those genes from population *pop* which are not dominated in $population\_best^{cl}$
6:     **until** $population\_best^{cl}$ contains at least $sizeBest$ elements
7: **end for**

the population $population\_best^{cl}$ which are meaningful for the class cl. To define fitness of an individual mg we assume the majority vote is performed to classify each data row and the number of correct answers is counted. Let $r = (\mathbf{x}, y)$ be a data row

$$which_i^{cl}(mg, \mathbf{x}) = \begin{cases} 1 & \text{if } mg[cl, i] = 1 \text{ and} \\ & \quad population\_best^{cl}(i) \text{ is true for } \mathbf{x} \\ 0 & \text{otherwise} \end{cases}$$

$$mv(mg, \mathbf{x}) = \max_{cl}(\sum_{i=1}^{sizeBest} which_i^{cl}(mg, \mathbf{x}))$$

and, finally

$$fitness(mg) = \frac{|r : \; mv(mg, \mathbf{x} = y|}{|dataset|}$$

Having defined individuals and fitness, standard genetic algorithms are applied to find one matrix mg best fitting the given dataset. This matrix is then used in testing where again, the majority vote is applied.

Algorithm 9. MV-cGEP second step (metalearning)

Require: training data with correct labels representing $|C|$ classes, integer $noIter$, population of $sizeBest$ genes for each class,
Ensure: best metagene mg

```
create random population popMet of metagenes
for i = 1 to noIter do
    calculate fitness of each metagene from popMet
    using roulette rule choose the metagenes for the next step,
    mutation,
    crossover,
end for
return the best metagene mg
```

3.5 Triplet Mass Function-Based Ensemble Classifier with GEP Learning (TMF-GEP)

The idea to apply mass functions is based on the Dempster-Shafer (DS) theory of evidence [34] where evidence is represented in terms of evidential functions and ignorance. Evidential functions include mass, belief and plausibility functions, each conveying the same information as any of the others. The DS theory formulates the reasoning process as pieces of evidence and hypotheses allowing to infer conclusions from the given uncertain evidence. In this section we define a method of attaching a triplet mass function to

Algorithm 10. MV-cGEP

Require: training data with correct labels representing $|C|$ classes, integer $sizePop$, $sizeBest$, testing data TS
Ensure: qc quality of the AdaBoost classifier
apply Algorithm 8 to define $population\_best$
apply Algorithm 9 to find metagene mg
{test mg for testing data TS}
$qc \leftarrow 0$
for all $(\mathbf{x}, y) \in TS$ **do**
  if $mv(mg, \mathbf{x}) = y$ **then**
    $qc \leftarrow qc + 1$
  end if
end for
$qc \leftarrow qc/|TS|$
return qc

a given classifier Cl and a given instance of data, then the algorithm of applying a set of triplets to get a class for the data, and finally the classification algorithm which makes use of the former one.

Given a classifier Cl as an output of algorithm INC, and a new instance of data d, the decision profile $DP(d)$ is a structure (not really a matrix because rows are of different length, not exceeding P)

$$DP(d) = \begin{pmatrix} g_1^1(d) & \dots & g_{j_1}^1(d) \\ \dots & \dots & \dots \\ g_1^i(d) & \dots & g_{j_i}^i(d) \\ \dots & \dots & \dots \\ g_1^{|C|}(d) & \dots & g_{j_{|C|}}^{|C|}(d) \end{pmatrix}$$

where the row $(g_1^i(d), \dots, g_{j_i}^i(d))$ is the support for class i. And $g_j^i(d) = 1$ if the expression tree for gene g_j^i is 'true' for instance d, otherwise $g_j^i(d) = 0$. To assign a class label i to instance d, the entire decision profile $DP(d)$ is considered. Let

$$m(i) = \frac{\sum_{k=1}^{j_i} g_k^i(d)}{\sum_{l=1}^{|C|} \sum_{k=1}^{j_l} g_k^l(d)} \tag{1}$$

Then the triplet (see [6] for the details) for the classifier Cl and instance d is an expression

$$T_{Cl}(d) = < u, v, C > \tag{2}$$

where

$$u = arg\max\{m(i) : 1 \le i \le |C|\}$$

$$v = arg\max\{m(i) : 1 \le i \le |C|, \; u \ne v\}$$

$$m(u) + m(v) + m(C) = 1$$

and m is the mass function defined in (1).

Any two triplets T_1, T_2 are combined to form a new triplet.

Definition 1. Let $T_1 =< x_1, y_1, C >$, $T_2 =< x_2, y_2, C >$ be two triplets with respective mass functions m_1, m_2. Then T is a new triplet with a mass function $m_1 \oplus m_2$. Consider the following possible cases:

1. two focal points equal, that is $x_1 = x_2$, $y_1 = y_2$, $x_1 \neq y_1$
   Then the mass functions are not in conflict if

$$m_1(x_1) \cdot m_2(y_2) + m_1(y_1) \cdot m_2(x_2) < 1$$

   and $T =< x_1, y_1, C >$ with

$$m_1 \oplus m_2(x_1) := K \cdot (m_1(x_1) \cdot m_2(x_2) + m_1(x_1) \cdot m_2(C) + m_1(C) \cdot m_2(x_2))$$

$$m_1 \oplus m_2(y_1) := K \cdot (m_1(y_1) \cdot m_2(y_2) + m_1(y_1) \cdot m_2(C) + m_1(C) \cdot m_2(y_2))$$

   where

$$m_1(C) = 1 - m_1(x_1) - m_1(y_1)$$

$$m_2(C) = 1 - m_2(x_2) - m_2(y_2)$$

$$K^{-1} = 1 - m_1(x_1) \cdot m_2(y_2) - m_1(y_1) \cdot m_2(x_2)$$

2. one focal point equal, that is $x_1 = x_2$, $y_1 \neq y_2$. Then the mass functions are not in conflict if

$$m_1(x_1) \cdot m_2(y_2) + m_1(y_1) \cdot m_2(y_2) + m_1(y_1) \cdot m_2(x_2) < 1$$

   Let

$$f(x_1) := K \cdot (m_1(x_1) \cdot m_2(x_2) + m_1(x_1) \cdot m_2(C) + m_1(C) \cdot m_2(x_2))$$

$$f(y_1) := K \cdot m_1(y_1) \cdot m_2(C)$$

$$f(y_2) := K \cdot m_1(C) \cdot m_2(y_2)$$

   where

$$K^{-1} = 1 - m_1(x_1) \cdot m_2(y_2) - m_1(y_1) \cdot m_2(y_2) - m_1(y_1) \cdot m_2(x_2)$$

   Let

$$x = arg\max(f(x_1), f(y_1), f(y_2))$$

$$y = arg\max\{f(t) : t \in \{x_1, y_1, y_2\}, t \neq x\}$$

   Then the new triplet

$$T =< x, y, C >$$

$$m_1 \oplus m_2(x) := f(x)$$

$$m_1 \oplus m_2(y) := f(y)$$

$$m_1 \oplus m_2(C) = 1 - m_1 \oplus m_2(x) - m_1 \oplus m_2(y)$$

3. completely different focal points, $x_1 \neq x_2$, $y_1 \neq y_2$. Then the mass functions are not in conflict if

$$m_1(x_1) \cdot m_2(x_2) + m_1(x_1) \cdot m_2(y_2) + m_1(y_1) \cdot m_2(x_2) + m_1(y_1) \cdot m_2(y_2) < 1$$

Let

$$f(x_1) := K \cdot m_1(x_1) \cdot m_2(C)$$
$$f(y_1) := K \cdot m_1(y_1) \cdot m_2(C)$$
$$f(x_2) := K \cdot m_1(C) \cdot m_2(x_2)$$
$$f(y_2) := K \cdot m_1(C) \cdot m_2(y_2)$$

where

$$K^{-1} = 1 - m_1(x_1) \cdot m_2(x_2) - m_1(x_1) \cdot m_2(y_2) - m_1(y_1) \cdot m_2(x_2) - m_1(y_1) \cdot m_2(y_2)$$

Let

$$x = arg\max(f(x_1), f(y_1), f(x_2), f(y_2))$$
$$y = arg\max\{f(t) : t \in \{x_1, y_1, x_2, y_2\}, t \neq x\}$$

And the new triplet

$$T =< x, y, C >$$
$$m_1 \oplus m_2(x) = f(x)$$
$$m_1 \oplus m_2(y) = f(y)$$
$$m_1 \oplus m_2(C) = 1 - m_1 \oplus m_2(x) - m_1 \oplus m_2(y)$$

Algorithm 11. TMF-GEP

Require: training data $TD = \{(\mathbf{x}_i, y_i) : y_i \in \{1, \ldots, |C|\}, i = 1, \ldots, N\}$, test dataset TS, integer T - number of triplets
1: call Algorithm 2 to receive classifier CL_1
2: curr_triplet← triplet for CL_1
3: **for** $i = 2$ to T **do**
4: call Algorithm 2 and receive classifier CL_i
5: new_triplet← triplet for CL_i as defined in (2)
6: new_triplet← combination of new_triplet and curr_triplet
7: **end for**
8: $CL \leftarrow$ the first focal point of curr_triplet
{testing Cl for the dataset TS similarly as in Algorithm 3 to receive qc}
9: **return** qc

4 Computational Experiment Results

To evaluate the proposed approach computational experiment has been carried out. The experiment involved the following datasets from the UCI Machine Learning Repository [2]: Wisconsin Breast Cancer (WBC), Diabetes, Sonar, Australian Credit (ACredit), German Credit (GCredit), Cleveland Heart (Heart), Hepatitis and Ionosphere. Basic characteristics of these sets are shown in Table 1. In the reported experiment classification tools described in Section 3 have been compared with the performance of 12 well-known

Table 1 Datasets used in the experiment

name	data type	attribute type	no. instances	no. attributes
WBC	multivariate	integer	699	11
Diabetes	multivariate, time-series	categorical, integer	768	9
Sonar	multivariate	real	208	61
ACredit	multivariate	categorical, integer, real	690	15
GCredit	multivariate	categorical, integer	1000	21
Heart	multivariate	categorical, real	303	14
Hepatitis	multivariate	categorical, integer, real	155	20
Ionosphere	multivariate	integer, real	351	35

Table 2 Comparison of classifier accuracy (%)

no	classifier	WBC	Diab.	Sonar	ACr.	GCr	Heart	Hep.	Ion.
1	Naive Bayes	95,99	76,30	67,78	77,68	75,40	83,70	84,51	82,62
2	Bayes Net	97,14	74,35	80,28	86,23	75,50	81,11	83,22	89,46
3	Logistic	96,56	77,21	73,08	85,22	75,20	83,70	82,58	88,89
4	RBF Network	95,85	75,39	72,11	79,71	74,00	84,07	85,80	92,31
5	AdaBoost M1	94,85	74,34	71,63	84,64	69,50	80,00	82,58	90,88
6	SVM	96,99	77,34	75,96	84,92	75,10	84,07	85,16	88,60
7	Ensemble Sel.	94,42	74,61	75,48	84,93	73,10	80,00	81,29	90,59
8	Bagging	95,56	74,61	77,40	85,07	74,40	79,26	84,52	90,88
9	Class. via clust.	95,71	64,84	54,32	74,06	56,60	77,04	74,19	70,94
10	Random Comm.	95,99	73,95	84,13	83,48	73,90	80,37	84,52	92,59
11	C4.5	94,56	73,82	71,15	86,09	70,50	76,66	83,87	91,45
12	Rotation Forest	96,99	76,69	84,13	87,25	74,80	80,74	82,58	93,73
13	AB-GEP	**98,57**	**78,10**	84,10	**89,31**	**80,53**	**88,00**	85,76	90,23
14	AB-cGEP	95,86	77,21	81,24	86,52	77,37	83,84	**87,13**	91,35
15	TMF-GEP	96,71	69,46	83,12	87,82	73,56	79,26	86,65	90,89
16	MV-GEP	97.80	75,20	83,65	86,23	75,20	86,50	85,38	90,67
17	MVC-GEP	96,70	76,70	**84,27**	86,79	77,54	84,40	84,15	**93,82**
18	MV-cGEP	95,58	76,99	80,79	87,39	76,27	80,24	86,46	91,73

classifiers from WEKA Environment for Knowledge Analysis v. 3.7.0 [40] including Naïve Bayes, Bayes Net, Logistic, RBF Network, Adaboost M1, SVM, Ensemble Selection, Bagging, Classification via Clustering, Random Committee, C4.5, and Rotation Forrest.

Computations involving GEP have been run with the following arbitrary parameter settings: population size – 100, number of iterations – 100. In cellular GEP, xmax = ymax = 10 and number of iterations – 50. In AdaBoost-based classifiers the number of AdaBoost iterations was set to 20. All probabilities of performing genetic operations in all cases have been set to 0.2. In all WEKA classifiers the default parameter settings have been applied.

Table 2 shows average classifier accuracies obtained over 10 repetitions of the 10-cross-validation scheme.

In tables 3 — 10 the extended range of the average performance measures, characterizing family of the discussed GEP-induced ensemble classifiers, is shown. Set of the performance measures include classifier accuracy, precision, recall, f-measure and area under the ROC curve calculated as the Wilcoxon-Mann-Whitney statistic (for the detailed description of these

Table 3 Average performance measures obtained in the experiment for the WBC dataset

Classifier	Accuracy (%)	Precision	Recall	F-measure	ROC
AB-GEP	**98.57**	**0.975**	**0.975**	**0.974**	**0.993**
AB-cGEP	95.86	0.947	0.938	0.943	0.978
TMF-GEP	96.71	0.949	0.949	0.951	0.972
MV-GEP	97.80	0.970	0.969	0.968	0.990
MVC-GEP	96.70	0.952	0.951	0.953	0.980
MV-cGEP	95.50	0.919	0.948	0.933	0.980
Bayes Net	97.14	0.972	0.971	0.972	0.992
SVM	96.99	0.970	0.970	0.970	0.968

Table 4 Average performance measures obtained in the experiment for the Diabetes dataset

Classifier	Accuracy (%)	Precision	Recall	F-measure	ROC
AB-GEP	**78.10**	0.736	0.760	0.748	0.802
AB-cGEP	77.21	0.860	**0.804**	0.831	0.820
TMF-GEP	69.46	0.712	0.716	0.720	0.732
MV-GEP	75.20	0.775	0.755	0.742	0.812
MVC-GEP	76.70	0.760	0.760	0.756	0.762
MV-cGEP	76.99	**0.908**	0.779	**0.838**	0.810
SVM	77.34	0.769	0.773	0.763	0.720
Logistic	77.21	0.772	0.772	0.765	**0.832**

Table 5 Average performance measures obtained in the experiment for the Sonar dataset

Classifier	Accuracy (%)	Precision	Recall	F-measure	ROC
AB-GEP	84.10	0.793	**0.859**	0.825	0.911
AB-cGEP	81.24	0.786	0.856	0.819	0.845
TMF-GEP	83.12	0.838	0.839	0.838	0.888
MV-GEP	83.65	0.841	0.749	0.792	0.881
MVC-GEP	**84.27**	0.833	0.843	0.838	0.875
MV-cGEP	80.79	0.786	0.778	0.776	0.822
Rot. Forrest	84.14	**0.843**	0.841	**0.841**	**0.925**
Random Com.	84.13	0.841	0.841	**0.841**	0.912

Table 6 Average performance measures obtained in the experiment for the Australian credit dataset

Classifier	Accuracy (%)	Precision	Recall	F-measure	ROC
AB-GEP	**89.31**	0.886	0.884	0.868	0.912
AB-cGEP	86.52	0.880	0.826	0.852	0.893
TMF-GEP	87.82	0.892	**0.892**	**0.891**	0.906
MV-GEP	86.23	0.876	0.856	0.864	0.910
MVC-GEP	86.79	0.886	0.873	0.874	**0.930**
MV-cGEP	87.39	**0.893**	0.832	0.861	0.890
Bayes Net	86.23	0.864	0.862	0.861	0.921
C4.5	86.09	0.861	0.861	0.861	0.887

Table 7 Average performance measures obtained in the experiment for the German credit dataset

Classifier	Accuracy (%)	Precision	Recall	F-measure	ROC
AB-GEP	**80.53**	**0.896**	0.842	**0.854**	**0.887**
AB-cGEP	77.37	0.892	0.806	0.846	0.872
TMF-GEP	73.56	0.728	0.728	0.736	0.766
MV-GEP	75.20	0.776	0.842	0.808	0.780
MVC-GEP	77.54	0.818	**0.857**	0.837	0.806
MV-cGEP	76.27	0.880	0.802	0.839	0.823
Bayes Net	75.50	0.746	0.755	0.749	0.780
Naive Bayes	75.40	0.743	0.754	0.746	0.787

measures see [13]). In addition, for each dataset covered in the experiment, the performance measures of the two best classifiers in terms of the classification accuracy, selected from the 12 classifiers used in the reported experiment, are shown.

Table 8 Average performance measures obtained in the experiment for the Cleveland Heart dataset

Classifier	Accuracy (%)	Precision	Recall	F-measure	ROC
AB-GEP	**88.00**	**0.856**	**0.858**	**0.848**	**0.902**
AB-cGEP	83.84	0.804	0.830	0.817	0.843
TMF-GEP	79.26	0.726	0.714	0.732	0.812
MV-GEP	86.50	0.809	0.809	0.809	0.877
MVC-GEP	84.40	0.819	0.776	0.797	0.845
MV-cGEP	80.24	0.681	0.844	0.754	0.880
RBF Network	84.07	0.841	0.841	0.841	0.893
SVM	84.07	0.841	0.841	0.840	0.837

Table 9 Average performance measures obtained in the experiment for the Hepatitis dataset

Classifier	Accuracy (%)	Precision	Recall	F-measure	ROC
AB-GEP	85.76	0.862	0.840	0.834	0.780
AB-cGEP	**87.13**	**0.930**	0.911	0.917	0.792
TMF-GEP	86.65	0.926	**0.926**	**0.922**	**0.924**
MV-GEP	85.38	0.811	0.808	0.808	0.862
MVC-GEP	84.15	0.680	0.772	0.735	0.830
MV-cGEP	86.46	0.920	0.910	0.915	0.920
RBF Network	85.80	0.852	0.858	0.854	0.835
SVM	85.16	0.847	0.852	0.849	0.756

Table 10 Average performance measures obtained in the experiment for the Ionosphere dataset

Classifier	Accuracy (%)	Precision	Recall	F-measure	ROC
AB-GEP	90.23	0.873	0.871	0.872	0.946
AB-cGEP	91.35	0.943	0.924	0.934	0.960
TMF-GEP	90.89	0.851	0.886	0.875	0.958
MV-GEP	90.67	0.841	0.915	0.882	0.952
MVC-GEP	**93.82**	0.888	0.920	0.906	**0.980**
MV-cGEP	91.73	**0.968**	0.907	0.936	**0.980**
Rot. Forrest	93.73	0.937	0.937	**0.937**	0.967
Random Com.	92.59	0.926	**0.926**	0.926	0.976

From the above tables it is clear that the family of the ensemble classifiers, constructed from GEP-induced expression trees, performs very well. Table 11 shows average ranks of all investigated classifiers with the classification accuracy as a criterion. For each data set covered by the experiment, the best classifier has been awarded value of 1 point, the second best value of 2 points, etc. Subsequently, classifiers ranks have been averaged over all datasets. Applying the Friedman's non-parametric test using ranks of the data, have confirmed that the null hypothesis stating that all of the 18 population distribution functions are identical should be rejected at the significance level of $0,05$.

Table 11 Average ranks of the investigated classifiers obtained in the reported computational experiment

classifier	rank	classifier	rank
AB-GEP	3,38	RBF Network	9,38
MVC-GEP	4,50	Logistic	9,75
AB-cGEP	5,63	Rand. Comm	9,75
MV-GEP	6,50	Naive Bayes	11,38
MV-cGEP	6,75	Bagging	11,50
Rotation F.	7,13	C4.5	13,75
SVM	8,38	Ens. Sel	13,88
TMF-GEP	8,88	AdaBoost M1	14,13
Bayes Net	9,13	Class. V. clust.	17,25

To support the above findings the Wilcoxon rank-sum test was used. Wilcoxon rank-sum test, (or Wilcoxon-Mann-Whitney test) is a non-parametric test for assessing whether two independent samples of observations come from the same distribution [7]. In the reported case a pairwise comparison of the classification accuracy rank of different classifiers has been carried out under the following hypotheses:

- Null Hypothesis H_0: There is no difference between rank medians. Two samples are drawn from a single population, and therefore their probability distributions are equal.
- Alternative Hypothesis H_1: Rank medians are statistically different. One sample is statistically greater.

In our case the critical value of the Wilcoxon T statistics is 2 (n = 81 = 7; two-tailed test) and the assumed significance level is 0. 05. AB-GEP has proven to produce statistically greater classification accuracy than 9 other classifiers out of the remaining 17 considered in the experiment.

Analysis of the precision, recall and F-measures provides further insight into the experiment results. Precision for a class is the number of true positives (i.e. the number of items correctly labeled as belonging to the positive class) divided by the total number of elements labeled as belonging to the positive class (i.e. the sum of true positives and false positives, which are items incorrectly labeled as belonging to the class). Recall in this context is defined as the number of true positives divided by the total number of elements that actually belong to the positive class (i.e. the sum of true positives and false negatives, which are items which were not labeled as belonging to the positive class but should have been). A measure that combines precision and recall is the harmonic mean of precision and recall, known as the F-measure derived by van Rijsbergen [32]. F-measure (denoted also as F1) assumes that recall and precision are evenly weighted. Comparing two classifiers performing best in each measure and belonging to the family of the proposed ensemble classifiers with two best from the remaining ones, gives the following results:

- Best GEP-induced ensemble classifier produce, on average, by 5.56% better precision values then best of the non GEP-induced classifiers
- Best GEP-induced ensemble classifier produce, on average, by 3.08% better recall values then best of the non GEP-induced classifiers
- Best GEP-induced ensemble classifier produce, on average, by 3.10% better F-measure values then best of the non GEP-induced classifiers
- Best GEP-induced ensemble classifier produce, on average, by 1.54% better area under the ROC curve values then best of the non GEP-induced classifiers

Comparing the proposed family of classifiers with the performance of several widely used and popular classifiers (some of them also built through applying collective computational intelligence tools) shows good potential of gene expression programming and cellular evolutionary algorithms, when applied to the field of machine learning. The visible advantage in terms of the classification accuracy can be attributed to their collaborative and synergetic features. The proposed ensemble classifiers offer performance comparable to the state-of-the-art heterogeneous classifier combination approaches. In Tab. 12,

Table 12 Comparison of the classification accuracy. Best combined classifiers obtained through selective fusion of heterogeneous classifiers as reported in [37] versus best ensemble classifiers proposed in the chapter

Dataset	Best combined with with voting [37]	Best combined with weighted voting [37]	Best GEP/cellular evolutionary
WBC	97.53%	97.03%	**98.57%**
Heart	76.66%	76.53%	**88.00%**
Hepatitis	84.96%	82.84%	**87.13%**
Ionosphere	35.75%	35.00%	**91.73%**
Diabetes	**93.31%**	**93.31%**	78.10%
ACredit	**95.25%**	**95.25%**	89.31%
GCredit	84.45%	**84.62%**	80.53%

the best classifiers induced using GEP and cellular evolutionary algorithm are compared with the best classifiers obtained in the process of selective fusion of heterogeneous classifiers proposed by Tsoumakas et al. [37].

5 Conclusions

Experiment results presented in the previous section confirm that next generation collective computational intelligence techniques like gene expression programming and cellular evolutionary algorithms, when applied to the field of machine learning, can be of help in producing excellent quality results through exploiting their collaborative and synergetic features. The proposed family of ensemble classifiers constructed from expression trees stands out as a convincing example showing effectiveness of the combinatorial effort of integrating collaborative data technologies for computational intelligence. The presented research allows also to draw the following conclusions with respect to application of the proposed techniques to the field of machine intelligence:

- Gene expression programming is a versatile and useful tool to automatically induce expression trees.
- Using GEP-induced expression trees allows for construction of a high quality ensemble classifiers competitive, in terms of classification accuracy, to many other approaches.
- High quality of the ensemble classifier performance can be attributed to the expression trees induced by GEP.
- GEP-induced expression trees can be easily converted into sets of rules easy to understand and interpret.
- GEP-induced ensemble classifiers should be used in applications where precision of the classification results is of primary concern to the user.

Future research should focus on extending the approach through integration with the data reduction algorithms to achieve better performance and reduce computation time required.

References

1. Alba, E., Dorronsoro, B.: Cellular Genetic Algorithms. Springer Science, New York (2008)
2. Asuncion, A., Newman, D.J.: UCI Machine Learning Repository. University of California, School of Information and Computer Science (2007), `http://www.ics.uci.edu/~mlearn/MLRepository.html`
3. Battiti, R., Colla, A.M.: Democracy in neural nets: Voting schemes for classification. Neural Networks 7(4), 691–707 (1994)

4. Bauer, E., Kohavi, R.: An empirical comparison of voting classification algorithms: Bagging, boosting, and variants. Machine Learning 36(1-2), 105–139 (1999)
5. Bennett, P.N.: Building Reliable Metaclassifiers for Text Learning, Ph.D. Thesis, School of Computer Science, Carnegie Mellon University, Pittsburgh (2006)
6. Bi, Y., Guan, J., Bell, D.: The combination of multiple classifiers using an evidentional reasoning approach. Artif. Intell. 172, 1731–1751 (2008)
7. Corder, G.W., Foreman, D.I.: Nonparametric Statistics for Non-Statisticians: A Step-by-Step Approach. J. Wiley, New Jersey (2009)
8. Dasarathy, B.V., Sheela, B.V.: Composite classifier system design: concepts and methodology. Proceedings of the IEEE 67(5), 708–713 (1979)
9. Drucker, H., Cortes, C., Jackel, L.D., LeCun, Y., Vapnik, V.: Boosting and other ensemble methods. Neural Computation 6(6), 1289–1301 (1994)
10. Duan, L., Tang, C., Zhang, T., Wei, D., Zhang, H.: Distance guided classification with gene expression programming. In: Li, X., Zaïane, O.R., Li, Z.-h. (eds.) ADMA 2006. LNCS (LNAI), vol. 4093, pp. 239–246. Springer, Heidelberg (2006)
11. Duda, R., Hart, P., Stork, D.: Pattern Classification. J. Wiley, New York (2001)
12. Engelbrecht, A.P.: Computational Intelligence. An Introduction. J. Wiley, Chichester (2007)
13. Fawcett, T.: ROC Graphs: Notes and Practical Considerations for Researchers, HP Labs Tech Report HPL-2003-4, Palo Alto, Ca (2003)
14. Ferreira, C.: Gene expression programming: a new adaptive algorithm for solving problems. Complex Systems 13(2), 87–129 (2001)
15. Ferreira, C.: Gene Expression Programming. SCI, vol. 21, pp. 337–380. Springer, Heidelberg (2006)
16. Freund, Y., Schapire, R.E.: Decision-theoretic generalization of on-line learning and application to boosting. Journal of Computer and System Science 55(1), 119–139 (1997)
17. Fulcher, J.: Computational Intelligence. An Introduction. SCI, vol. 115, pp. 3–78. Springer, Heidelberg (2006)
18. Gama, J.: Local cascade generalization. In: Proceedings of the 15th International Conference on Machine Learning, pp. 206–214 (1998)
19. Hartigan, J.A., Wong, M.A.: A k-means clustering algorithm. Applied Statistics 28(1), 100–108 (1979)
20. Ho, T.K., Hull, J.J., Srihari, S.N.: Decision combination in multiple classifier systems. IEEE Transactions on Pattern Recognition and Machine Intelligence 16(1), 66–75 (1994)
21. Huang, C.-L., Chen, M.-C., Wang, C.-J.: Credit scoring with a data mining approach based on support vector machines. Expert Systems with Applications 33, 847–856 (2007)
22. Jedrzejowicz, J., Jedrzejowicz, P.: GEP-induced expression trees as weak classifiers. In: Perner, P. (ed.) ICDM 2008. LNCS (LNAI), vol. 5077, pp. 129–141. Springer, Heidelberg (2008)
23. Jedrzejowicz, J., Jedrzejowicz, P.: A Family of GEP-induced ensemble classifiers. In: Nguyen, N.T., Kowalczyk, R., Chen, S.-M. (eds.) ICCCI 2009. LNCS, vol. 5796, pp. 641–652. Springer, Heidelberg (2009)

24. Jedrzejowicz, J., Jedrzejowicz, P.: Two ensemble classifiers constructed from GEP-induced expression trees. In: Jędrzejowicz, P., Nguyen, N.T., Howlet, R.J., Jain, L.C. (eds.) KES-AMSTA 2010. LNCS, vol. 6071, pp. 200–209. Springer, Heidelberg (2010)
25. Karakasis, V.K., Stafylopatis, A.: Data mining based on gene expression programming and clonal selection. In: Proc. IEEE Congress on Evolutionary Computation, pp. 514–521 (2006)
26. Kuncheva, L.I.: Classifier ensembles for changing environments. In: Roli, F., Kittler, J., Windeatt, T. (eds.) MCS 2004. LNCS, vol. 3077, pp. 1–15. Springer, Heidelberg (2004)
27. Lam, L., Suen, C.Y.: Optimal combination of pattern classifiers. Pattern Recognition Letters 16(9), 945–954 (1995)
28. Last, M., Maimon, O.: A compact and accurate model for classification. IEEE Transactions on Knowledge and Data Engineering 16(2), 203–215 (2004)
29. Li, X., Zhou, C., Xiao, W., Nelson, P.C.: Prefix gene expression programming. In: Proc. Genetic and Evolutionary Computation Conference, Washington, pp. 25–31 (2005)
30. Pena Centeno, T., Lawrence, N.D.: Optimising kernel parameters and regularisation coefficients for non-linear discriminant analysis. Journal of Machine Learning Research 7, 455–491 (2006)
31. Polikar, R.: Ensemble based systems in decision making. IEEE Circuits and Systems Magazine 3, 22–43 (2006)
32. Van Rijsbergen, C.V.: Information Retrieval, 2nd edn. Butterworth, London (1979)
33. Schapire, R.E., Freund, Y., Bartlett, P., Lee, W.S.: Boosting the margin: a new explanation for the effectiveness of voting methods. The Annals of Statistics 26(5), 1651–1686 (1998)
34. Shafer, G.: A Mathematical Theory of Evidence. Princeton University Press, Princeton (1976)
35. Srinivasa, K.B., Singh, A., Thomas, A.O., Venugopal, K.R., Patnoik, L.M.: Generic feature extraction for classification using fuzzy c-means clustering. In: Proc. Intelligent Sensing and Information Processing Conference, pp. 33–38 (2005)
36. Torre, F.: Boosting Correct Least General Generalizations, Technical Report GRAppA-0104, Grenoble (2004)
37. Tsoumakas, G., Angelis, L., Vlahavas, I.: Selective fusion of heterogeneous classifiers. Intelligent Data Analysis 9(6), 511–525 (2005)
38. Wang, W., Li, Q., Han, S., Lin, H.: A preliminary study on constructing decision tree with gene expression programming, In: Proc. First International Conference on Innovative Computing, Information and Control, vol. 1, pp. 222–225 (2006)
39. Weinert, W.R., Lopes, H.S.: GEPCLASS: a classification rule discovery tool using gene expression programming, In: Li, X., Zaïane, O.R., Li, Z.-h. (eds.) ADMA 2006. LNCS (LNAI), vol. 4093, pp. 871–880. Springer, Heidelberg (2006)
40. Witten, I.H., Frank, E.: Data Mining: Practical Machine Learning Tools and Techniques, 2nd edn. Morgan Kaufmann, San Francisco (2005)

41. Wolpert, D.H.: Stacked generalizatio. Neural Networks 5, 241–259 (1992)
42. Statlog Datasets: comparison of results, `http://www.is.umk.pl/projects/datasets.html#Cleveland` (accessed on December 27, 2007)
43. Zeng, T., Xiang, Y., Chen, P., Liu, Y.: A model of immune gene expression programming for rule mining. Journal of Universal Computer Science 13(7), 1239–1252 (2007)
44. Zhou, C., Xiao, W., Tirpak, T.M., Nelson, P.C.: Evolving accurate and compact classification rules with gene expression programming. IEEE Transactions on Evolutionary Computation 7(6), 519–531 (2003)

Chapter 8
Self-Organized Load Balancing through Swarm Intelligence

Vesna Šešum-Čavić and Eva Kühn

Abstract. The load balancing problem is ubiquitous in information technologies. New technologies develop rapidly and their complexity becomes a critical issue. One proven way to deal with increased complexity is to employ a self-organizing approach. There are many different approaches that treat the load balancing problem but most of them are problem specific oriented and it is therefore difficult to compare them. We constructed and implemented a generic architectural pattern, called SILBA, which stands for "self-initiative load balancing agents". It allows for the exchanging of different algorithms (both intelligent and unintelligent ones) through plugging. In addition, different algorithms can be tested in combination at different levels. The goal is to ease the selection of the best algorithm(s) for a certain problem scenario. SILBA is problem and domain independent, and can be composed towards arbitrary network topologies. The underlying technologies encompass a black-board based communication mechanism, autonomous agents and decentralized control. In this chapter, we present the complete SILBA architecture by putting the accent on using SILBA at different levels, e.g., for load balancing between agents on one single node, on nodes in one subnet, and between different subnets. Different types of algorithms are employed at different levels. Although SILBA possesses self-organizing properties by itself, a significant contribution to self-organization is given by the application of swarm based algorithms, especially bee algorithms that are modified, adapted and applied for the first time in solving the load balancing problem. Benchmarks are carried out with different algorithms and in combination with different levels, and prove the feasibility of swarm intelligence approaches, especially of bee intelligence.

1 Introduction

The IT-industry continuously faces a rapid increase in the complexity of software systems. New requirements, a large number of interacting components with

---

Vesna Šešum-Čavić · Eva Kühn
Technical University Vienna, Institute of Computer Languages,
Argentinierstrasse 8, 1040 Wien, Austria
{vesna,eva}@complang.tuwien.ac.at

N. Bessis, F. Xhafa (Eds.): Next Generation Data Technologies for CCI, SCI 352, pp. 195–224.
springerlink.com

internal states defined by many thousands of parameters, applications that rely on other unreliable systems, and many components tied together are only a few reasons that impose the necessity of finding new approaches for software systems. Main factors that determine software complexity are:

- Huge amounts of distributed components that interplay in a global solution,
- Problem size like number of computers, clients, requests, size of queries etc.,
- Heterogeneity,
- Autonomy of organizations, and
- Dynamic changes in the environment.

Distributed software systems are forced to integrate other software systems and components that are often not reliable, exhibit bad performance, and are sometimes unavailable. These challenges are so fundamental that, the usually taken approach to control distributed components across enterprise boundaries through one central and predefined coordinator software, reaches its technical and conceptual limits. The huge number of unpredictable dependencies on participating components cannot be coped with any more in the traditional way, namely through one central coordinator that implements the entire business logic and that possesses the complete picture of the distributed environment and all possible exceptions. A very useful concept in the adaptation of complex systems is *self-organization*. Certainly, self-organizing systems will not be able to adapt to all possible events, but they have proven to pose a good perspective to deal with complexity through self-organization, self-repairing, self-configuring, self-grouping, self-learning, self-adaptation, etc.

In this chapter, we consider the problem of load balancing (LB) in the light of the above mentioned challenges of today's systems. LB can be described as finding the best possible workload (re)distribution and addresses ways to transfer excessive load from busy (overloaded) nodes to idle (under-loaded) nodes. Dynamic LB should improve the performance of the overall distributed system and achieve the highest level of productivity.

1.1 Related Approaches

There are many different approaches that cope with LB. The first group consists of different conventional approaches without using any kind of intelligence, e.g.: Sender Initiated Negotiation and Receiver Initiated Negotiation [33], Gradient Model [22], Random Algorithm [38], and Diffusion Algorithm [7]. In Sender algorithm, LB is initiated by the over-loaded node. This algorithm has a good performance for low to moderate load levels while in Receiver algorithm, LB is initiated by the under-loaded node and this algorithm has a good performance for moderate to heavy load levels. Also the combination of these two algorithms

(Symmetric) is possible. The Gradient Model is based on dynamically initiated LB requests by the under-loaded node. The result of these requests is a system wide gradient surface. Overloaded nodes respond to requests by migrating unevaluated tasks down the gradient surface towards under-loaded nodes. In Random Algorithm each node checks the local workload during a fixed time period. When a node becomes over-loaded after a time period, it sends the newly arrived task to a randomly chosen node without taking in consideration whether the target node is over-loaded or not. Only the local information is used to make the decision. The principle of diffusion algorithms is keeping the process iterate until the load difference between any two processors is smaller than a specified value. The *second* group includes theoretical improvements of LB algorithms using different mathematical tools and estimations [2] without focusing on implementation and benchmarks. The *third* group contains approaches that use intelligent algorithms like evolutionary approaches [5], and ant colony optimization approaches [12]. Evolutionary approaches use the adjustment of some parameters specific for evolutionary algorithms to achieve the goal of LB. Ant colony optimization is used in [12] for a graph theoretic problem formulated from the task of computing load balanced clusters in ad hoc network. The intelligent algorithms from the last group showed promising results. However, they still need improvement concerning experience in the tuning of algorithms, the quality of solution they provide, scalability, the provisioning of a general model, and flexibility. In [21], non-pheromone-based (bee intelligence) versus pheromone-based algorithms are compared. Their conclusion is that the former are significantly more efficient in finding and collecting food.

These approaches mainly try to improve only one of the components of the whole LB infrastructure, namely the LB algorithm itself. A comprehensive classification of different LB approaches is given in [19], where we refer to the problem as the lack of a general framework, autonomy, self-* properties, and arbitrary configurations, and introduced a LB pattern, i.e. a software building block that abstracts the LB problem and that can be re-used in many different situations by simply configuring it termed SILBA (self-initiative load balancing agents) that addresses the following issues:

General Framework: Existing LB approaches are very problem specific. As there is no "one-fits-all" solution, in order to find the best solution for a problem, a general framework is needed that allows for testing and tuning of different LB algorithms for a given problem and environment. The SILBA architectural pattern is agile [24] and supports an easy and dynamic exchange of pluggable algorithms as well as combinations of different algorithms with the goal to ease the selection of the best algorithm for a certain problem scenario under certain conditions. Note that a framework itself doesn't solve the LB problem but serves as necessary basement for testing LB algorithms.

Autonomy and Self- Properties:* Increased complexity of software systems, diversity of requirements, and dynamically changing configurations imply a necessity to find new solutions that are e.g. based on self-organization, autonomic computing and autonomous (mobile) agents. Intelligent algorithms require autonomous agents which are advantageous in situations that are characterized by high dynamics, not-foreseeable events, and heterogeneity.

Arbitrary Configurations: LB can be required to manage the load among local core processors on one node, as well as in a network (intranet, internet, cloud). A general LB framework must be able to cope with all these demands at the same time and offer means to abstract hardware and network heterogeneities.

Our research focuses on a new conception of a self-organizing coordination infrastructure that suggests a combination of coordination spaces, self-organization, adaptive algorithms, and multi-agent technologies[1]. Each of these technologies has some form of self-organization in its incentive. In this chapter, after explaining the SILBA pattern in its basic form that supports LB between nodes in one subnet and briefly describing the obtained results, as a further step, SILBA is extended:

1. to support load balancing on several levels, i.e. not only between agents of the same node, but also between agents of different nodes and possible in different subnets and
2. to allow for combinations of different algorithms on different levels (e.g., swarm intelligence on each level, or swarm intelligence combined with unintelligent algorithms).

Our contributions are summarized in the following points where (1.) concerns our previous work on basic SILBA, and (2.)-(5.) concern extended SILBA:

1. Implementation of different algorithms, fine tuning of parameters and comparison of unintelligent versus intelligent algorithms, by plugging them into the SILBA pattern and benchmarking them: For the intelligent algorithms, we: a) adapted and implemented two ant algorithms, b) adapted and for the first time implemented the concepts of bee intelligence to the LB problem. The novelty includes the mapping and implementation of bee intelligence for the LB problem to improve the quality of the solution and scalability.
2. Realization of LB by extending it to several levels.
3. Construction of different combinations of algorithms for LB.
4. Investigation which combination of algorithms fits best for a particular network topology, and which topologies profit the most from the application of swarm intelligence.
5. Achievement of self-organization through different methods (like swarm intelligence, autonomous agents) in combination.

---

[1] It is assumed that a reader is familiar with the basic concepts of multi-agent technologies (see, for example [34], for an overview).

2 SILBA Framework

The SILBA framework is based on multi-agents technology and space-based computing. Space-based computing (SBC) is a powerful concept for the coordination of autonomous processes, an easy to use solution that handles the complexity of the interplay of autonomous components in a heterogeneous environment through a high abstraction of the underlying hardware and operating system software [20]. The processes communicate and coordinate themselves by simply reading and writing distributed data structures in a shared space. Although SBC is mainly a data-driven coordination model, it can be adapted and used according to control-driven coordination models. A space offers many advantages: a high level abstraction for developers that allows for hiding complexity, reliable communication, transactions, asynchrony, near-time event notification, scalability and availability [20].

SILBA uses a space-based architecture, called XVSM (extensible virtual shared memory) [16]. It generalizes Linda tuple based communication [11] as well as several extensions to it like reactions [30], programmable behaviour [3], and further coordination laws like priority and user defined match makers [29]. Comparable to Linda, a container represents a shared data space that can be accessed by the operations *read*, *take*, and *write*. Beyond that it can be addressed via an URL, can reference other containers, and is extensible through aspects [15], i.e., code fragments that react to certain events and serve to build higher level behaviour and interfaces on top of a container thus forming more complex coordination data structures. This asynchronous and blackboard based communication model is advantageous to for collaboration of autonomous (multi)agents as it avoids coupling through direct interactions [13] between the agents, especially when mobile agents are assumed [3], [29].

2.1 Basic SILBA

Basic SILBA supports the exchange of algorithms (both unintelligent and intelligent ones) as a test bed to ease the evaluation of the best algorithm for a certain problem scenario under certain conditions. It is based on decentralized control. Self-organization is achieved by using a blackboard based style of collaboration to build up a shared view on the current state. The SILBA pattern is domain independent and can be used at different levels:

- *Local* node level: allocating load to several core processors of one computer – the determining factor for load distribution is the balanced utilization of all cores.
- *Network* level: distributing load among different nodes. This includes load balancing within and between different subnets. One must take into consideration

the time needed for transferring data from a busy node to an idle node and estimate the priority of transferring, especially when the transfer itself requires more time to complete than the load assignment.

The basic components of SILBA are clients, autonomous agents, tasks and policies. Clients request tasks to be executed, i.e. load originates from clients. Different types of autonomous agents operate in a peer-to-peer manner and decide on their own when to pick up or push back work, assuming that the amount of work is changing dynamically. A task can be described as a tuple of the form "(priority, job, description, properties, timeout, answer space)", denoting the priority of a task in absolute terms, the job in a standard format like XML or WSDL, an optional (semantic) description, properties (e.g., whether task's execution mode is "at-most-once" or "best-effort"), a timeout, and a URL of an Internet addressable resource where to write the result of the execution back, that we term answer space to make the protocol stateless and to support not always connected networks. SILBA puts emphasis on two main policies termed transfer policy and location policy. The transfer policy determines whether and in which form a resource participates in load distribution and in that sense determines the classification of resources [33] into the following categories: under-loaded (UL), ok-loaded (OK) and overloaded (OL). The transfer policy is executed by a worker agent autonomously. A worker agent may reject a task it has started and re-schedule it. The location policy determines a suitable partner of a particular resource for LB [33]. The SILBA pattern is composed of three sub-patterns [19]:

The **local node pattern** is responsible for the execution of requests by local worker agents actively competing for work. The basic components of this pattern are: clients, worker agents, load space, and answer space. Load space is a place where new requests are put by clients and information about all worker agents' registrations and the current load status (UL, OK, OL) of a node are maintained. Requests are accessible in either the order they arrived, or by means of other criteria like their priority, the required worker role, or their expiration date. Answer spaces are places where the answers computed by worker agents are put directly (not routed) and where they can be picked up by the corresponding clients.

The **allocation pattern** redirects load between load spaces of different local nodes. The basic components of the allocation pattern are: load space, allocation agents, policies, and allocation space. There are three kinds of allocation agents: arbiter agents, IN agents, and OUT agents. Arbiter agents query the load of the load space and decide about re-distribution of work. They publish this information to the routing space in form of routing requests. Both IN and OUT agents read routing information from the allocation space and pull respectively push work from/to another node in a network to which the current node has a connection. The

IN and OUT allocation agents assume that the information about the (best) partner to/from which to distribute load can be queried from the allocation space. The allocation space holds information about partner nodes as computed by the location policy. This information is queried by the allocation agents and can be either statically configured or dynamically computed by routing agents.

The **routing pattern** executes the location policy according to a particular LB algorithm. The basic components of the routing pattern are: allocation space, routing agents, and routing space. Routing agents perform the location policy by implementing a certain LB algorithm and by communicating with other routing agents of the same type forming a dynamically structured overlay network. The collaboration between routing agents of different nodes is carried out via the corresponding routing spaces of this type. Each kind of routing agents has its own routing space where specific information, required by the applied algorithm, is stored and retrieved (e.g. pheromones for ants, or duration of waggle dance for bees). Eventually, the information about the best or suitable partner nodes is stored in the allocation space where a corresponding IN or OUT allocation agent grabs this information and distributes the load between the local node and its partner node.

The above described patterns can be composed towards more complex patterns by "hooking" them via shared spaces. They must agree on the format of entries stored in these spaces, and on the interaction patterns on these. With SILBA patterns, bi-directional control flows are possible and arbitrary logical network configurations can be easily and dynamically be constructed. Example in Fig. 1 shows four subnets A-D that have different relationships to each other. Nested subnets are allowed and two (or more) subnets might overlap, i.e. have in their intersection 0, 1 or more nodes. Therefore, nodes can belong to one or more subnets, e.g., nodes N1 and N2 are part of one subnet each, whereas N3 belongs to two subnets.

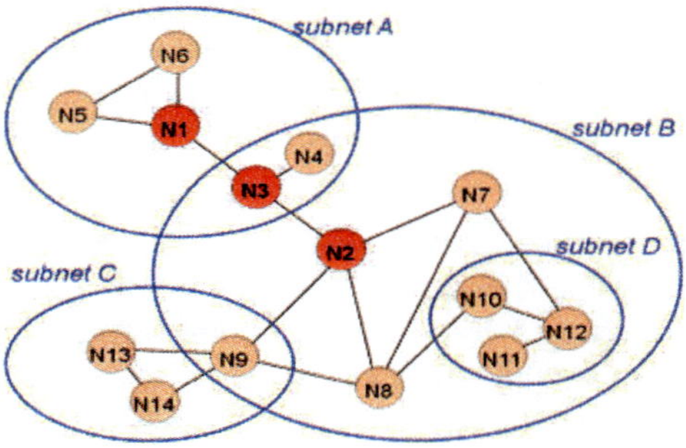

Fig. 1 Topology example.

The XVSM shared data space, which has already been successfully applied in several agents based projects [17], [18] serves as the coordination middleware for SILBA. Shared data structures maintain collaboration information and other LB relevant parameters to tune the algorithms. This indirect communication allows for a high autonomy of agents. Concurrent agents either retrieve, or subscribe to this information being notified in near-time about changes, or modify it. Clients continuously put tasks to any node in the distributed network.

2.2 Extended SILBA

SILBA is designed so that it can be extended towards the remote load balancing as sketched in the following. The main point is that the routing sub-patterns for different levels of load balancing are the same and simply composed towards the desired topology by "connecting" them via shared spaces; all sub-patterns can be parameterized by different algorithms. Each level can apply a different algorithm and load balancing in the entire network occurs through the combination of all algorithms. Fig. 2 represents the realization of nodes N1, N2 and N3 from Fig. 2. Boxes represent processes (clients or agents), and circles represent shared spaces. For simplicity, the three roles of arbiter, IN and OUT agent are represented by one box in the allocation pattern. Sub-patterns are edged with dotted lines. A composition occurs where sub-patterns overlap, in that they jointly access a shared space. SILBA can be composed to support an arbitrary amount of subnets.

LB within a subnet. In this case, the behaviour of the routing agent must be implemented. E.g., in Fig. 2, node N3 belongs to two different subnets. In one subnet, routing agents are of type 1 (e.g. implementing ants based LB algorithm) and in the other one they are of type 2 (e.g. implementing bee based LB algorithm). In order to collaborate with nodes from both subnets, N3 must possess both types of routing agents including both kinds of routing spaces that hold the information specific for each respective LB algorithm. The collaboration between different types of routing agents at N3 goes through its allocation space. It holds the information about partner nodes as computed by the continuously applied location policy. The IN and OUT allocation agents assume that the information about best/suitable partners to/from which to distribute load can be queried any time from the allocation space.

LB between subnets. This level of LB requires a further extended behaviour of routing agents for inter-subnet routing. Note that spaces are represented by XVSM containers that are referenced by URLs. For inter-subnet routing, each routing space is published under a public name using the JXTA based peer-to-peer lookup layer[2] of XVSM so that the routing agents can retrieve the foreign routing spaces in the network. This way, routing within a subnet uses the same pattern as routing between one or more subnets.

[2] http://www.sun.com/software/jxta

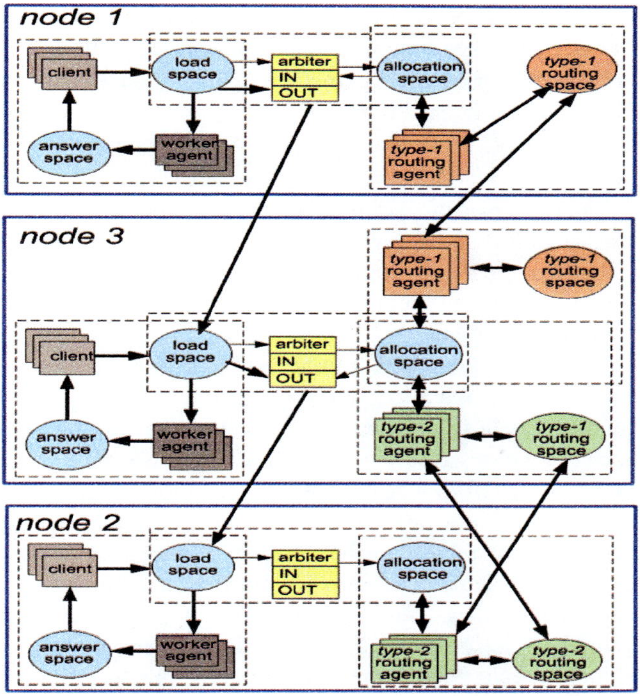

Fig. 2 N1, N2, N3 implementation.

3 Swarm Based Algorithms

The main obstacle in solving the combinatorial optimization problems is that they cannot be solved (optimally) within the polynomial bounded computational time. Therefore, in order to solve large instances, the approximate algorithms (heuristics) have to be used. These algorithms obtain near-optimal solutions in a relatively short time [37]. A set of algorithmic concepts, that can be used to define heuristic methods applicable to a wide set of different problems, was emerged. This new class of algorithms, the so-called metaheuristics, increases the ability of finding very high quality solutions to hard combinatorial optimization problems in a reasonable time [37]. Generally, swarm intelligence describes the collective behavior of fully decentralized, self-organized systems from nature. Particularly successful metaheuristics are inspired by swarm intelligence [10]. This concept belongs to the area of artificial intelligence. Swarm intelligence algorithms refer to a specific set of metaheuristics, adaptive algorithms. Adaptive algorithms usually manipulate with a population of items. Each item is evaluated by means of a figure of merit and its adequacy for the solution. The evaluation is done by using the so-called fitness function. When searching for the adequate solution, exploration and exploitation of a search space are mixed. The exploration investigates unknown areas of the search space, whereas exploitation makes use of accumulated knowledge. A good trade-off between these two contradictory requirements leads to finding a global optimum.

In this section, two types of swarm intelligent algorithms are presented. Bee algorithms are adopted for the LB problem and implemented to this problem for the first time [31]. Although ant algorithms have been applied previously to LB [12], we adapted and implemented them in order for comparison with non-pheromone based swarm intelligence (bees).

Different dynamic processes characterize the LB scenario. Nodes join and leave dynamically, information about load changes permanently, and tasks are dynamically added and continuously processed. A structured peer-to-peer (P2P) network [1] has an overlay topology that is controlled. There is a simple mapping of content to location, and therefore it scales well. On the other side, the support of dynamics is not so good. Queries can only be simple key/value mappings, i.e., exact match queries instead of more complex queries. For these reasons, they are not suitable for the LB problem. In an unstructured P2P network, a placement of information can be done independently of an overlay topology, but the content must be localized explicitly, e.g., through brute force mechanisms or flooding. It is very well suitable for dynamic populations, and complex queries are possible. Therefore, an unstructured P2P network fits better to our problem. The negative point is that it does not scale so well, which is the starting point for improvements. In order to point out the arguments for the potential of using bees for the LB problem, we give a short comparison of Gnutella and swarm-based systems:

Gnutella [1] operates on an unintelligent query flooding based protocol to find a particular node. For communication between peers ping (discover hosts), pong (reply to ping), query (search request), and query hit (reply to query) messages are used. It needs many concurrent agents for one (exhaustive) search, as for each branch, a new agent is required.

Bees search the network and build up routes as overlays. If a bee finds the required information, it directly flies back to its hive and informs the "starting place" of the search directly in a P2P way. Bounding the number of bees is possible, which is an indication that bees can scale better than Gnutella. However, in the first iteration step, there is no guarantee of finding a solution, but one will find a solution in upcoming iterations through learning. Knowledge distribution takes place in the own hive. Bees of different hives do not communicate with each other.

Ants leave information (pheromones) at all nodes on their backward trips. Their forward trip is comparable to the bees' forward movement (navigation), but their backward trip is different as ants do not directly contact the "starting place" in a P2P way but must go the entire way back.

3.1 Bee Algorithm

3.1.1 Bee Behaviour in Nature

A bee colony consists of bees with different roles: foragers, followers, and receivers. This natural intelligence performs self-organization through two main strategies: navigation and recruitment. Navigation means searching for nectar in an unknown landscape. A forager scouts for a flower with good nectar, returns to the hive, unloads nectar, and performs a recruitment strategy, meaning that it communicates the knowledge about the visited flowers to other bees. The best known way of bee communication is the so-called waggle dance which is the main part of the recruitment strategy. Using this "dance language", a bee informs its hive mates about direction, distance and quality of the food found. A follower randomly chooses to follow a forager and visits the flower that has been "advertised" without own searching. A forager can choose to become a follower in the next step of navigation, and vice versa. A receiver always stays in the hive and processes the nectar. Autonomy, distributed functioning, and self-organization characterize the biological bee behaviour [4].

Bee-inspired algorithms have been applied to several computer science problems like travelling salesman [36], scheduling jobs [6], [28], routing and wavelength assignment in all-optical networks [23], training neural networks for pattern recognition [27], and computer vision and image analysis [26]. Although some of these applications deal with some kind of job scheduling, they differ from our general and domain independent approach as they use a simplified version of a scheduling problem by including the limitations given in advance, e.g., a single machine supplies jobs, and each job needs only one operation to be executed.

3.1.2 Algorithm

In [31], the principals for usage of bee intelligence for LB are proposed. Software agents represent bees at the particular nodes. A node contains exactly one hive and one flower with many nectar units. A task relates to one nectar unit. A hive has a finite number of receiver bees and outgoing (forager and follower) bees. Initially, all outgoing bees are foragers. Foragers scout for a location policy partner node of their node to pull/push nectar from/to it, and recruit followers. The goal is to find the best location policy partner node by taking the best path which is defined to be the shortest one. A suitability function δ (see below) defines the best location policy partner. A navigation strategy determines which node will be visited next and is realized by a state transitions rule [36]:

$$P_{ij}(t) = \frac{[\rho_{ij}(t)]^{\alpha} \cdot [\, 1/d_{ij} \,]^{\beta}}{\sum_{j \in A_i(t)} [\rho_{ij}(t)]^{\alpha} \cdot [\, 1/d_{ij} \,]^{\beta}} \tag{1}$$

where $\rho_{ij}(t)$ is the arc fitness from node i to node j at time t, d_{ij} is the heuristic distance between i and j, α is a binary variable that turns on/off the arc fitness influence, and β is the parameter that controls the significance of a heuristic distance. In the calculation of the arc fitness values, we differentiate:

(1) **Forager:** A bee behaves in accordance with the state transition rule and $\rho_{ij} = 1/k$, where k is the number of neighbouring nodes of node i. A forager can decide to become a follower in the next cycle of navigation.
(2) **Follower:** Before leaving the hive, bee observes dances performed by other bees and randomly chooses to follow one of the information offered through these dances. This information contains the set of guidance moves that describes the tour from the hive to the destination previously explored by one of its hive mates. This is the so-called *preferred path* [36]. When a bee is at node i at time t, two sets of next visiting nodes can be derived: the set of allowed next nodes, $A_i(t)$ and the set of favoured next nodes, $F_i(t)$. $A_i(t)$ contains the set of neighbouring nodes of node i, whereas $F_i(t)$ contains a single node which is favoured to reach from node i as recommended by the preferred path. The arc fitness is defined in

$$\rho_{ij}(t) = \begin{cases} \lambda & \text{if } j \in F_i(t) \\ \dfrac{1-\lambda \cdot |A_i(t) \cap F_i(t)|}{|A_i(t)| - |A_i(t) \cap F_i(t)|} & \text{if } j \notin F_i(t) \end{cases} \quad \forall j \in A_i(t),\ 0 \le \lambda \le 1 \tag{2}$$

where $|S|$ is the sign of the cardinality (i.e., the number of elements) of some set S. So, $|A_i(t) \cap F_i(t)|$ can be either 0 or 1, as $A_i(t)$ and $F_i(t)$ may have either none element or only one element in their intersection.

A recruitment strategy communicates obtained knowledge about path and quality of solution to bees. From this we can derive a fitness function for bee i,

$$f_i = \frac{1}{H_i}\delta \tag{3}$$

where H_i is the number of hops on the tour, and δ is the suitability function. The colony's fitness function f_{colony} is the average of all fitness functions (for n bees)

$$f_{colony} = \frac{1}{n}\sum_{i=1}^{n} f_i \tag{4}$$

If bee i finds a highly suitable partner node, then its fitness function, f_i obtains a good value. After a trip, an outgoing bee determines how "good it was" by comparing its result f_i with f_{colony}, and based on that decides its next role [25].

Pseudocode1: Bee Colony Optimization (BCO) metaheuristic [36].

```
procedure BCO_MetaHeuristic
  while(not_termination)
        ObserveWaggleDance()
        ConstructSolution()
        PerformWagledance()
  end while
end procedure
```

Each node can start the location policy. If the node is UL, its bee searches for a suitable task belonging to some OL node and carries the information about how complex the task the node can accept. If the node is OL, its bee searches for a UL node that can accept one or more tasks from this OL node. It carries the information about the complexity of tasks this OL node offers and compares it with the available resource of the current UL node that it is visiting. Therefore, the complexity of the task and the available resources at a node must be compared. For this purpose, we need the following definitions: task complexity *c*, host load *hl* and host speed *hs* [8]. *hs* is relative in a heterogeneous environment, *hl* represents the fraction of the machine that is not available to the application, and *c* is the time necessary for a machine with $hs = 1$ to complete a task when $hl = 0$. We calculate the argument $x = (c/hs)/(1 - hl)$ of suitability function δ and define it as $\delta = \delta(x)$ (cf. Table 2). If $x = 1$, then the situation is ideal. The main intention is to find a good location policy partner. For example, when a UL node with high resource capacities is taking work from an OL node, a partner node offering tasks with small complexity is not a good partner as other nodes could perform these small tasks as well. Taking them would mean wasting available resources. A detailed description about bee algorithm for LB can be found in [31].

3.1.3 Implementation Parameters

In our implementation, we introduced one parameter, the so-called *search mode* that is configurable and determines which nodes in the network (according to their load status) will trigger a load balancing algorithm.

Table 1 Search Modes.

SM1	the algorithm is triggered from UL nodes, OK nodes (in a situation when it's likely that the node will become OL, but is not yet heavily loaded) and consequently OL nodes
SM2	the algorithm is triggered from UL nodes
SM3	the algorithm is triggered from OK nodes (in a situation when it's likely that the node will become OL, but is not yet heavily loaded) and consequently OL nodes; the computation of x argument for δ(x) suitability is slightly changed[3]
SM4	the algorithm is triggered from OL nodes
SM5	the algorithm is triggered from UL and OL nodes
SM6	the algorithm is triggered from OK nodes (in a situation when it's likely that the node will become OL, but is not yet heavily loaded) and consequently OL nodes.

[3] If a node is in OK state, the algorithm is triggered, and searching for a suitable node among the neighbor nodes is started (afterwards, this information about the most suitable node is stored locally). As soon as the node gets OL, the tasks get re-routed to this target node. To achieve this à priori searching for a suitable node (when the information about a task is still unavailable, i.e., the task complexity c is yet unknown), we compute argument x only on the basis of host speed and host load parameters.

For *suitability* function δ, we implemented the following functions:

Table 2 Suitability Functions.

SF0	one linear function: if (x = 1.0) $\delta(x) = n$, else $\delta(x) = 5x$ (if the number of nodes $\leq n$)
SF1	an exponential function: $\delta(x) = 10^x$
SF2	a polynomial function: $\delta(x) = 10x^3$
SF3	another linear function: if (x < 1.0) $\delta(x) = 4nx$, else $\delta(x) = 5n$ (if the number of nodes $\in [5n\text{-}4, 5n]$)

The *fitness* function *f* is computed from the suitability function of the found node and the number of hops to this node using the following combinations:

Table 3 Fitness Functions.

FF0	f(x) = δ(x) / number_of_hops
FF1	f(x) = δ(x) • (quality_of_links / number_of_hops)
FF2	f(x) = δ(x) / sqrt(number_of_hops)
FF3	similar to FF0, only the local node is excluded from the comparison and the rest of neighbouring nodes are taken in consideration.

3.2 Ant Algorithms

The basic requirements - to find the *best* location policy partner node by taking the *best* path - are the same as in the bee case. The best location policy partner is defined by the maximum amount of pheromones left on the path. The Ant Colony Optimization metaheuristic (ACO) has been inspired by the real ant colonies. The ants' behaviour is characterized by indirect communication between individuals in a colony via pheromone. A software agent plays the role of an ant. The natural pheromone is stigmergic information that serves as the communication among the agents. Ants make pure local decisions and work in a fully distributed way. In ACO, ants construct solutions by moving from the origin to the destination, step by step, according to a stochastic decision policy. After that, the aim of the pheromone update is to increase the pheromone values associated with good solutions (deposit pheromones) and decrease those associated with bad ones [10].

Pseudocode2: Ant Colony Optimization (ACO) metaheuristic [10].

```
procedure ACO_MetaHeuristic
  while(not_termination)
        ConstructSolutions()
        pheromoneUpdate()
        daemonActions()
  end while
end procedure
```

The most popular variations and extensions of ACO algorithms are: Elitist AS, Rank-Based AS, MinMax Ant System (MMAS), and Ant Colony System. AntNet [9] is a network routing algorithm based on ACO. It is an algorithm for adaptive routing in IP networks, highly adaptive to network and traffic changes, robust to agent failures and provides multipath routing. AntNet algorithm supports adding and removing network components.

The following is a brief "tutorial" about ant algorithms [10], needed for explanation of results and clarification of benchmarks parameters: MinMax [10] is an improvement of the initial Ant System algorithm. In each Ant System algorithm, there are two phases: ants' tour (solution) construction and pheromone update. In the 1st phase, m artificial ants concurrently build their solutions starting from randomly chosen nodes and choosing the next node to be visited on their trips by applying a random proportional rule:

$$p_{ij}^k = \frac{[\tau_{ij}]^\alpha [\eta_{ij}]^\beta}{\sum_{l \in N_i^k} [\tau_{il}]^\alpha [\eta_{il}]^\beta}, \text{ if } j \in N_i^k \tag{5}$$

where τ_{ij} is a pheromone trail on (i,j)-arc, $\eta_{ij} = 1/d_{ij}$ is a heuristic value (available à priori), α and β are two parameters that determine the influence of the pheromone trail and the heuristic information, and N_i^k is the set of cities that ant k has not visited yet. In the 2nd phase, the pheromone trails are updated. The pheromone value on all arcs is decreased by a constant factor:

$$\tau_{ij} \leftarrow (1-\rho)\tau_{ij} \tag{6}$$

where $0 < \rho \leq 1$ is the pheromone evaporation rate. After evaporation, the additional amount of pheromones is deposited on the arcs that have being crossed in the ants' constructions of solutions:

$$\tau_{ij} \leftarrow \tau_{ij} + \sum_{k=1}^{m} \Delta\tau_{ij}^k \tag{7}$$

where $\Delta\tau_{ij}^k$ is the amount of pheromones ant k deposits on arcs it has visited.

In the MinMax algorithm, the following modifications are done [10]:

- Best tours found are strongly exploited.
- Possible range of pheromone trail values are limited to the interval $[\tau_{min}, \tau_{max}]$.
- Pheromone trails are initialized to the upper pheromone trail limit.
- Pheromone trails are reinitialized each time the system approaches any kind of stagnation.

So, the 1st phase is the same as in the initial Ant System algorithm, but the 2nd phase is modified – the update of pheromone trails is implemented as follows:

$$\tau_{ij} \leftarrow \tau_{ij} + \Delta\tau_{ij}^{best} \tag{8}$$

where $\Delta\tau_{ij}^{best} = 1/C^{best}$ and C^{best} can be either the length of the iteration's best tour or the length of the best tour so far.

AntNet algorithm [9] is similar to all Ant Algorithms, i.e., has two phases: solution construction and data structures update. The necessary data structures used in this algorithm are: an artificial pheromone matrix τ_i and a statistical model M_i of the traffic situation over the network. Both matrices are associated with each node *i* of the network. Two sets of artificial ants exist: forward ants and backward ants. Generally, ants have the same structure, but their actions differ:

- Forward ant, $F_{s\to d}$, travels from the source node *s* to a destination node *d*.
- Backward ant, $B_{s\to d}$, travels back to the source node *s* using the same path as $F_{s\to d}$ but in the opposite direction; it uses the information collected by $F_{s\to d}$ in order to update routing tables of the visited nodes.

In the 1st phase, each $F_{s\to d}$ starts its travel from the source node *s* and chooses its destination *d* according to this probabilistic rule:

$$p_{sd} = \frac{f_{sd}}{\sum_{i=1}^{n} f_{si}} \tag{9}$$

where f_{sd} is some measure of data flow. The ant constructs the path on this way:

a) An ant that is currently at node *i* chooses the next node *j* to be visited by applying the following probabilistic rule:

$$P_{ijd} = \frac{\tau_{ijd} + \alpha\eta_{ij}}{1+\alpha(|N_i|-1)} \tag{10}$$

where τ_{ijd} is an element of the pheromone matrix τ_i that indicates the learned desirability for an ant in node *i* with destination *d* to move to node *j*, $|N_i|$ is a number of neighbours of node *i*, η_{ij} is a heuristic value that takes into account the state of the jth link queue of the current node *i*:

$$\eta_{ij} = 1 - \frac{q_{ij}}{\sum_{l=1}^{|N_i|} q_{il}} \tag{11}$$

The parameter α from Eq.(10) weighs the importance of the heuristic values with respect to the pheromone values stored in the pheromone matrix.

b) When $F_{s\to d}$ comes to destination node *d*, it generates $B_{s\to d}$, transfers to it all of its memory and is deleted.
c) $B_{s\to d}$ travels back to the source node *s* using the same path as $F_{s\to d}$ but in the opposite direction. It uses the information collected by $F_{s\to d}$ in order to update the routing tables of the visited nodes.

The 2nd phase considers updating matrices τ_i and M_i. In the pheromone matrix τ_i, values that suggest choosing neighbour *f* when destination is *d*, are incremented:

$$\tau_{ifd} \leftarrow \tau_{ifd} + r\cdot(1-\tau_{ifd}) \tag{12}$$

The other pheromone values are decremented:

$$\tau_{ijd} \leftarrow \tau_{ijd} - r\cdot\tau_{ijd} \quad j\in N_i, \;\; ; j\neq f \tag{13}$$

There are several ways to determine and assign r values: from the simplest way of setting $r = constant$ to a more complex way that defines r as a function of the ant's trip time and the parameters of the statistical model M_i.

Remodelling of these ant algorithms for a location policy comprises the following changes. What does "Construct Solution" mean in our case? The ant made a path and found the data on that path. We are not only interested in the best path, but also in the quality of the data found. Therefore, *DepositPheromone* procedure is changed as follows. If an ant on its trip:

1. Found exact data, it deposits pheromone;
2. Found acceptable data with the accuracy/error rate < ε, (ε is a parameter given in advance related to the definition of δ), it deposits less amount of pheromone,
3. Did not find data, then skips depositing pheromones on its trip (i.e., the values on arcs it traversed will be the same as the values on the rest of unvisited arcs in the network).

A different amount of pheromones is deposited according to the quality of the solution found. The suitability function $\delta = \delta(x)$ describes how good (acceptable) the found solution is, $\delta \in [0,1]$. In case of changing the type of δ, its value can be scaled into the same segment [0,1]. DepositPheromone procedure is changed:

1. For MinMax algorithm: $\Delta\tau = 1/MC^{best}$ where $M=1/\delta$;
2. For AntNet algorithm: $\tau := r \cdot (1-\tau) \cdot \delta$.

4 Benchmarks

This section describes the benchmarks performed in the SILBA framework. As a detailed explanation of the basic SILBA benchmarks can be seen in [32], emphasis is put on more sophisticated benchmarks in the extended SILBA. Therefore, the basic SILBA benchmarks, i.e., their conclusion are mentioned briefly here.

4.1 Basic SILBA Benchmarks

The tests are constructed on the basis of the following criterions [32]:

- Find out the best combination of parameter settings for each intelligent algorithm: Bee Algorithm, MinMax and AntNet Ant Algorithms,
- Compare these optimally tuned swarm based algorithms with several well-known algorithms: Round Robin, Sender, Adapted Genetic Algorithm (GA).

The benchmarks demonstrate: the agility of the SILBA pattern by showing that algorithms can be easily exchanged, and the promising approach of bee algorithms.

The load is generated by one single client. For performing test examples, an arbitrary topology is used in which a full connection between all nodes is not required. All benchmarks are carried out on a cluster of 4 machines, and on the Amazon EC2 Cloud. As the figure of merit, the absolute execution time and scalability of the solutions are used. The values of the suitability function help to discern the usefulness of intelligent algorithms and emphasize the correctness of properly chosen partner nodes, i.e., the methodology to determine the best partner node. This function reflects how a good solution is chosen and the degree of self-organization of the used swarms. The average x value is 1, meaning the best node is always chosen.

In the basic SILBA, the best combination of feasible parameters for each algorithm is identified and, under these conditions, the advantages of using bee swarm intelligence in the context of load balancing are presented. The obtained results show that the bee algorithm behaves well, does not impose an additional complexity and outperforms all other test candidates [32].

4.2 Extended SILBA Benchmarks

As the extended SILBA supports the multi-level LB strategy, the goal was to exchange the algorithms on each level. In the considered case, there are 2 levels on which LB is realized concurrently: between several subnets and inside each subnet. Also, the success of a particular combination depends on a network topology. The tests are performed on the basis of the following criterions:

- Find the best combination of algorithms for each of the well-known topologies (chain, full, ring, star) used in these benchmarks.
- Compare and analyze the best obtained combinations.
- After obtaining the best combinations, perform the benchmarks on different network (and subnets) dimensions and evaluate the scalability issue.

The benchmarks demonstrate: 1) the flexibility of the SILBA pattern by showing that the LB problem could be easily treated in more complex network structures with several subnets, 2) detection of those topologies which profit most of swarm intelligent algorithms (particularly bee algorithms).

4.2.1 Test Examples and Test Environment

Test examples are constructed taking into account the following issues: the combination of algorithms, the different number of subnets and the number of nodes per subnets, the increased number of clients per each subnet, different topologies:

Combinations (36) of all algorithms on two levels (6 algorithms on 2 levels): Level 1 denotes the used algorithm inside a subnet, whereas level 2 denotes the used algorithm between subnets; the *values* of the respective parameters are described in Table 4 and reused from basic SILBA.

Table 4 Combinations of algorithms.

level1 / level2	Bee Alg.	MinMax	AntNet	adaptedGA	Sender	Round Robin
Bee Alg.	1	2	3	4	5	6
MinMax	7	8	9	10	11	12
AntNet	13	14	15	16	17	18
adaptedGA	19	20	21	22	23	24
Sender	25	26	27	28	29	30
Round Robin	31	32	33	34	35	36

Different number of subnets and number of nodes per subnets:

Table 5 Distribution of nodes in subnets.

total number of nodes	number of subnets	number of nodes in each subnet
16	4	4
16	8	2
32	4	8
32	8	4

Increased number of clients per each subnet:
In the basic SILBA, only one client is responsible for putting the tasks into the network. This leads to a light to moderate loaded network. In extended SILBA, the number of clients is drastically increased, i.e., for a subnet of *n* nodes, the assigned number of clients is *n/2*. The number of clients per subnet is increased until the subnet becomes fully loaded. Each client supplies the same number of tasks. Clients are symmetrically positioned in order to have fairly loaded subnet. The same parameter is used for all test runs.

Different topologies:
The well-known topologies, ring, star, full, chain, are chosen in order to define which combination of algorithms fits the best to a particular topology. Fig. 3 depicts one example of each topology. Subnets can be with intersections and without intersections, but in both cases at least one node from each subnet must possess two types of routing agents in order to allow for the realization of different types of load balancing algorithms (inside a subnet, between subnets).

Two different *test environments* are used: a cluster of 4 machines, and the Amazon EC2 Cloud[4]. Each machine of the cluster had the following characteristics:

[4] http://aws.amazon.com/ec2/

2*Quad AMD 2,0GHz with 16 GB RAM. We simulated a network with 16 (virtual) nodes. Each test run began with a "cold start" and all nodes were being UL. On Amazon Cloud, we used standard instances of 1.7 GB of memory, 1 EC2 Compute Unit (1 virtual core with 1 EC2 Compute Unit), 160 GB of local instance storage, and the 32-bit platform.

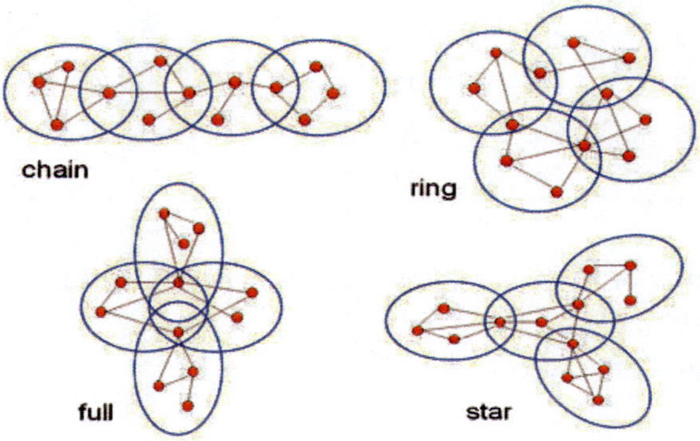

Fig. 3 Topology examples.

4.2.2 Raw Result Data

The next figures (Fig. 4 – Fig. 7) show all combinations of algorithms on different topologies, searching for the best combination in each topology. The presented results demonstrate a 4*4 structure, i.e., 4 subnets and 4 nodes in each subnet. In each subnet, each client supplies 200 tasks, giving a total of 1600 tasks.

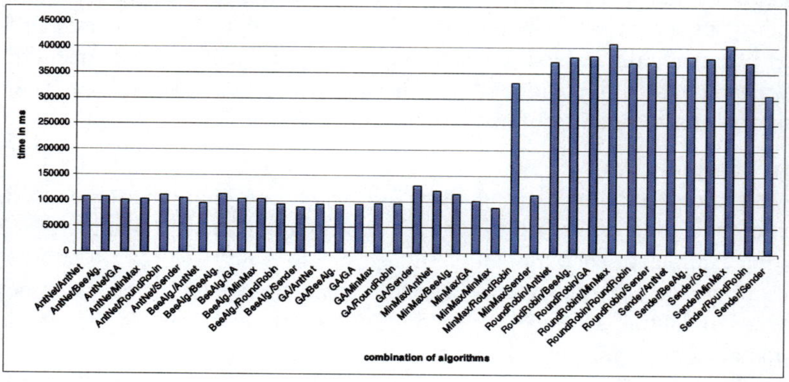

Fig. 4 Combination of algorithms in chain topology.

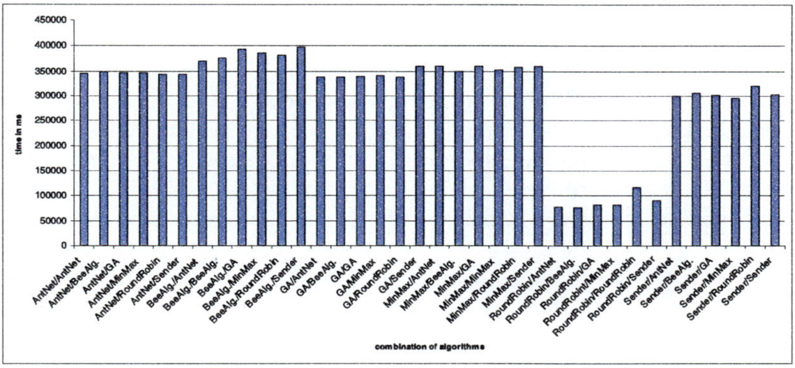

Fig. 5 Combination of algorithms in full topology.

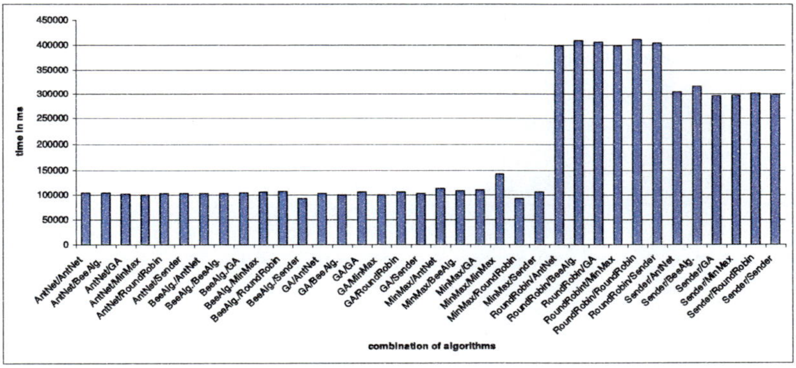

Fig. 6 Combination of algorithms in ring topology.

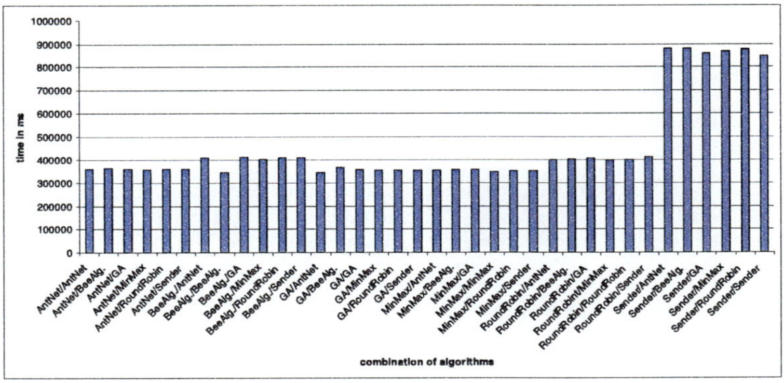

Fig. 7 Combination of algorithms in star topology.

On the basis of the results obtained (Fig. 4 – Fig. 7), the overall comparison is done (Table 6). Many appearances of the same topology in Table 6 denote that the respective combinations are equally good (e.g., both combinations Bee Alg./Sender and MinMax/MinMax are equally good in a chain topology).

As can be noticed, in each topology (except star) the best combination is made by one intelligent and one unintelligent algorithm. Although these combinations are not real hybrid algorithms (each pure algorithm works either inside a subnet or between subnets), the overall load distribution in the entire network is realized through their synergy. Intelligent algorithms find good starting solutions (quality and fastness), while unintelligent algorithms improve these solutions (fastness).

Table 6 Overall comparison of the best results in all topologies.

topology	combination of algor.	time (ms)
chain	BeeAlg./Sender	88000
chain	MinMax/MinMax	88000
full	RoundRobin/BeeAlg.	76000
ring	BeeAlg./Sender	93000
ring	MinMax/RoundRobin	93000
star	BeeAlg./BeeAlg.	346000
star	GA/AntNet	346000

The results from Table6 are graphically presented in Fig.8.

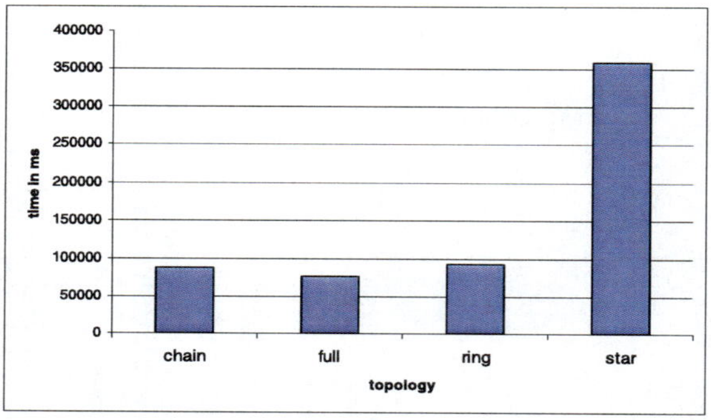

Fig. 8 Results of the best combinations for each topology.

After obtaining the best combination for each topology, the benchmarks with the best combinations are performed on larger network dimensions. Table7 summarizes these results and shows that the results are stable as the same combination(s) of algorithms are obtained as the best ones for each of different dimensions (4*4, 8*2, 4*8, 8*4).

Table 7 Results (time in ms) of the best combinations in different network dimensions.

total number of nodes	number of subnets	number of nodes in each subnet	chain	full	ring	star
16	4	4	88000	76000	93000	346000
	8	2	374000	384000	359000	365000
32	4	8	420000	556000	582000	388000
	8	4	406000	455000	484000	356000

Extended SILBA offers better and more powerful solutions than basic SILBA. The situations that can benefit from extended SILBA are the following:

1. Subnets are physically required.
2. Extremely large networks with a high number of nodes where building subnets and applying the extended SILBA strategy helps transferring the load between very distant nodes. Load need not be transferred via a number of hops from one node to another one, but can be transferred by using a shortcut, "jumping" from the original node's subnet to the distant destination node's subnet.

4.2.3 Overall Evaluation

The *absolute execution time* is used as metric for the benchmarks. According to the obtained results (see section 4.2.2.), the behaviour of a particular combination of algorithms depends on a topology. The questions to be analyzed are:

1. How much is the best combination (in each topology) better than "extreme" combinations: the worst one and the combination of the second best one?
2. What is the "behaviour" of the other combinations, i.e., how much do they deviate from the best solution? What is the "collective behaviour" of algorithm combinations and the used SILBA pattern in each topology?

For *chain* topology, the best result is obtained by both BeeAlgorithm/Sender and MinMax/ MinMax. They are equally good, and 5.4% better than the combination in the second place, GA/Bee Algorithm, 78% better than the worst combination, and 56% better than the average of all combinations. The additional measurements, the interval of variation and the root mean square deviation (RMSD) are introduced in order to examine the behaviour of the other combinations, i.e., how much they deviate from the best solution. The interval of variation is defined as the difference between the maximum value of the used metric (time) and its minimum value: $t_{max} - t_{max}$ and is equal to 320000ms. The used RMSD is a quantitative measure (a decimal number) that tells how many good combinations in a particular topology exist, i.e., how far from the best solution the data points (the rest of the combinations) tend to be (smaller RMSD means more good combinations). For chain topology, the value of RMSD is 172121.

The combination RoundRobin/BeeAlgorithm shows the best results in the *full* topology. This combination is 1.3% better than the combination in the second place, RoundRobin/AntNet, 80.9% better than the worst combination, and 74.9% better that the average of all combinations. The interval of variation, $t_{max} - t_{max}$, is 322000ms and the RMSD is 248227.7.

Both BeeAlgorithm/Sender and MinMax /RoundRobin are equally good in the *ring* topology. They are 1.4% better than the combination in the second place, MinMax/RoundRobin, 60.7% better than the worst combination, and 24.3% better that the average of all combinations. The interval of variation, $t_{max} - t_{max}$, is 535000ms and the RMSD is 216194.9.

For the *star* topology, the combinations BeeAlgorithm/BeeAlgorithm and GA/AntNet are the best with the same resulting value. They are 6.1% better than the combination in the second place, AntNet/MinMax, 77.4% better than the worst combination, and 50.1% better that the average of all combinations. The interval of variation, $t_{max} - t_{max}$, is 319000ms and the RMSD is 153859.9.

From these results we can conclude that bee algorithms play an important role in almost every topology. The best obtained results in each topology are based on bee algorithms used either inside or between subnets, or in both. Also, the rest of intelligent algorithms give good results in all topologies. The exception is the full topology where the best results are obtained when round robin algorithm is used inside subnets and combined with all others algorithms (except the combination Round Robin/Round Robin).

The RMSD shows that the greatest deviation is reached in full topology, i.e., the majority of the other combinations differentiate a lot (they are worse in a significant extent) comparing to the best obtained combination. The smallest deviation is in star topology, so the combinations behave evenly in this topology. If we analyze how good response will be obtained by plugging any (random) combination of algorithms into SILBA, the equally good results are obtained in star topology. So, SILBA is very stabile (without peaks in results) in star topology. At the other side, Fig.8 shows that the results of the individual combinations of the SILBA pattern are successful for chain, full and ring topologies, whereas the results obtained for star topology are not so good.

In the next table, the behaviour of the swarm intelligence algorithms' combinations is extracted as these algorithms are promising ones and not exploited so much. Table8 shows how much they deviate from the best solution in each of the used topologies. For example, the set of all combinations that use bee algorithms inside subnets is denoted in the table as "bee/others". According to these results, all the combinations from this set deviate slightly from the best combination in the chain topology (that are BeeAlgorithm/Sender and MinMax/MinMax), whereas the combinations from this set deviate more from the best combination in star topology, although the best combination is BeeAlgorithm/BeeAlgorithm.

Table 8 Deviation swarm based algorithms' combinations from the best solution.

	chain	full	ring	star
RMSD (Bee/Others)	35171.0	755211.9	25337.7	141470.8
RMSD (Others/Bee)	417868.4	600503.1	387401.6	537938.7
RMSD(AntNet/Others)	44899.9	659335.3	25869.2	35787.1
RMSD(Others/AntNet)	404891.3	608559.9	372385.6	541636.4
RMSD(MinMax/Others)	249164.6	686738.7	58813.3	21725.6
RMSD(Others/MinMax)	450334.3	603189.0	371052.6	526899.4

Additionally, scalability is analyzed. Here, we focus on the issue of *load scalability*. A very general definition of scalability is taken into account [14], [35]. According to [14], a general family of metrics can be based on the following definition:

$$\psi = \frac{F(\lambda_2, QoS_2, C_2)}{F(\lambda_1, QoS_1, C_1)} \tag{14}$$

where F evaluates the performance, λ evaluates the rate of providing services to users, QoS is a set of parameters which evaluate the quality of the service seen by users, and C reflects the cost. Further, [14] establishes the scaling strategy by means of a scaling factor k and a set of scaling variables which are functions of k. They express the strategy as a scaling path in a space in which they are the coordinates. Fig.9 shows how $\psi(k)$ behaves in different situations:

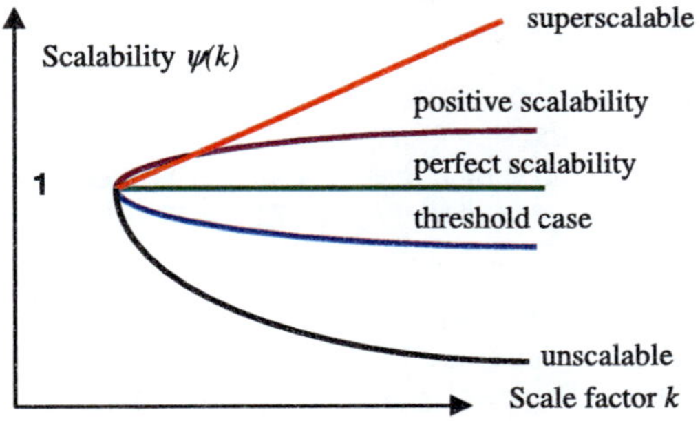

Fig. 9 Scaling behavior [14].

We specialize it to a simplified version of interest to our problem in terms of load, resources and performance measure. This restricted aspect of scalability can be quantitatively described on the basis of the computational resources available (R), load of the system (L) and some performance measure (P). Then scalability can be quantified by means of a "scalability ratio" r_{scal} for a given constant k

$$r_{scal} = \frac{P(L,R)}{P(kL,kR)} \quad (15)$$

Usually, performance P is the function of load L and resources R. A certain aspect of scalability is described by the answer to the question of how P is affected when more resources (larger R) have to compensate for more load (larger L). A constant remaining value of P when simultaneously increasing L and R by the same factor leads to the "ideal" scalability ratio of 1. In our test examples, this interpretation of load scalability is applied. We analyze the increasing of load with the increasing of the resources. By comparing results (Table 12), it is easy to see that the best chosen combinations based on bee algorithm scale well [14]. Load and resources are increased *twice* for consecutive test runs.

Scalability in basic SILBA

For example in the cluster environment, load and resources are increased twice for consecutive test runs, i.e., they are increased by 2^n compared with the starting test run (4 nodes, 50 tasks). The values of r_{scal} are 2.9, 3.0, 3.4, 3.2 (rounded to one decimal) for consecutive bee test runs, i.e., 2.9, 9.1, 31.0, 100.8 compared with the starting test run (4 nodes, 50 tasks). These values converge to positive scalability. Such behaviour is even better in a more real environment, i.e., on the Cloud. Almost the similar situation occurs with AntNet algorithm.

Scalability in extended SILBA

For *chain* topology: a) If the number of subnets is increased and the number of nodes inside a subnet is the same, i.e., **4***4, **8***4, r_{scal} is 4.6 (rounded to one decimal), that leads to positive scalability. b) If the number of nodes in a subnet is increased and the number of subnets is the same, i.e., 4***4**, 4***8**, r_{scal} is 4.8; 8***2**, 8***4**, r_{scal} is 1.1, that converges to perfect scalability.

For *full* topology: a) If the number of subnets is increased and the number of nodes inside a subnet is the same, i.e., **4***4, **8***4, r_{scal} is 5.98; that leads to positive scalability. b) If the number of nodes in a subnet is increased and the number of subnets is the same, i.e., 4***4**, 4***8**, r_{scal} is 7.3; 8***2**, 8***4**, r_{scal} is 1.8; that leads to positive scalability.

For *ring* topology: a) If the number of subnets is increased and the number of nodes inside a subnet is the same, i.e., **4***4, **8***4, r_{scal} is 5.2; that leads to positive scalability. b) If the number of nodes in a subnet is increased and the number of subnets is the same, i.e., 4***4**, 4***8**, r_{scal} is 6.2; 8***2**, 8***4**, r_{scal} is 1.3; that converges to perfect scalability.

For *star* topology: a) If the number of subnets is increased and the number of nodes inside a subnet is the same, i.e., **4***4, **8***4, r_{scal} is approximately 1; that leads to perfect scalability. b) If the number of nodes in a subnet is increased and the number of subnets is the same, i.e., 4***4**, 4***8**, r_{scal} is approximately 1; 8***2**, 8***4**, r_{scal} is approximately 1; that leads to perfect scalability.

5 Conclusion

In this chapter, the problem of dynamic load balancing is investigated and treated. First, the generic load balancing architectural pattern SILBA is introduced and shortly explained. It allows the plugging and easy exchanging of a variety of algorithms. First, SILBA is developed in its basic form, which refers to load balancing within one network. Later, SILBA is extended in a way that load balancing can be done through different levels (between nodes in one network, between subnets in one network, between several networks) and this can be done concurrently. Different load balancing algorithms can be plugged into SILBA.

In the basic SILBA, the advantages of using bee swarm intelligence in the context of load balancing are presented. Besides bee swarm intelligence, two further intelligent algorithms are adapted based on MinMax and AntNet ant algorithms. For these algorithms, the best combination of feasible parameters is identified, and they are compared with three well-known algorithms: Round Robin, Sender, and Adapted Genetic Algorithm. The load is generated by one single client, and as a performance parameter the absolute execution time is used. Under these conditions, the obtained results show that the bee algorithm outperforms all other test candidates. All benchmarks are carried out on a cluster of 4 machines, and on the Amazon EC2 Cloud.

The extended SILBA shows the advantages of using bee swarm intelligence in the combination with the other algorithms (both intelligent and unintelligent) for load balancing in more complex network structures that consist of different subnets which might overlap or be nested: investigating different network topologies, the combinations that are based on swarm algorithms show the best results in the chain, ring and full topologies. The best combinations in all topologies are based on bee algorithms. The load is generated by many clients, positioned symmetrically in subnets. The benchmarks are also carried out on a cluster of 4 machines, and on the Amazon EC2 Cloud. The performance measure is the absolute execution time, expressed in milliseconds. The best obtained combinations scale well in all investigated topologies.

Future work will concern the following issues:

- Except execution time and scalability, different metrics will be used for the evaluation of results and analysis: communication delay, utilization, stability, fairness across multi-user workloads, robustness in the face of node failure, adaptability in the face of different workloads, etc. Also, it will be investigated under which circumstances, each metric is most appropriate.
- Benchmarking of very large instances and specific examples of the state of the art in the real world.
- Investigation of the impact of load injection in different places in the network.
- Although enlarging their parameter space is the part of the future work, the other way of investigation will play a role, i.e., a shrinking of the parameter space, with more samples and more determinism so that the nature of swarm intelligent algorithms (especially bee intelligence) can be better understood.

- Developing of a recommendation system for a given problem, e.g., the determination of the best topology, algorithm combinations, and parameters tuning for a particular problem.
- Application of the results to distribute load in collaborative security scenarios like distributed spam analysis and intrusion detection.

Acknowledgments. The work is partially funded by the Austrian Government under the program FIT-IT (Forschung, Innovation und Technologie für Informationstechnologien), project 825750 Secure Space - A Secure Space for Collaborative Security Services. We would also like to thank Deguang Sea and Fabian Fischer for implementation and benchmarking of SILBA.

References

[1] Androutsellis-Theotokis, S., Spinellis, D.: A survey of peer-to-peer content distribution technologies. ACM Computing Surveys 36(4), 335–371 (2004)
[2] Bronevich, A.G., Meyer, W.: Load balancing algorithms based on gradient methods and their analysis through algebraic graph theory. Journal of Parallel and Distributed Computing 68(2), 209–220 (2008)
[3] Cabri, G., Leonardi, L., Zambonelli, F.: Mars: A programmable coordination architecture for mobile agents. IEEE Internet Computing 4(4), 26–35 (2000)
[4] Camazine, S., Sneyd, J.: A model of collective nectar source selection by honey bees: Self-organization through simple rules. Journal of Theoretical Biology 149, 547–571 (1991)
[5] Chen, J.C., Liao, G.X., Hsie, J.S., Liao, C.H.: A study of the contribution made by evolutionary learning on dynamic load-balancing problems in distributed computing systems. Expert Systems with Applications 34(1), 357–365 (2008)
[6] Chong, C.S., Sivakumar, A.I., Low, M.Y., Gay, K.L.: A bee colony optimization algorithm to job shop scheduling. In: Proceedings of the Thirty-Eight Conference on Winter Simulation, pp. 1954–1961 (2006)
[7] Cortes, A., Ripolli, A., Cedo, F., Senar, M.A., Luque, E.: An asynchronous and iterative load balancing algorithm for discrete load model. Journal of Parallel and Distributed Computing 62(12), 1729–1746 (2002)
[8] Da Silva, D.P., Cirne, W., Brasileiro, F.V.: Trading Cycles for Information: Using Replication to Schedule Bag-of-Tasks, pp. 169–180. Applications on Computational Grids, Proceeding of European Conference on Parallel Processing (2003)
[9] Di Caro, G., Dorigo, M.: AntNet: Distributed Stigmergetic Control for Communications Networks. Journal of Artificial Intelligence Research 9, 317–365 (1998)
[10] Dorigo, M., Stuetzle, T.: Ant Colony Optimization. MIT Press, Cambridge (2005)
[11] Gelernter, D., Carriero, N.: Coordination languages and their significance. ACM Communication 35(2), 97–107 (1992)
[12] Ho, C., Ewe, H.: Ant colony optimization approaches for the dynamic load-balanced clustering problem in ad hoc networks. In: Proceeding of Swarm Intelligence Symposium, IEEE/SIS 2007, pp. 76–83 (2007)
[13] Janssens, N., Steegmans, E., Holvoet, T., Verbaeten, P.: An agent design method promoting separation between computation and coordination. In: Proceedings of the 2004 ACM Symposium on Applied Computing, SAC 2004, pp. 456–461 (2004)

[14] Jogalekar, P., Woodside, C.M.: Evaluating the Scalability of Distributed Systems. IEEE Transactions on Parallel and Distributed Systems 11(6), 589–603 (2000)
[15] Kühn, E., Mordinyi, R., Schreiber, C.: An extensible space-based coordination approach for modelling complex patterns in large systems. In: Proceedings of the Third International Symposium on Leveraging Applications of Formal Methods, pp. 634–648 (2008)
[16] Kühn, E., Riemer, J., Lechner, L.: Integration of XVSM spaces with the Web to meet the challenging interaction demands in pervasive scenarios. Ubiquitous Computing and Communication Journal - Special issue of Coordination in Pervasive Environments 3 (2008)
[17] Kühn, E., Mordinyi, R., Keszthelyi, L., Schreiber, C.: Introducing the Concept of Customizable Structured Spaces for Agent Coordination in the Production Automation Domain. In: Proceedings of the Eighth International Conference on Autonomous Agents and Multiagent Systems, AAMAS 2009, pp. 625–632 (2009)
[18] Kühn, E., Mordinyi, R., Lang, M., Selimovic, A.: Towards Zero-delay Recovery of Agents in Production Automation Systems. In: Proceedings of the 2009 IEEE/WIC/ACM International Joint Conference on Web Intelligence and Intelligent Agent Technology, vol. 2, pp. 307–310 (2009)
[19] Kühn, E., Sesum-Cavic, V.: A space-based generic pattern for self-initiative load balancing agents. In: Aldewereld, H., Dignum, V., Picard, G. (eds.) ESAW 2009. LNCS (LNAI), vol. 5881, pp. 17–32. Springer, Heidelberg (2009)
[20] Kühn, E.: Virtual Shared Memory for Distributed Architectures. Nova Science Publishers (2001)
[21] Lemmens, N., De Jong, S., Tuyls, K., Nowé, A.: Bee behaviour in multi-agent systems. In: Tuyls, K., Nowe, A., Guessoum, Z., Kudenko, D. (eds.) ALAMAS 2005, ALAMAS 2006, and ALAMAS 2007. LNCS (LNAI), vol. 4865, pp. 145–156. Springer, Heidelberg (2008)
[22] Lin, F.C., Keller, R.M.: The gradient model load balancing method. IEEE Transactions On Software Engineering 13(1), 32–38 (1987)
[23] Markovic, G., Teodorovic, D., Acimovic-Raspopovic, V.: Routing and wavelength assignment in all-optical networks based on the bee colony optimization. AI Communications 20(4), 273–285 (2007)
[24] Mordinyi, R., Kühn, E., Schatten, A.: Towards an Architectural Framework for Agile Software Development. In: Proceedings of the Seventeenth International Conference and Workshops on the Engineering of Computer-Based Systems, pp. 276–280 (2010)
[25] Nakrani, S., Tovey, C.: On honey bees and dynamic server allocation in the Internet hosting centers. Adaptive Behaviour 12(3-4), 223–240 (2004)
[26] Olague, G., Puente, C.: The Honeybee Search Algorithm for Three-Dimensional Reconstruction. In: Rothlauf, F., Branke, J., Cagnoni, S., Costa, E., Cotta, C., Drechsler, R., Lutton, E., Machado, P., Moore, J.H., Romero, J., Smith, G.D., Squillero, G., Takagi, H. (eds.) EvoWorkshops 2006. LNCS, vol. 3907, pp. 427–437. Springer, Heidelberg (2006)
[27] Pham, D.T., Soroka, A.J., Ghanbarzadeh, A., Koç, E., Otri, S., Packianather, M.: Optimising neural networks for identification of wood defects using the Bees Algorithm. In: Proceedings of the IEEE International Conference on Industrial Informatics, pp. 1346–1351 (2006)
[28] Pham, D.T., Koç, E., Lee, J.Y., Phrueksanant, J.: Using the Bees Algorithm to schedule jobs for a machine. In: Proceedings of the Eighth International Conference on Laser Metrology, pp. 430–439 (2007)

[29] Picco, G.P., Balzarotti, D., Costa, P.: Lights: a lightweight, customizable tuple space supporting context-aware applications. In: Proceedings of the ACM Symposium on Applied Computing, SAC 2005, pp. 413–419 (2005)
[30] Picco, G.P., Murphy, A.L., Roman, G.C.: Lime: Linda meets mobility. In: Proceedings of the IEEE International Conference on Software Engineering, pp. 368–377 (1999)
[31] Šešum-Čavić, V., Kühn, E.: Instantiation of a generic model for load balancing with intelligent algorithms. In: Hummel, K.A., Sterbenz, J.P.G. (eds.) IWSOS 2008. LNCS, vol. 5343, pp. 311–317. Springer, Heidelberg (2008)
[32] Šešum-Čavić, V., Kühn, E.: Comparing configurable parameters of Swarm Intelligence Algorithms for Dynamic Load Balancing. In: Proceedings of the Fourth IEEE International Conference on Self-Adaptive and Self-Organizing Systems, Workshop Self-Adaptive Network, SASO/SAN, pp. 255–256 (2010)
[33] Shivaratri, N.G., Krueger, P.: Adaptive Location Policies for Global Scheduling. IEEE Transactions on Software Engineering 20, 432–444 (1994)
[34] Shoham, Y., Leyton-Brown, K.: Multiagent Systems: Algorithmic, Game-Theoretic, and Logical Foundations. Cambridge University Press, Cambridge (2009)
[35] Van Steen, M., Van der Zijden, S., Sips, H.J.: Software Engineering for Scalable Distributed Applications. In: Proceedings of the Twenty-Second International Computer Software and Applications Conference, COMPSAC, pp. 285–293 (1998)
[36] Wong, L.P., Low, M.Y., Chong, C.S.: A Bee Colony Optimization for Traveling Salesman Problem. In: Proceedings of the Second Asia International Conference on Modelling & Simulation, AMS, pp. 818–823. IEEE, Los Alamitos (2008)
[37] Yang, X.: Nature-Inspired Metaheuristic Algorithms. Luniver Press (2008)
[38] Zhou, S.: A trace-driven simulation study of dynamic load balancing. IEEE Transactions on Software Engineering 14(9), 1327–1341 (1988)

Glossary

FF	fitness function
GA	genetic algorithm
LB	load balancing
LP	location policy
MINMAX	min-max ant system algorithm
OK	ok-loaded
OL	overloaded
SF	suitability function
SILBA	self initiative load balancing agents
SM	search mode
TP	transfer policy
UL	under-loaded
XVSM	extensible virtual shared memory

Chapter 9
Computational Intelligence in Future Wireless and Mobile Communications by Employing Channel Prediction Technology

Abid Yahya[1], Farid Ghani[1], Othman Sidek[2], R.B. Ahmad[1], M.F.M. Salleh[3], and Khawaja M. Yahya[4]

Abstract. This work presents a new scheme for channel prediction in multicarrier frequency hopping spread spectrum (MCFH-SS) system. The technique adaptively estimates the channel conditions and eliminates the need for the system to transmit a request message prior to transmit the packet data. The new adaptive MCFH-SS system employs the Quasi-Cyclic low density parity check (QC-LDPC) codes instead of the regular conventional LDPC codes. In this work performance of the proposed MCFH-SS system with adaptive channel prediction scheme is compared with the fast frequency hopping spread spectrum (FFH-SS) system. The proposed system has full control of that spectrum; it plans for the system to keep off unacceptable adjacent channel interference. When an interferer suddenly changes its carrier, the set of appropriate channels has a large return and resultantly the adjacent channel interference between the systems is reduced. It has been shown from results that the signal power in FFH system exceeds the average by at least 6.54 dB while in the proposed MCFH-SS system signal power exceeds the average only 0.84 dB for 1% (correct use) of the time. The proposed MCFH-SS system is more robust to narrow band interference and multipath fading than the FFH-SS system, because such system requires more perfect autocorrelation function.

Abid Yahya · Farid Ghani · R.B. Ahmad
School of Computer and Communication Engineering,
Universiti Malaysia Perlis, Perlis, Malaysia

Othman Sidek
Collaborative Microelectronic Design Excellence Center
Universiti Sains Malaysia
14300 Nibong Tebal, Pulau Penang, Malaysia

M.F.M. Salleh
School of Electrical and Electronic Engineering Universiti Sains Malaysia
14300 Nibong Tebal, Pulau Penang, Malaysia

Khawaja M. Yahya
Department of Computer Systems Engineering, NWFP University of
Engineering & Technology Peshawar, Pakistan

N. Bessis, F. Xhafa (Eds): Next Generation Data Technologies for CCI, SCI 352, pp. 225–250.
springerlink.com

1 Introduction

The application of Computational Intelligence (CI) techniques play an important role in the wireless and mobile communications industry, which is a fast growing, stimulating, and critically vital research and development field. CI-related schemes have been incorporated into wireless and mobile communications systems at all levels of hierarchical information processing, such as, from a sensor signal to a symbolic information level. They are commonly employed in components and merged with other technologies in hybrid systems. Multiple input modalities are employed to develop devices and services with enhanced ergonomic user interfaces. Nevertheless, this challenging area is a perfect field for novel applications, since inherent strengths of CI algorithms such as robustness, adaptivity and multi-source information processing. To smooth the progress of practical use, we need a secure and efficient system coupled with newly channel coding scheme incorporated with channel prediction technique.

2 Spread Spectrum Communications

In spread spectrum (SS) the transmission bandwidth employed in a communication system is much greater than the minimum bandwidth required to transmit the information. Spread spectrum takes a narrow-band signal and spreads it over a broader portion of the radio-frequency band.

In Spread Spectrum unintentional noise, like Additive White Gaussian Noise (AWGN), will not interfere the signal so much. But, in case of intentional noise when the signal is spread, the jammer (intentional noise) will either spread its band limited PSD over the new bandwidth, which will reduce its effect on signal, or to continue at its original *BW*, which will induce it to affect only a portion of data. Such effect might be further reduced by error correction coding at the receiver end.

During the World War II spread spectrum technology is first used by the military, since spread spectrum offers low interference and much-needed security. Most of the applications of spread spectrum techniques previously are in the fields of military applications such as radar and communication systems. The history of spread spectrum covers over six decades and may provide as a subject of separate study. From 1920 through World War II, thorough research in radar spread spectrum systems had been taken on in Germany, the USA, the UK and the USSR. In advancements in technology, various solid theoretical analyses had been carried on into the accuracy and signal resolution of radar.

It is true that a great deal of information on new practical developments in spread spectrum radar and navigation was assorted for a long time, since military and intelligence services managed the great majority of projects. Nevertheless, many ideas were extensively recognized as soon as they were accomplished in systems of large-scale employment.

The commercial spread spectrum era started around the late 1970s, at the time when the mobile telephone began its dominant invasion of the world. The first proposal for CDMA cellular networks in the USA and Europe (1978–1980) granted to alternative projects, which later acquired into the GSM and DAMPS standards. The 2G standard IS-95 was proposed in the mid 1990s, resting on a fully spread spectrum/CDMA platform. At a large pace, networks of this standard attained wide identification in America, Asia and the former Soviet Union countries. The great accomplishment of IS-95, as well as careful analysis and additional experiments, had lead to acceptance of the spread spectrum/CDMA philosophy as the basic platform for the major 3G mobile radio specifications. These 3G mobile radios are UMTS and cdma2000, both of them are now the core mobile communication instruments.

Surveys published in the West usually report only a little on Soviet research since the Soviet cold war strict limits on the contacts of Soviet specialists with their foreign colleagues and publications abroad. However, it is an open secret that Soviet advancement in the spread spectrum field between the 1950s and the 1990s was very up-to-date and quite competitive with developments in the USA and Europe.

In 1985, the US Federal Communications Commission (FCC) allocated three spectrum frequencies, which soon headed to an explosion of spread spectrum use among business communities [1] .In the field of communications the SS technique is used in mobile networks communication and wireless local area network (WLAN) [2]. The use of spread spectrum technique in communication consist of anti-jamming, anti-interference, low probability of intercept (LPI), message privacy, multiple-users access, and selective addressing.

Spread spectrum is categorized into two major techniques: Direct sequencing and Frequency Hopping Spread Spectrum.

2.1 Direct Sequencing Spread Spectrum

Direct sequencing (DS) is the most common form of spread spectrum sequencing. DS technique spreads its signal by expanding it over a broad portion of the radio band [3]. This happens when the information is divided into small blocks or data packets that are spread over the transmitting bandwidth and results in a low power density. The low power density comes across as a white noise and lowers the possibility of interference. At the receiver, in order to recover the original message, the incoming waveform is multiplied by an identical synchronized spreading waveform. The signal contains the frequency changes in the spectrum in accordance to the code. The rest of the frequencies in the spectrum that are not in the signal are transmitted as noise. Each time, if another user is added, the noise level increases. Direct sequence is used in broadband bandwidths.

Preferably, a direct-sequence signal with binary phase-shift keying (PSK) data modulation can be represented by [4].

$$x(t) = Am(t)s(t)\cos(2\pi f_c t + \theta) \tag{1}$$

where, the parameter A is the signal amplitude, $m(t)$ denotes the data modulation, $s(t)$ represents the spreading waveform, f_c is the carrier frequency, and θ is the phase at t=0.

2.2 Frequency Hopping Spread Spectrum (FHSS)

The other method for converting the baseband data stream into larger bandwidth signal is the FHSS technique. In FHSS technique the transmission bandwidth W Hertz is divided into q non-overlapping frequency slots. After the signal is modulated to an intermediate frequency, the carrier frequency is hopped periodically according to some pre-designated code (a pseudo-random sequence) [5].

A patent Hedy Lamarr and music composer George Antheil [5] for a "Secret Communication System," in 1942, is based on the frequency hopping concept, with the keys on a piano representing the different frequencies and frequency shifts used in music. In that year, the technology could not be realized for a practical implementation. Lemarr and Antheil incurred a patent for their idea soon after the expiry of the original patent. Then the U.S applied the FHSS technique for military communication systems onboard ships [6].The use of FHSS systems has then increased dramatically since 1962. Figure 1 shows the block diagram of FHSS technique.

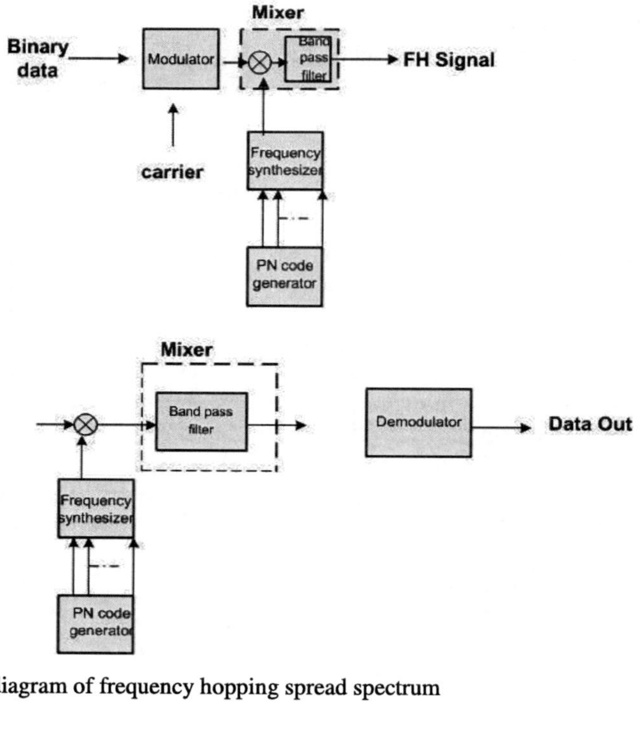

Fig. 1 Block diagram of frequency hopping spread spectrum

The benefit of a frequency hopping signal is that it intercepts resistant. This feature is extensively used in military communications where the risk of signal jamming or intercept is higher. Nowadays, it is used in the mobile communication industry as a multiple access technique. The frequency hopping communication systems are utilized to handle high capacity data in an urban setting [5]. Frequency hopping communication systems play an important role in military communications strategy. FH communication systems offer an enhancement in the performance when subjected by hostile interference. FH communication systems also reduce the ability of a hostile observer to receive and demodulate the communications signal. FH communication systems are susceptible to a number of jamming threats, such as noise jammers and narrowband, single or multitone jammers.

If all frequency hops are to be jammed, the jammer has to divide its power over the entire hop band. Thus, it needs to lower the amount of received jamming power at each hop frequency. Unfortunately, if the tone jamming signal has a significant power advantage, reliable communications will not be possible, even when the jamming tones experience fading [7,8]). If the FH signal has an ample hop range, received jamming power will be negligible. If a tone jammer is concentrated on a particular portion of the FH bandwidth, its power may adversely impact communications. A likely antijamming strategy is used as a narrow bandstop filter to remove the tone from the received signal spectrum [5]. Another method based on the undecimated wavelet packet transform (UWPT) isolates the narrowband interference using frequency shifts to confine it to one transform subband [9]. This technique is a robust to avoid interference and is suitable for FHSS systems.

2.2.1 Frequency Hopping over Direct Spread Spectrum

The fundamental difference between direct spread and frequency hop is that the instantaneous bandwidth and the spread bandwidth are identical for a direct spread system. While for a frequency hop system the spread bandwidth can be and is typically far greater than the instantaneous bandwidth. For the frequency hop system, all the anti-jamming (AJ) systems' processing gain depends upon the number of available frequency slots. The spread bandwidth is generally equal to the frequency excursion from the lowest available frequency slot to the highest available frequency slot. The instantaneous bandwidth, in turn, is determined by the hopping rate or symbol rate, whichever is greater.

For the direct spread system, the spread bandwidth is limited to the instantaneous bandwidth or the rate at which the PN sequence is clocked. Theodore (2001) at Virginia Tech conducts a research project comparing the effects of interference on wireless LANs using DSSS and FHSS technology with the 802.11 and 802.11b standards. The experimental results prove that FHSS is superior to DSSS in high interference settings. Experimental analysis shows that DSSS systems are suffered a performance degradation of 27-45% while FHSS systems are degraded only by 7-8%.

The following characteristics show that FHSS overcomes DSSS [6, 10, 11].

- **Throughput:** Point-to-point throughput is variable between both DSSS and FHSS products. Protocols for DSSS throughput sacrifice mobility and roaming performance, but FHSS provides greater power, signal efficiency, mobility, and immunity from multipath interference.
- **Interception:** DSSS data is easier to intercept than FHSS data. Constant hopping of FHSS signals make it less susceptible to interception and interference.
- **Power:** FHSS radios use less power than DSSS.
- **Efficiency:** FHSS can provide up to four times more network capacity than DSSS.
- **Mobility:** FHSS products provide better mobility, are smaller, lighter, and consume less power. Unlike DSSS, FHSS incorporates roaming without sacrificing throughput and scalability.
- **Immunity from Multipath Interference:** Multipath interference is caused by signals that bounce off from the walls, doors, or other objects so that signals arrive at the destination at different time. This problem is automatically avoided by FHSS. FHSS simply hops to a different frequency whenever the channel is attenuated. DSSS however, is not capable of overcoming this effect.

Both spread spectrum methods carry large volumes of data, but FHSS is superior. FHSS is a very robust technology, it is scalable, mobile, secure, can accommodate overlapping networks, resistance to interference, and with little influence from noises, reflections, other radio stations or other environment factors [6,11]. Furthermore, the number of simultaneously active systems in the same geographic area is considerably higher than the equivalent number for DSSS systems.

2.2.2 Multicarrier Frequency Hopping Spread Spectrum (MCFH-SS) Systems

Multicarrier frequency hopping spread spectrum (MCFH-SS) systems have received great attention because they take advantage of both multicarrier modulation and the FH concept and can be implemented coherently at the receiver when appropriately and specifically designed [12].

Authors in [13] propose a modified multicarrier (MC) direct-sequence code division multiple-access (DS-CDMA) system with adaptive frequency hopping for use over slow multipath fading channels with frequency selectivity in the reverse link transmission of a cellular network. Rather than transmitting data substreams uniformly through subchannels, data substreams hop over subchannels with the hopping patterns adaptively adjusted to the channel fading characteristics. Authors design an efficient algorithm, based on the water-filling (WF) principle to determine the optimal hopping pattern and show that the performance, in terms of average bit-error probability (BEP) is substantially better than that of single carrier RAKE receiver systems, conventional MC-CDMA systems applying moderate error protection, or diversity systems with different combining schemes. The proposed work illustrates that such an enhancement can be directly translated into an increase in CDMA system capacity. A similar, but more general, framework for applying the FH concept to multicarrier DS-CDMA schemes is proposed in [14]. Nonlinear constant-weight codes are introduced in the proposed scheme, in order to control the associated FH patterns and to competently share the system's frequency resources by each user. Furthermore, constant-weight codes are employed with different weights, in order to activate a number of subcarriers to support multirate services. Performance of the proposed system is evaluated by using a coherent RAKE receiver with maximum ratio combining (MRC) for demodulation and compares with that of corresponding single-carrier DS-CDMA and MC DS-CDMA systems, in a multipath Nakagami fading environment. It is observed from simulation results that the proposed SFH/MC DS-CDMA is competent of interworking with the existing 2G and 3G CDMA systems, while providing an evolutionary path for future unlicensed and broadband radio access networks (BRAN) without stiff and unnecessary spectrum fragmentation. Authors in [15] propose a truncated adaptive transmission scheme for the hybrid multicarrier CDMA/FDM system in forward link under single and multiple-cell environment. In the proposed research work, a substream of information is transmitted over the subchannels of which the channel benefits are greater than a specified threshold, established on the feedback data from the mobile station. The proposed scheme outperforms the adaptive FH/DS system as well as the conventional MC DS/CDMA system, in the single-cell environment, when orthogonal signature sequences are used. Authors emphasize on the orthogonality between users in order to eliminate the multiuser interference. It is found in the proposed scheme that by transmitting signals over good subchannels, the received signal energy is increased, while the interference from other cell base stations does not increase. The proposed scheme has better performance characteristics than the adaptive FH/DS system, in the multiple-cell environment when orthogonal or random codes are employed as spreading sequences.

A new allocation algorithm to overcome the limitations of WF algorithm in the MC-CDMA system with adaptive FH is proposed in [16]. In the proposed system signal to interference and noise ratio (SINR) is used instead of BER as the performance measure, and concentrate on the performance of the substream to maximize the SINR with the lowest SINR, since the error events are linked with that substream dominate the error rate. At the receiver end of MC-CDMA system, linear decorrelating detector is employed in order to enhance the spectral efficiency. Authors investigate that the linear decorrelating detector that employs the proposed allocation algorithm is very effective in mitigating MAI, with performance approaching the single user bound for MC-CDMA system with adaptive FH.

Against the background of the extensive development of the Internet and the continued dramatic increase in demand for high-speed multimedia wireless services, there is an urgent requirement for flexible, bandwidth-efficient transceivers. Multi-standard operation is also an essential demand for the future generations of wireless systems. Authors in [14] demonstrate the possible implementation of the proposed FH/MC DS-CDMA scheme by software-defined radios, and its competence in handling multirate services. The FH/MC DS-CDMA exhibits a high grade of flexibility in the context of system design and parameter reconfiguration especially, in the existing second- and third-generation CDMA system bands [14].

Taking the advantage of a bandwidth-efficient multicarrier on-off keying (MC-OOK) modulation, authors in [17] propose an efficient modulation method for frequency-hopped multiple-access (FHMA) communications in order to furnish a higher immunity against multiple-access interference in FHMA systems. Bit error probability of the proposed scheme is examined in slow frequency non-selective Rayleigh fading channels with background noise, while authors in [18] analyze the same system but with FFH. The former system shows that MC-OOK/FHMA provides a lower interference over MFSK/FHMA for E_b / N_o greater than a threshold (interference-limited region), but the opposite is found to be true at low E_b / N_o. Experimental results indicate that the capacity gain that MC-OOK/FHMA system provides over MFSK/FHMA system in an interference-limited region is more than 2.5dB, when the modulation alphabet size *M* is set to 8, and becomes higher for larger *M*. Authors in [19] propose a multicarrier direct sequence slow FH CDMA system with similar properties to that of conventional multicarrier DS-CDMA system, except that the main frequency subbands in the proposed scheme are divided into a number of hopping frequency dwells. A similar FH technique is applied in [20] to a conventional multicarrier CDMA system, allowing for the narrowband frequency subcarriers of a user to hop within some groups of frequency slots. The proposed scheme is examined in an uncoded multi-access environment by utilizing a Gaussian assumption for the MAI.

Authors in [21] propose a multicarrier M-*ary* frequency shift keying (MFSK)/FH-CDMA, which utilizes FH patterns with cross correlation, not greater than one and derives at reducing mutual interference. BER of the proposed system is investigated in multiuser non-fading and fading channels with noncoherent reception, and has found to be decreasing function of the number of subcarriers. The performance of coded MC-FH multiple access schemes is examined in [22] in the presence of jamming in static and fading channels. It is shown that Gaussian approximation for the MAI is unacceptable for most cases. Simulation results show that optimal weights for the receiver soft outputs in countering smart jamming, give rise to worst jamming. Stronger MAI or greater SJR reduces the gain achieved by the optimal weighting in the proposed investigation. It is observed from the proposed work that if the receiver has JSI it should opt a weight factor greater than unity based on an adaptive method.

Wireless communication systems have to be planned in order to meet the required error protection levels. The construction of forward error correction (FEC) codes generally comprises of choosing a fixed code with a definite code rate, encoding/decoding complexity, and error-correcting capacity.

3 Channel Coding

Channel coding is one of the major means that boost the transmission consistency at higher data rates. For practical applications in wireless communication systems, the channel coding scheme with low complexity and shorter length is preferred. The rediscovery of Low Density Parity Check (LDPC) codes, which are originally proposed by [23] and afterward are extrapolated by MacKay [24] have acquired considerable attention. The performance of LDPC codes is investigated, at many events of interests and is encountered to outperform turbo codes with good error correction [24,25,26].

LDPC codes are generally designed to be linear and binary block codes. In this case there is a generator matrix $\boldsymbol{G}$ that converts a message vector $\boldsymbol{m}$ into a code vector $\boldsymbol{c}$ by means of a matrix multiplication. The corresponding parity check matrix H has the property that it is constructed with linearly independent row vectors that form the dual subspace of the subspace produced by the linearly independent row vectors of $\boldsymbol{G}$. This means that every code vector satisfies the condition $\boldsymbol{H} \circ \boldsymbol{c} = \boldsymbol{0}$.

3.1 Representations for LDPC Codes

Basically there are two different underlying theories to represent LDPC codes. Like all linear block codes, LDPC codes can be described using matrices. Alternatively, it can be described using a graphical representation.

3.1.1 Matrix Representation

LDPC codes are in the category of linear codes. They cater near capacity performance on a large data transmission and storage channels. LDPC codes are rendered with probabilistic encoding and decoding algorithms. LDPC codes are designated by a parity check H matrix comprising largely 0's and has a low density of 1's. More precisely, LDPC codes have very few 1's in each row and column with large minimum distance. In specific, a (n, j, k) low-density code is a code of block length n and source block length k . The number of parity checks is delimitated as $m = n - k$. The parity check matrix weight (number of ones in each column or row) for LDPC codes can be either regular or irregular. LDPC can be regular if the number of ones is constant in each column or row and gets irregular with a variable number of ones in each column or row. A regular LDPC code is a linear block code whose parity-check matrix H constitutes exactly J 1's in each column and exactly $k = j\left(n/m\right)$ 1's in each row, with the code rate $R = 1 - j/k$.

3.1.2 Graphical Representation

Tanner considered LDPC codes and showed how to represent these codes effectively by a bipartite graph, now call a Tanner graph [27].Tanner graphs provide a complete representation of the code and also help to explain the decoding algorithm.

In Tanner graphs the nodes of the graph are separated into two typical sets and edges connecting nodes of two different types. The two types of nodes in a Tanner graph are called variable nodes (v -nodes) and check nodes (c -nodes). The Tanner graph of a code is drawn according to the following rule: It consists of m check nodes (the number of parity bits) and n variable nodes (the number of bits in a codeword).Check node y_i is connected to variable node y_j if the element h_{ij} of H is a 1. Tanner graph corresponding to H is shown in Figure 2.

$$H = \begin{bmatrix} 1\,1\,1\,0\,0\,0\,0\,0\,0\,0 \\ 1\,0\,0\,0\,1\,1\,1\,0\,0\,0 \\ 0\,1\,0\,0\,1\,0\,0\,1\,1\,0 \\ 0\,0\,1\,0\,0\,1\,0\,1\,0\,1 \\ 0\,0\,0\,1\,0\,0\,1\,0\,1\,1 \end{bmatrix} \qquad (2)$$

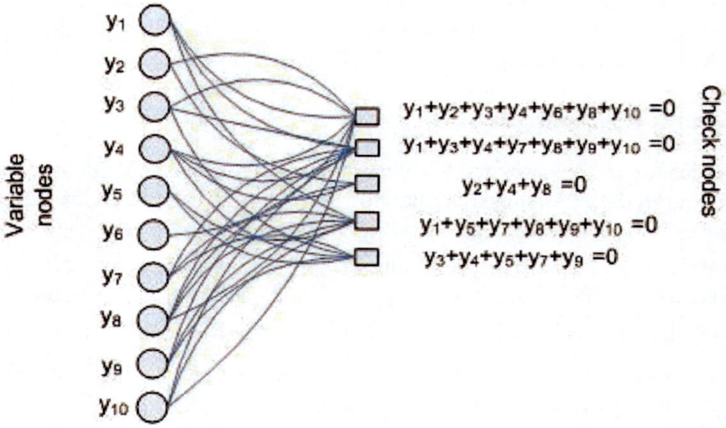

Fig. 2 Tanner Graph

A new necessary and adequate condition for determining the girth of QC-LDPC codes is derived in [28] based on the theory of adjacency matrices, without an explicit enumeration of cycles. It is shown from simulation results that the obtained codes are often with performance comparable to the LDPC codes constructed by progressive-edge-growth (PEG) algorithm [29,30]. The performance error for regular QC-LDPC and PEG-LDPC codes are plotted with iterative sum-product decoding [26] , with same code rate and observes that the PEG-LDPC codes are frequently with smaller girth than the proposed QC-LDPC code by adopting concentrated parity-check degree distribution. There is considerable work on optimizing girth in LDPC codes. Another method for constructing large girth QC-LDPC codes from graphical models is proposed in [31]. The proposed QC-LDPC codes based on circulant permutation matrices with girths 16 and 18, by employing a simple quadratic congruential equation.

Authors of this book Chapter keeping in mind the aforesaid systems, have proposed a novel construction of QC-LDPC code which not only reduces encoding complexity but also improves the decoding part of the system [32, 33, 34, 35]. In order to simplify the hardware implementation, the proposed codes incorporate some form of structured decoder interconnections. In the proposed algorithm, the restructuring of the interconnections is invented by splitting the rows with the group size. Such a division guarantees a concentrated node degree distribution and reduces the hardware complexity. The new codes offer more flexibility in terms of high girth, multiple code rates and block length.

This work presents a new scheme for channel prediction in multicarrier frequency hopping spread spectrum (MCFH-SS) system. The technique adaptively estimates the channel conditions and eliminates the need for the system to transmit a request message prior to transmit the packet data. The new adaptive MCFH-SS system employs the Quasi-Cyclic low density parity check (QC-LDPC) codes instead of regular conventional LDPC codes. In this work performance of the proposed MCFH-SS system with adaptive channel prediction scheme is compared with the fast frequency hopping spread spectrum (FFH-SS) system. In the aforesaid systems QC-LDPC codes are utilized as a forward error correction (FEC) scheme.

4 Proposed Channel Prediction Scheme

The performance of the MCFH-SS system can be enhanced by incorporating the new proposed channel prediction scheme to the system. This means the MCFH-SS system responds to the noise and fading by avoiding the channel that is unfit for communication. Usually, a channel is banned only after it has been used to transmit data, which results in retransmission. The principle of blocking the poor channels is based on the prospect that these channels will remain in poor condition for transmission in quite some time. However, the banned channels may fit inadequately to the actual noise and fading.

In a severe condition, the system may end up employing the low quality channels regularly and would ban the good ones too frequently and disturbs the performance. Apparently, it is attractive to mitigate such undesirable consequences. If the system is designed in such a way that it attempts to forecast and ignores the poor channels, better performance can be incurred. This furnishes the need for attempting to forecast the quality of channels. In order to predict the fit channels the system transmits short test packets on channels. If the test packet arrives is readable, the channel will be occupied with a Pseudonoise (PN) code and use for transmission or else, it will be banned.

4.1 The Algorithm

The channel prediction algorithm flow chart is shown in Figure 3 and explains as in the following steps:

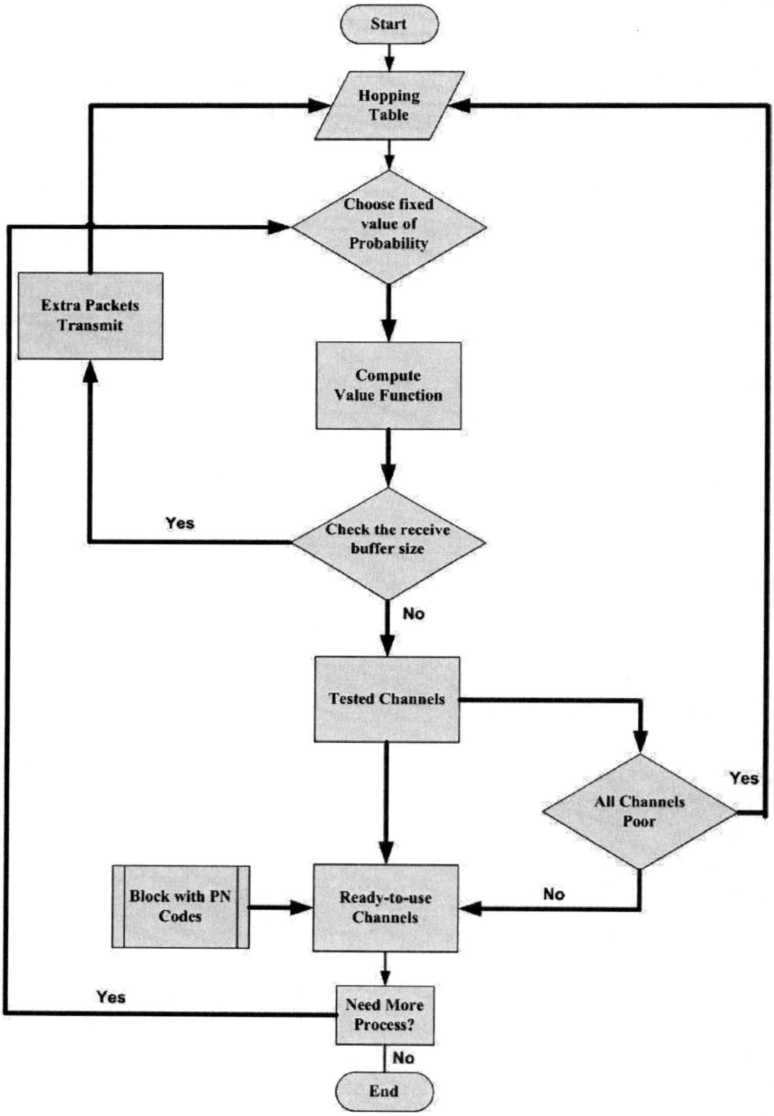

Fig. 3 Flowchart of proposed channel prediction

1. The probability of a poor channel p must be determined.

$$P = \sum_{i=1}^{h} p(b_0 | i, b_q) \tag{3}$$

where the parameters,
q is the number of test packets/channel test
h is the hopping table
b is the maximum number of packets in the receiver buffer
b_0 is the actual number of packets in the receiver buffer

2. Determine the Value function for each possible channel test.

$$E_{h,q}(b) = \min_{h} P\left\{S + e^{-rT} E_{h,q}(b_0)\right\} \tag{4}$$

where , $E_{h,q}(b)$ represents the Value function, the other parameters are mentioned as;

q is the number of test packets/channel test
h is the hopping table
b is the maximum number of packets in the receiver buffer
b_0 is the actual number of packets in the receiver buffer
S is the stochastic variable
e^{-rT} is the discount factor, and $0 \leq \gamma \leq 1$

3. Following step 2, check the receive buffer size and the number of best possible extra packets.
4. If all the q tested channels are poor, it is proposed that the radio shall simply use the next untested channel from the hopping table to transmit the frame.
5. Block the ready-to-use channels with PN code in order to keep them off from intentional and unintentional jammers such as narrow band jammers.
6. Update ready-to-use channels list.

5 Spectrum Analysis of the Proposed MCFH-SS System

In this Section the spectrum analysis of the proposed MCFH-SS system (MCFH-SS system coupled with adaptive channel prediction scheme together with proposed QC-LDPC codes) is investigated under slow fading frequency channel with partial band noise jamming environment.

5.1 Complementary Cumulative Distribution Function (CCDF)

Complementary Cumulative Distribution Function (CCDF) measurements provide important information in the field of design and manufacturing of the system's components used in spread spectrum modulation that result in large throughput for cellular services. The CCDF curve is generated for the normal distribution starting with the familiar normal probability density function (PDF).

5.1.1 Performance Comparison of CCDF Curves

A technique to determine the suitable vector range is to measure the CCDF of the digitally-modulated signal. CCDF as shown in Figure 4, is a statistical characterization that plots power level on the x-axis and probability on the y-axis of a graph. Each point on the CCDF curve shows what percentage of time a signal spends at or above a given power level. The power CCDF curves illustrate only information from average power on up. The spectrum analyzer result in Figure 4 shows the overshoot or trajectory vector length versus statistical frequency. The power level is expressed in dB relative to the average signal power level. This work finds a percentage of time on the ordinate, then read across the graph and ascertain the signal's consequent number of dB above average power for that time.

For comparison, the performance of CCDF curve is shown in Figure 5. However, the proposed MCFH-SS system in Figure 4 consists of channel prediction scheme coupled with QC-LDPC codes as FEC codes. Parameters are set for the proposed MCFH-SS system in Table 1. There is a significant difference between the CCDF curves of the two signals. The position of the CCDF curves indicates the degree of peak-to-average deviation, with more stressful signals further to the right. It shows from the spectrum analyzer results in Figure 4 and 5 respectively, employing DPSK modulation, that when $t = 1\%$ on the y-axis, the corresponding peak to average ratio is 0.84 dB and 6.54 dB respectively. This means the signal power in Figure 5 exceeds the average by at least 6.54 dB while in Figure 4 the signal power exceeds the average only 0.84 dB for 1% (correct use) of the time. The proposed MCFH-SS system is more robust to narrow band interference and multipath fading than the FFH-SS system, because such system requires more perfect autocorrelation function.

Table 1 Parameters setup for CCDF measurement of the proposed MCFH-SS system

Code	Length	Rate	Iterations	Modulation
(3,6)	1536	0.5	8	DPSK

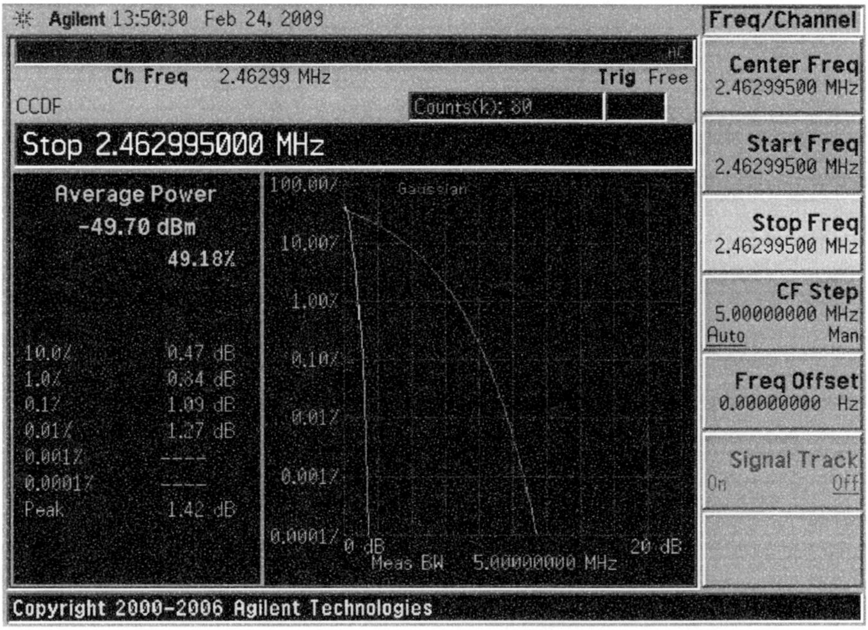

Fig. 4 Complementary Cumulative Distribution function of the proposed MCFH-SS system

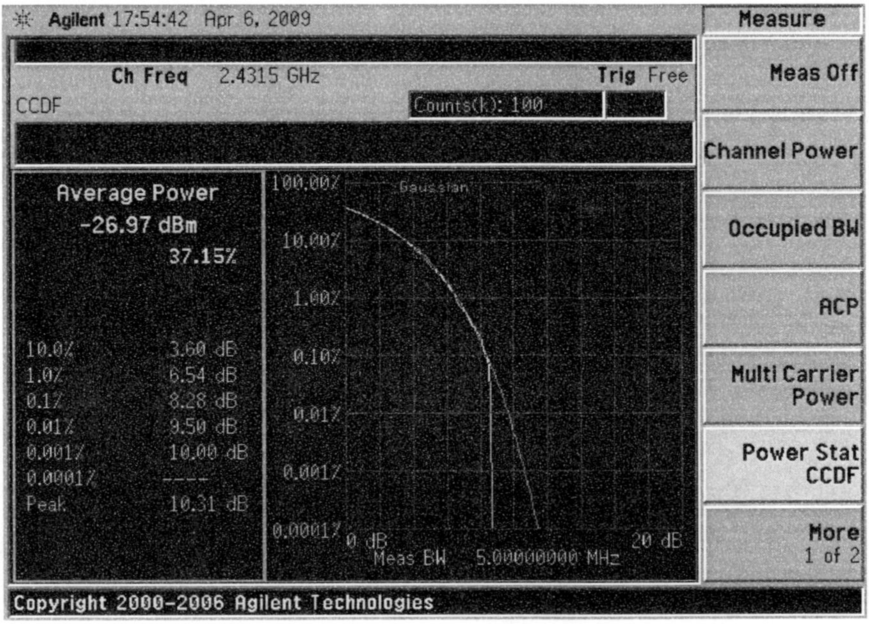

Fig. 5 Complementary Cumulative Distribution Function of the FFH-SS system

5.2 Adjacent Channel Power

In general, engineers are most interested in the relative difference between the signal power in the main channel and the signal power in the adjacent or alternate channel. The measurement application is set up to measure the adjacent channel on either side of the main channel. Figure 6 shows the practical implementation of an Adjacent Channel Power Ratio (ACPR) measurement. In order to determine ACPR, first the channel power is measured by integrating the power over a 2.5 MHz bandwidth. Then, the power at the expected offset frequency is integrated over 1.5 MHz to measure the power in a particular adjacent channel. The term integration is used freely in this case. The power measurements are more accurately described as the summation of the various power samples taken by the spectrum analyzer describes more precisely the power measurement over the channel or adjacent channel bandwidths. The goal is to reduce the ACPR as much as possible, so as not to cause interference in the adjacent channel.

Resolution bandwidth has to be set narrow as compared to the channel bandwidth in order not to widen the channel bandwidth. The channel becomes widen due to the high shape factor of the spectrum analyzer filters. When the resolution bandwidth of the signal as shown in Figure 6, is taken as 24 kHz, this gives an effective number of 6000 samples in the transmit channel and 3600 samples in the adjacent channels. It can be seen that the overall adjacent channel power ratio results are subjugated by the ambiguity linked with the 600 samples in the adjacent channel. These results are obtained, as shown in Figure 6, with a sweep of only 100 milli seconds. Table 2 shows the parameters setup for ACP measurement of the proposed MCFH-SS system. The typical measurement application provides in Table 3 showing the integrated power in the main channel and the integrated and relative powers in the adjacent channels and compares with Standard IEEE 802.11g ACPR. The comparison results show that the proposed MCFH-SS system has only -11.25 dBc ACPR as compared to the Standard IEEE 802.11g -63.8 dBc ACPR.

For comparison with FFH-SS system, the traces in the Figure 7, employing DPSK modulation show that the amount of energy in the neighboring channels is steadily increases as the input signal increased. The ACP results in a decline in the number of active users which can be operated at the same time and also distortions leading an increase in BER. When the system has more than one ready-to-use channel available, it may decide to increase the spacing between channels on its own network in order to allow a small guard band between itself and the neighboring system. In the proposed MCFH-SS system, the suitability of one channel is measured independent from all other channels else the transceiver would need to find optimal action polices while being in an operation. The proposed system has full control of that spectrum; it plans for the system to keep off unacceptable adjacent channel interference. When an interferer suddenly changes its carrier, the set of appropriate channels has a large return and resultantly the adjacent channel interference between the systems is reduced as shown in Figure 6.

Table 2 Parameters setup for ACP measurement of the proposed MCFH-SS system

Code	Length	Rate	Iterations	Modulation
(3,6)	1536	0.5	8	DPSK

Table 3 Trace properties of adjacent channel power

		Figure 6	Figure 7
1	**Channel Frequency**	2.431 GHz	2.431 GHz
2	**ACP Low**	-38.37 dBm	-75.48 dBm
3	**ACP UP**	-38.27 dBm	-76.97 dBm
4	**ACPR Low**	-11.35 dBc	-55.28 dBc
5	**ACPR Up**	-11.25 dBc	-56.77 dBc
6	**Carrier Power**	-27.02 dBm	-20 dBm
Standard IEEE 802.11g ACPR =-63.8 dBc			

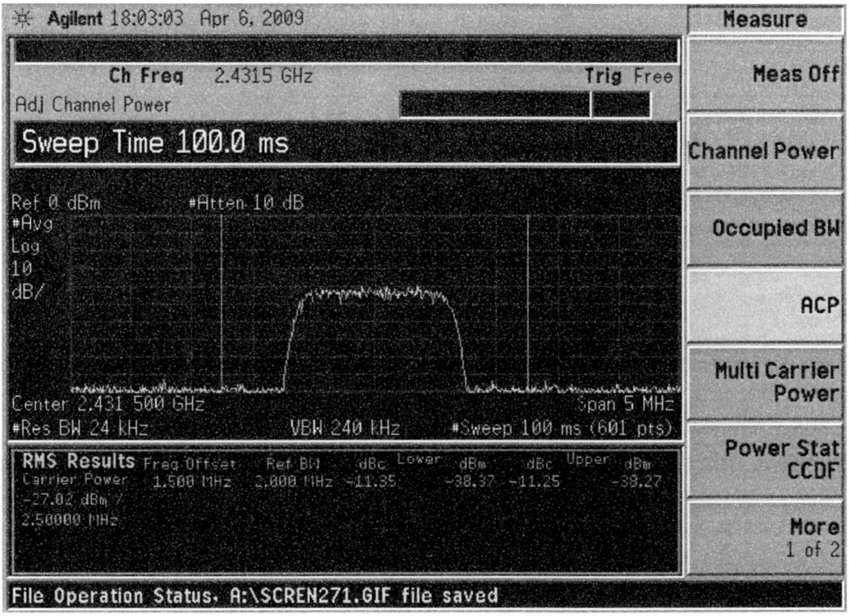

Fig. 6. Adjacent Channel Power of the proposed MCFH-SS system

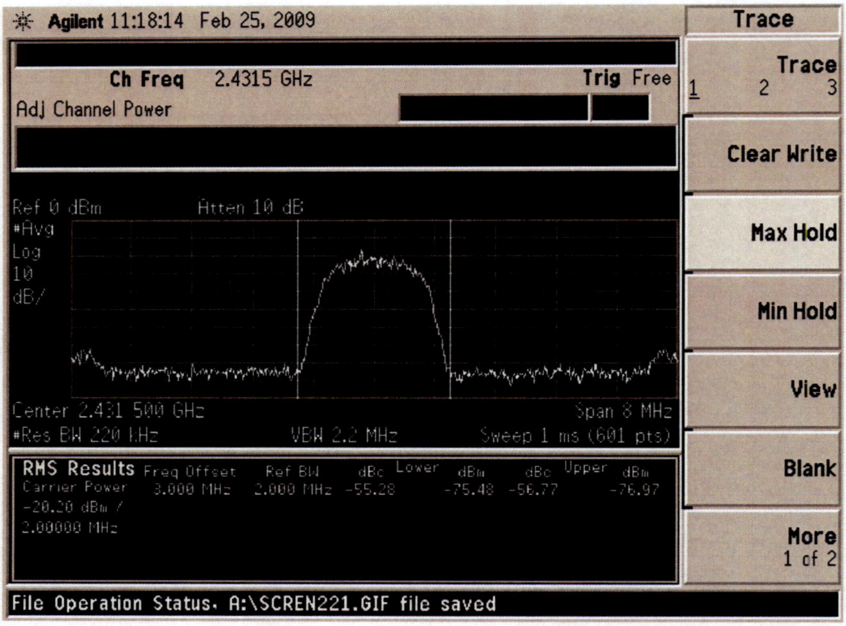

Fig. 7 Adjacent Channel Power of the FFH-SS system

6 Interference Analysis of the Proposed MCFH-SS System

The proposed MCFH-SS system's Transmitter setup is shown in Figure 8 (a). The workstation installed with Matlab program and VSA software is connected to signal generator by means of a GPIB bus. It is then connected to a hardware platform and spectrum analyzer. The hardware platform output is connected to the spectrum analyzer RF in and serially connects to the workstation. The whole system is fully controlled by a workstation.

The Anritsu MS2721B Spectrum Master Handheld Spectrum Analyzer is the perfect portable tool for trouble-shooting interfering signals in the field due to its lightweight and calibrated measurement receiver. Since the interfering process generally implies overlapping frequencies, a spectrum analyzer is the instrument that easily catches and exhibits such signal overlaps. The MS2721B approximately demonstrates the same sensitivity as a typical affected system receiver. The receiver setup of the proposed MCFH-SS system using Anritsu spectrum analyzer is shown in Figure 8 (b).

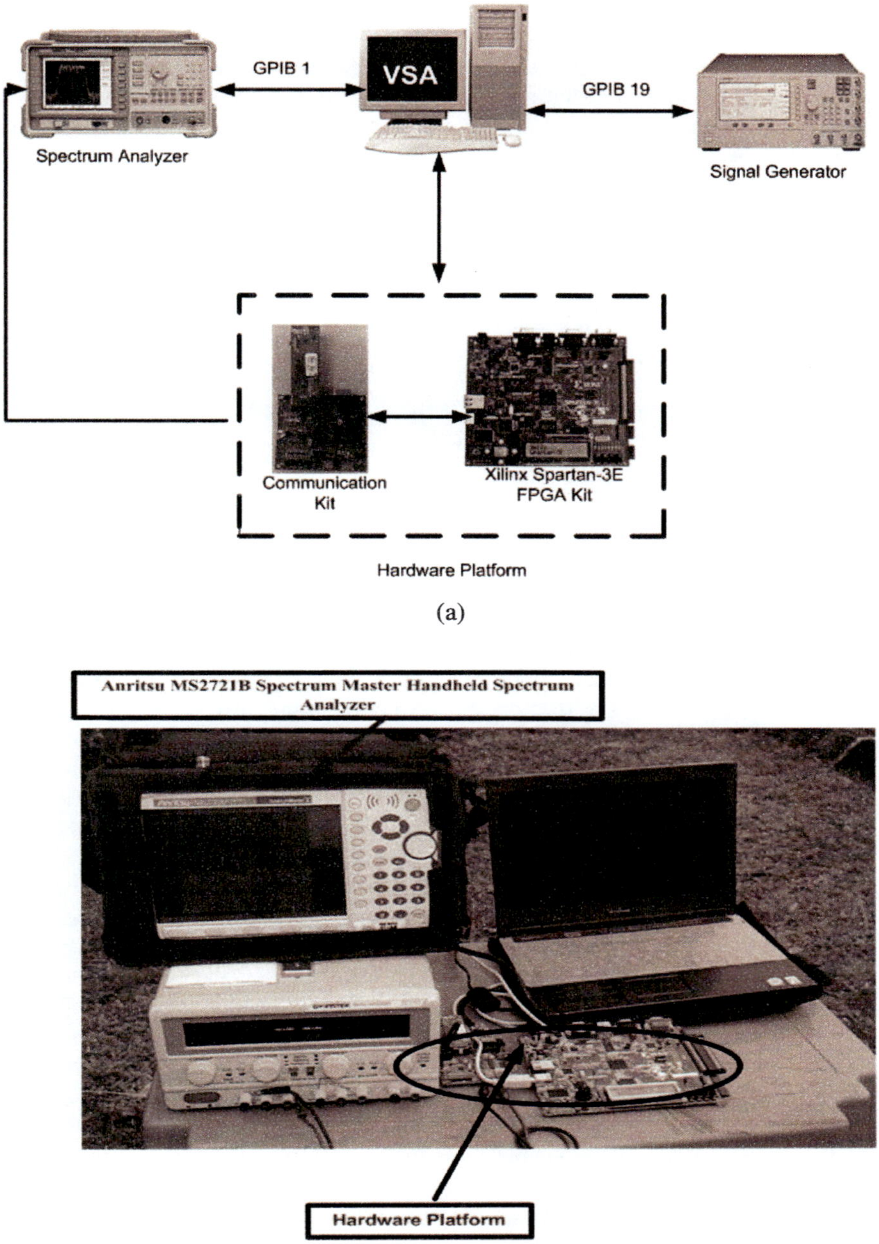

(b)

Fig. 8 Proposed MCFH-SS system (a) Transmitter and (b) Receiver setup

6.1 Interference Analyzer

Signal interference occurs in the wireless networks when unwanted signals are received at signal strengths that make the receiver insensitive. In the past, solving the interference problems were usually the most time consuming. Most of the time was spent in collecting the data of nearby transmitters and their identification, which wasted time and money. Today, with the increased number of transmitters and other communication devices, a deliberate process is required to figure out interference.

In order to examine the performance of the transmitted signals and the presence of adjacent frequencies, spectrum analyzers are generally used. These equipments are also used to figure out the possible sources of interference.

6.2 Spectrogram and Received Signal Strength

High-quality signal transmission and reception are ascertained by the interference analysis option. The system is able to collect data for up to 72 hours within a defined frequency range.

The Spectrogram is a valuable tool for tracking the source of the interfering signal. The power at a certain frequency (in dBm) is displayed along with the spectrogram. The spectrogram display provides signal's activity while exhibiting frequency, time, and power information. The spectrogram display is ideal for location of periodic interference. Since spectrogram shows the desired portion of the spectrum to see changes in the spectrum over time, mutually with its consequent frequency and power level.

The Anritsu spectrum analyzer is switched into an interference analyzer mode. Meanwhile the spectrum display at the bottom of the spectrogram is analyzed to trace the suspected interfering signal. The spectrogram saves the time sets by the time interval feature between two consecutive measurement points. The spectrogram display in Figures 9 and 10 show signals over time with various colors corresponding to the signal strength in the presence of worst condition of partial band noise jamming. The blue color shows the weakest signal while the brown color shows the strongest signal. As the signal strength increases, the color on the spectrogram changes accordingly. The proposed MCFH-SS system is shown in Figure 9 presents the strongest signal which is marked with an arrow displayed in brown color. FFH-SS system shows some weak at the start represented by arrow 1 and then with the passage of time shows the noise level of the signal represented by arrow 2 as presented in Figure 10. The error correcting often employed to remove the effect of a single error-causing peak in the system. The permutation of proposed QC-LDPC bit nodes coupled with new channel prediction scheme enhances the burst erasure correction and results in a permutation-equivalent code that retains all the random error correcting properties of the original code. The bounds for maximum burst erasure correction capability under iterative decoding has been provided by the proposed system which makes the system robust against hostile jamming.

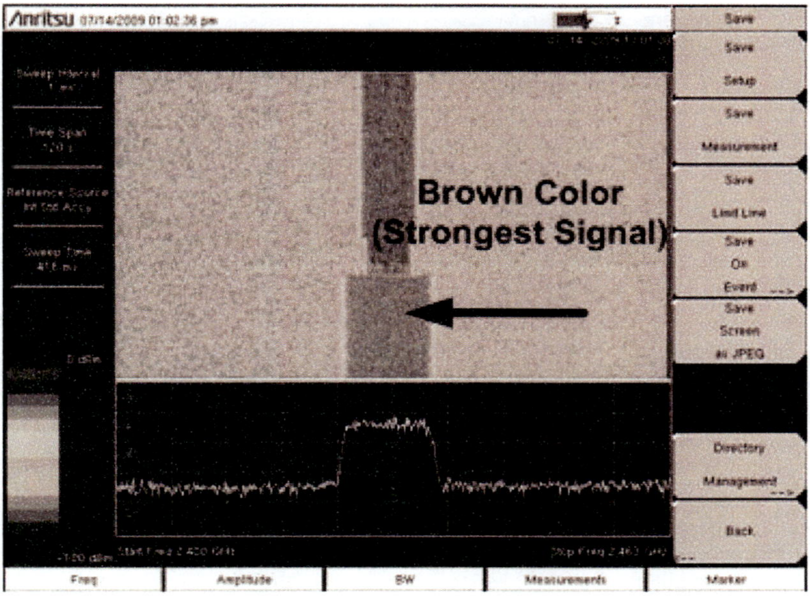

Fig. 9 Spectrogram display of the proposed MCFH-SS signals over time with color corresponding to the signal strength

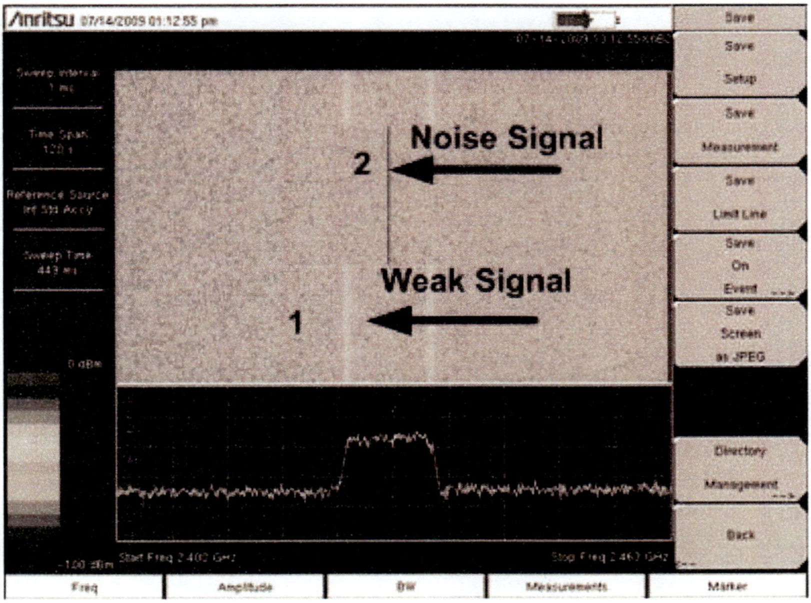

Fig. 10 Spectrogram display of the FFH-SS signals over time with color corresponding to the signal strength

7 Conclusions

In this work QC-LDPC channel coding scheme has been adapted.The new codes offer more flexibility in terms of large girth, multiple code rates and large block lengths. In this method of code construction, the rows are used to form as the distance graph. They are then transformed to a parity-check matrix in order to acquire the desired girth. In this work a new scheme for channel prediction in MCFH-SS system has been proposed. The technique adaptively estimates the channel conditions and eliminates the need for the system to transmit a request message prior to transmitting the packet data. The proposed MCFH-SS system uses PN sequences to spread out frequency spectrum reduce the power spectral density and minimize the jammer effects.

In this Chapter, the hardware implementation of the proposed system is described as well. The hardware platform consists of communication development kit that is interfaced with Xilinx Spartan-3E development board. The user can easily program the device with custom protocols for use in their end product or for general product development. The workstation is installed with Matlab and VSA software which is connected to a signal generator by means of a GPIB bus. It is also then connected to the hardware platform and spectrum analyzer.

If technological advances in the areas of sensing, computation and wireless communications continue at their current rates, then it will not be long before hand-held devices could be used for different applications throughout the world. This is an interesting area of future research, as it brings together different fields and has the potential to help billions of people.

References

[1] SSS Online. Spread spectrum history page, `http://sss-mag.com/shistory.html`

[2] Tanner, R., Woodard, J.: WCDMA - Requirements and Practical Design. Wiley, Chichester (2004)

[3] Simon, M., Omura, J., Scholtz, R., Levitt, B.: Spread Spectrum Communications handbook. McGraw–Hill Inc., New York (1994) (revised edition)

[4] Dixon, R.C.: Spread Spectrum Systems with Commercial Applications. John Wiley & Sons, Chichester (1994)

[5] Don, R.: Principles of Spread Spectrum Communication System. Springer, New York (2005)

[6] Hoffman, M.A.: IEEE History Center website (2002), `http://www.ieee.org/organizations/history_center/lamarr.html`

[7] Robertson, R.C., Sheltry, J.F.: Multiple Tone Interference of Frequency Hopped Noncoherent MFSK Signals Transmitted Over Ricean Fading Channels. IEEE Transactions on Communications 44(7), 867–875 (1996)

[8] Katsoulis, G., Robertson, R.C.: Performance Bounds for Multiple Tone Interference of Frequency-hopped Noncoherent MFSK Systems. In: Proceedings IEEE Military Communications Conference, pp. 307–312 (1997)

[9] Pérez, J.J., Rodriguez, M.A., Felici, S.: Interference Excision Algorithm for Frequency Hopping Spread Spectrum Based on Undecimated Wavelet Packet Transform. Electronics Letters 38(16), 914–915 (2002)

[10] Baer, H.P.: Interference Effects of Hard Limiting in PN Spread-Spectrum Systems. IEEE Transactions on Communications 5, 1010–1017 (1992)

[11] Cheun, K., Stark, W.E.: Performance of FHSS Systems Employing Carrier Jitter against One-Dimensional Tone Jamming. IEEE Transaction on Communications 43(10), 2622–2629 (1995)

[12] Lance, E., Kaleh, G.K.: A diversity scheme for a phase-coherent frequency-hopping spread-spectrum system. IEEE Transactions on Communications 45(9), 1123–1129 (1997)

[13] Chen, Q., Sousa, E.S., Pasupathy, S.: Multicarrier CDMA with adaptive frequency hopping for mobile radio systems. IEEE Journal on Selected Areas in Communications 14(9), 1852–1858 (1996)

[14] Yang, L.L., Hanzo, L.: Slow Frequency-Hopping Multicarrier DS-CDMA for Transmission over Nakagami Multipath Fading Channels. IEEE Journal on Selected Areas in Communications 19(7), 1211–1221 (2001)

[15] Kim, H.J., Song, I., Lee, J., Kim, S.Y.: A truncated adaptive transmission scheme for hybrid multicarrier CDMA/FDM systems in forward link. IEEE Transaction on Vehicular Technology 54(3), 967–976 (2005a)

[16] Jia, T., Duel-Hallen, A.: Subchannel Allocation for Multicarrier CDMA with Adaptive Frequency Hopping and Decorrelating Detection. In: Military Communications Conference, MILCOM 2006, pp. 1–7 (2006)

[17] Kim, S.H., Kim, S.W.: Frequency-Hopped Multiple-Access Communications with Multicarrier On–Off Keying in Rayleigh Fading Channels. IEEE Transactions on Communications 48(10), 1692–1701 (2000)

[18] Sharma, S., Yadav, G., Chaturvedi, A.K.: Multicarrier on-off keying for fast frequency hopping multiple access systems in Rayleigh fading channels. IEEE Transactions on Wireless Communications 6(3), 769–774 (2007)

[19] Wang, J., Huang, H.: MC DS/SFH-CDMA systems for overlay systems. IEEE Transactions on Wireless Communications 1(3), 448–455 (2002)

[20] Elkashlan, M., Leung, C., Schober, R.: Performance analysis of channel aware frequency hopping. IEE Proceedings on Communication 153(6), 841–845 (2006)

[21] Hong, C.F., Yang, G.C.: Multicarrier FH codes for multicarrier FH-CDMA wireless systems. IEEE Transactions on Communications 48(10), 1626–1630 (2000)

[22] Nikjah, R., Beaulieu, N.C.: On Antijamming in General CDMA Systems–Part II: Antijamming Performance of Coded Multicarrier Frequency-Hopping Spread Spectrum Systems. IEEE Transactions on Wireless Communications 7(3), 888–897 (2008)

[23] Gallager, R.G.: Low-Density Parity-Check Code. MIT Press, Cambridge (1963)

[24] Mackay, D.J.: Good error-correcting codes based on very sparse matrices. IEEE Transactions on Information Theory 45(2), 399–431 (1999)

[25] Berrou, C., Glavieux, A., Thitimajshima, P.: Near Shannon limit error correcting coding and decoding: turbo-codes. In: IEEE ICC 1993, Geneva, Switzerland, pp. 1064–1070 (1993)

[26] Chung, S.Y., Richardson, T.J., Urbanke, R.L.: Analysis of Sum-Product Decoding of Low-Density Parity-Check Codes Using a Gaussian Approximation. IEEE Transactions on Information Theory 47(2), 657–670 (2001)

[27] Tanner, R.M.: A recursive approach to low complexity codes. IEEE Transactions on Information Theory IT-27, 533–547 (1981)

[28] Wu, X., You, X., Zhao, C.: A necessary and sufficient condition for determining the girth of quasi-cyclic LDPC codes. IEEE Transactions on Communications 56(6), 854–857 (2008)
[29] Hu, X.Y., Eleftheriou, E., Arnold, D.M.: Irregular progressive edge growth (PEG) Tanner graphs. In: Proceedings IEEE International. Symposium on Information Theory, Lausanne, Switzerland, p. 480 (2002)
[30] Hu, X.Y., Eleftheriou, E., Arnold, D.M.: Progressive edge-growth Tanner graphs. In: IEEE Global Telecommunications Conference, pp. 995–1001 (2001)
[31] Huang, C.M., Huang, J.F., Yang, C.C.: Construction of Quasi-Cyclic LDPC Codes from Quadratic Congruences. IEEE Communications Letters 12(4), 313–315 (2008)
[32] Abid, Y., Othman, S., Salleh, M.F.M., Farid, G.: Lower Computation and Storage Complexity of QC-LDPC Codes in Rayleigh Fading Channel. International Journal of Computer Theory and Engineering (IJCTE) 1(2), 115–118 (2009a) ISSN: 1793-821X (online version); 1793-8201 (print version)
[33] Abid, Y., Othman, S., Salleh, M.F.M., Farid, G.: A New Quasi-Cyclic Low Density Parity Check Codes. In: IEEE Symposium on Industrial Electronics and Applications (ISIEA 2009), Kuala Lumpur, Malaysia, October 4-6, pp. 329–342 (2009c)
[34] Abid, Y., Othman, S., Salleh, M.F.M., Farid, G.: Row Division Method to Generate QC-LDPC Codes. In: IEEE Proceedings on Fifth Advanced International Conference on Telecommunications, Venice/Mestre, Italy, May 24-28, pp. 183–187 (2009d)
[35] Abid, Y., Othman, S., Salleh, M.F.M., Sardar, A.: An Efficient Encoding-Decoding of Large Girth LDPC Codes Based on Quasi-Cyclic. Australian Journal of Basic and Applied Sciences 3(3), 1734–1739 (2009b) ISSN 1991-8178

Glossary of Terms and Acronyms

1 *Code Size*

The code size defines the dimensions of the parity check matrix $(M \times N)$. Occasionally the term code length is used referring to *n*. usually a code is defined employing its length and row column weight in the form (N, j, k). M can be deduced from the code parameters N, j and k. It has been determined that long codes execute better than shorter codes but need more hardware implementation resources.

2 *Code Weights and Rate*

The rate of a code R, is the number of information bits over the total number of bits transmitted. Higher row and column weights result in extra computation at each node because of many incoming messages. Nevertheless, if many nodes contribute in estimating the probability of a bit the node accomplishes a consensus faster. Higher rate indicate fewer redundancy bits. Namely, more information data is transmitted per block resulting in high throughput. Though, low redundancy implies less protection of bits and thus less decoding performance or higher error rate. Low rate codes have more redundancy with less throughput. More redundancy results in more decoding performance. But, low rate may have poor performance with a small number of connections.

LDPC codes with column-weight of two have their minimum distance increasing logarithmically with code size as compared to a linear increase for codes with column-weight of three or higher. Column weights higher than two are generally employed but carefully designed irregular codes could have better performance.

3 Code Structure

The structure of a code is ascertained by the pattern of connection between rows and columns. The connection pattern ascertains the complexity of the communication interconnect between check and variable processing nodes in an encoder and decoder hardware implementations. Random codes do not chase any predefined or known pattern in row-column connections. Structured codes on the hand have a known interconnection pattern.

4 Complementary Cumulative Distribution Function

Complementary Cumulative Distribution Function (CCDF) measurements provide important information in the field of design and manufacturing of system components used in spread spectrum modulation that result in large throughput for cellular services. The CCDF curve is generated for the normal distribution starting with the familiar normal probability density function (PDF).

5 Cycle

A cycle in a distance graph is formed by a path edges or vertices starting from a vertex V_x and ending at V_x. No more than two vertices forming the cycle belong to the same column.

6 Girth

The girth g, is the smallest cycle in the graph. A cycle of length of g in the graph corresponds to a cycle of length 2g in the matrix form.

7 Minimum Distance

The Hamming weight of a codeword is the number of 1's of the codeword. The Hamming distance between any two code words is the number of bits with which the words differ from each other, and the minimum distance of a code is the smallest Hamming distance between two code words. The larger the distance the better the performance of a code. Very long and large girth LDPC codes tend to have larger minimum distances. A better code could be ascertained by employing minimum distance as a measure.

8 Spectrogram

The Spectrogram is a valuable tool for tracking the source of the interfering signal. The power at a certain frequency (in dBm) is displayed along with the spectrogram. The spectrogram display provides signal's activity while exhibiting frequency, time, and power information. The spectrogram display is ideal for location of periodic interference.

Chapter 10
Decision Making under Synchronous Time Stress among Bilateral Decision Makers

Hideyasu Sasaki

Abstract. In this chapter, we discuss a theory on process for decision making under time stress which is common or synchronous constraint among two or bilateral decision makers. Its typical application is found in a system that analyzes the pricing mechanism of group-or-collaborative behaviors among multi-party using multi-agents. The problem on time stress is to evaluate cost of time or value of the entire duration of a certain process for decision making. We propose a formula to compute cost of time by introduction of opportunity cost to its evaluation. We also propose a formula on strategic points for decision making to minimize cost of time for a certain process for decision making under time stress as synchronous constraint. A strategic point is always located at the one-third of the entire or remaining duration for decision procedures, instead of a heuristic point of its half time. We have conducted a feasibility check on the proposed formulas on cost of time and strategic points in their applications to a case study.

Keywords: Decision making, Cost of time, Strategic point, Time constraint, Synchronous, Bilateral

1 Introduction

In this chapter, we discuss a theory on process for decision making under time stress which is common or *synchronous* constraint among two or *bilateral* decision makers.

Hideyasu Sasaki
Associate Professor of Computer Science, Department of Information Science and Engineering, Ritsumeikan University, 6-4-10 Wakakusa, Kusatsu, Shiga 525-0045 Japan (Correspondence Address);
Visiting Senior Researcher, Keio Research Institute at SFC, Keio University, 5322 Endo, Fujisawa, Kanagawa 252-0882 Japan
e-mail: hsasaki@uchicago.edu

N. Bessis, F. Xhafa (Eds.): Next Generation Data Technologies for CCI, SCI 352, pp. 251–269.
springerlink.com

The proposed theory for decision making under synchronous time constraint is applicable in a variety of fields like autonomous agents and multi-agent systems, in information search but also environments where collective intelligence can be harvested.

Its typical application is found in a system that analyzes the pricing mechanism of group-or-collaborative behaviors among multi-party using multi-agents. In general, economic analysis in transactions relies on given heuristics based on certain autonomy among participants however the proposed theory provides a rational basis on evaluation of cost in transactions as a clear solution to the pricing analysis using autonomous agents.

For giving the above basis, we introduce a generally-accepted concept of the evaluation of cost that relates to time constraint in the mathematical formulation of our theory.

1.1 Motivations

Web 2.0 provides a useful digital environment to support a variety of decision making [7]. The community of intelligent systems discusses process for decision making in a variety of implementations: information systems management [22], agent-based systems and evolutionary game approaches for e-commerce [33,2], advanced financial systems for derivatives transaction [5] and e-mediation systems [14].

Not a large number of researches have examined process for decision making under *time stress* or time constraint [20]. The impact of time constraint is however discussed in information search strategies [34] and real-time decision making by multi-agents [19], *et cetera*. We have also discussed the impact of time constraint in process for decision making under time constraint which is synchronous among bilateral decision makers in our previous studies [26,27,30,28,29].

Our purpose is to give a formula on the process for decision making under synchronous time constraint among bilateral decision makers. The problem on time constraint is to evaluate *cost of time* or value of the entire duration of certain process for decision making [24]. Its well-known concept of computation is *opportunity cost* which is value of the next-best alternative use of that time as generally-accepted in economics for evaluating the cost of time constraint [23]. We propose a formula to compute the cost of time by introduction of opportunity cost to its evaluation.

Opportunity cost allows us to discuss a *strategic point* for decision making, which is a point to minimize cost of time for a certain process for decision making under synchronous time constraint from not a heuristic but rational viewpoint. The subjects to time constraint rather than other factors often choose irrational strategies in their decision making [18].

Typical irrationality under time constraint is found in the *a priori* applications of a *heuristic point* or half time to process for decision making. *Half time* is the half entire duration of a certain process for decision making. That heuristic point of half time is popular and generally accepted in various empirical studies on human decision making. Farm employees work only half of their expected time-load [10].

People accept it as a certain psychological anchor on human beliefs [25]. Half time works in an economic model for making fallacy in decision making [6]. A number of product management systems explicitly or implicitly include the heuristic point in their constraint management mechanisms [31].

Instead of the heuristic point, we propose a formula on strategic points for decision making under synchronous time constraint among bilateral decision makers. The strategic points under synchronous time constraint are always located at the one-third entire duration and the one-third remaining duration of a certain process for decision making.

1.2 Definitions

We give brief definitions to following concepts which are introduced in this chapter, as below.

First, bilateral decision makers have only two options to leave from or to stay in their process for decision making [8].

Second, bilateral decision makers face two types of process for decision making or *games*, among which a *repeated game* consists of some number of repetitions of some base game but a *single stage game* is a non-repeated game [16].

When two respective decision makers select their strategy of collaboration, they stick to their current games at the final minute for avoiding escalation moves in their process for decision making. That process for decision making is considered as a non-repeatable or single game with *finite* or single stake for decision making. On the contrary, when the same decision makers select their strategy of competition, they have discretion to stay in or leave from their current process for decision making to other process at the final minute for avoiding escalation moves. That process for decision making is considered as a repeatable game with *infinite* or substitutable stake for decision making.

Third, value of *gain* in a certain process for decision making is subject to a certain hypothetical wage rate of possible work which is taken by decision makers instead of their current task [1]. Although other values may be used where appropriate, virtually from empirical studies, any measure of the cost of a worker's time, however, will be closely linked to a worker's wage. Wage is substitutable with price of individual items which are available in a certain process for decision making [3]. Gain takes the form of linear curve on its function [17]. Gain and cost discussed in this chapter are considered as equivalent to certain wage rate of possible work taken by decision makers instead of their current task which is committed to their decision making so that gain and cost curves are regarded to be linear.

Fourth, *free access to information* regarding any gain or price of individual items in a certain process for decision making is indispensable to computation of cost of time [3].

Cost of time in a repeated game is computed on the basis of opportunity cost, because a single stage game always takes unique values on prices of its individual items, however, a repeated game does not [23]. Fig. 1 of Page 254 describes its

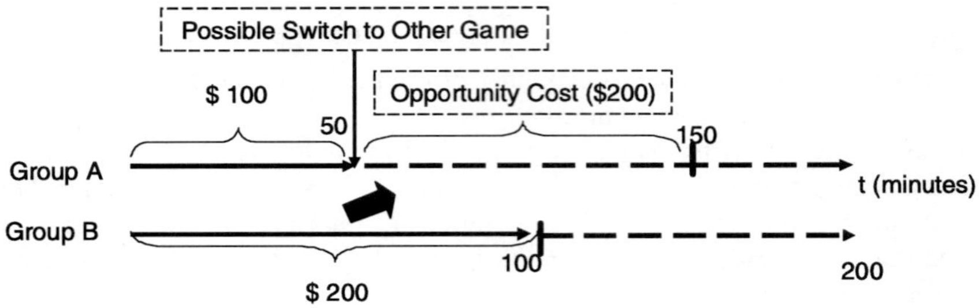

Fig. 1 The computation of opportunity cost.

basic idea on computation of opportunity cost in a repeated game by using prices of individual items in a single stage game.

Suppose that Group A gained $100 at 50 minutes during its entire 150 minutes in its repeated game. Group B gained $200 at 100 minutes during its entire 200 minutes in its single stage game. Thus, Group A dismissed its opportunity of the larger gain, here $200, so that its remaining time, here 100 minutes, was evaluated as equivalent to the same amount of monetary value, $200, as its opportunity cost.

Thus, opportunity cost of time for acquiring certain gain is equivalent to the additive of prices of individual items which are available during a certain process for decision making.

1.3 Contributions and Limitations

This chapter contributes to computation of cost of time for decision making and formulation of strategic points under synchronous time constraint, instead of the heuristic point. Especially, the proposed formulas are to minimize cost of time and to accelerate time-sensitive decision making which is indispensable to any solutions in collective computational intelligence.

We rely on following findings in behavioral sciences, especially cognitive science, as limitations to this chapter as below.

First, we introduce several findings in *transitional games* or process for decision making under time constraint. Decision makers under time constraint do not always behave rationally in transitional games as introduced in the transaction analysis (TA) of psychology, though several types of decision making are discussed in strategic games and coalitional games [4].

Contrast with the above strategic games and coalitional games, transitional games are different from those conventional games in following points. The decision makers do not always behave rationally, but behave more like real people who respond to the actions of their counter parties so that their selected strategies are sometimes not rational [21]. The property of transitional games is desirable to design a model of human behavior under time constraint, as found in an agent-based negotiation system [9].

Second, decision makers under time stress are more risk averse or conservative at lower risk levels [11].

Third, people under time stress face only two outcomes, *i.e.*, all or nothing, here, leave or stay, and can always account for certain equilibrium across any range of their games [12]. In transitional games, it is assumed that decision makers do not always rationally estimate either the best timing to close their current process for decision making or their possible gain and loss, although they are accessible to those kinds of information.

The above assumption allows decision makers to stay in their process for decision making even after they have selected strategy of competition before their final minute for avoiding escalation moves, *i.e.* actions to leave from current amicable situations to hostile stages [16]. Its typical case occurs, when their counter partners who selected strategy of collaboration and evaluated their final minute for avoiding escalation moves at the earlier stage than when those decision makers have assumed to close their process for decision making hostile. In that case, those decision makers are forced to stay in their current process for decision making.

On the other hand, those decision makers are forced to leave from their process for decision making even after they have selected strategy of collaboration before their final minute for avoiding escalation moves. Its typical case occurs, when their counter partners who selected strategy of competition and evaluated their final minute for avoiding escalation moves at the earlier stage than when those decision makers have assumed to close their process for decision making amicable. In that case, those decision makers are forced to leave from their current process for decision making.

Fourth, opportunity cost of time for acquiring certain gain is the equivalent to the additive of prices of individual items which are available in a certain process for decision making [23].

We also apply an assumption on *reinforcement learning* as a limitation to this chapter as below. Decision makers without prior information in their current process for decision making can be aware of information regarding the works of their environments, such as gain and cost of time through their first decision making, however they can only carry it out to their succeeding or future process for decision making [32].

1.4 Structure

The remaining of this chapter is organized as follows. In Section 2, we give a formula to evaluate cost of time, and deduct from it a formula on strategic points under synchronous time constraint, respectively. In Section 3, we conduct a feasibility check on those formulas in their applications to a case study. In Section 4, we give various analyses of the case study with contributions and limitations of the proposed formulas. In Section 5, we conclude this chapter with our future work.

2 Formulas

In this section, we give a formula to evaluate cost of time for decision making, and then deduct from it two formulas on strategic points to minimize cost of time for a certain process for decision making under synchronous time constraint.

2.1 Cost of Time

Cost of time is equivalent to the additive of prices of individual items which are available during a certain process for decision making.

We introduce two assumptions, as below:

Assumption 1. *Any price of an individual item is always given as certain static value in a single stage game;*

Assumption 2. *Cost of time is in proportion to ratio of elapsed time to the entire duration of a certain process for decision making.*

The assumptions 1 and 2 assure that prices of individual items in a single stage game and the entire duration of a certain process for decision making are given as static value, respectively.

We give a formula to evaluate its cost of time in a single stage game as $C^s_{(t)}$, as below:

Definition 1

$$C^s_{(t)} \equiv \frac{t}{\tau} \cdot \sum_{k=1}^{n} p_k. \tag{1}$$

s.t. t represents its elapsed time in a certain process for decision making, $t \in \mathscr{R}$; τ *is given as certain static value to the entire duration of a certain process for decision making,* $\tau \in \mathscr{R}$; p_k *represents each price of the k-th individual item from* 1 *to n in a single stage game,* $p \in \mathscr{R}$, $k, n \in \mathscr{N}$.

The definition 1 assures that cost of time for decision making is equivalent to gain or additive of the prices of individual items which are available in its single stage game. Its gain in a repeated game does not always have unique value in a variety of its next-best alternatives so that we apply opportunity cost to its computation in order to identify its unique value.

We introduce two more assumptions, as below:

Assumption 3. *Any repeated game is to be expected to spend at least the same duration of its previous game in its next game;*

Assumption 4. *Any function on cost of time always takes certain single equivalent value in its transition from a repeated game to a single stage game.*

Suppose that a repeated game transits into a single stage game at half time or the half entire duration; Its new single stage game takes that same half time again so that both its repeated game and single stage game face certain single equivalent value at its half time, $t = \frac{\tau}{2}$.

Those assumptions 1 to 4 allow us to deduct a lemma on its gain in a repeated game which is to be certain static value, P^r.

Lemma 1

$$P^r = \frac{1}{2}\sum_{k=1}^{n} p_k. \tag{2}$$

Proof

$$\frac{\frac{\tau}{2}}{\tau}\cdot\sum_{k=1}^{n} p_k = \lfloor\frac{\tau}{\frac{\tau}{2}}\rfloor\cdot\frac{\frac{\tau}{2}}{\tau}\cdot P^r. \quad \because C^s_{(\frac{\tau}{2})} = C^r_{(\frac{\tau}{2})}.$$

s.t. $\lfloor\cdot\rfloor$ is floor function; $\lfloor\frac{\tau}{t}\rfloor$ represents the number of possible times of repetitions of some base game to acquire the next-best alternatives as opportunity cost. □

We give a formula to evaluate cost of time in a repeated game using the lemma 1 as $C^r_{(t)}$, as below:

Definition 2

$$C^r_{(t)} \equiv \lfloor\frac{\tau}{t}\rfloor\cdot\frac{t}{\tau}\cdot\frac{1}{2}\sum_{k=1}^{n} p_k. \tag{3}$$

2.2 Strategic Points

Here, we give a formula on strategic points under synchronous time constraint.

Suppose that a repeated game transits into a single stage game at the heuristic point of half time, discussed as above. Bilateral decision makers in this game *a priori* accept the heuristic point of half time as their point for decision making. In this case, we give a function on the ratio of gain to cost of time for decision making, $\frac{P}{C_{(t)}}$, as below:

Definition 3

$$\frac{P}{C_{(t)}} = \begin{cases} \frac{P^r}{C^r_{(t)}} = \frac{2\tau}{\lfloor\frac{\tau}{t}\rfloor\cdot t} \; if \quad 0 \le \frac{t}{\tau} \le \frac{1}{2}, \\ \frac{P^s}{C^s_{(t)}} = \frac{\tau}{t} \quad else\ if \;\; \frac{1}{2} \le \frac{t}{\tau} \le 1. \end{cases} \tag{4}$$

Fig. 2 describes that the function always takes its largest value 3.00 at the one-third entire duration of its process for decision making before the heuristic point of half time. Its peak is the strategic points to minimize its cost of time for decision making.

We prove this finding as a theorem, as below:

Theorem 1

$$\arg\max_{[0\le t\le\frac{\tau}{2}]}\frac{P}{C_{(t)}} = \lim_{t\to\frac{\tau}{3}+0}\frac{P^r}{C^r_{(t)}} = 3.00. \tag{5}$$

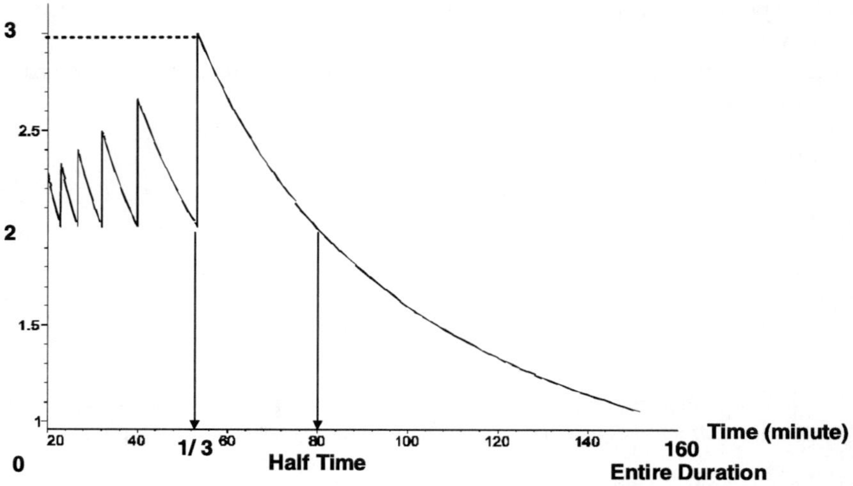

Fig. 2 The function on the ratio of gain to cost of time, $\frac{P}{C_{(t)}}$: A repeated game transits into a single stage game at half time.

Proof

$$\frac{P^r}{C^r_{(\frac{\tau}{3}+0)}} = 2\cdot\frac{\frac{\tau}{\frac{\tau}{3}}}{\lfloor\frac{\tau}{(\frac{\tau}{3}+0)}\rfloor} = 3 > \frac{P^r}{C^r_{(\frac{\tau}{4}+0)}} = 2.67 > \frac{P^r}{C^r_{(\frac{\tau}{5}+0)}} = 2.5 > \cdots .$$

$$\because \lim_{t\to\frac{\tau}{3}+0}\lfloor\frac{\tau}{t}\rfloor = 2;\ \lim_{t\to\frac{\tau}{4}+0}\lfloor\frac{\tau}{t}\rfloor = 3;\ \lim_{t\to\frac{\tau}{5}+0}\lfloor\frac{\tau}{t}\rfloor = 4;\cdots . \qquad \square$$

Therefore, a strategic point under synchronous time constraint is always located at the one-third entire duration of a certain process for decision making before the heuristic point of half time.

After half time, decision makers have two options: leave from or stay in their current process for decision making. The former option allows decision makers in their new process for decision making to scale down by half on the duration of their current process for decision making and to apply a function on the ratio of gain to cost of time for decision making as described in the equation (4).

In this option, its function takes the following equation:

Definition 4

$$\frac{P}{C_{(t)}} = \begin{cases} \frac{P^r}{C^r_{(t)}} = \frac{\tau}{\lfloor\frac{\tau}{2t-\tau}\rfloor\cdot(t-\frac{\tau}{2})} & \text{if } \frac{1}{2} \le \frac{t}{\tau} \le \frac{3}{4}, \\ \frac{P^s}{C^s_{(t)}} = \frac{1}{\frac{2t}{\tau}-1} & \text{else if } \frac{3}{4} \le \frac{t}{\tau} \le 1. \end{cases} \tag{6}$$

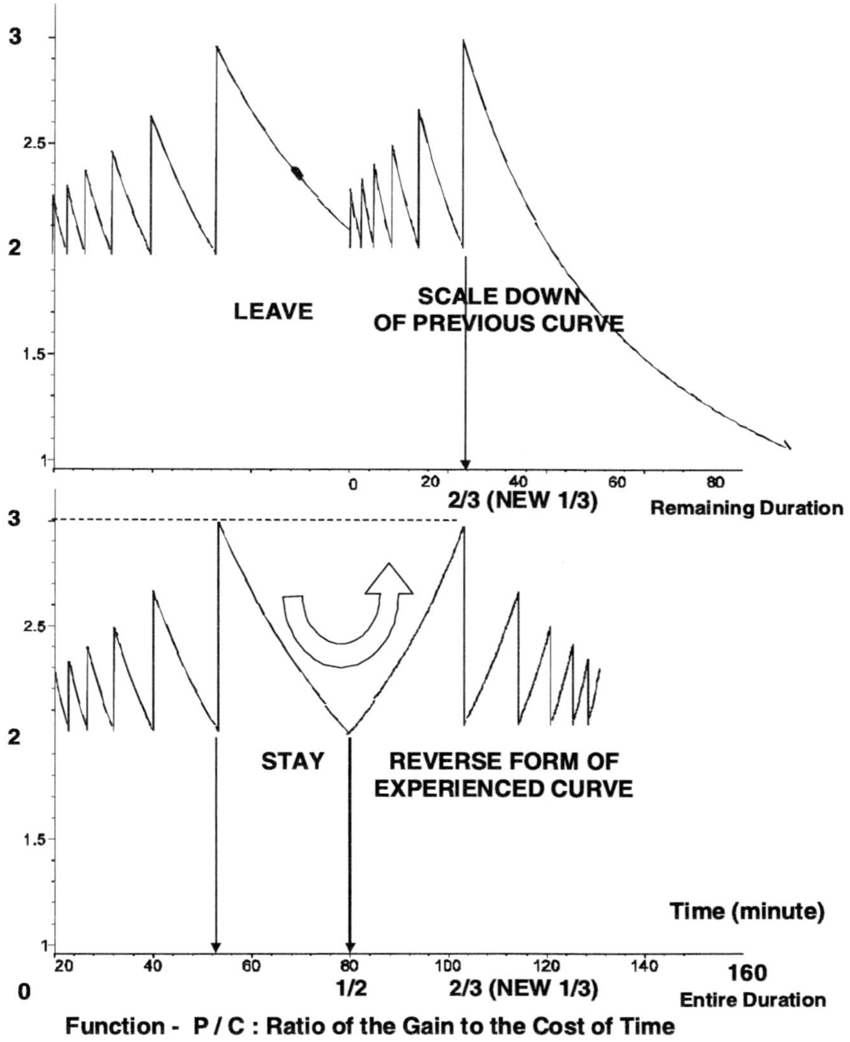

Fig. 3 The strategic points after leave and stay.

The upper of Fig. 3 of Page 259 describes that the function in the second repeated game always takes its largest value 3.00 at the two-thirds entire duration or the one-third remaining duration of its current process for decision making. Its peak is the strategic point to minimize cost of time.

On the contrary, decision makers who select strategy to stay in its current process for decision making face two types of functions: One function is the same with the above equation (6); The other function takes reverse or backward move from a point of half time to a point of the one-third entire duration of its current process for decision making. In this option, its function takes the following equation:

Definition 5

$$\frac{P^r}{C^r_{(t)}} = \frac{2\tau}{\lfloor\frac{\tau}{\tau-t}\rfloor\cdot(\tau-t)} \quad (\frac{1}{2} \le \frac{t}{\tau} \le 1). \tag{7}$$

The lower of Fig. 3 of Page 259 describes that the function in the second repeated game always takes the largest value 3.00 at the two-thirds entire duration or the one-third remaining duration of its current process for decision making. Its peak is the strategic point to minimize cost of time.

We prove this finding as a theorem, as below:

Theorem 2

$$\arg\max_{[\frac{1}{2}\le t\le 1]} \frac{P}{C_(t)} = \lim_{t\to\frac{2\tau}{3}+0} \frac{P^r}{C^r_{(t)}} = \lim_{t\to\frac{2\tau}{3}-0} \frac{P^r}{C^r_{(t)}} = 3.00. \tag{8}$$

Proof

$$\frac{P^r}{C^r_{(\frac{2\tau}{3}+0)}} = \frac{\tau}{\lfloor\frac{\tau}{2(\frac{2\tau}{3}+0)-\tau}\rfloor\cdot(\frac{2\tau}{3}-\frac{\tau}{2})} = 3 >$$

$$\frac{P^r}{C^r_{(\frac{5\tau}{8}+0)}} = \frac{8}{3} = 2.67 > \cdots >$$

$$\frac{P^r}{C^r_{(\frac{3\tau}{5}+0)}} = \frac{P^s}{C^s_{(\frac{3\tau}{5}+0)}} = \frac{5}{2} = 2.5 > \cdots >$$

$$\frac{P^r}{C^r_{(\frac{3\tau}{4}+0)}} = \frac{P^s}{C^s_{(\frac{3\tau}{4}+0)}} = 2;$$

$$\frac{P^r}{C^r_{(\frac{2\tau}{3}-0)}} = \frac{2\tau}{\lfloor\frac{\tau}{\tau-(\frac{2\tau}{3}-0)}\rfloor\cdot(\tau-\frac{2\tau}{3})} = 3 >$$

$$\frac{P^r}{C^r_{(\frac{5\tau}{8}-0)}} = \frac{8}{3} = 2.67 > \cdots >$$

$$\frac{P^r}{C^r_{(\frac{3\tau}{5}-0)}} = \frac{P^s}{C^s_{(\frac{3\tau}{5}-0)}} = \frac{5}{2} = 2.5$$

$$\frac{P^r}{C^r_{(\frac{3\tau}{4}-0)}} = \frac{P^s}{C^s_{(\frac{3\tau}{4}-0)}} = 2.$$

$$\because \lim_{t\to\frac{2\tau}{3}+0} \lfloor\frac{\tau}{2t-\tau}\rfloor = \lim_{t\to\frac{2\tau}{3}-0} \lfloor\frac{\tau}{\tau-t}\rfloor = 2. \qquad \square$$

Therefore, another strategic point under synchronous time constraint is always located at the one-third remaining duration of a certain process for decision making after the first heuristic point of half time and before the other heuristic point of half time in both a single stage game and a repeated game.

Those theorems 1 and 2 assure that respective strategic points under synchronous time constraint are always located before the heuristic points of the half entire duration and the half remaining duration of a certain process for decision making.

In the next section, we apply the above formulas on cost of time for decision making and strategic points for decision making to a case study.

3 Case Study

In this section, we apply the formulas on cost of time for decision making and strategic points to the following case study for their feasibility check.

3.1 Case

The Multisearch Software Case is an introductory practice for decision making in American business schools [13]. That case is business alliance on certain planning for software development between bilateral decision makers: a developer and a company.

3.2 Participant Record

We have three different groups for their trial sessions. All six decision makers of A and D, B and C, and E and F constitute two separate groups, respectively:

1. Those decision makers are divided in two groups, a single party of one-to-one players and a multi-party of two-to-two players; And,
2. Respective groups negotiate over a single case once, respectively.

3.3 Entire Duration of a Certain Process for Decision Making

The developer accepts 3 to 6 months and the company does 3 to 4 months for the release of their final product. The individual duration of its current process for decision making is scaled down in practice to 120 minutes for Group 1 and 160 minutes for Group 2, respectively.

3.4 Elapsed Time Record

The processes for decision making took 50 minutes in Group 1 of A and B and 95 minutes in Group 2 of C-D and E-F.

Table 1 The results on the case study

Groups	Group 1	Group 2
Company Side	A	C & D
Developer Side	B	E & F
Prices of Individual Items (*given*)	Single Party	Multi-party
1. Royalty to Developer	10%, 4 years	8%, 5 years
2. Advance to Developer	\$250,000-	\$150,000-
3. Promotion for Sales	\$1,000,000-	\$1,100,000-
4. The Additional to Developer	N/A	\$1,000,000- (5 years)
5. Commitment by Developer to Company: Abandon of Side Work	\$300,000- (2 years)	\$750,000- (5 years)
6. Developer's Independent Gain	\$150,000- /year after the 3rd year	N/A
Additive	P^{s1} = \$2,400,000-	P^{s2} = \$3,700,000-
Entire Duration (*given*)	$\tau^1 = 120$ minutes	$\tau^2 = 160$ minutes
Elapsed Time	$t^1 = 50$ minutes	$t^2 = 95$ minutes
Selection of Strategy	Stay	Stay

3.5 Prices of Individual Items

Table 1 describes the prices of individual items in detail which are given as static values. Annual revenue from sales of software product is estimated as \$1,000,000 among those groups. The monetary value of stock option and pension plan are estimated as \$300,000 in total.

4 Discussions

In this section, we discuss the applications of the proposed formulas on cost of time for decision making and strategic points with their contributions and limitations.

The primary issue is whether the proposed formulas are to suggest any improvement in a certain process for decision making of the respective groups regarding the results of the case study.

4.1 Results

Group 1 passed its first strategic point at 40 minutes and closed its process for decision making at 50 minutes, which was before the heuristic point of half time, 60 minutes.

Group 1 faced the equation (4) as its function on the ratio of gain to cost of time, as described in Fig. 4 of Page 263. The ratio of gain to cost of time at 50 minutes

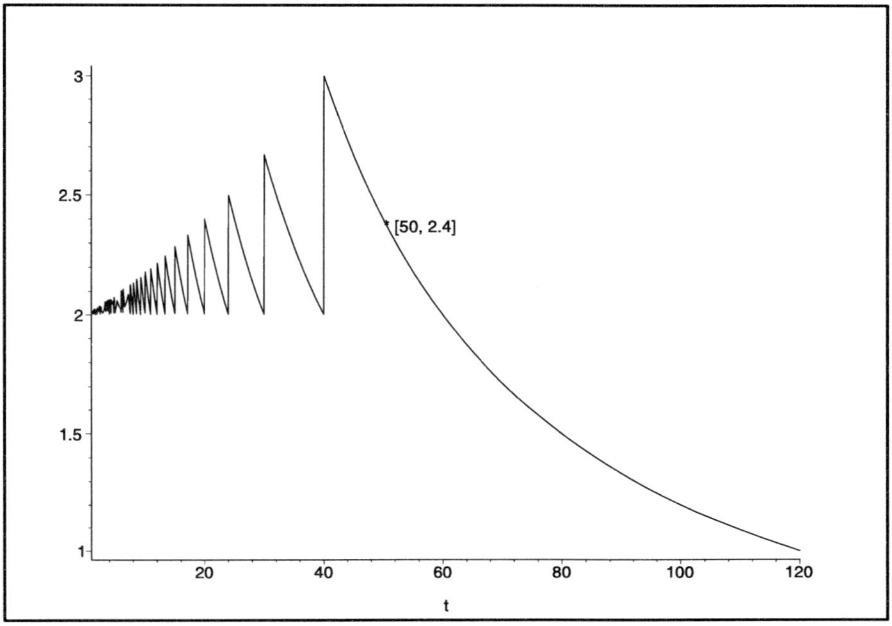

Fig. 4 Group 1's ratio of gain to cost of time (X: t (minutes); Y: $\frac{P^1}{C^1_{(t)}}$): Equation (4).

was 2.4 ($\because \frac{P^1}{C^1_{(50)}} = \frac{2\cdot 120}{\lfloor \frac{120}{50} \rfloor \cdot 50}$). That ratio of 2.4 was smaller than the ratio of 3.0 at its strategic point for decision making and even another ratio of 2.67 at its previous second-best peak point.

Group 2 passed its first strategic point at 53.3 minutes and closed its process for decision making at 95 minutes, which was located before its second heuristic point of half time of its remaining duration at 120 minutes, but quite close to another strategic point at 107 minutes.

Group 2 faced the equations (4), (5) and (6) as a series of three types of its functions on the ratio of gain to cost of time, as described in Fig. 5. The ratios of gain to cost of time at 95 minutes were 1.68 ($\because \frac{160}{95}$) on the equation (4), 2.13 ($\because \frac{160}{\lfloor \frac{160}{2\cdot 95-160} \rfloor (95-\frac{160}{2})}$) on the equation (5) and 2.46 ($\because \frac{2\cdot 160}{\lfloor \frac{160}{160-95} \rfloor (160-95)}$) on the equation (6), respectively. The ratio of 2.46 was smaller than the ratio of 3.0 at its strategic point for decision making and even another ratio of 2.67 at its previous second-best peak point.

Groups 1 and 2 faced their respective differentiated functions on the ratio of gain to cost of time before half time, respectively, as described in Fig. 6 of Page 265. Any amplitude of its differentiated function on Group 1's ratio of gain to cost of time was the larger than any amplitude on Group 2's, because Group 1's entire duration for its process for decision making was given with the smaller value, initially.

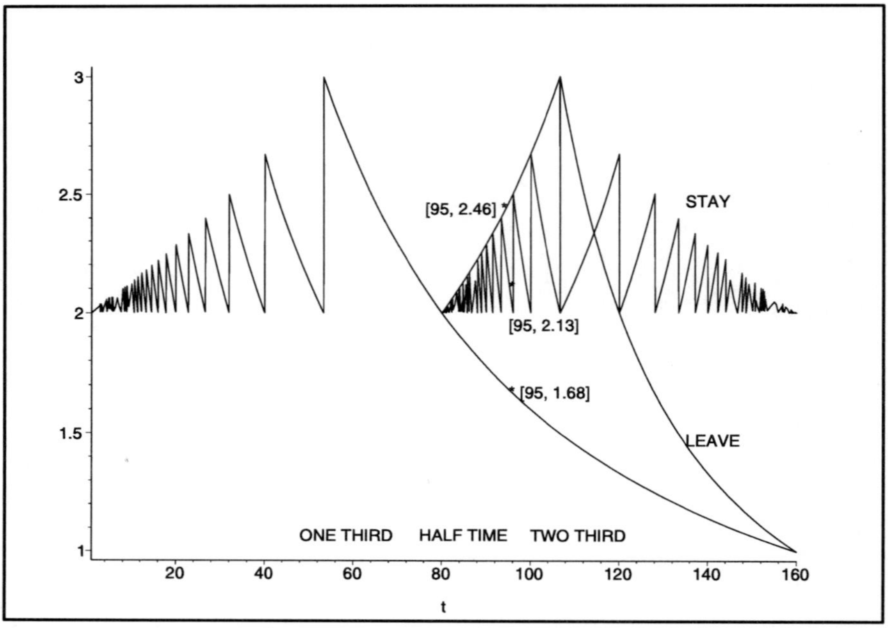

Fig. 5 Group 2's ratio of gain to cost of time (X: t (minutes); Y: $\frac{P^2}{C^2_{(t)}}$): Equations (4), (5) and (6).

Groups 1 and 2 faced their respective functions on cost of time, as described in Fig. 7 of Page 266. Here, Group 1's gain or its additive of prices of individual items in its single stage game was given as 1 so that Group 2's gain in its single stage game was to be 1.54 ($\because \frac{3{,}700{,}000}{2{,}400{,}000}$). Its initial value on Group 2's cost of time was the larger than Group 1's, because Group 2's gain in its single stage game was given with the larger amount, initially.

4.2 Elements

We discuss elements on selection of strategic points under synchronous time constraint.

First, the entire duration of a certain process for decision making is to be long enough. Its longer duration allows decision makers to face the more moderate slope of the function on the ratio of gain to cost of time. Group 2 allocated the longer duration for its process for decision making rather than Group 1 did so that Group 2 enjoyed its moderate amplitude of the slope of its ratio of gain to cost of time.

Second, gain or additive of prices of individual items is to be balanced in proportion to the given entire duration of a certain process for decision making. The larger gain to the longer duration allows decision makers to face the more appropriate initial value in the function on cost of time. Group 2 gave its larger amount of gain

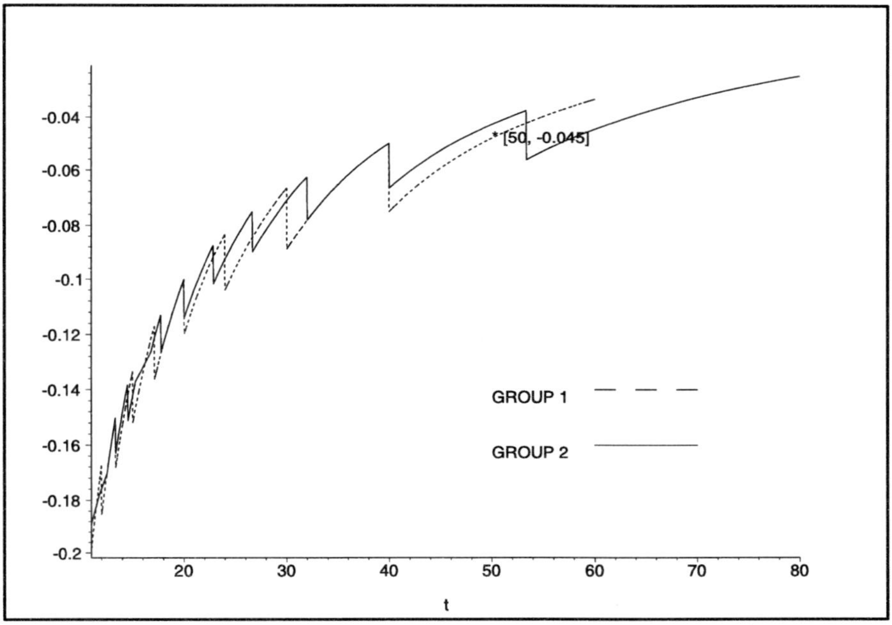

Fig. 6 Groups 1 and 2's differentiated ratio of gain to cost of time before half time (X: t (minutes); Y: $\frac{d\frac{P}{C_{(t)}}}{dt}$): Differentiation of the equation (4).

to its longer duration ($2.31 \times 10^4 = \frac{\$3{,}700{,}000}{160\ minutes}$) rather than Group 1 did ($2.0 \times 10^4 = \frac{\$2{,}400{,}000}{120\ minutes}$) so that Group 2 enjoyed its appropriate proportion of gain to cost of time.

Finally, the timing of decision making is to be proper around strategic points. The better timing allows decision makers to face a variety of options from the first strategic point to the second strategic point, if possible, even after the first heuristic point of half time. Group 2 stayed in after its first half time and closed its process for decision making at a point which was close to the one-third of its remaining duration, which was its second strategic point for decision making.

The contributions of the proposed formulas in this chapter are found as below.

First, the concept of cost of time or monetary value of the entire duration of a certain process for decision making allows us to locate strategic points or better timing for decision making to minimize cost of time for decision making in a transparent way of evaluation. The proposed formula is to be applicable to black-box process for decision making in human-machine intelligent systems, *e.g.*, design of time-out threshold in search engines.

Second, strategic points are to accelerate time-sensitive decision making, instead of the heuristic point of half time. Strategic points of the one-third of the entire or

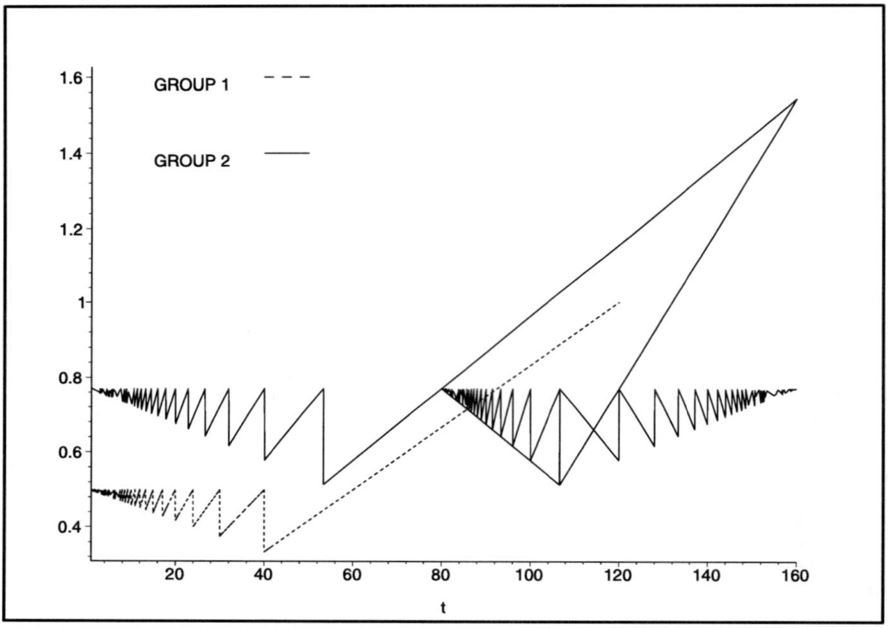

Fig. 7 Groups 1 and 2's cost of time (X: t (minutes); Y: $C_{(t)}$): P^{1s} is given as 1.

remaining duration for decision procedures allows those in decision making to take advantages of time resources more effectively.

A remaining limitation to our study in this chapter is that the proposed formulas still accept given initial values on prices of individual items. Those given values should be replaced with a certain formula regarding cost of time for decision making in our future work.

5 Conclusions and Future Work

In this chapter, we have proposed a formula to compute cost of time by introduction of opportunity cost to its evaluation under synchronous time constraint.

We have also proposed a formula on strategic points for decision making to minimize cost of time under synchronous time constraint among bilateral decision makers. The proposed strategic points are always located at the one-third of the entire and remaining duration for decision making, instead of the heuristic point of half time.

We have referred to a variety of concepts which are accepted in the state of the arts and the literatures: opportunity cost, transitional games and reinforcement learning.

We have conducted a feasibility check on our proposed formulas in their applications to the case study.

The proposed formulas contribute to transparent design of decision making and advanced management of time resources which are to be implemented in human-machine intelligent systems. The proposed strategic points are to accelerate time-sensitive decision making which is indispensable to any solutions in collective computational intelligence.

The proposed theory harvests a variety of fields which relate to collective intelligence. Among them, its promising field is pricing mechanism of group-or-collaborative behaviors among multi-party using multi-agents.

With the proposed concept of opportunity cost that is generally-accepted in economics as a concept of the evaluation of cost of time constraint, the proposed theory provides a rational basis on evaluation of cost in transactions as a clear solution to the pricing analysis using autonomous agents instead of given heuristics based on certain autonomy among participants.

In our future work, we would implement our proposed formulas on process for decision making under time constraint which is asynchronous among multilateral decision makers in agent-based intelligent systems.

Acknowledgements. The Author of this chapter is supported financially in part by the Grant-in-Aid for Scientific Research ("KAKENHI") of the Japanese Government: No. 21,700,281 (FY 2009-2011), No. 22,240,023 (FY 2010-2012) and by the Moritani Scholarship Foundation (FY 2010-2013). The Author would like to thank an anonymous reviewer who gives insight comments on this chapter.

References

1. Ashenfelter, O., Greenstone, M.: Using mandated speed limits to measure the value of a statistical life. Journal of Political Economy 112(2, part 2), 226–267 (2004)
2. Ba, S., Whinston, A.B., Zhang, H.: The dynamics of the electronic market: An evolutionary game approach. Information Systems Frontiers 2(1), 31–40 (2000)
3. Baird, D.G., Gertner, R.H., Picker, R.C.: Game theory and the law, 3rd edn. Harvard University Press, Cambridge (1998)
4. Berne, E.: Games people play: The psychology of human relationships. Grove, New York (1964)
5. Bichler, M.: Trading financial derivatives on the Web: An approach towards automating negotiations on OTCmarkets. Information Systems Frontiers 1(4), 401–414 (2000)
6. Brunnermeier, M.K., Papakonstantinou, F., Parker, J.A.: An economic model of the planning fallacy: NBER working paper (No. 14228). National Bureau of Economic Research, Cambridge (2008)
7. Bull, G., Thompson, A., Searson, M., Garofalo, J., Park, J., Young, C., Lee, J.: Connecting informal and formal learning: Experiences in the age of participatory media. Contemporary Issues in Technology and Teacher Education 8(2), 100–107 (2008)
8. Chen, J.H., Chao, K.M., Godwin, N., Soo, V.W.: Combining cooperative and non-cooperative automated negotiations. Information Systems Frontiers 7(4/5), 391–404 (2005)

9. Cheng, Z., Capretz, M.A.M., Osano, M.: A model for negotiation among agents based on the transaction analysis theory. In: Proceedings of the Second International Symposium on Autonomous Decentralized Systems, pp. 427–433. IEEE Computer Society, Silver Spring (1995)
10. Deininger, K.W.: Cooperatives and the breakup of large mechanized farms theoretical perspectives and empirical evidence: World Bank discussion papers (No. 218). The International Bank for Reconstruction and Development/The World Bank, Washington, DC (1993)
11. Dror, I.E., Busemeyer, J.R., Basola, B.: Decision making under time pressure: An independent test of sequential sampling models. Memory & Cognition 27(4), 713–725 (1999)
12. Goeree, J.K., Holt, C.A., Palfrey, T.R.: Risk averse behavior in asymmetric matching pennies games. Games and Economic Behavior 45(1), 97–113 (2003)
13. Gould, E.C., Easter, M.: The multisearch software case. Program on Negotiation, Harvard Law School (1998)
14. Greenstein, S.: The commercialization of information infrastructure as technological mediation: The Internet access market. Information Systems Frontiers 1(4), 329–348 (2000)
15. Kriesberg, L.: Timing and the initiation of de-escalation moves. In: Breslin, J.W., Rubin, J.Z. (eds.) Negotiation Theory and Practice, pp. 223–231. Harvard University Press, Cambridge (1991)
16. Kriesberg, L.: Timing and the initiation of de-escalation moves. Negotiation Journal 3(4), 375–384 (2007)
17. Lee, M.S., Ratchford, B.T., Talukdar, D.: The impact of the Internet on information search for automobiles. Review of Marketing Science 1(2, Working Paper 1), 1–47 (2002)
18. Lehner, P., Seyed-Solorforough, M.-M., O'Connor, M.F., Sak, S., Mullin, T.: Cognitive biases and time stress in team decision making. IEEE Transactions on Systems, Man and Cybernetics, Part A: Syst. and Humans 27(5), 698–703 (1997)
19. Noh, S., Gmytrasiewicz, P.J.: Flexible multi-agent decision making under time pressure. IEEE Transactions on Systems, Man and Cybernetics, Part A: Syst. and Humans 35(5), 697–707 (2005)
20. O'Grady, L.A., Witteman, H., Wathen, C.N.: The experiential health information processing model: Supporting collaborative Web-based patient education. BMC Medical Informatics and Decision Making 8(58), 1–22 (2008)
21. Olson, D.L.: Rationality in information systems support to decision making. Information Systems Frontiers 3(2), 239–248 (2001)
22. Osei-Bryson, K.M., Ngwenyama, O.: Decision models for information systems management. Information Systems Frontiers 10(3), 277–279 (2008)
23. Payne, J.W., Bettman, J.R., Luce, M.F.: When time is money: Decision behavior under opportunity cost time pressure. Organizational, Behavior and Human Decision Process 66, 131–152 (1996)
24. Russell, S., Wefald, E.: On optimal game-tree search using rational meta-reasoning. In: Proceedings of 11th Joint Conference on Artificial Intelligence, pp. 334–340. Morgan Kaufmann, San Francisco (1989)
25. Sanna, L.J., Parks, C.D., Chang, E.C., Carter, S.E.: The hourglass is half full or half empty: Temporal framing and the group planning fallacy. Group Dynamics: Theory, Research, and Practice 9(3), 173–188 (2005)

26. Sasaki, H.: An evaluation method for strategic decision making on group collaboration under temporary constraints. In: Proceedings of the First IEEE International Symposium on Advanced Management of Information for Globalized Enterprises (AMIGE 2008), vol. (10), pp. 1–5. IEEE Computer Society Press, Silver Spring (2008)
27. Sasaki, H.: Strategic decision making on group collaboration under temporary constraints. In: Proceedings of the Fifth IEEE/ACM International Conference on Soft Computing as Transdisciplinary Science and Technology (CSTST 2008), pp. 343–349. ACM Press, New York (2008)
28. Sasaki, H.: Decision making under time constraint: From heuristics to strategy. In: Proceedings of the IEEE International Conference on Systems, Man, and Cybernetics (SMC 2010), pp. 2935–2940. IEEE Computer Society Press, Silver Spring (2010)
29. Sasaki, H.: A Study on strategic points for decision making under time constraint. In: Proceedings of the First International Workshop on Emerging Data Technologies for Collective Intelligence (EDTCI 2010) in conjunction with the First International Conference on P2P, Parallel, Grid, Cloud and Internet Computing (3PGCIC 2010), pp. 308–313. IEEE Computer Society Press, Silver Spring (2010)
30. Sasaki, H.: A computing theory for collaborative and transparent decision making under time constraint. Information Systems Frontiers (in press), doi:10.1007/s10796-009-9189-5
31. Schragenheim, E.: A systematic approach to common and expected uncertainty. Journal of System Improvement 1(2), 1–8 (1997)
32. Sutton, R.S., Barto, A.G.: Reinforcement learning: An introduction. MIT Press, Cambridge (1998)
33. Szirbik, N.: A negotiation enabling agent based infrastructure: Composition and behavior. Information Systems Frontiers 4(1), 85–99 (2002)
34. Weenig, M.W.H., Maarleveld, M.: The impact of time constraint on information search strategies in complex choice tasks. Journal of Economic Psychology 23(6), 689–702 (2002)

Glossary of Terms and Acronyms

- Cost of time: A value of the entire duration of a certain process for decision making.

- Opportunity cost: A value of the next-best alternative use of that time as generally-accepted in economics as a concept of the evaluation of cost of time constraint.

- Repeated game: A game consisting of some number of repetitions of some base game with infinite or substitutable stake for decision making.

- Single stage game: A non-repeated game with finite or single stake for decision making.

- Strategic point: A point minimizing cost of time for a certain process for decision making under synchronous time constraint from not a heuristic but rational viewpoint.

Chapter 11
Augmenting Human Intelligence in Goal Oriented Tasks

Ana Cristina Bicharra Garcia

Abstract. *Tell me and I forget, teach me and I remember, involve me and I learn. (Benjamin Franklin).* The world is an increasingly complex with problems that require swift resolution. Although knowledge is widely available, be it stored in companies databases or spread over the Internet, humans have intrinsic limitations for handling very large volumes of information or keeping track of frequent updates in a constantly changing world. Moreover, human reasoning is highly intuitive and potentially biased due to time pressure and excess of condence. Computer systems that manage knowledge by thoroughly exploring the context and range of alternatives may improve human decision-making by making people aware of possible misconceptions and biases. Computer systems are also limited in their potential usage due to the frame problem. Systems are not aware of their ignorance, thus they cannot substitute human intelligence; however, they may provide a useful complement. The objective of this chapter is to present the AGUIA model for amplifying human intelligence, utilizing agents technology for task-oriented contexts. AGUIA uses domain ontology and task scripts for handling formal and semiformal knowledge bases, thereby helping to systematically (1) explore the range of alternatives; (2) interpret the problem and the context; and (3) maintain awareness of the problem. As for humans, knowledge is a fundamental resource for AGUIA performance. AGUIAs knowledge base remains in the background and keeps updating its content during interaction with humans, either through identied individuals or through anonymous mass contribution. The feasibility and benets of AGUIA were demonstrated in many different elds, such as engineering design, fault diagnosis, accident investigation and online interaction with the government. The experiments considered a set of criteria including: product cost, number of explored alternatives, users problem understanding and users awareness of problem context changes. Results indicate that AGUIA can actually improve human problem solving capacity in many different areas.

Ana Cristina Bicharra Garcia
Instituto de Computação, Universidade Federal Fluminense - Niteroi - Rio de Janeiro, Brasil
e-mail: bicharra@ic.uff.br

N. Bessis, F. Xhafa (Eds.): Next Generation Data Technologies for CCI, SCI 352, pp. 271–299.
springerlink.com

1 Introduction

Intelligence works as a mechanism for optimizing human interaction with the environment, by allowing man to adapt to it, change it, or leave it for a better suitable context. Analytical, creative and practical abilities work together so that individuals may reach their goals within their environment. However, the lack of time and of a good problem understanding may lead individuals to reasoning bias. It is not unusual for inexperience, imperfect understanding, and overconfidence in familiar options to result in wrong conclusions, and consequently in unwise decisions [42].

Carrying out goal oriented tasks involves analyzing and choosing alternatives within a range of possibilities considering the available information and time constraints. This view of bounded rationality [38] emphasizes the pragmatic limitations of human decision-making process. Alternatives are generated and analyzed in light of knowledge about a certain domain and task.

Articial intelligence, since its formalization in the mid-1940s, has aimed to create devices that reproduce intelligent human behavior. The idea is to improve collective capacity by adding softwares that can substitute man in certain activities. This view, which placed articial intelligence in competition with man, created high expectations for these systems performance, as well as fears that another industrial revolution could actually render man obsolete. The 1980s revealed the limitations of these so-called intelligent systems, which did not yet show signs of becoming smarter than their creators. However, the area's technological advances can be applied to amplify human intelligence, working in partnership with people to rationally reach superior solutions to increasingly complex problems.

In order to function, intelligent systems need knowledge about a domain. Knowledge can be found in books, norms and standards, but also scattered in technical reports or even in the memories of individuals experiences. Since automatic knowledge acquisition methods based on these sources has not been successful, the effort is now on overcoming this limitation by using collective human knowledge. This effort can be put into two categories: 1) explicit knowledge acquisition, for which trained individuals are hired to maintain and generate knowledge[17, 21, 26], and implicit knowledge acquisition, for which users provide knowledge as a subproduct of another task they want to perform [45, 39, 4]. The challenge of acquiring knowledge from individuals involves aligning their interests, especially with regard to time allocation, with those of the society, which would benefit overall from an extensive and updated knowledge base. One approach to this challenge is to focus on the psychological aspect of the interaction [45], designing an incentive mechanism that encourages individuals to make a social contribution by satisfying their immediate interests. An efficient knowledge acquisition tool should have the following characteristics: ease of use with minimum training required; compatibility with the proposed incentives; a single mechanism applicable to all users; and virtually no extra effort requirements.

This research addresses the human challenge to produce better solutions to increasingly complex problems. Since it is a very broad problem, this research has focused on goal oriented tasks, considering following assumptions:

- the more knowledge one has, the more well-grounded his decision will be;
- the more a problem is understood, the greater the chances of properly solving it;
- the higher the number of alternative solutions considered for a problem, the greater the chances of finding a superior solution;
- given a set of options, the best one can be defined based on a set of known criteria;
- knowledge evolves, and therefore must be expanded constantly.

The objective of this chapter is to introduce the AGUIA model [13] for amplifying human intelligence, based on problem solving through cooperation between humans and computers. The AGUIA utilizes domain ontology and task scripts for handling formal or semiformal knowledge bases, thereby helping (1) to explore the range of alternatives; (2) to interpret the problem and the context; and (3) to maintain "awareness" of the problem. AGUIA's agents should also have strategies for amplifying their knowledge in order to remain useful. These agents learn incrementally through interaction, amplifying human decision-making capacity for goal oriented tasks. In addition to presenting the AGUIA model, this chapter describes five different implementations of the AGUIA model as applied in different domain areas. In some domains, experiments objectively confirmed that solutions produced by the human-AGUIA partnership were significantly superior.

The upcoming section presents key concepts that will be used throughout the chapter, such as: agents, ontology and rational decision-making. Section 3 presents the AGUIA model for amplifying human intelligence, a model composed of three basic elements: agents that use knowledge to enhance user perception of the range of alternatives; agents that elicit user knowledge and develop the knowledge base; and the knowledge itself represented formally, semiformally or informally. Section 4 presents examples of agents that aid users in understanding the problem, analyzing, evaluating and selecting alternative solutions, as well as keeping users aware of changes in context. Section 5 presents AGUIA agents which are, directly or indirectly, responsible for eliciting knowledge from users in order to maintain AGUIA's good performance. Section 6 presents related work, and lastly Section 7 presents the conclusions, underscoring this work's contributions and limitations, as well as future projects.

2 Theoretical Foundation

The software industry has undergone considerable expansion, becoming more diversified and focusing on agents designed for entertainment purposes. However, computer agents also have a very important role in helping people reach their goals in an efficient manner. Yet this proactive role is limited because the existing knowledge needed for computed processing is often not formalized [35]. Computers are not able to process human's natural language and humans do not easily express knowledge in a formal language [5]. This communication mismatch creates a barrier to effective understanding and reproduction of the human evaluation process by computer agents. A computer agent is limited to the scope of its knowledge and thus cannot be expected to find the optimum solution. However, it is reasonable to look for better solutions from an already recognized partnership between human creativity and computer expedited exploration of alternatives. AGUIA's model is based on building agents that understand the world through an ontology. These agents need to elicit knowledge to keep their performance updated. They act according a a rational decision-making process. This section presents an overview of the key concepts underlying AGUIA's model such as: ontology, knowledge acquisition and agents.

2.1 Knowledge Representation Ontology

Ontology has many different meanings, from the philosophical notion defining existence to the pragmatic computer idea of being a specification of a conceptualization [16]. Even Gruber's definition, however, which is quite broad, does not include the role of the ontolgy builder, nor does it address the purpose of the ontology or how it will evolve [36]. Ontologies have a history of helping people and computer programs share data, information and knowledge. Their importance in Web information retrieval systems [36] is due to the ease with which both computers and people can interpret them.

Ontologies are usually represented as semantic graphs, in which nodes represent concepts and arcs represent the relationships between them. Each concept or relationship has a name, rules for being that concept, list of exceptions, list of assumptions and a validity timeframe. An ontology designer creates any meaning he considers necessary to describe a domain, though there are some relations with already consolidated semantics such as:

- Is-a: suggests the relationship of set-subset. The properties of the higher set are inherited by its subsets. For example, a cat is a mammal. Once this relationship is expressed, we know that all mammal's properties will also be a cat's properties (but not vice-versa).
- Part-of: suggests the relationship of composition. A given concept will be formed by the composition of its parts. For example, an engine, the chassis and tires are integral parts of a car. A car does not exist without an engine.
- Is-an-attribute-of: suggests a relationship that is weaker than a characterization, such as: color is an attribute of cars. It is also an attribute of many other concepts, for example houses and clothes.
- Cause: suggests a relationship of consequence between two concepts, such as: Eating spoiled food causes sickness.
- Temporal Sequence (before, after, in between): suggests a temporal relationship between two concepts, such as: Mary will attend the U2 concert after buying the ticket.

An essential rule to follow when building an ontology is the annotation principle [40]: "One cannot use the same symbol to represent different things; one cannot represent equal things using different symbols." Common examples of annotation errors are caused by poor mapping of natural language's polysemic symbols into the formal language. For example, we could say that a car has both a color and an engine, but the meaning of the verb *to have* is different with regards to the concept *car color* and *car engine*. In the first context, the verb relates an object and its characteristics: *to have* characterizes the object according to the property of color. In the second context, the verb relates objects that together compose another object: an engine is *part of* a car. This distinction is important when knowledge is being formalized: A computer agent does not identify the two different semantic meanings if the same symbol is used in both scenarios.

2.2 Knowledge Acquisition

If making knowledge explicit is a complex matter, due to communication noise and faulty representation, formalizing it for computational use is even more complicated. This undertaking is often postponed due to the extra time and effort it requires. Another discouraging factor is that the person who actually describes the knowledge is very often not the one who will use it [24].

Keeping knowledge based systems updated is an enormous challenge. When knowledge is completely formalized, the challenge is even greater, as this normally requires specially trained professionals. This creates a dependency on knowledge engineers that could be problematic in maintaining consistency in the knowledge base updates.

On the other hand, systems based on knowledge fragments articulated by metadata, such as in argumentation by gIBIS hypertext [8], MIKROPOLIS [11] and wiki, make users codesigners and coresponsible for maintaining the base. These systems are not domain dependent and require a low level of knowledge formalization; however, they leave the knowledge incomprehensible to computational processing.

For Shipmam III and McCall [35], the dilemma, of wanting to use formalized knowledge to provide more efficient computer assistance, but not wanting to spend time formalizing it, is the result of the tradeoffs between an all-or-nothing understanding of the need and an all-at-once processing approach to knowledge acquisition.

One approach to incremental knowledge acquisition, as suggested in [12], mitigates this problem by diluting the acquisition process over the time a system takes to perform a task. Girgensohn [15] proposes a similar approach by making the acquisition process like the model extension process, but with users changing the representation directly.

There are two approaches to knowledge acquisition: explicit, where people consciously contribute, and implicit, where knowledge is acquired as a subproduct of another task that the contributing user is interested in performing.

Explicit approaches to knowledge acquisition engage people tin a conscious effort to build and maintain an information base. This requires a considerable contribution on the part of the people involved, who altruistically or based on monetary compensation dedicate their time to the task. Web content indexers, used by Google and generators of general ontologies, as proposed in the CYC (short for enCYClopedia) project [22], are examples of this type of explicit knowledge. The (ambitious) goal of the CYC project was to build a repository that contained all of the world's knowledge, creating a comprehensive ontology that established common sense knowledge, or "everything that everybody knows", that could be used to leverage any intelligent system. However, since knowledge evolves and expands, the task of feeding the CYC seems endless. Use of the CYC has been limited to small systems for specific domains, for example natural language processing [9] and speech recognition [3]. Even though the CYC knowledge is considered to be high quality, it still represents knowledge from a limited number of sources that could thus skew inferences [4].

Open Mind program [41, 39] has the specific aim of collecting a certain category of knowledge from altruistic users who deliberately contribute to the system, as with the Open Source Initiative [19]. One example is the Open Mind Common Sense program, which tries to get users to supply the possible continuation of a given scenario and builds connections based on the pieces of information obtained.

These initiatives make it possible to collect knowledge on a broad range of subjects. One problem with the Open Mind concerns knowledge quality, since nothing ensures that entries will help expand the knowledge model. The strength of the CYC is the quality of the knowledge it collects, because contributors are qualified users. However, this fact may also cause biased reasoning.

Implicit knowledge acquisition involves collecting data without users' awareness of this task. Many confidential services are accessible online, for example banking information. Systems for secure access to information bases need to confirm that the user accessing them is human and not a computer agent, the latter being utilized largely for malicious intent. Traditional ways of accessing information bases via passwords are considered fragile, because illicit access can be gained through trial and error [44].

One approach for verifying user identity is the CAPTCHA (Completely Automated Public Turing test to tell Computers and Humans Apart) [44] based on the Turing test [43]. This test, developed in 1950, determines whether a system is intelligent based on whether or not it performs a specific task like a human. Consider a group of observers that interact with two agents, one human and the other computational. The interaction is carried out exclusively through text messages in a specific domain. If the evaluators are unable to distinguish between the human and the computational agent at the conclusion of the interaction, the computer agent is considered intelligent. With CAPTCHA, the aim is to establish a task that exaggerates this difference. Furthermore, the task should be easy for the human so as not to prevent him from achieving his objective, in this case accessing the desired information bases.

This opportunity for knowledge acquisition has already been noted by many researchers who have been developing CAPTCHAs for different knowledge acquisition needs [18]. We consider this to be an excellent incentive mechanism, as it requires no extra effort.

Efforts to access a database can be channeled so that access is secured at the same time the meaning of the test's images (non-processable symbols) is acquired. Since users have interest in accessing the base, they are not likely to provide false information, under penalty of exerting additional and unnecessary effort.

2.3 Agents

Agents are computer systems aimed at achieving goals autonomously to determine actions in a dynamic environment [23]; they can act in isolation or within a community of what are called multiagent systems. An agent perceives the environment and reacts or reasons about what to do, choosing the most appropriate action applicable in that environment. It has three components: sensors for perceiving the environment; reasoning to decide what to do; and actuators to change the environment [33].

An agent is a special type of computational system that can be classified based on the following principals:

- cognition: an agent has explicit representation of the environment and other agents; it can reason about past actions and plan future actions (cognitive agent); or it reacts to environment stimuli, without memory of capacity to project the future (reactive agent);
- focus: an agent can be physically or behaviorally similar to humans;
- action: an agent can act in isolation or in community;
- environment: an agent act on an environment, such as in a company's Intranet or in the Internet.

The architecture of a computer agent is characterized by its internal processes and its interaction with the environment. It can be as simple as a process that only reacts when it recognizes a certain situation in the environment. Agents can also have complex reasoning processes, the most common involving the generation and evaluation of alternatives based on a set of preferences. Finally, it is important to underscore that agents must learn as they go. This evolution helps agents to adapt to changes in the environment [33].

3 AGUIA: An Intelligence Amplification Model Based on Agents Active on Demand

The AGUIA model for human intelligence amplification integrates users and autonomous agents in the decision making process, with each performing different functions. Autonomous agents expand perception of a problem's context and systematically generate and evaluate possible solutions, while users develop creative solutions and decide what action to take under the circumstances. As Figure 1 shows, the model has two types of agents: those that help people explore the range of options to reach better solutions, and those that collect knowledge from the users to enrich AGUIA's knowledge base. Users are both consumers and producers of the knowledge that will be organized, stored and made available to decision-makers.

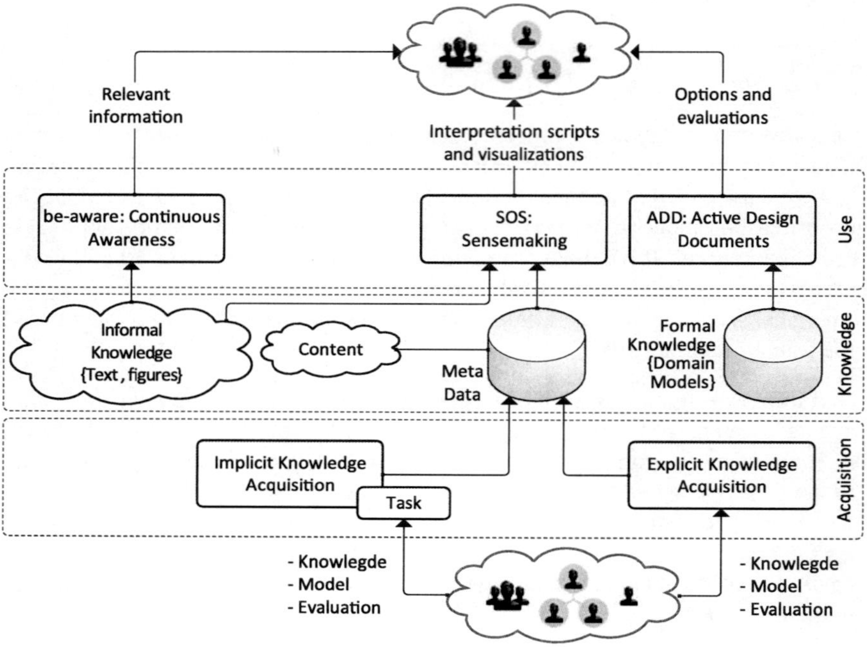

Fig. 1 The AGUIA model for intelligence amplification: knowledge use and acquisition. Arrows indicate the flow of data/information/knowledge.

The model is based on three assumptions: (1) more and better investigation of the range of possibilities lead to better solutions; (2) solution space's exploration depends on knowledge about the domain and the task to be performed; and (3) humans have limited capacity to generate, evaluate and compare a great number of alternative solutions and could benefit from computational help.

According to the AGUIA model, three types of partnerships are possible: Exploration guided by models, Sensemaking [32], guided by scripts and ongoing Awareness. In the first type, computer agents manipulate formal knowledge to generate, compare and justify alternatives in order to provide a suggestion according to a rational decision-making model . The speed with which a computer system systematically processes many different scenarios allows the range of alternatives to be explored much more efficiently than humans. However, it only operates provided there are formalized models of the task and domain from which it can reason upon.

Since knowledge is essential for AGUIA agents good behavior, our approach also includes special agents for keeping AGUIAs knowledge base updated. AGUIA included two types of knowledge amplification agents: explicit acquisition and implicit acquisition. The challenge of explicit acquisition is obtaining reliable knowledge. Knowledge acquisition is more problematic when information is contributed anonymously. This requires incentive mechanisms that align individual interests with those of the community. The objective is to encourage honest contributions from knowledgeable people (*maven*[1])

Implicit knowledge acquisition involves gathering knowledge from users as subproduct while they are performing other computational tasks. In this research, users contribute by attributing meaning to images when accessing secure databases that apply tests to filter out computer agents. The acquired knowledge is formal, including for example descriptors of images and the relationships between them, and therefore it can be used by computers.

4 AGUIA for Amplifying Human Intelligence

Users must choose between three types of partnerships with AGUIA: Exploration guided by models, Sensemaking guided by scripts or maintaining Awareness. In the first type of partnership, appropriate standard solutions are obtained with the help of agents that manipulate formal models. The second is suitable for investigating innovative solutions, usually in domains for which there is no standard process for reaching a solution . In this scenario, users manipulate in knowledge fragments guided by an interpretative script designed to enhance reflective reasoning without biases. Lastly, AGUIA agents may keep users informed about changes that may jeopardize previous decisions.

4.1 ADD: AGUIA for Exploration of Options Guided by Models

ADD(Active Design Documents) [12] is the most powerful agent for amplifying human intelligence, because it allows users to evaluate a much greater number of alternative solutions than they could evaluate on their own leading to a better quality final solution. ADD agent uses logical reasoning and is capable of generating alternatives and evaluating them based on domain models. Its results are precise: evaluated options and a suggestions so that the user can choose the best alternative. In addition, since this agent has a formal model it can help people evaluate creative alternatives and understand the rationale for them. This task-oriented agent provides unsurpassed assistance. Its strength is not only in the solution it provides, but rather in the partial solutions that users can improve upon, and conversely, in its ability to test user provided creative solutions.

[1] A Yiddish word often used to identify the best experts in a field.

4.1.1 The ADDModel

The ADD model, as illustrated in Figure 2, has three main elements: Interfaces, a design knowledge base and reasoning components.

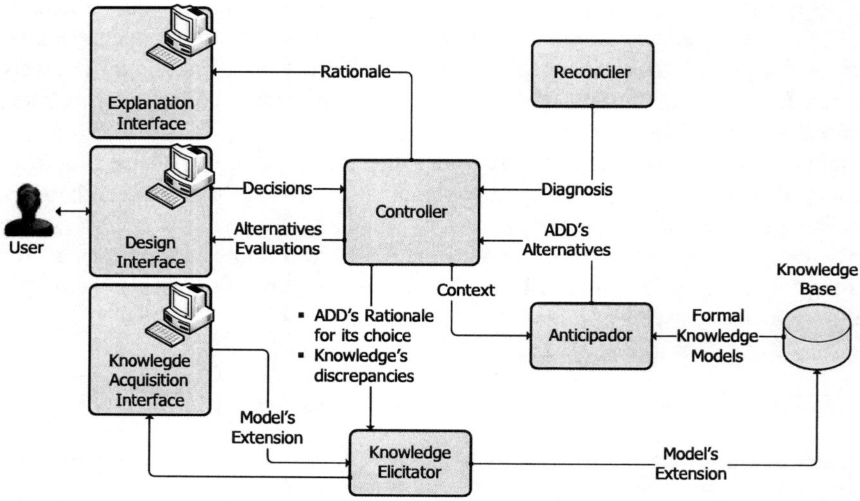

Fig. 2 The ADDmodel of active design documentation.

Designers interact with the ADD agent through design interfaces. The design interface allows users to naturally develop their design project while providing the system with means to understand the users' design choices. The explanation interface allows users to retrieve the rationale for the final design, that is, the justifications for the decisions that eventually conducted to the solution.

The knowledge base contains a description of the domain and the rules that guide inferencing in that domain in terms of heuristics, physical laws, artifact's functioning model or even a history of previously solved cases. This base serves as input for producing and evaluating alternative solutions.

The reasoning components determine the strategy for generating and evaluating alternatives. The Controller, the Anticipator and the Reconciler are the three ADD's reasoning components. The Controller is the central component that orchestrate the entire process. It monitors users' actions on the design task, activating the Anticipator, the Reconciler and the Knowldge Elicitor whenever needed. The Anticipator is the reasoning component that generates and evaluates the range of alternative solutions based on its knowledge base. The Anticipator sends the Controller its decision's expectations that are confronted to the user's chosen option by the Reconciler. When the Reconciler considers the two options similar, the Controller deduces that the ADD knowledge base has sufficient knowledge to justify that user's decision. On the other hand, whenever the Reconciler identifies significant discrepancies between the two solutions, it alerts the Controller about the incompatibility. Upon noticeable discrepancy between the human and computer models, the Controller activates the Knowledge Elicitor, which shows the user the ADD's expectation and requests user for more elements to help ADD reach the same alternative solution or to accept ADD's suggestion.

Therefore, ADD works as both an expert and an apprentice. As an expert, ADD is reminding users of bad choices. As an apprentice, it is showing its expectations and letting users adjust its knowledge base. These accounts of design rationale will be used to respond to subsequent requests for decisions' rationale.

4.1.2 Experiments

To evaluate ADD feasibility and benefits, applications were built for different engineering domains, such as air conditioning design for offshore oil platforms, oil process plant platform and oil pipeline layout. All applications were successful, however the application for the air conditioning design, called ADDVAC(Active Design Documents for Ventilation and Air Conditioning systems), allowed further studies as described here. The analyzed cases were randomly chosen from pre-existing projects, but unknown to the participants. The average cost of an air conditioning system on offshore platforms was around $1 million and took one to three months to complete. Table 1 presents a summary of the experiment.

Table 1 ADDVAC Experiment.

	Description
Participants	There were two participants: experienced air conditioning systems designers for offshore platforms. They were also familiar with computational drawing systems.
Material	Six past design cases chosen at random from the company's project library and unknown to the two participants. The cases had similar complexity. Although none of the cases required complex solutions, they were intricate and involved many interrelating details.
Method	Participants were expected to input the design case specification into the ADDVAC system and to develop a design solution with ADD assistance.

4.1.3 Results Analysis

As Table 2 shows, the results were significant. In all examined cases, the ADDVAC, in partnership with the designer, examined more options and reached better solutions. The two designers were satisfied not only with the solutions generated, but also with the process of exploring options and with the easy access to alternative solutions' comparison. We also tested the ADDVAC on a recently completed project designed by one of the participants. Given the positive results of the trials, he agreed to subject his design, which was already in the bidding process for construction, to the test. In three days, the designer - ADDVAC partnership yielded a result which was 30% cheaper while respecting all other criteria, such as safety.

Table 2 Results of using the ADDVAC. Experiments were conducted considering 6 different design cases with 6 different participants

	Design made without ADD	ADDVAC Effect
Average cost of the device designed	$1 million	20-40% less
Number of options evaluated	1-2	10-20
Time to develop design	1-3 months	1-2 weeks

4.2 AGUIA for Script-Based Ontology Sensemaking

Sensemaking agents facilitate human reflection using specific scripts for exploring and interpreting specific knowledge in order to enhance people's efficiency and awareness. This assistance to human sensemaking (SM) [32] is the purpose of our AGUIA's agent for scenarios in which there is a call for innovation. Our AGUIA's SM agent, called SOS (Sensemaking based on Ontology and Storytelling), is based on storytelling and ontology techniques. Domain ontology provides an unified vocabulary to investigate and describe a problem, while storytelling technique provides guidance to explore the solution space.

4.2.1 The SOSModel

The SOS model reflects a psychoanalytical view of the reflective reasoning process. By reconstructing stories, people see details that help them understand the context, the problem and even the solution. A domain ontology plays an important role in generating shared understanding of the stories among participants. The SOS model contains three main components:

- Script: defines the order in which information is requested and provided within a context.
- Domain ontology: defines the vocabulary for writing the stories.
- Anchor: provides evidence that increase credibility of the story.

Using storytelling for knowledge elicitation is not new. Recently, storytelling techniques have been used to investigate crimes: using a reasoning script based on textual evidence, criminal reports are created to shed light on legal cases [2]. Storytelling techniques have already been used to settle disputes [1], to share perspectives in collaborative work [14] and for tacit knowledge acquisition [20], which is recorded as accounts of experience. In the latter context, users tell and retell stories in a search for evidence to support conclusions.

4.2.2 Experiments

The SOS model was implemented in the DMWizard (short for Data Mining Wizard) system to help investigators find the root causes of accidents on offshore platforms. Figure 3 illustrates DMWizard's main interface.

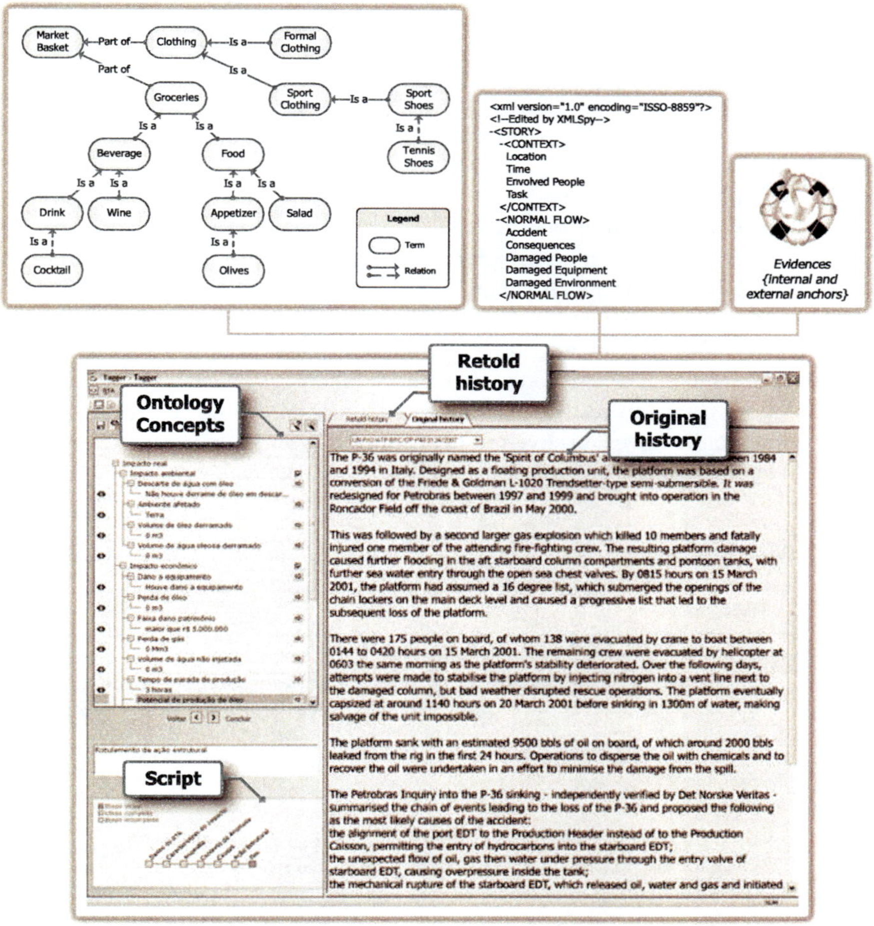

Fig. 3 SOS model for sensemaking of the context and the problem.

A story is structured into sections, organized according to the script. The right hand side of the interface displays notes and evidences from which an accident story can be reconstructed. The left hand side is divided into two panels: the bottom part shows the sections of the story and the top part displays a detailed view of the elements contained in each section. The system asks questions concerning the accident, and answers can be provided by the user in the top panel. Users answer these questions using a restricted vocabulary provided by the domain ontology. According to the specified story script, the same question may be asked again in a different context to improve the chances that users will reflect on the information provided.

Stories about accidents are based on events connected in time that may or may not have a causal relationship. Evidence is needed to give credibility to the story. Additionally, expectations are that multiple stories about the same theme would contain relationships between similar events.

The script includes the following elements:

- Context: include information about the physical location, timing, actors involved and actions performed.
- Turning Points: are important events that alter the flow of the story. For example, a sudden explosion in the workplace.
- Actions: are responses to events. In the aforementioned exemple, a factual action in response to the explosion would be evacuating the premises and calling the police and fire departments.
- Reversal: are actions that alter the stituation. For example, an action of an under-trained fireman that worsened the accident situation.
- Resolution: means to resolve the undesired situation. In the case of accidents, the resolution involves discovering the root causes and subsequent actions that would have prevented the undesired incident from occurring.

Table 3 Analysis scenario

	Description
Participants	28 workers from four different business units of a major oil company participated in the experiment. They had experience in accident investigation, but not experts at the task. Experts could be called upon should a participant feel incapable of recounting an accident story.
Material	Original stories, retold stories and evaluation surveys
Method	Each participant received basic, three-day training in the use of DMWizard, and was given the task of reading and interpreting the textual reports describing the accidents. For each report, a story would be told using the limited vocabulary defined by the domain ontology and enriched by evidences found in the original accident annotation or other sources to support the final story. Answering questions according to evidence promoted reflection about the accident annotations, which were often considered to be incomplete and inconsistent. The participants reanalyzed and retold the stories of accidents that occurred between 2006 and 2008. A total of 3,145 stories were retold over a period of six months. In the process they sought additional evidence and reanalyzed the accidents. The product was a structured information base that could be mined.

The domain ontology includes concepts such as accident events, economic or human impacts, equipment taxonomy and work force description. Evidence is anything that enforces a root cause hypothesis including pictures, text or lab results. Table 3 described the experiment using DMWizard.

Following the storytelling period, a survey was given to assess whether DMWizard helped the investigations by prompting them to look for evidence of all the items on the questionnaire. Users from the D unit did not respond before the deadline.

4.2.3 Results Analysis

Results, presented in Table 4, show significant gains in the report quality and comprehension when DMWizard was used. Comprehension gains were directly related to the amount of information users were able to extract from different sources and explicitly report in DMWizard. This degree of completion was calculated by the ratio between the amount of information the user entered and the total information considered essential for the story. The retold stories' level of completion reached 95% with a considered good quality when compared to experts' interpretations.

Table 4 Completeness of reanalysis with assistance from the DMWizard (SOS)

Business Unit	Number of participants	Number of analyzed accident reports	Degree of completeness on story's reevaluation
A	10	1827	90%
B	11	624	85%
C	5	630	95%

The company has already included the DMWizard in their risk assessment procedure, but, due to confidentiality issues, the data is no longer available for academic purposes. The company's technology adoption seems a strong indication of AGUIA's ability to assist people to better understand and investigate accident events by inducing reflective reasoning.

4.3 AGUIA for Keeping User Awareness of Changes in Context

Often people make decisions as soon as they become aware of a specific information. Although we may know where to find the needed information, time restrictions may prevent us from continuously accessing it. For example, if a person is notified of a traffic violation, he will probably pay it if he considers it warranted. The time lapse between notification and action is extremely important given the consequences of not acting. This sections presents an agent to take care of awareness in dynamic scenarios.

4.3.1 The Be-Aware Model

The Be-aware is a special type of user configured AGUIA agent that constantly monitors changes to recognized information bases available online. Its design is based on specific information from personal data records, with information indicated based on user interest. The research was conducted in the domain of electronic government.

The Be-aware model, as shown in Figure 4, consists of a personal agent configured to meet the requirements of each user's desired needs for information and an e-agency that coordinates the agents which continuously access government databases to find relevant data that may affect users.

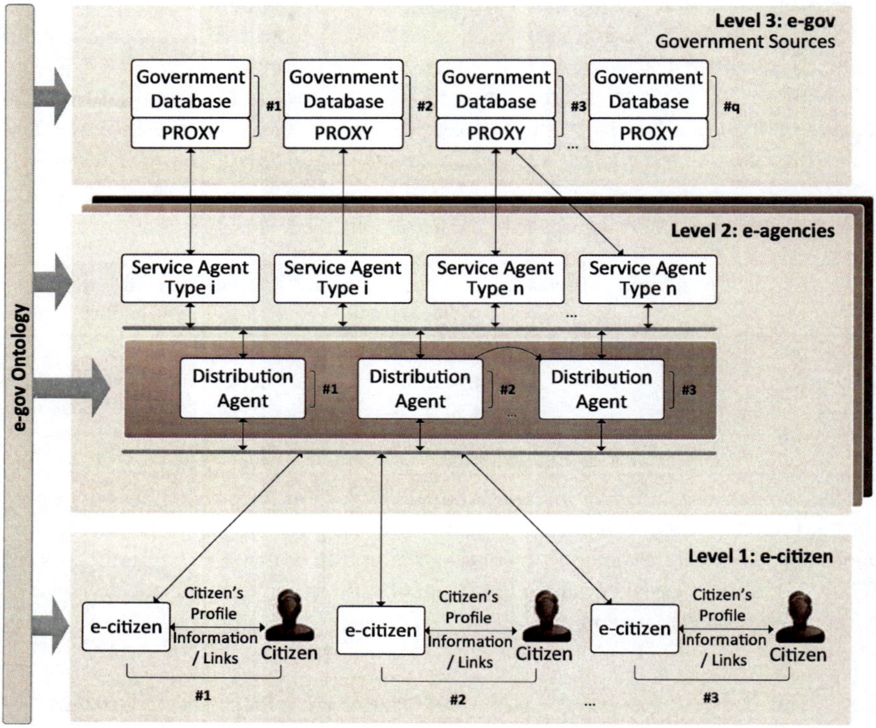

Fig. 4 The Be-aware model for maintaining user awareness.

These agencies are activated, taking into consideration an optimum service load for their functioning. At this level, the agency itself initiates a balancing process, replicating agents with heavy service loads and eliminating any that remain idle for a long period of time. It is in each agent's interest to find work so that it is not eliminated. In each round they bid for services in an auction-like process that distributes the work to the one that is most efficient and available.

This architecture, developed by Nogueira [28] [2], allows each agent (Be-aware) to represent one user virtually to resolve requests independently in order to achieve the desired objectives. None of the agents have sufficient resources, information or capacity to solve the entire problem on its own. Each one has knowledge and expertise, but it is their combined abilities that make it possible to produce the desired result.

4.3.2 Experiments

The e-citizen system was developed based on the Be-aware model to test the feasibility and usefullnes of maintaining context awareness by keeping citizens updated on government changes that directly or indirectly impact them.

Table 5 Be-aware pilot study.

	Description
Participants	A group of 20 volunteer evaluators was formed, among them IBGE (Brazilian Institute of Geography and Statistics) employees and UFF (Fluminense Federal University) students and employees. All participants were skilled Internet users with their own e-mail accounts.
Material	Participants' testimonies and personal data.
Method	The experiment lasted 15 consecutive days, beginning September 1, 2007. The process was explained through an explanatory bulletin that gave an overview of the e-citizen project. The e-citizen system was designed with agents capable of interacting with two official government information sources, one concerning the State Fire Department Tax and the other concerning the municipal property tax (Rio de Janeiro City).
Result Analysis	e-citizen sent relevant news to users concerning their relationship with government. It was discovered that one of the participants owed a Fire Department Tax debt for the years 2003, 2004 and 2005, and another participant owed property tax debt for the year 2006. All participants evaluated e-citizen as providing relevant assistance.

Three experiments were conducted [28]. As described in Table 5, a small pilot project was designed to evaluate the potential benefit of the e-citizen system. The second experiment, described in Table 6, involved 200 participants. Two groups were formed: a control group and a treatment group, each with 100 participants. The third experiment involved 700 participants, all of them using e-citizen's assistance. The third experiment ratified findings from the second experiment. 93% evaluated e-citizen as providing GOOD to VERY GOOD assistance.

[2] PhD thesis advised by this chapter's author.

Table 6 Be-aware 200-participants experiment.

	Description
Participants	IBGE employees and UFF students and employees were invited to participate. 200 were chosen at random from the list of those who signed up. All the participants had over two years experience using the Internet and virtually all of them (94.3%) had Internet access at home.
Material	Each participant filled out a questionnaire to define his interest profile with regard to government interaction, his experience using the Internet and his familiarity with the electronic government context.
Method	The experiment lasted 60 consecutive days, beginning on November 1, 2007. Each participant filled out a questionnaire to define his interest with regard to government information. During the 60-day experiments, they were expected be kept informed of government's actions that might affected them. Participants answered questions about their interest in issues related to the government domain and how they normally obtained information. After the 60-day period, participants were asked to answer another questionnaire concerning their current knowledge on government issues that might have affected them in that 60-day period.
Result Analysis	An evaluation questionnaire was distributed, with a return rate of 82%. Results indicated the participants rated the assistance as: Good or Very Good. Moreover, group using e-citizen's assistance felt more aware of government's actions than the other group. Participants without e-citizen claimed their interest in visiting government web site faded with time.

4.3.3 Results Analysis

The data collected during the pilot study showed the difficulty of maintaining awareness about information that affects us, no matter how important. The second experiment underscored how people initially accessed government bases, but soon stopped consulting them regularly. Additionally, they did not visit all the bases, even when these were listed on the task completion form. The third experiment confirmed the perceived benefits the e-citizen provided by keping people aware of e-government changes.

5 AGUIA for Knowledge Amplification

A knowledge model is essential for developing any intelligent system to help people to efficiently perform tasks. Amplifying knowledge as quickly and extensively as needed is as challenging as it is important. Knowledge can be collected through explicit or implicit knowledge acquisition methods. Explicit knowledge acquisition involves eliciting knowledge while it is being used, as explained with the HYRIWYG model. This research involved gathering information from a mass of people and addressed the major issue of information reputation. On the other hand, knowledge can be acquired implicitly as a side effect of other activity, such as described with the KA-CAPTCHA model.

5.1 HYRIWYG: How You Rate Influences What You Get

The HYRIWYG is a new reputation mechanism [31] for providing users the right incentive when they voluntarily supply reliable evaluations that become the bases for recommender system (RS) suggestions. It is therefore an indirect way of increasing the RS's reliability and also adjusting the RS's inference mechanism. Although the long-term benefits of providing evaluations for building a knowledge base and adjusting the RS may be very clear, users do not see these gains immediately. The HYRIWYG incentive pays users immediately, but only with a form of compensation that is as good as the quality of the RS being evaluated. This way, HYRIWYG encourages users and the RS to act correctly.

5.1.1 Description of the HYRIWYG Mechanism

Assuming a set of evaluators $I = \{1,2,...,N\}$; each evaluator i has a profile of preferences θ_i. Associated with each item available for evaluation is a vector $x = \{K, \Im_1, ..., \Im_i\}$, in which K is the aggregated evaluation of the aggregated product and $\Im$ is the incentive for each one of the N product evaluators. The K amounts vary according to the evaluation model of the RS, such as "Good/Bad" or 1-5 stars.

The agent i aims to maximize its utility function (u_i). Therefore, it should be compensated in proportion to its workload. In the absence of incentive, an RS requires that people act out of altruism. The incentive, or social function, has a critical role of moving the benefit of adjusting the RS to the time that a person provides a recommendation. Sometimes people acknowledge the indirect benefits of acting sincerely by offering their opinion on a product.

The HYRIWYG proposes a social function that includes incentives for each individual that adjusts the RS for his profile. As shown in the compensation function (see Eq. (1)), the incentive is a constant C, corresponding to the points, coupons, awards or any other reward, depending on the product's marketing strategy. C is adjusted by the contribution of the agent i for improving the RS.

Compensation Function

The compensation function of user i becomes:

$$(\tau_i = C * (1 + \alpha * \mid v'(RS(\theta_i)) - v_i \mid) \quad (1)$$

where:

- $v'(RS(\theta_i))$ is the expectation created by the RS with regard to the evaluation (concept normalized to be always in the interval $[0,1]$) that will be provided by user i, considering his projected profile θ_i;
- v_i is the evaluation (concept normalized to be always in the interval $[0,1]$) actually issued by the agent i;
- $\mid v'(RS(\theta_i)) - v_i \mid$ represents the contribution of the agent i to system adjustment;
- C is a constant defined by the product's marketing or business model;
- α is a constant to adjust the importance of obtaining diversity. Based on a scale from 0 to 1, it should be as close to 1 as possible when diversity of opinion is desirable or very close to 0, when homogeneity of opinion is preferred.

Rule for Redeeming:
The incentives are cumulative, shown in the redeeming benefits rule, Eqs. (2) (3). An agent i may only redeem award $\Im$ when this value is greater than a threshold T. In addition, $\Im$ functions only with the selected products, which is to say, with the products selected by the RS for the profile of the agent i.

$$\$_{product} = \$_{product} - f(\Im) \tag{2}$$

$$Reward(\$_{product}, \theta_i) = \$_{product} * Product(\theta_i) \tag{3}$$

where:

- $\$_{product}$ is the value of the reward, $\$_{product} = 0$ when $\Im < T$
- $f(\Im)$ is the discount associated with $\Im_i$
- $Product(\theta_i) \in$ to the products recommended by the RS for a user with profile θ_i

The agent i may only apply the incentive earned to products that correspond to the profile of the agent i, according to the RS. Thus there is positive incentive for users to express their opinion, as well as to adjust the RS. Therefore, providing an honest evaluation is in the individual's best interest. Incentives can be translated into cumulative points or discount coupons. The main point is to ensure that only products consistent with the evaluator's profile, according to the RS, are redeemed. This is a simple system that encourages people to tell the truth from their own viewpoint, since the reward will only be as good as the quality of the RS. For example, a user who does not like children's movies and lies about it to accumulate rewards could earn a quantity of coupons to "The Princess and the Frog" movie, and will have no way of selling it.

5.1.2 HYRIWYG: Empirical Evaluation

The experiments, as presented in Table 7 were divided into three phases, [6]. Each experiment lasted three weeks. They were conducted sequentially to avoid intersecting. Announcements were made using posters in the participating video stores and an electronic banner on the rental services' homepages.

Six video rental stores in the city of Juiz de Fora, Brazil, participated in the experiments. They were divided into two groups. Three of the stores participated in phases one and two, while the other three, which had similar characteristics and were located in the same neighborhoods, participated in phase three. We chose to work with two different groups of stores so that the customers who registered under phase 2 were not aware of the conditions that applied to phase 3, and vice-versa. The evaluation was made on the website by an acquisition system. Users were allowed to enter evaluations for one or more products. Around 650 people participated in the experiments. Participants of all three experimente were customers of the participating video rental stores.

Table 7 HYRIWYG experiment.

Experiment	Method
Phase 1	no incentive: no prize was awarded to the evaluators.
Phase 2	free DVD rentals were distributed in proportion to the number of films a participant evaluated.
Phase 3	free HYRIWYG's selected DVD rentals were distributed in proportion to the number of films a participant evaluated.

5.1.3 HYRIWYG: Result Analysis

As expected, when offering no rewards, the number of participations was small. During the period of freely chosen awards, the weekly registration rate was the highest. When incentives were generated by the system's recommendation (phase 3), the number of registered more than doubled the "No reward participation", but remained far below that of phase 2, in which awards were freely chosen.

Users' opinions on the recommendations generated were also analyzed. A high rate of satisfaction signalled that the collaborators' evaluations were largely honest. For this analysis, a comparison was made between the scores of users' movie evaluations and the system's suggested movie. The questionnaire asked those who did not claim the award film (46 out of 100) what motivated this. The responses indicated that only 20% were not satisfied with the award chosen by the RS. However, 30% did not respond. Even assuming that the 30% who did not respond were dissatisfied with the choice, this would be attributed to the human learning process, which requires time for people to realize the consequences, perhaps losses, of their actions. These people either will no longer participate or will do so more honestly in the future.

Web users are potential knowledge contributors. However, it is not reasonable to expect them to devote time and effort altruistically sharing their knowledge without any immediate benefit from it [30]. Even if they will benefit later from this knowledge base, they do not have a sense of being compensated. Therefore, implicit acquisition methods can be more efficient for ongoing knowledge acquisition, as discussed below.

5.2 *KA-CAPTCHA for Implicit Knowledge Acquisition*

KA-CAPTCHA is an AGUIA agent for implicit knowledge acquisition that takes advantage of the scenario configured by the CAPTCHAs. An application developed according to this model was programmed for the domain of image indexing and was tested by student Bruno Silva in his Master's thesis[3]. Results showed feasibility and efficiency of KA-CAPTCHA to implicit KA [37].

Normally, CAPTCHA [44] generates tests by retrieving data from a public base and distorting the image. The KA-CAPTCHA, an AGUIA agent for knowledge acquisition, retrieves data from public bases that are semantically poor to formulate a CAPTCHA exercise to users. Users provide information that will be used to build meaning to images. An individual contribution is weak to provide semantic to an image, but when mass contributions go toward the same meaning, there is a great chance this meaning is an acceptable one.

The KA-CAPTCHA requires an ontology of concepts from which images receives a classification. The support and reliability of each semantic relationship will be calculated so that reliable information can be distinguished from noise. The support specifies the frequency with which users identify a given relationship as suitable in the CAPTCHA database. High support indicates that many users considered a specific relationship between symbols as meaningful and valid. Reliability is related to the relative frequency in which a symbol and a meaning were associated. Reliability varies from zero (to date, no user has acknowledged that relationship as valid) to one (all users acknowledge it).

5.2.1 KA-CAPTCHA: Experiment

In this experiment, the task of the KA-CAPTCHA is to associate semantic labels with image. Initially, a label from the ontology is chosen randomly. Then, images previously correlated with the selected label, whether positively or negatively, are also chosen randomly. Correlations are considered positive when they pass the support and confidence threshold, and negative otherwise. Some images are also selected from which relationship with the chosen labels is unknown. Then the images are mildly distorted to prevent them from being compared automatically.

Controlled experiments were conducted in three stages, as described in Table 8. The first phase evaluated the interaction of 147 volunteers with the KA-CAPTCHA . Initially, evaluators measured the average number of attempts for a user to pass the test and access the desired database. They then measured the information precision and recall when the labeled base was used to retrieve information from the Web. The results were compared with those from the Google image search engine. Finally, the results extracted from the previous stages of the test were evaluated. There was no intersection among users to ensure that noise did not affect the test; each one participated in only one phase of the experiment. Each of the test's four columns contained five images. Support was considered the equivalent of getting at least two figures right, and confidence the value of 80%.

[3] Master's thesis, with this chapter author as advisor.

Table 8 KA-CAPTCHA experiment.

Experiment	Participants	Method
Phase 1	two undergraduate students from Engineering	Researchers collected 15 of the most common questions to the Google Press Center made in the month of September 2006. Each question was presented as a label to be associated with images; volunteers were asked to retrieve 10 images from the Internet that they considered closely related to the question, and 10 they considered totally unrelated. Volunteer 1 retrieved 101 images, while volunteer 2 retrieved only 62. The groups of images were classified as Base 1 and Base 2, respectively.
Phase 2	143 undergraduate students from CS, Engineering and Physics, randomly selected	The students could only access their grades from the mid-term exams using the site created with the KA-CAPTCHA. Students didn't known they were collaborating by associating test labels and images. There were no other way to retrieving their grades. This experiment was conducted twice: with the midterm and final grades.
Phase 3	Two graduate students from Law and Veterinary	Participants were presented with all the classifications and labels from Phases 1 and 2. They were instructed to identify which label best described the images selected in the earlier experiments.

5.2.2 KA-CAPTCHA: Results Analysis

The first evaluation concerned the users' ability to pass the Turing test. In this experiment only 2% of users had to make two attempts to access the desired data.

Base 1 was populated with 101 image-label associations after the students accessed it to view the mid-term grade. Of these, the system considered 38 to be valid, 21 false and 42 inconclusive. Base 2, populated by students accessing their final grade, was composed of 63 associations, showing 29 considered to be valid, 20 false and 14 inconclusive.

To finalize the evaluation, search using the populated image database was compared to that of Google Images as a way to evaluate precision and recall. Three participants were asked to choose, for each label in our base, if one of the 180 Google results would best match it. The precision of the KA-CAPTCHA reached 98% with the same data, indicating potential improvement over Google's results. However, recall was low, in the range of 108/164=0.658. This poor result was expected, because the experiment took place in a narrow window of time, when very little was registered; use over time should offset this factor.

The strength of the KA-CAPTCHA is in the fact that the acquisition task is embedded in a commonplace and necessary user task. One of its limitations is that it requires an initial knowledge base and an ontology that needs populating.

6 Related Work

As discussed throughout the text, amplification of human intelligence is seen here as an improvement in the process of choosing alternatives in the context of performing goal oriented tasks.

Thus, we can amplify human reasoning capacity without having to change the brain's physical structure. In this sense, users benefit greatly from knowledge manipulation and extraction and from systematic examination of the range of options.

The AGUIA model can be compared to decision support systems, since both are designed to help in the decision making process. In this regard, AGUIA offers support to different scenarios examined by decision support systems [29], such as: communication of information and knowledge among participants in a collaborative decision making process; manipulation of large volumes of data; manipulation of document management and interpretation; and by searching and supporting the generation and evaluation of alternatives.

RSs can be seen as decision support systems specifically targeting product suggestion. RSs warrant special note due to their suggestion precision and their growing popularity. The AGUIA model can also be compared with argumentation systems designed to organize the decision making process in an individual or group argumentation system. This organization allows ideas to be structured.

However, the AGUIA model is broader than a decision support system, largely distinguished by its active work in partnership with people, and not in passive work substituting human reasoning. It not only offers solutions, but it leads users to solutions, at the same time it allows users to amplify computational knowledge. This is perhaps the greatest difference: embedding knowledge acquisition in its processes. Even an AGUIA that manipulates formalized knowledge has an incremental and contextualized knowledge acquisition mechanism [35]. The AGUIA model considers all assistance targets, from manipulating large volumes of data to navigating online information. Users add content over a span of time and activity. Users also organize content by identifying labels that index the information. This organization provides search engines with more efficient access. Production and Use are combined in the same platform. Examples of successful application of this technology include software on the Internet, social networks, wikis, blogs and folksonomies.

The semantic web[34] is designed to allow computer systems to process existing content more effectively. In this regard, the semantic web means enriching content connections using a domain ontology to help computer agents find the desired result, in contrast to blind navigation. The semantic web allows agents to be more active in helping people conduct searches.

Examples of applications that use the semantic web are Swoogle [10] and BING [25]. Swoogle searches the structured content in an ontology, e.g., OWL (Web Ontology Language) data [46]. This improves computer agents' ability to manipulate and locate the knowledge sought.

BING [25] is Microsoft's newest search engine, which operates on the semantic web and is designed especially for indexing and searching for images and videos. It is marketed as a decision support engine, and it outperforms traditional visual and video search engines. Results are organized based on a taxonomy of logical concepts, which should expand human perception about what is sought. The AGUIA operates similarly when it provides its suggestions using domain ontologies as a background and schema for organizing results.

Decision support systems, RSs and the semantic web are approaches to enhance human perception. Applications specific for a domain, as suggested by the ADDAGUIA agents, can provide more precise and efficient assistance, but they require formal knowledge. This is a trade-off, always in play, when choosing which type of AGUIA agent to use. The quality of the assistance has earned attention and inspired activity in other domains, such as the semantic web. The interconnection between knowledge acquisition and knowledge usage may be the key element for AGUIA's successful applications.

7 Conclusion

The popularization of the Internet and the ease with which knowledge can be shared have facilitated the solution of complex problems by providing easy access to information. On the other hand, the solution itself has become a problem. An information overload bogs down the process and makes it difficult to identify what is applicable and relevant to the specific task at hand. Man has limits for processing large quantities of information and knowledge and for determining information accuracy. The synergy of a work group can mitigate the effects, but not necessarily solve the problem. And although human beings are creative and highly intuitive, their rationality is limited by excess information to process, which may result in biased solutions.

Computer agents can be used to amplify human intelligence in problem solving. Partnerships between computer agents and people result in better solutions, with lower costs, less time spent on the task and higher final quality. These partnerships allow humans to learn and computer agents to expand their knowledge bases.

This chapter detailed a model for amplifying human intelligence, called AGUIA. It investigated the technical feasibility, costs and benefits of building AGUIA agents for amplifying human intelligence for efficient execution of many different tasks. Concrete studies were conducted involving engineering designs, fault diagnosis, accident investigation and electronic interaction with the government. The main results were reported herein.

This research has shown the feasibility of expanding human intelligence with intelligent agents that act in partnership with people and learn from this interaction, in different domains. The diverse cases studied, all of them with industrial or pre-industrial prototypes, highlighted the impact in final product cost and quality as well as decision making process improvement.

AGUIA's active intelligent agents manipulate and acquire knowledge and can:

- enhance perception of the context and the problem
- enhance examination of more and better solution alternatives
- enhance awareness of the environment and the context changes, and thereby
- amplify human intelligence, enabling human capacity to make better decisions.

There is still much to study about the use of AGUIA agents to amplify human intelligence. The AGUIA model integrates knowledge acquisition into its use, but handles knowledge formalized in a variety of manners. In this regard, there is still a need for bridges between the parts of knowledge represented differently. In fact, there are no systems that operate like an AGUIA to help people build these bridges. For example, early discussions about a project contain many knowledge fragments that are not necessarily formalized. However, bridges such as argumentation networks could connect discussions [7], and the models generated based on these could be represented by ontologies [16].

There are limitations associated with required formalized representation. The ADDagents require an initial knowledge base represented by models (formal representation) in order to operate. ADDacquisition in done by expanding an initial base. However, generation of this initial base happens outside the process through meetings with experts in the field. Thus acquisition of formal initial models continues to be an obstacle to successful application of the ADDagents. On the other hand, SOS does not require formalized knowledge, but its assistance is superficial since it does not understand what the user is doing. Be-aware can be embedded in both ADD and SOS agents. HYRIWYG and KA-CAPTCHA acquire both formal and semi-formal knowledge. Both are very incremental and it takes time and great number of people to make the effort worthwhile.

Many areas do not yet have the maturity necessary to create a formal initial knowledge model. However, there are knowledge fragments that can be organized and eventually formalized. In this regard, mechanisms are needed for interaction between agents and individuals to help structure increasingly formalized knowledge. These mechanisms must allow mapping between formal and informal knowledge.

Lastly, it is worth noting that sometimes no knowledge is available, but collective intelligence can be built of small fragments of information dispersed throughout a community. Numerous efforts, albeit in their early stages, are moving in this direction [27]. Developing the means to direct these efforts could help design an AGUIA that generates creative solutions from the collective intelligence. In this sense, AGUIA works as a knowledge catalyst to solve problems that do not yet have solutions.

Intelligent systems are far from substituting human's performance. Nevertheless, humans frequently decide upon biased view of the problem, with little time to properly explore the solution space. Though, combining human creativity with computer systems' systematic evaluation might boost final results.

Acknowledgements. This research was conducted with the precious help of my Ph.D. and M.Sc. students that developed most of the software I used in my studies. Many thanks to Adriana Vivacqua, Jose Luis Nogueira, Leandro Ciuffo, Bruno Silva e Thiago Cortat. I also would like to thank Petrobras, the Brazilian Petroleum Company, for allowing me observe the improvement of its employees' decision making process while using an AGUIA component.

References

1. Antunes, P., Relvas, S., Borges, M.: Alternative dispute resolution based on the storytelling technique. In: Proceedings of the 13th International Conference on Groupware: Design Implementation, and Use, pp. 15–31 (2007)
2. Bex, F., Van den Braak, S., Van Oostendorp, H., Prakken, H., Verheij, B., Vreeswijk, G.: Sense-making software for crime investigation: how to combine stories and arguments? Law, Probability and Risk 6(1-4), 145–168 (2007)
3. Brewster, C., O'Hara, K.: Knowledge representation with ontologies: the present and future. IEEE Intelligent Systems 19(1), 72–73 (2004)
4. Chklovski, T.: Deriving quantitative overviews of free text assessments on the web. In: Proceedings of the 2006 International Conference on Intelligent User Interfaces, pp. 155–162 (2006)
5. Chommsky, N.: Three models for the description of language. IRE Transactions on Information Theory 2, 113–124 (1956)
6. Ciuffo, L.: Um estudo de caso para verificar a suscetibilidade a incentivos de avaliadores de produtos na web. Master's thesis, CS Department, Universidade Federal Fluminense, Brazil (2005)
7. Conklin, J., Burgess Yakemovic, K.C.: A process-oriented approach to design rationale. Human-Computer Interaction 6(3), 357–391 (1991)
8. Conklin, J., YakemBegemanovic, M.: gibis: A hypertext tool for exploratory policy discussion. ACM Transactions on Office Information Systems 6(4), 303–331 (1988)
9. Curtis, J., Cabral, J., Baxter, D.: On the application of the cyc ontology to word sense disambiguation. In: Proceedings of the 19th International Florida Artificial Intelligence Research Society Conference, pp. 652–657 (2006)
10. Ding, L., Finin, T., Joshi, A., Pan, R., Cost, S., Peng, Y., Reddivari, P., Doshi, V., Sachs, J.: Swoogle: a search and metadata engine for the semantic web. In: Proceedings of the 13th ACM Conference on Information and Knowledge Management (2004)
11. Fischer, G., Lemke, A.C., McCall, R., Morch, A.I.: Making argumentation serve design. Human-Computer Interaction 6(3), 393–419 (1991)
12. Garcia, A.C.B.: A new approach for supporting documentation during preliminary routine design. PhD thesis, Stanford University (1992)
13. Garcia, A.C.B.: AGUIA: agentes-guia para ampliar a inteligencia humana em tarefas orientadas a metas. PhD thesis, Full professorship context, Universidade Federal Fluminense (2009)
14. Garcia, A.C.B., Carretti, C.E.L., Ferraz, I.N., Bentes, C.: Sharing design perspectives through storytelling. Artificial Intelligence for Engineering Design, Analysis and Manufacturing 16(3), 229–241 (2002)
15. Girgensohn, A.: Modifier: improving an end-user modifiable system through user studies. In: Grechenig, T., Tscheligi, M. (eds.) VCHCI 1993. LNCS, vol. 733, pp. 141–152. Springer, Heidelberg (1993)
16. Gruber, T.: A translation approach to portable ontologies. Knowledge Acquisition 5(2), 199–220 (1993)
17. Guha, R.V., Lenat, D.B.: Cyc: Enabling agents to work together. Communications of the ACM 37(7), 127–142 (1994)
18. Hacker, S., von Ahn, L.: Matchin: eliciting user preferences with an online game. In: Proceedings of the 27th International Conference on Human Factors in Computing Systems, pp. 1207–1216 (2009)
19. Open Source Initiative, `http://www.opensource.org/` (retrieved November 22, 2010)

20. LeBlanc, S., Hogg, J.: Storytelling in knowledge management: an effective tool for uncovering tacit knowledge. In: Proceedings of the STC Atlanta Currents Conference (2006)
21. Lenat, D., Feigenbaum, E.: On the thresholds of knowledge. Artificial Intelligence 47 (1991)
22. Lenat, D.B.: Cyc: A large-scale investment in knowledge infrastructure. Communications of the ACM 38(11), 33–38 (1995)
23. Maes, P.: Agents that reduce work and information overload. Communications of the ACM 37(7), 30–40 (1994)
24. Marshall, C., Shipman, F.: Searching for the missing link: discovering implicit structure in spatial hypertext. In: Proceedings of 1993 ACM Hypertext Conference, pp. 217–230 (1993)
25. Microsoft. Bing, `http://www.microsoft.com` (retrieved January 20, 2010)
26. Miller, G.A.: Wordnet: a lexical database for english. Communications of the ACM 38(11), 39–41 (1995)
27. MIT-CCI. Mit center for collective intelligence, `http://cci.mit.edu/news/index.html` (retrieved November 22, 2010)
28. Nogueira, J.L.: e-Cidadao: agentes para auxiliar cidadaos com e-gov. PhD thesis, CS Department, Universidade Federal Fluminense (2008)
29. Power, D.J.: A brief history of decision support systems, `http://DSSResources.COM/history/dsshistory.html` (retrieved January 10, 2008)
30. Resnick, P., Sami, R.: The information cost of manipulation resistance in recommender systems. In: Proceedings of the ACM Recommender Systems Conference (2008)
31. Resnick, P., Zeckhauser, R., Friedman, E., Kuwabara, K.: Reputation systems. Communications of the ACM 43(12), 45–48 (2000)
32. Russell, D.M., Stefik, M.J., Pirolli, P., Card, S.K.: The cost structure of sensemaking. In: Proceedings of the 1993 ACM INTERACT and CHI Conference on Human Factors in Computing Systems (INTERCHI 1993), pp. 269–276 (1993)
33. Russell, S., Norvig, P.: Artificial intelligence: a modern approach, 2nd edn. Prentice Hall, Upper Saddle River (2003)
34. Shadbolt, N., Hall, W., Berners-Lee, T.: The semantic web revisited. IEEE Intelligent Systems 21(3), 96–101 (2006)
35. Shipman III, F.M., McCall, R.: Supporting knowledge-base evolution with incremental formalization. In: Proceedings of 1994 ACM Conference on Human Factors in Computing Systems (CHI 1994), pp. 285–291 (1994)
36. Shirky, C.: Ontologies are overrated: categories, links, and tags, `http://www.shirky.com/writings/ontology_overrated.html` (retrieved November 20, 2010)
37. Silva, B.N.: Ka-captcha: An opportunity for knowledge acquisition on the web. Master's thesis, CS Department, Universidade Federal Fluminense, Brazil (2007)
38. Simon, H.: Bounded rationality and organizational learning. Organization Science 2(1), 125–134 (1991)
39. Singh, P.: The public acquisition of commonsense knowledge. In: Proceedings of AAAI Spring Symposium on Acquiring (and Using) Linguistic (and World) Knowledge for Information Access (2002)
40. Stefik, M.J.: Introduction to knowledge systems. Morgan Kaufmann, San Francisco (1995)
41. Stork, D.G.: The open mind initiative. IEEE Expert Systems and Their Applications, 16–20 (1999)

42. Thaler, R.H., Sunstein, C.R.: Nudge: improving decisions about health, wealth, and happiness. Yale University Press, New Haven (2008)
43. Turing, A.M.: Computing machinery and intelligence. Mind, 433–460 (1950)
44. Von Ahn, L., Blum, M., Hopper, N., John Langford, J.: Captcha: using hard ai problems for security. In: Biham, E. (ed.) EUROCRYPT 2003. LNCS, vol. 2656, pp. 294–311. Springer, Heidelberg (2003)
45. von Ahn, L., Dabbish, L.: Labeling images with a computer game. In: ACM Conference on Human Factors in Computing Systems (CHI 2004), pp. 319–326 (2004)
46. W3C. Owl web ontology language guide: W3c recommendation, `http://www.w3.org/` (retrieved November 20, 2010)

Glossary

Agent is a computational component with autonomous behavior that perceives the environment and acts according to its goals and objectives.

Artificial intelligence is an area of computer science for studying and developing artifacts that suggest human intelligent behavior.

Collective intelligence is a form of distributed intelligence that raises from effective mobilization of individual skills with no rigid control leading to new knowledge.

Design is a plan for a construction of an artifact.

Knowledge base is a special type of database that holds knowledge of a domain in terms of heuristic rules, first-principle equations or even a history of known cases, that must be understood by computational processes..

Ontology is a description of a domain to be used for a specific purpose and described according to the view of a group of users.

Acronyms

ADD	Active Design Documents
ADDVAC	ADD for the design of Ventilation and Air Conditioning systems
AGUIA	Agent Guidance for Human Intelligence Amplification
CAPTCHA	Completely Automated Public Turing test to tell Computers and Humans Apart
DMWizard	Data Mining Wizard system
e-Gov	Electronic Government
HYRIWYG	How You Rate Influences What You Get
KA-CAPTCHA	Implicit Knowledge Acquisition using CAPTCHA applications
SOS	Script-based Ontology Sensemaking

Part III
Advanced Applications

Chapter 12
Clouds and Continuous Analytics Enabling Social Networks for Massively Multiplayer Online Games

Alexandru Iosup and Adrian Lăscăteu

Abstract. Many of the hundreds of millions Massively Multiplayer Online Games (MMOGs) players are also involved in the social networks built around the MMOGs they play. Through these networks, these players exchange game news, advice, and expertise, and expect in return support such as player reports and clan statistics. Thus, the MMOG social networks need to collect and analyze MMOG data, in a process of continuous MMOG analytics. In this chapter we investigate the use of CAMEO, an architecture for Continuous Analytics for Massively multiplayEr Online games on cloud resources, to support the analytics part of MMOG social networks. We present the design and implementation of CAMEO, with a focus on the cloud-related benefits and challenges. We also use CAMEO to do continuous analytics on a real MMOG community of over 5,000,000 players, thus performing the largest study of an online community, to-date.

1 Introduction

Massively Multiplayer Online Games (MMOGs) have emerged in the past decade as a novel Internet application with a rapidly growing, world-wide user base of tens of

Alexandru Iosup
Parallel and Distributed Systems Group,
Faculty of Electrical Engineering, Mathematics, and Computer Science,
Delft University of Technology,
Mekelweg 4, 2628CD, The Netherlands,
e-mail: A.Iosup@tudelft.nl

Adrian Lăscăteu
Computer Science Department,
Faculty of Automatic Control and Computer Science,
Politehnica University of Bucharest,
Spl. Independentei 311, Bucharest, Romania
e-mail: Adrian_Lascateu@yahoo.com

N. Bessis, F. Xhafa (Eds.): Next Generation Data Technologies for CCI, SCI 352, pp. 303–328.
springerlink.com

millions of players. There exist now hundreds of MMOGs providers with thousands of MMOGs currently in operation; from these, FarmVille and World of Warcraft number each over 10,000,000 constant players. Around each game or groups of similar games, third-parties such as volunteers and small businesses have built online communities (social networks) that inform and entertain the players. These communities use MMOG analytics to improve visitor experience with player reports, progress charts, clan statistics, etc. While the analysis may differ, the data collection and analysis (collectively, the *game analytics*) can benefit from recent advances in the availability of on-demand resources through cloud computing services such as Amazon's Elastic Compute Cloud (EC2). In this chapter we present an architecture for MMOG analytics in the cloud.

Cloud computing has emerged in the past few years as a new paradigm for IT, in which the infrastructure, the platform, and even the software are outsourced services. These services have fixed cost and general Service Level Agreements and, most importantly, can be used when and for how long they are needed. There currently exist hundreds of commercial cloud service providers, such as Amazon, Microsoft, FlexiScale, and NewServers.

CAMEO is our architecture for MMOG analytics in the cloud. CAMEO mines information from the Web, collects information using Web 2.0 interfaces provided by various MMOG operators and their collaborators, integrates the information into comprehensive and time-spanning MMOG datasets, analyzes the datasets, and presents application-specific results. The resources used by CAMEO can be provisioned from commercial or enterprise clouds. Through its use of computational intelligence, CAMEO addresses the main challenges faced by system designers addressing the problem of continuous analytics enabling MMOG social networks, including the understanding of the user community needs; the use of distributed and collaborative technologies such as clouds; the data management, data growth, and data storage; and building a system with high performance, scalability, and robustness. Thus, CAMEO can readily be used by a variety of MMOG communities, and may provide a good step forward in supporting other next generation, data-oriented organizations.

Online data crawling and analysis has often been employed in the past to determine the stationary and dynamic characteristics of Internet-based communities. However, the focus of the research community has been either in making the crawling process more parallel [2, 8, 23], or analyzing the acquired data using more scalable parallel or distributed algorithms [34, 3]; both these approaches assume that enough resources are available for the task. Recently, data analysis in the cloud has received much attention, and generic programming models such as MapReduce [9] and Dryad [21, 41] are now supporting large-scale processing both for the general public and for companies such as Google, Facebook, Twitter, and Microsoft. In contrast to this body of previous work, which addresses a wide range of issues, in this chapter we focus on a domain-specific application, MMOG analytics, for which we design and build an integrated solution. Extending our previous work on CAMEO [15], our main contribution is four-fold. First, we introduce a number of MMOG-specific tools to better understand and support the user community needs.

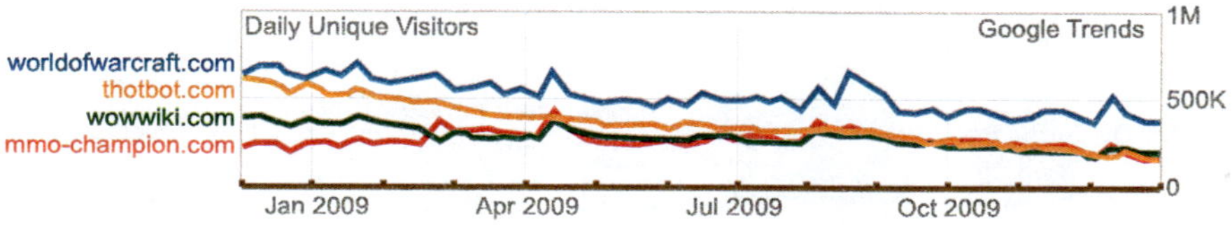

Fig. 1 Size of four online communities built around World of Warcraft. Except for worldofwarcraft.com, the communities are not maintained by the MMOG operator.

Second, we extend the cloud-specific part of CAMEO to support data acquisition even under traffic limitation. Third, we extend the data-specific part of CAMEO to support more efficient data storage and transfer. Fourth, we show evidence that CAMEO can be used in practice by performing continuous analysis on RuneScape, a popular MMOG, on resources provisioned from either commercial or free-to-use clouds.

The remainder of this chapter is structured as follows. Section 2 formulates the problem of continuous analytics for MMOGs. Section 3 contrasts this work and related research. The CAMEO architecture is introduced in Section 4. The next two sections focus on experiments using CAMEO: Section 5, which introduces the experimental setup, and Section 6, which presents the experimental results. Section 7 discusses future research directions and concludes the chapter.

2 Mission: Continuous Analytics for MMOGs

MMOGs generate data that need to be analyzed at various levels of detail and for various purposes, from high-level analysis of the number of players in a community, to the detailed analysis of the users' mouse-clicking behavior. As in our previous work on CAMEO [15], we define *continuous analytics for MMOGs* as the process through which relevant MMOG data are analyzed in such a way that prevents the loss of important events affecting the data; the relevance of the data is application- and even community-of-users-specific. In this section we present our goal in supporting continuous analytics for MMOGs.

2.1 Goal

Around each popular MMOG, the dedicated players and even commercial interests have created social networks that support active communities. Through collaborative paradigms such as Wikis, Data Mashups, and Web Services, these *MMOG social networks* aggregate information about the MMOG in the form of encyclopedic reports, tutorials, videos, and even player-customized information. Many of the MMOG social networks are built with the volunteered contributions of common players, who may in return get social rewards such as community recognition.

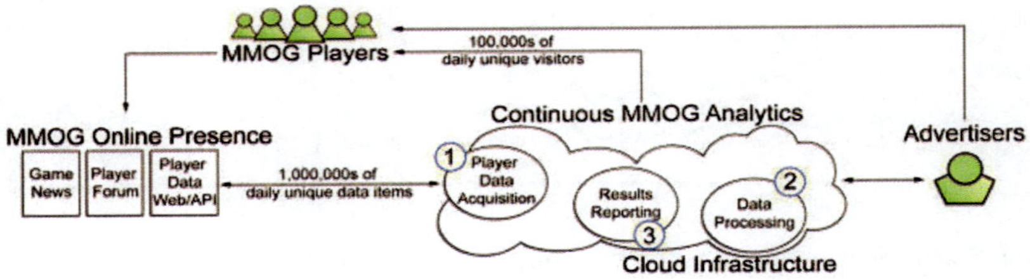

Fig. 2 The MMOG/players/advertisers/third-party communities ecosystem, for a single MMOG.

The MMOG social networks have to support large communities with hundreds of thousands of unique daily visitors. Often, the popularity of the social networks built "by the community, for the community" reaches or even exceeds the popularity of the communities built by the game's operator. Figure 1 shows the size over time of four communities built around World of Warcraft: `worldofwarcraft.com`, which is built and maintained by the game operators, and the community-built `mmo-champion.com`, `thottbot.com`, and `wowwiki.com`. The four communities total over one million daily unique visitors at the beginning of 2010; the community-built `thottbot.com` has had at the beginning of 2009 a size equal to `worldofwarcraft.com`'s.

The business model for MMOG social networks built by third-parties, that is, organizations not sponsored by the MMOG operator, is to obtain revenue from advertisements or from selling virtual goods and services. For this business model to function, the third-party web sites need to retain their visitors, which are only a subset of all the MMOG players; a big step toward visitor retention is to use MMOG analytics to improve visitor experience with player reports, progress charts, clan statistics, etc. These rely, in turn, on continuous MMOG analytics.

Our goal is to design a generic architecture for continuous analytics in support of social networks for MMOGs. We envision such a system operating in the players/MMOG operator/advertisers ecosystem depicted in Figure 2. To materialize this vision, we introduce in Section 4 CAMEO, our cloud-based continuous MMOG analytics architecture, which provides the services labeled 1 through 3 in the figure.

2.2 *Challenges*

We identify four main challenges in achieving our goal: understanding user community needs, enabling and using distributed and collaborative technology, the combined data challenge, and several MMOG-specific challenges relating to system design. We describe each of the challenges, in turn.

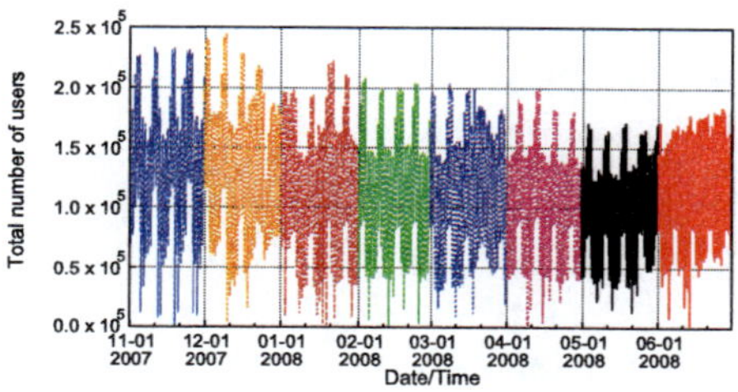

Fig. 3 The number of Active Concurrent Players in RuneScape, between Nov 1, 2007 and Jul 1, 2008. Each month of data is depicted with a different type of line. (Source: our previous work [16].)

2.2.1 Understanding User Community Needs

MMOG communities vary by type and size, and the same community will vary in size over time (see also Section 2.2.2). The communities for casual and constant (hard-core [11]) gamers are very different, with leads to many types of analysis. For example, a hard-core gamer community could be interested in showing each player's evolution over time and in identifying top-performing players; a casual gamer community could try to identify specific classes of players, for example players with unique combination of skills, and form groups of complementary players. The consumers of the analysis results may be the users or the community operators, which entails again different analysis needs. For example, a third-party company providing content for the game could base its decisions on the distribution of playing skills and achievements. We conclude that the main challenge for this topic is to **understand the specific and dynamic needs of each community**.

2.2.2 Enabling and Using Distributed and Collaborative Technology

The traditional approach to supporting online communities is shared-nothing, as companies build and operate their own infrastructure. This approach is unattractive for MMOG analytics, because the number of potential users varies greatly on both short and long time periods, and the variation is difficult to predict [29]. For example, consider the evolution over time of Zynga's popular MMOG FarmVille. FarmVille achieved over 1 million daily players after 4 days of operation, and over 10 million after 2 months, exceeding any other MMOG before and after it (until mid-2010). Eventually, FarmVille reached its peak of around 28 million daily players and 75 million monthly players in Feb 2010, nine months after launch, but it stands in July 2010 at only 16 million daily players and 62 million monthly players. For shorter time scales, consider the hourly evolution of the number of concurrent players in RuneScape, another popular MMOG (see Figure 3). The visible daily pattern in the figure indicates a strong variation in the number of players over the course of each day. Thus, for MMOGs such as FarmVille and RuneScape it is not worth provisioning resources in advance for continuous analytics. Instead, the challenge for next generation of platforms is to **provision resources on-demand**, that is, when needed and only for as long as they are needed.

2.2.3 Data Management, Data Growth, and Data Storage

Data Management: Web 2.0, Mash-ups, and Provenance As Web 2.0 has become more pervasive, MMOG operators have started to provide access to non-sensitive player data through Web 2.0 interfaces. For example, Jagex's RuneScape provides access to player scores, and so did the now defunct Dungeon Runners by NCsoft. However, Web 2.0 is not common among MMOGs, and most games still provide only traditional Web pages with selected player information. The presence of online APIs allows the creation of data mash-ups. For example, most operators provide information for specific player identities, but do not list the identities themselves; instead, the identities can be provided by other web sites. We have used this two-step process to collect data about an online bridge community [33]. Thus, the main challenge related to data management is to be able **to use and aggregate both Web (legacy) and Web 2.0 information**. Finally, the management of the data also involves the ability to trace back the results to their data source and production process, that is, **to record data provenance information** [27, 28].
Data Growth and Storage. The number of players across all MMOGs has increased exponentially over the past decade [38, 29]; the trend also applies to the first operational months of the popular MMOGs. Additionally, many games increase in complexity over time by adding more characteristics and abilities to the player. Thus, the amount of data produced by these games increases quickly, matching the general IT trend of the past decade to produce more data than there is (or will be) storage for them [5]. The main challenge related to data growth is to **filter out irrelevant data**. The main challenge related to data storage is to **employ data storage systems that can store the predicted amount of data**.

2.2.4 Performance, Scalability, and Robustness

Performance and Scale: MMOGs pose unique data scale and rate challenges. The population size of successful commercial MMOGs ranges from a few thousands of players to a tens of millions of players [29]. Popular MMOGs generate massive amounts of information; for example, the database logging user actions for Everquest 2, a popular MMOG, stored over 20 new terabytes (TB) of data per year for each of the peak years of the game. Other IT projects, such as CERN's Large Hedron Collider or the Sloan Digital Sky Survey, produce data orders of magnitude larger than MMOGs, but these projects are using large and pre-provisioned (expensive) computational and data infrastructure that game companies cannot afford. Furthermore, the data production rate for these other projects is stable over time for spans of days or even weeks, whereas for MMOGs the daily user activity has peaks and may even change hourly (see Figure 3). The main challenge related to performance is that **the system must operate well at MMOG scale**. The main challenge related to scale is that **the system needs to be able to to scale up and down quickly**.

Robustness: MMOGs require robustness to retain and grow their communities. Platform failures are common at large scale [22]; for MMOG analytics, the presence of failures is increased by traffic shaping and failures in the analytics middleware.

However, not responding to failures in the continuous analytics process can quickly lead to community shrinking; we have experienced before [29] *en-masse* departure of gamers, based on rumored or real system problems. Thus, the main challenge related to robustness is to build a system that **is robust**, that is, it minimizes the chances of failure, but also **is able to respond to failures quickly**.

2.3 Other Applications

Building a system for continuous MMOG analytics does not benefit only MMOG communities. Instead, it enables next generation application for the gaming industry and for other domains. We describe in the remainder of this section the most important applications of our research.

Within the gaming industry, the main applications are to audit the group of companies that develop, operate, and distribute the MMOG; to understand the play patterns of users and support future investment decisions; to detect cheating and prevent game exploits; to provide user communities with data for ranking players; to broadcast gaming events; and to produce data for advertisement companies and thus increase the revenue stream for the MMOG owners.

In other domains, the applications may include studying emergent behavior in complex systems (systems theory), understanding the emergence and evolution of the contemporary society [35] (social sciences) and economy [7] (economics), uncovering the use of MMOGs as cures and coping mechanisms [37] (psychology), investigating disease spread models [4] (biology), etc.

3 Related Work

Our goal of designing a generic architecture for continuous analytics in support of social networks for MMOGs places our work at the intersection between large-scale data collection and mining, and data processing using on-demand resources. We survey in this section these two fields, in turn.

3.1 General Data Collection and Mining

Data Collection (crawling) The interest generated by Web search, promoted by companies such as Google and Yahoo, lead to many general web crawling approaches [2, 8, 26, 23]. Garcia-Molina et al. [2, 8] proposed a generic parallel crawling architecture, where the crawler operates via many concurrent crawling processes, either intra-site (parallel) or distributed, with various degrees of coordination between crawling processes (independent, dynamic or static); they also analyzed different crawling modes for static assignment. The topical (focused) crawlers, introduced and studied by Menczer et al. [26], adjust dynamically their crawl according to a rich context provided at start-up, such as topics, queries, user profiles. IRLbot [23] is a generic web crawler designed to scale web crawling to billions of pages using limited resources. IRLbot was used to crawl over 6 billion valid HTML pages while sustaining an average download rate of over 300 MB/s and almost 2,000 pages/s during an experiment that lasted 41 days. Many of these techniques can be employed for acquiring MMOG data, but need adaptation for domain-specific data collection.

Data Mining. The general data mining community has focused on parallel or distributed analytics since the end of the 1990s [34, 25, 3]. For example, Provost and Kolluri [34] examine many basic techniques for scaling up inductive algorithms. *FacetNet* [25] is a framework for analyzing communities and their evolutions in dynamic temporal networks. Unlike traditional two-stage techniques, which separate community extraction and extraction of community evolution, the FacetNet framework combines these two tasks into a unified extraction process. Many of these techniques can be employed for MMOG analytics, but need adaptation for large-scale data processing and on-demand resource use.

More recently, the ability to process data in many small tasks has been pioneered by the industry, with companies such as Google, Yahoo, Microsoft, Facebook, and Twitter starting general data processing frameworks. Google's MapReduce [9] and Yahoo's open-source variant Hadoop support generic data processing at large scale through a data flow model comprised of a much simplified programming abstraction and a runtime environment for this abstraction. Using only map and reduce commands, the user can execute complex data mining processes on distributed computing platforms; however, the application has to keep track of its own objects and data. Domain-specific programming languages such as Pig and Facebook's Hive hide the complexity of this programming model, but are not oriented towards processing MMOG data. Microsoft's Dryad [21] and the language DryadLINQ [41] provide together a data flow execution model, similar to MapReduce, that is most suitable for a wide range of problems from large-scale graph analysis and machine learning. Both MapReduce and Dryad scale and are fault-tolerant. These general tools can be used for continuous MMOG analytics, but would need specific adaptation for efficiently combining the data collection part with data processing.

3.2 Data Collection and Mining Using On-Demand Resources

With the growth of infrastructure capacity, and the trend of commercialization of computing and storage infrastructure, companies and organizations have started to use on-demand resources instead of the traditional self-owned infrastructure. Two main types of infrastructure are currently available for on-demand resource use, grids and clouds. Grids [10] are collections of resources ranging from clusters to supercomputers, shared by various organizations as a computing and data platform that is ubiquitous, uninterrupted, and has uniform user access–in this sense, similar to the power grid. Desktop grids are a flavor of grid computing in which the resources are provided by volunteers, for example through a resource manager such as BOINC. Cloud computing is an infrastructure similar to grid computing, but with simplified access and strict service guarantees; access to resources is paid. We survey in the following the use of each of these platforms for data mining.

Grid Computing. Many grid projects, such as CERN's Large Hedron Collider and the Sloan Digital Sky Survey, have focused on data processing on on-demand resources. For example, Natarajan et al. [30] show that large-scale data mining can benefit from using on-demand resources such as grids, when the processing modules (kernels) minimize the data transfer between grid clusters, and between the compute and storage resources within a cluster. However, grids were not designed to support MMOG analytics workloads. As many of the grid data processing workloads were compute-intensive [18], that is, data processing usually took much longer than the data input and output, grids were optimized for long-running, compute-intensive jobs. Furthermore, with few exceptions, grids were not designed to crawl large amounts of small files, a typical scenario in MMOG analytics.

Cloud Computing. The increase in the number of commercial clouds led to an increase in general data processing projects that can use cloud resources. Google has designed many core tools for data storage, processing, and collaboration, such as the Google Files System [12], MapReduce [9] (provided as a service on on-demand cloud resources by Amazon), and the Google Fusion Tables [13], respectively. The Google Fusion Tables enables users to upload and share tabular data and the results of filtering and aggregating them. Thus, it enables collaboration between users, albeit it currently supports only datasets of up to 100MB. The users still have to write their processing code as queries using a subset of SQL. The *Sector storage cloud* and the *Sphere compute cloud* [14] were designed to perform high performance data mining, scaling up to wide area networks, while using a cloud-based infrastructure.

4 The CAMEO Architecture

In this section we present the CAMEO architecture for continuous MMOG analytics. The CAMEO architecture is built around the idea of enabling continuous MMOG analytics while using resources only when needed. To achieve this goal, it acquires and releases dynamically computational and storage resources from clouds such as Amazon Web Services.

4.1 Overview

The core dataset in CAMEO is the *snapshot*, that is, a read-only dataset which has been extracted from original data (often provided by the MMOG operator). The CAMEO architecture is built around the idea of obtaining representative snapshots, which is a classic problem of creating replicas. Similar to other cases of information replicas in distributed systems, creating exact copies of the data for analysis purposes may not be only expensive, but also unnecessary [40]. Instead of ensuring that the replicas are strongly consistent, our goal is to maintain information replicas whose difference is bounded and the bound is under the control of the analyst.

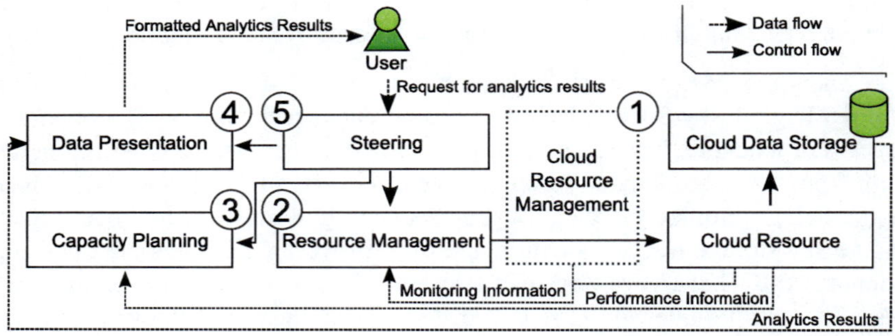

Fig. 4 The CAMEO architecture.

This goal stems from traditional work on quasi-copying [1] and on continuous consistency of information replicas with deviation in the staleness of information [40]. The former model assumes that data writing has a single possible source, which allows for looser consistency guarantees: data in a replica may differ from the original, but the current value of the replica must have been the value of the original at some (earlier) point in time, and the largest delay between an original and the replica can be controlled by the distributed system. The latter model assumes the presence of multiple data writers, and considers inconsistency along multiple axes; the goal here is to bound the absolute inconsistency between replicas.

The CAMEO architecture is designed to collect snapshots of MMOG data and to perform analytics operations on snapshots; all data storage and computation are designed to make use of on-demand resources. The five main components of the CAMEO architecture are depicted in Figure 4. The *Cloud Resource Management* component (component 1 in Figure 4) provides access to the computational and storage resources of the cloud computing environment, and is maintained by the cloud owner. The *Resource Management* component (#2) acquires and releases resources from the cloud and runs the analytics applications. It also uses the monitoring information provided by the cloud resource management and the resources as input for further management actions, such as transparent fault tolerance through application instance replication. The *Capacity Planning* component (#3) is responsible for deciding how many resources must be acquired for the analytics process. The decisions are based on the system's capability to produce results, analyzed during the course of the analytics process, and on the accuracy and cost goals of the process. The *Data Presentation* component (#4) formats and presents the results of the analytics process to the user. The *Steering* component (#5) is responsible for coordinating the analytics process. Towards this end, it takes high-level decisions, expressed through the configuration of each other's component process.

Except for the specific support for MMOGs and for the use of cloud computing resources, our architecture uses a traditional approach. For example, CAMEO's crawling can be classified according to Garcia-Molina et al.'s taxonomy [2, 8] as intra-site and/or distributed (depending on the location of the machines they run on), using static assignment (each process downloads the pages assigned at start), and operating in firewall mode (each process downloads only the pages within its partition). However, the components have unique features specific to the targeted application. We describe in the remainder of this section three distinctive features of CAMEO.

4.1.1 Resource Management Mechanisms

The triggering of the analytics process depends on the nature of the application and on the system status. On the one hand, the nature of the application may allow the system analyst to design a stable analysis process such as a daily investigation of the whole community of players. On the other hand, special analysis may be required when the system is under unexpectedly heavy load, or when many players are located in the same area of the virtual world. To address this situation, we design the Resource Management component to provide two mechanisms for using cloud resources: one static and one dynamic. The *steady analytics*[1] mechanism allows running a periodic analytics operation on cloud resources. The *dynamic analytics* mechanism allows running a burst of analytics operations on cloud resources. Optimizing the allocation of resources for static analytics or for mixed static-dynamic analytics is a target for this component, but beyond the scope of this work. Similarly, the case when the cost of data transfers is significant, that is, similar or higher to the cost of the computational resources, is left for future work.

4.1.2 Steering through Snapshots of Different Size

The analytics process includes collecting the necessary information from the data source. We further *complete snapshot* a snapshot that includes data for all the players managed by the MMOG, and contrast it to a *partial snapshot*. Taking snapshots complies with the continuous analytics definition introduced in Section 2.1.

Depending on the goal of the analysis, it may be possible to obtain meaningful results through continuous analytics based on partial snapshots; for example, when the goal is to obtain statistical information about the player community it may suffice to continuously analyze a randomly chosen group of players of sufficient size. We design the Steering component to be able to perform a two-step analytics process in which first complete snapshots are taken from the system with low frequency, and partial snapshots are acquired often.

4.1.3 Controlling the Process

Taking a snapshot takes time, which depends on the performance of the cloud resources and also on the limitations set by the owners of the original data; to prevent denial-of-service attacks and to improve scalability with the number of requests, it is common for the data owners to limit the network bandwidth available for an individual resource (IP address).

[1] We do not use the term "static" to underline that this is a continuous process.

Assume that a single machine can acquire a new snapshot every T time units (seconds). Then, we can achieve linear scaling (to a certain degree) in the number of acquired snapshots by installing new machines; K machines can acquire K snapshots every T time units. We can then control either how many snapshots we acquire every T time units, or the minimal performance that has to be delivered by each machine to acquire exactly one snapshot every T time units.

4.2 Workflow

In this section we present the flow of data and tasks in CAMEO. Our motivating scenario focuses on a community built around the popular MMOG RuneScape (`http://runescape.com`). The goal of this community is to process player data and post online the results of MMOG analytics.

4.2.1 Data Collection

This data collection process has two parts, identifying players and using player identifiers to obtain data for each player.

There are very few MMOGs or third-party services that offer player identifiers. Instead, these identifiers can be obtained through crawling and then processing web pages made available by either MMOGs or the third-party services, such as high-score lists or forums. RuneScape offers open access to its high-score lists, but not to its forums; several third-party World of Warcraft communities offer open access to high-score lists and forums. RuneScape's high-score lists include the player identifiers of the top 2,000,000 players for over 30 ranking criteria (by each of the 24 skills in the game, by the sum of all skills, by specific in-game achievements, etc.) The crawling system often needs to access hundreds of thousands of web pages to create a comprehensive list of player identifiers; in RuneScape, each list of 2,000,000 players is accessible through web pages that present small chunks of a few tens of players each. Since crawling these pages is not supported through any API, data acquisition of player identifiers can lead to traffic limitation and thus to a serious performance bottleneck. In RuneScape, our crawler was banned for a few hours if more than about 20,000 names were acquired within a few minutes.

Player data can often be obtained individually for each player identifier through a Web 2.0 API. RuneScape and NCsoft's Dungeon Runners offer each an API for collecting player data in this way; for World of Warcraft, several third-party services offer such functionality. Most APIs are explicitly designed to allow collecting data about a single player per request, which leads to millions of requests being necessary to collect data for all players of an MMOG. During this second data collection phase, data is stored, for preservation and to enable later (re-)analysis. For data collected for each player in RuneScape include a triplet ($rank$, $level$, $experience\ points$) for each skill in the game. While acquiring data, CAMEO's crawlers may fail to retrieve information either because it is not offered anymore by the MMOG, or because the server to which the requests are made is overloaded. The latter errors, distinguished by socket-level (thus, API-independent) error messages, are recorded and a second attempt is made later to gather their corresponding data.

4.2.2 Data Processing

Data processing is application-specific and applied to the original or derivative datasets. The original datasets have been collected directly by CAMEO's crawlers. The derivative datasets can be obtained (automatically) by extracting specific pieces of information from the original datasets. For example, a derivative dataset may contain only the player overall skill, which is the sum of all individual skills.

CAMEO already supports two generic types of applications to process any of these two types of datasets: single-snapshot and multi-snapshot statistical analysis of players. Single-snapshot analysis is based on a single snapshot acquired by CAMEO. Examples of results that can already be investigated by CAMEO include: ranking players according to one or more skills (which extends the functionality of the current RuneScape web site); extracting the statistical properties of a skill for the whole community, including the extraction of empirical cumulative distribution functions (CDFs); etc.

It has recently become attractive for system analysts to investigate the evolution of large-scale systems such as peer-to-peer file-sharing [36] and social [24] networks. CAMEO enables such analysis for MMOGs through multi-snapshot analysis. For this type of analysis, multiple datasets acquired by CAMEO at different times are analyzed together, and the timestamp of each dataset can influence the results of the analysis. Thus, analyzing single player evolution and the evolution of the complete community are both possible through multi-snapshot analysis. Examples of multi-snapshot analysis already implemented in CAMEO are: extracting the characteristics of players for the players who improved most during a period; computing the average evolution of the Top-k best players during a period, for arbitrary values of k; etc.

4.3 Addressing the Challenges of Continuous MMOG Analytics

In the remainder of this section we show, in turn, how the CAMEO architecture addresses each of the four challenges of continuous MMOG analytics introduced in Section 2.2.

4.3.1 Understanding User Community Needs

CAMEO addresses the main challenge of understanding the user community needs, to understand the specific and dynamic needs of each community, in two ways.

As shown in Section 2.2.2, the number of players in an MMOG is highly dynamic, both over short and long time periods. CAMEO adapts to population dynamics by identifying the number of players dynamically. In the example presented in Section 4.2, CAMEO first extracts the active player identifiers from the high-score lists, and only then collects detailed information about each player. This mechanism can also be used to support communities with a specific focus, for example by obtaining player identifiers only for the best players featuring a specific skill or achievement and pruning out the other players.

CAMEO already supports a wide variety of specific types of analysis. First, it can analyze various pieces of player information, such as skill, experience points, and rank given by the MMOG. Second, as explained in Section 4.2.2, CAMEO can process information from a single or from multiple snapshots, allowing for single time point and evolution analysis. Third, there are many types of specific analysis that CAMEO already implements: ranking players according to one or more skills (which extends the functionality of the current RuneScape web site); extracting the statistical properties for the whole community for one or more skills; extracting the characteristics of the k players who improved most during a period (e.g., week); computing the (average) evolution of the Top-k best players during a period; identifying specific players with unique combined skills; etc. We leave as future work the task of supporting more types of analysis.

4.3.2 Enabling and Using Distributed and Collaborative Technology

The CAMEO architecture is built around the use of on-demand resources, provisioned mainly from the cloud. In practice, we have used CAMEO on top of the Amazon Web Services cloud and our own, locally-installed, Eucalyptus-based [31] cloud.

One of the fine points of supporting MMOG analytics is the ability to collect and process millions of small web pages. For this workload, the data collection time is dominated by Internet latencies. The collection time can be greatly reduced if the resource collecting data is located near the data server. There currently exist hundreds of cloud providers, and several of them, including Amazon, have multiple sites spread across the world. While we did not implement a location-aware cloud selection mechanism, we show in our experiments in Section 6.2 that collection time can be reduced greatly through location-aware resource selection. For the same workload, data processing time can be reduced by caching the data, pipelining the use of data (for example through multi-threading), or by grouping data and storing it in larger chunks; the latter technique is employed by Google's File System [12].

Using on-demand resources can help alleviate the traffic limitations imposed by some web sites, per IP address. For example, RuneScape will limit the amount of player identifiers that a single IP address can access during an interval of a few minutes to about 20,000 players, but will not limit in any way the amount of player data; only the latter is accessed through a Web 2.0 API. We associate this behavior, which is typical for several MMOGs, with poor practice in designing the access APIs. Bypassing this limitation by leasing new cloud resources when the old resources are banned can be very costly. Since the typical charging interval is an hour, using a leased resource for only a few minutes before its traffic becomes limited is impractical. Instead, CAMEO uses cloud services that allow the IP address of a leased resource to change in time, such as the Elastic IP Address (EIP) service provided by Amazon EC2. An EIP is a static IP address that is owned by the cloud provider

Table 1 Comparison of storage solutions. The aspects are rated through a 7-level Likert scale, from "– – –" to "X" to "+ + +", where the "–", "X", and "+" signs denote negative, neutral, and positive appreciation, respectively.

	Centralization	Traffic Speed	Reliable	Small Load	Medium load	Large load
Cloud machine	–	+ + +	+ + +	+ + +	+ +	–
Local storage	+ +	– – –	– –	+ + +	+ + +	+ +
Cloud storage	+ + +	+ +	+ + +	+ + +	+ + +	+ + +

and can be leased for use. An EIP can be attached to any instance that is currently running, replacing the old IP address it exposed outside the cloud. EIP addresses are associated with an account and not a particular instance; thus, they are just another resource to be managed by CAMEO.

4.3.3 Data Management, Data Growth, and Data Storage

CAMEO can use and aggregate both Web (legacy) and Web 2.0 information by design. Since there are no standards in MMOG player data presentation, each community will have to write its own data parsers. CAMEO already provides parsers for all the player data offered by the RuneScape operators.

To simplify data management and recording data provenance information, CAMEO stores data centrally, that is, using a single storage administrator (e.g., Amazon S3). For management, CAMEO interacts automatically with the storage administrator.

Three main solutions to store data are available to CAMEO: store the data on the same machine that acquires or generates it; store the data outside the cloud; and store the data using the dedicated cloud storage services. Transferring data in/out of the cloud incurs additional costs and has much lower performance than intra-cloud data transfers. The dedicated cloud storage services are much more reliable than any machine in the cloud. We analyze in the following each solution; Table 1 compares the solutions for three load levels, small, medium, and large. Storing data on the same machine means that the machine gathering the data preserves the data for further local processing, or for being used by another machine; this forces the data processing application to keep track of data location, and effectively breaks the CAMEO assumption that data are stored under a central administrator. This solution also has limited storage capacity (per machine) and limited reliability. Storing data outside the cloud has the downside of low data transfer performance, with data transfers into the cloud being particularly slow. The reliability of the storage solution is another issue for this solution. Storing the data using the dedicated cloud services is a solution that keeps the data inside the cloud as much as possible, and transfers outside the cloud only the results that are needed. Cloud storage is centralized, more reliable than storing data on a cloud machine, and faster than transferring data from outside the cloud to the processing machine; however, cloud storage can be expensive.

CAMEO stores the data using the dedicated cloud storage services, by default. Thus, the ability of CAMEO to store MMOG data matches the ability of the cloud infrastructure. For example, Amazon S3 is used as the storage solution in the experiments using Amazon EC2 for computation. In S3, users store data in labeled "buckets" (for our purpose, the equivalent of directories, with provenance support). We have already stored several terabytes of data using Amazon S3.

CAMEO can filter out irrelevant data to the extent by which the MMOG API supports this feature, for example by pruning out players (described in Section 4.3.1) or by allowing the acquisition of specific data fields for each player.

4.3.4 Performance, Scalability, and Robustness

The performance of CAMEO is upper-bounded by the performance of the leased cloud resources, and by the location of the data.

The scalability and robustness challenges are eased through the use of cloud resources. Scalability-wise, cloud resource allocations are designed to scale up and down quickly; our previous evaluations of Amazon EC2 scaling capability [32, 20] reveal that the resource allocation time is below two minutes when allocating a single resource of the type used throughout this work. Robustness-wise, in our experience with this work and with benchmarking four clouds [19], clouds are much more robust than grids (see the grid failure data in the Failure Trace Archive [22]).

In the experiments presented in Section 6.1, the reference CAMEO implementation was able to scale and perform without failures the largest MMOG analytics experiment to-date.

5 Experimental Setup

To test CAMEO in practice we follow the scenario introduced in Section 4.2. In this scenario, CAMEO performs continuous analytics on RuneScape, a popular MMOG, and uses resources leased from one of two clouds, the commercial Amazon Web Services and the Eucalyptus-based cloud installed locally. We describe in the remainder of this section the experimental setup we have used.

5.1 CAMEO Implementation

The reference CAMEO implementation was written in Python, which makes it portable for a wide range of platforms but with a lower performance than can be achieved in other programming languages, such as C. We have written RuneScape-specific web crawlers for the data collection process.

For enabling S3 and EIPs support we have used boto [6], an integrated interface to services offered by Amazon Web Services such as the Elastic Compute Cloud (EC2) and the Simple Storage Service (S3).

We have also used two additional tools for debugging purposes. Hybridfox is a Firefox extension for managing an Amazon EC2 account. With Hybridfox, launching, stopping, connecting to, and monitoring machines leased from clouds can be done via a graphical interface as opposed to standard command line interface.

```
0     Herccrazy  4306  2321  325506318  2127   99   42880737     24922  99
      15801687   24171 99    19000663   5746   99   34673693     5755   99
      20804786   13596 99    13057112   26763  99   13892161     47528  99
      13270010   54709 99    13162562   13561  99   14118083     178680 82
      2568328    33054 99    13052685   8879   99   13052970     15712  90
      5365614    46742 85    3403281    17862  90   5366501      16882  85
      3499380    10024 99    13167150   7451   99   13155100     11879  93
      7390556    214   99    23932428   57488  82   2421684      7746   90
      5361562    6636  99    13069715   106769 40   37870 14673  1830   -1
      -1    -1   -1    207371     1276  -1     -1   20899 1906   6645
      2424  32885 1138 69240 1544  -1   -1
```

Fig. 5 Player data example.

Table 2 The resource characteristics for the instance types offered by Amazon EC2.

Resource Type	Cores (ECUs)	RAM [GB]	Architecture [bit]	I/O Performance	Disk [GB]	Cost [$/h]
m1.small	1 (1)	1.7	32	Med	160	0.085
m1.large	2 (4)	7.5	64	High	850	0.34
m1.xlarge	4 (8)	15.0	64	High	1,690	0.68
c1.medium	2 (5)	1.7	32	Med	350	0.17
c1.xlarge	8 (20)	7.0	64	High	1,690	0.68
...						

Hybridfox can also be used with the private cloud at UPB. We have also used the S3 Organizer as a visual interface for managing the contents of the S3 bucket.

5.2 MMOG Case Study: RuneScape

RuneScape is a popular MMOG, ranking in Aug 2008 as second by number of players in the US and European markets among MMORPG (about 7 million active players, second to World of Warcraft), and number one by number of opened accounts (over 135 million). RuneScape offers detailed player data through a Web 2.0 interface. Figure 5 shows the data gathered for one player. Besides the index (0) and the name ("Herccrazy"), the data includes a triplet $(rank, level, experiencepoints)$ for each skill available to RuneScape players, and other information (see Section 4.2.1).

5.3 Cloud Infrastructure

We have used in our experimental setup two cloud platforms: Amazon EC2 and a Eucalyptus-based private cloud. We describe the two platforms in the following.

Amazon Web Services We have used the commercial cloud Amazon Web Services. Amazon EC2 provides the computational resources to acquire and process Runescape data. The EC2 user can use any of a number of resource (*instance*) types currently available on offer, the characteristics of which are summarized in Table 2.

An ECU is the equivalent CPU power of a 1.0-1.2 GHz 2007 Opteron or Xeon processor. The theoretical peak performance can be computed for different instances from the ECU definition: a 1.1 GHz 2007 Opteron can perform 4 flops per cycle at full pipeline, which means at peak performance one ECU equals 4.4 giga float operations per second (GFLOPS). Throughout the experiments conducted for this work we have used the `m1.small` instances; extending the Capacity Planning module with the ability to use multiple instance types is left as future work. We have also used Amazon S3 for storage and the Elastic IP services to alleviate traffic limitations imposed by the MMOG data provider (see Section 4.3.2).

Table 3 The resource characteristics for the instance types offered by the UPB cloud.

Resource Type	Compute Units	RAM [GB]	Architecture [bit]	Disk [GB]
m1.small	1	192	64	4
m1.large	1	256	64	4
m1.xlarge	2	512	64	4
c1.medium	2	1024	64	4
c1.xlarge	4	2048	64	4

UPB A private cloud was set up at Politehnica University, Bucharest, Romania (UPB) is based on Ubuntu Enterprise Cloud, which in turns relies on the open-source cloud middleware Eucalyptus. This cloud offers Amazon EC2-like services allowing us to run the same experiments as on the Amazon cloud. Five instance types are available with characteristics detailed in Table 3. One compute unit in this case is the equivalent of an Intel Xeon processor at 2 Ghz. Although the private cloud is able to support elastic IP assignment we did not use this service because we wanted to conduct an experiment without EIP support for comparison purposes. Generated data is stored inside the storage attached to the private cloud.

6 Experimental Results

Using CAMEO, we have taken and analyzed several complete snapshots of the state of RuneScape over a period of one and a half years. We have also also taken partial snapshots of the state of RuneScape in quick succession, which enabled us to study the short-term dynamics of the RuneScape community. In this section we show evidence of CAMEO being able to fulfill in practice the challenges of continuous MMOG analytics. To this end, we present sample results rather than perform an in-depth performance evaluation or a detailed analysis of the RuneScape community. For an in-depth performance evaluation we refer to our previous work [42].

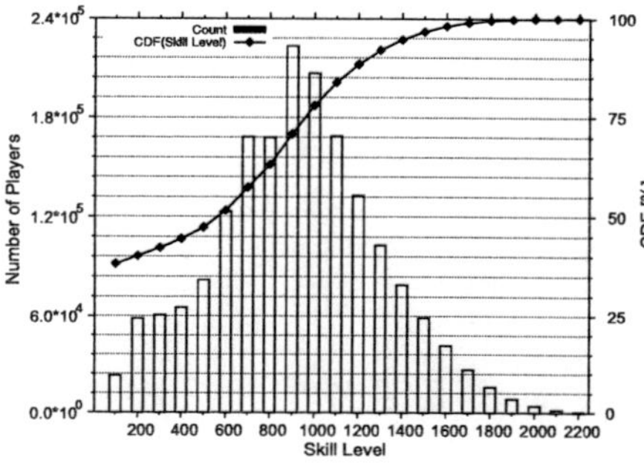

Fig. 6 Pareto graph, that is, combined PDF (left vertical axis) and CDF (right vertical axis) depiction of the skill level of the RuneScape player population. Each bar represents a range of 100 levels. CDF stands for cumulative distribution function; $CDF(x)$ is the total number of players with skill level up to and including x. Note that the left vertical axis is not linear. See text for why the CDF of the skill level does not start at 0%.

6.1 Understanding User Community Needs

We exemplify in this section the ability of the reference CAMEO implementation to adapt to the specific and dynamic user community needs.

Using CAMEO, we analyzed the skill level of millions of RuneScape players, which shows evidence that CAMEO can be used for measurements several orders of magnitude larger than the previous state-of-the-art [39]. CAMEO collected in August 2008 official skill level data for 2,899,407 players. We have repeated the experiment in 2009 and 2010; for these measurements, the population size has increased to over 3,000,000 active players, despite a high rate of player abandonment (more than 25% of the players identified in 2008 are not present in the 2010 experiments). The success of this experiment across multiple years also demonstrates the ability of CAMEO to adapt to long-term changes in the population size.

CAMEO automatically analyzes the player skill distribution; from the 2008 dataset, 1,817,211 (over 60%) players had an overall skill level above 100; the maximum overall skill level was 2280 in 2008. The values for players with skill level below 100 include application-specific noise (mostly starting players) and are therefore polluted. Thus, we present here only data for all players with skill above 100, and for a single measurement. Figure 6 depicts the overall skill level of RuneScape players, with bins of 100 levels. The number of players per bin is well characterized by a skewed normal-like distribution; the majority of the players are of average skill or below, the most populated skill level bins are those corresponding to the middle skill level, and the number of high-level players is significant. We have explored the implications of the overall skill level distribution in our previous work on automatic content generation [16, 17].

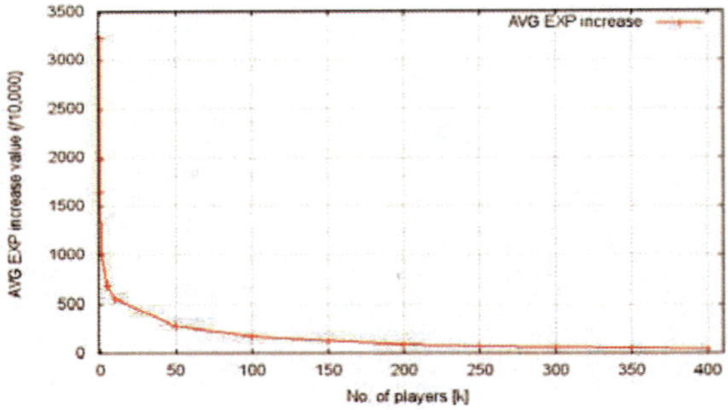

Fig. 7 Average experience increase over a period of 11 days for the Top-*k* RuneScape players. The values of *k* are in in 1,000, e.g., $k = 400$ refers to 400,000 players.

Table 4 The Top-15 players by increase of total experience points, over an 11-day period.

Top-15 Rank	RuneScape Overall Rank	Experience Points gained
1	272	49,880,094
2	28	39,233,769
3	101,092	37,635,063
...		
11	364,344	23,710,650
12	357,856	23,002,277
13	65	21,061,414
14	127,681	20,308,110
15	53,922	20,181,985

We show in the following two types of analysis enabled by CAMEO; both use multiple datasets and follow the evolution RuneScape players over a period of 11 days. We exemplify with a use scenario each of the two types.

Often, MMOG designers cannot account for real (emerging) gameplay, which leads to playing difficulty that is too high or too low in comparison with design expectations. Several large player communities asked and obtained from the designers of RuneScape to revert particular design changes [29]. The average experience increase is a type of analysis that can show, for a community of players, if their advancement rate is similar to the design expectation. Figure 7 shows the average experience increase for the Top-*k* players for various values of *k*. The decrease in the average is abrupt, with a Pareto-like shape; a few top players dominate the others in the amount of experience obtained. Depending on what the community wishes, the game designers may be asked to "even out" the difficulty of the game.

Advancing only one skill or a very reduced set of skills is called *skilling*, and can lead to good rewards in MMOGs, because players are often rewarded by overall skill instead of their best skill. However, players who are skilling (*skillers*) cannot perform the skilling activity alone for very long periods of time unless they find a group of players that needs the skill, because skilling is a highly repetitive activity. Thus, identifying active skillers can be used to form groups of highly-effective, yet lowly-ranked, players and at the same time prevent skillers from becoming bored and leaving the game. CAMEO can find skillers by analyzing the characteristics of individual players, and can find *active* skillers by observing the evolution of a player's characteristics (stats) over time. Table 4 depicts the Top-15 players, ranked by the number of experience points gained over the 11 day period of our observation. Four out of the top fifteen players are ranked 100,000 or lower in RuneScape's overall skill classification, and two of these four are even ranked lower than rank 350,000 by the same criterion. This gives evidence that even lowly-ranked players can quickly advance in experience, using their best skill or group of skills; these players are active skillers that would help most both themselves and the group of players they will join.

6.2 Enabling and Using Distributed and Collaborative Technology

To demonstrate the capability of CAMEO to perform both dynamic and steady analytics, and to monitor the process, we show in Figure 8 the evolution of the cumulative number of consumed CPU hours over time. The dynamic analytics are based on uneven bursts of activity, of which the burst during March 10 is the most prominent. The steady analytics part of the experiments reveals an even use of resources over time, with the steps indicating a new work cycle.

To give a first estimation on the cost of continuous MMOG analytics, we use a simple analytics process that acquires partial snapshots and only browses the data in memory during the processing phase. Figure 9 shows the total cost incurred by the continuous analytics process over the course of one month. For this simple analysis process the cost is below $500 per month. It is not our intention to argue that the cost of continuous analytics for an MMOG can be this low; much more complex

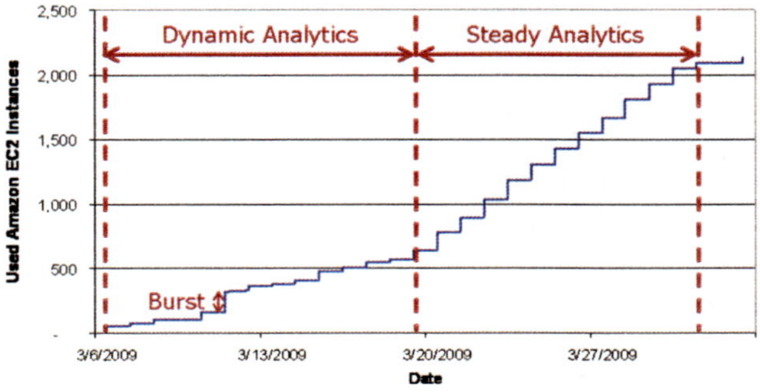

Fig. 8 Resource consumption in the two analytics modes: dynamic and static.

Billing Statement: April 1, 2009

Billing Cycle for this Report: March 1 - March 31, 2009

Expand All | Collapse All

Rate	Usage	Totals
Amazon Elastic Compute Cloud View/Edit Service		
Amazon EC2 running Linux/UNIX		
$0.10 per Small Instance (m1.small) instance-hour (or partial hour)	2,097 Hrs	209.70
Amazon EC2 Bandwidth		
$0.100 per GB Internet Data Transfer - all data transfer into Amazon EC2	611.005 GB	61.10
$0.170 per GB Internet Data Transfer - first 10 TB / month data transfer out of Amazon EC2	507.121 GB	86.21
Taxes		**67.83**
Charges due on April 1, 2009†		**424.85**

Fig. 9 Putting a cost on continuous analytics for MMOGs.

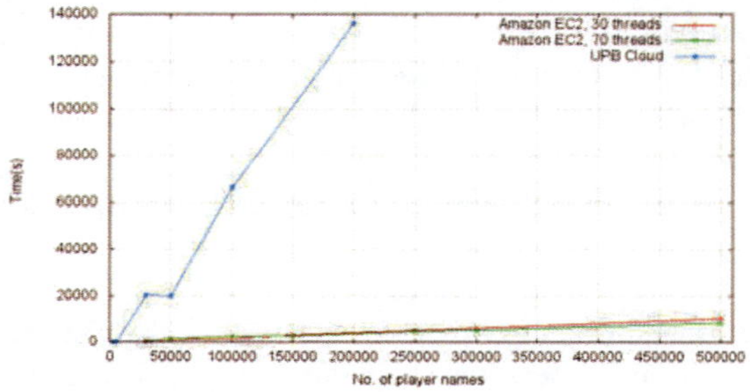

Fig. 10 Performance of collecting player identifiers for three configurations, Amazon EC2 with 30 and 70 collection threads, and UPB cloud with 30 collection threads.

analytics taking many more computational hours are performed for any of the applications presented in Sections 2.3 and 6.1.

To demonstrate the benefits of using EIPs, we collect player identifiers from the RuneScape servers located in the US (West Coast) using a single cloud machine. We use, in turn, the closely located resources of Amazon and the remote resources of UPB; besides a different location, UPB does not use EIPs. We also vary the number of threads for the faster Amazon location, which allows us to estimate the benefit of using EIPs in comparison with using traditional parallelism through multithreading. Figure 10 shows the time spent by each approach in acquiring player identifiers for a number of players ranging from 10,000 to 500,000. As expected, for UPB the time needed to collect the same amount of player names without EIP support increases quickly with the number of player identifiers, because of the repeated bans. In contrast, the Amazon EC2 uses the EIP mechanism to alleviate traffic limitations imposed by the MMOG data provider (see Section 4.3.2). This way, a single machine can be used effectively to collect the player identities in a reasonable amount of time. The number of threads does not affect the performance similarly to the number of EIPs, and may lead to data collection loss as many concurrent requests can lead to the requesting IP address being banned (not shown here).

7 Conclusion and Future Work

The expanding world of Massively Multiplayer Online Games (MMOGs) fosters important derivative online applications and raises interesting new challenges to the distributed computing community. In this work we present CAMEO, an architecture for MMOG analytics in the cloud that mines MMOG data from the Web, collects information using the Web 2.0 interfaces provided by various MMOG operators and their collaborators, integrates the information into comprehensive and time-spanning MMOG datasets, analyzes the datasets, and presents the results. CAMEO hides the complexity of detailed resource allocation and use from the user, and operates on top of clouds to provision resources on-demand, that is, only when and for long they are needed.

We have implemented and deployed CAMEO in practice, and were able to perform various large-scale analysis processes on data provided by the popular MMOG RuneScape over a period of over two years. Using resources provisioned on-demand from Amazon Web Services, a commercial cloud, we have analyzed the characteristics of almost 3,000,000 players, and followed closely the progress of 500,000 players for over a week. Our results give evidence that cloud computing resources can be used for continuous MMOG data acquisition and analysis. Finally, we have provided a first cost estimation for the continuous MMOG analytics process.

The analysis capabilities already implemented in CAMEO already cover a wide range of community uses. However, we identify as future research directions the use of these results to find and reward good players, to single out potential cheaters, to identify good matches between players and to make grouping recommendations based on them, etc. Each of these uses of MMOG analytics comes with distinct design, implementation, and testing challenges, but will potentially impact the lives of a large number of people.

Acknowledgments

We would like to acknowledge the help of Prof. Dr. Nicolae Ţăpuş, from Politehnica University of Bucharest, who enabled this collaboration.

References

1. Alonso, R., Barbará, D., Garcia-Molina, H.: Data caching issues in an information retrieval system. ACM Trans. Database Syst. 15(3), 359–384 (1990)
2. Arasu, A., Cho, J., Garcia-Molina, H., Paepcke, A., Raghavan, S.: Searching the web. ACM Trans. Internet Technol. 1(1), 2–43 (2001), http://doi.acm.org/10.1145/383034.383035

3. Bayardo, R.J., Ma, Y., Srikant, R.: Scaling up all pairs similarity search. In: International Conference on the World Wide Web (WWW), pp. 131–140 (2007)
4. BBC NEws, Virtual game is a 'disease model'. News Item (2009), http://news.bbc.co.uk/2/hi/6951918.stm
5. Berman, F.: Got data?: a guide to data preservation in the information age. Commun. ACM 51(12), 50–56 (2008)
6. Boto: Boto - a python interface to amazon web services, http://code.google.com/p/boto/
7. Castronova, E.: On virtual economies. Game Studies 3(2) (2003)
8. Cho, J., Garcia-Molina, H.: Parallel crawlers. In: International Conference on the World Wide Web (WWW), pp. 124–135 (2002)
9. Dean, J., Ghemawat, S.: MapReduce: simplified data processing on large clusters. Commun. ACM 51(1), 107–113 (2008)
10. Foster, I.T., Kesselman, C., Tuecke, S.: The anatomy of the grid: Enabling scalable virtual organizations. Int'l. J. of High Performance Computing Applications 15(3), 200–222 (2001)
11. Fritsch, T., Voigt, B., Schiller, J.H.: Distribution of online hardcore player behavior (how hardcore are you?). In: Workshop on Network and System Support for Games (NETGAMES), p. 16 (2006)
12. Ghemawat, S., Gobioff, H., Leung, S.T.: The Google File System. SIGOPS Oper. Syst. Rev. 37(5), 29–43 (2003)
13. Gonzalez, H., Halevy, A.Y., Jensen, C.S., Langen, A., Madhavan, J., Shapley, R., Shen, W., Goldberg-Kidon, J.: Google Fusion Tables: web-centered data management and collaboration. In: SIGMOD 2010: Proceedings of the 2010 International Conference on Management of Data, pp. 1061–1066. ACM, New York (2010)
14. Grossman, R.L., Gu, Y.: Data mining using high performance data clouds: Experimental studies using sector and sphere. CoRR (2008), http://arxiv.org/abs/0808.3019
15. Iosup, A.: CAMEO: Continuous analytics for massively multiplayer online games on cloud resources. In: Lin, H.-X., Alexander, M., Forsell, M., Knüpfer, A., Prodan, R., Sousa, L., Streit, A. (eds.) Euro-Par 2009. LNCS, vol. 6043, pp. 289–299. Springer, Heidelberg (2010)
16. Iosup, A.: POGGI: Puzzle-based Online Games on Grid Infrastructures. In: Sips, H., Epema, D., Lin, H.-X. (eds.) Euro-Par 2009. LNCS, vol. 5704, pp. 390–403. Springer, Heidelberg (2009)
17. Iosup, A.: POGGI: generating puzzle instances for online games on grid infrastructures. Concurrency and Computation: Practice and Experience (2010) (accepted April 2010, in print), doi: 10.1002/cpe.1638
18. Iosup, A., Dumitrescu, C., Epema, D.H.J., Li, H., Wolters, L.: How are real grids used? the analysis of four grid traces and its implications. In: IEEE/ACM International Conference on Grid Computing (GRID), pp. 262–269. IEEE, Los Alamitos (2006)
19. Iosup, A., Ostermann, S., Yigitbasi, N., Prodan, R., Fahringer, T., Epema, D.: Performance analysis of cloud computing services for many-tasks scientific computing. IEEE Trans. on Parallel and Distrib. Sys. (2010) (accepted September 2010, in print)
20. Iosup, A., Yigitbasi, N., Epema, D.: On the performance variability of production cloud services. Tech.Report, TU Delft (2010), pds.twi.tudelft.nl/reports/2010/PDS-2010-002.pdf

21. Isard, M., Budiu, M., Yu, Y., Birrell, A., Fetterly, D.: Dryad: distributed data-parallel programs from sequential building blocks. In: EuroSys, pp. 59–72. ACM, New York (2007)
22. Kondo, D., Javadi, B., Iosup, A., Epema, D.: The Failure Trace Archive: Enabling comparative analysis of failures in diverse distributed systems. In: IEEE/ACM International Symposium on Cluster Computing and the Grid (CCGRID), pp. 398–407 (2010), Archive data available: `http://fta.inria.fr`
23. Lee, H.T., Leonard, D., Wang, X., Loguinov, D.: Irlbot: Scaling to 6 billion pages and beyond. ACM Transactions on the Web (TWEB) 3(3) (2009)
24. Leskovec, J., Backstrom, L., Kumar, R., Tomkins, A.: Microscopic evolution of social networks. In: ACM SIGKDD International Conference on Knowledge Discovery and Data Mining (KDD), pp. 462–470. ACM, New York (2008)
25. Lin, Y.R., Chi, Y., Zhu, S., Sundaram, H., Tseng, B.L.: Facetnet: a framework for analyzing communities and their evolutions in dynamic networks. In: Huai, J., Chen, R., Hon, H.W., Liu, Y., Ma, W.Y., Tomkins, A., Zhang, X. (eds.) International Conference on the World Wide Web (WWW), pp. 685–694. ACM, New York (2008)
26. Menczer, F., Pant, G., Srinivasan, P.: Topical web crawlers: Evaluating adaptive algorithms. ACM Transactions on Internet Technology (TOIT) 4(4), 378–419 (2004)
27. Miles, S., Groth, P.T., Deelman, E., Vahi, K., Mehta, G., Moreau, L.: Provenance: The bridge between experiments and data. Computing in Science and Engineering 10(3), 38–46 (2008)
28. Muniswamy-Reddy, K.K., Macko, P., Seltzer, M.I.: Making a cloud provenance-aware. In: Workshop on the Theory and Practice of Provenance. USENIX (2009)
29. Nae, V., Iosup, A., Podlipnig, S., Prodan, R., Epema, D.H.J., Fahringer, T.: Efficient management of data center resources for massively multiplayer online games. In: ACM/IEEE Conference on High Performance Networking and Computing (SC). IEEE/ACM (2008)
30. Natarajan, R., Sion, R., Phan, T.: A grid-based approach for enterprise-scale data mining. Future Generation Comp. Syst. 23(1), 48–54 (2007)
31. Nurmi, D., Wolski, R., Grzegorczyk, C., Obertelli, G., Soman, S., Youseff, L., Zagorodnov, D.: The eucalyptus open-source cloud-computing system. In: IEEE/ACM International Symposium on Cluster Computing and the Grid (CCGRID), pp. 124–131 (2009)
32. Ostermann, S., Iosup, A., Yigitbasi, M.N., Prodan, R., Fahringer, T., Epema, D.: An early performance analysis of cloud computing services for scientific computing. In: CloudComp. LNICST, vol. 34, pp. 1–10. Springer, Heidelberg (2009)
33. Posea, V., Balint, M., Dimitriu, A., Iosup, A.: An analysis of the BBO Fans online social gaming community. In: RoEduNet International Conference (RoEduNet), pp. 218–223. IEEE, Los Alamitos (2010)
34. Provost, F.J., Kolluri, V.: A survey of methods for scaling up inductive algorithms. Data Min. Knowl. Discov. 3(2), 131–169 (1999)
35. Steinkuehler, C., Williams, D.: Where everybody knows your (screen) name: Online games as "third places". In: DIGRA Conf. (2005)
36. Stutzbach, D., Rejaie, R., Sen, S.: Characterizing unstructured overlay topologies in modern p2p file-sharing systems. IEEE/ACM Trans. Netw. 16(2), 267–280 (2008)
37. Williams, D., Yee, N., Caplan, S.: Who plays, how much, and why? debunking the stereotypical gamer profile. Journal of Computer-Mediated Communication 13(4), 993–1018 (2008)
38. Woodcock, B.S.: An analysis of mmog subscription growth. Online Report (2006), `http://www.mmogchart.com` (November 2008)

39. Yee, N.: The demographics, motivations, and derived experiences of users of massively multi-user online graphical environments. Presence 15(3), 309–329 (2006)
40. Yu, H., Vahdat, A.: Efficient numerical error bounding for replicated network services. In: VLDB 2000: Proceedings of the 26th International Conference on Very Large Data Bases, pp. 123–133. Morgan Kaufmann Publishers Inc., San Francisco (2000)
41. Yu, Y., Isard, M., Fetterly, D., Budiu, M., Erlingsson, Ú., Gunda, P.K., Currey, J.: Dryadlinq: A system for general-purpose distributed data-parallel computing using a high-level language. In: OSDI, pp. 1–14. USENIX (2008)
42. Iosup, A., Lascateu, A., Tapus, N.: CAMEO: Enabling Social Networks for Massively Multiplayer Online Games through Continuous Analytics and Cloud Computing. In: Workshop on Network and System Support for Games (NETGAMES), vol. 7, pp. 1–6. IEEE Press, Piscataway (2010)

Chapter 13
Adaptive Fuzzy Inference Neural Network System for EEG and Stabilometry Signals Classification

Pari Jahankhani, Juan A. Lara, Aurora Pérez, and Juan P. Valente

Abstract. The focus of this chapter is to study feature extraction and pattern classification methods from two medical areas, Stabilometry and Electroencephalography (EEG).

Stabilometry is the branch of medicine responsible for examining balance in human beings. Balance and dizziness disorders are probably two of the most common illnesses that physicians have to deal with. In Stabilometry, the key nuggets of information in a time series signal are concentrated within definite time periods are known as events.

In this chapter, two feature extraction schemes have been developed to identify and characterise the events in Stabilometry and EEG signals. Based on these extracted features, an Adaptive Fuzzy Inference Neural network has been applied for classification of Stabilometry and EEG signals.

The model constructs its initial rules by a hybrid supervised/unsupervised clustering scheme while its final fuzzy rule base is optimised through competitive learning. A two-stage learning methodology is applied to this Neuro-Fuzzy structure, by incorporating gradient descent and recursive least squares estimations. The proposed modelling scheme is characterised by its high performance accuracy, high training speed and provides an efficient solution to the "curse of dimensionality" problem inherited in traditional neuro-fuzzy schemes.

In order to classify Stabilometric time series, a set of balance-related features have been extracted according to the expert's criteria.

The proposed Stabilometric medical diagnostic system is based on a method for generating reference models from a set of time series.

The experimental results validated the proposed methodology.

Pari Jahankhani
University of East London, School of Computing, Information Technology and Electronic, London E16 2RD, United Kingdom
e-mail: `pari@uel.ac.uk`

Juan A. Lara · Aurora Pérez · Juan P. Valente
Technical University of Madrid, School of Computer Science, Campus de Montegancedo, 28660, Boadilla del Monte, Madrid, Spain
e-mail: `j.lara.torralbo@upm.es, {aurora, jpvalente}@fi.upm.es`

N. Bessis, F. Xhafa (Eds.): Next Generation Data Technologies for CCI, SCI 352, pp. 329–355.
springerlink.com

1 Introduction and Background

The first medical diagnoses made by humans were based on what ancient physicians could observe with their eyes and ears. Most sophisticated diagnostic tools and techniques such as the thermometer for measure temperature and the stethoscope for measuring hear rate were not used until the end of 19th century. In the 19th century, diagnostic tools, including the microscope and X-ray helped provide hard data independent of subjective judgement and anecdote [1].

Number of medical diagnostic decision support (DS) systems based on computational intelligence methods has been developed for medical diagnosis. The medical diagnostic knowledge can be automatically derived from the description of cases solved in the past.

In medicine data mining is often used to complement and expand the work of the clinician and researcher by expanding knowledge rather than providing new knowledge as is the trend in other domains.

Data mining is the research area involving powerful processes and tools that allow an effective analysis and exploration of usually large amounts of data. Data mining techniques have found application in numerous different scientific fields with the aim of discovering previously unknown patterns and correlations.

Classifiers can play an intermediate role in multilevel filtering systems. In medical practice, the collection of patient data is often expensive, time consuming and harmful for the patients. Therefore it is necessary to have a classifier that is able to reliable diagnose with a small amount of data, also the process of determining the right subset of data may be time consuming as it is essentially a combinatorial problem.

The derived classifier can then be used either to assist the physician when diagnosing new patients in order to improve the diagnostic speed and accuracy.

Some medical diagnosis systems based on computational intelligence methods use expert systems (ESs) [2,3], fuzzy expert systems (FESs) [4,5,6], neural networks (NNs) [7,8,9,10] and genetic algorithms (GAs) [11]. Fuzzy Logic (FL) [12] is a "language", which uses syntax and local semantics where we can imprint any qualitative knowledge about the problem to be solved.

With the continuously growing demand for models of complex systems inherently associated with nonlinearity, high-order dynamics, time-varying behaviour, and imprecise measurements there is a need for a relevant modelling environment. Efficient modelling techniques should allow for a selection of pertinent variables and a formation of highly representative datasets. The models should be able to take advantage of the existing domain knowledge (such as a prior experience of human observers or operators) and augment it by available numeric data to form a coherent data knowledge modelling entity.

Fuzzy systems accept numeric inputs and convert these into linguistic values (represented by fuzzy numbers) that can be manipulated with linguistic IF-THEN rules and with fuzzy logic operations, such as fuzzy implication and composition rules of inference. However, at present there is no systematic procedure for the design of a fuzzy system. Usually the fuzzy rules are generated by converting human operators' experience into fuzzy linguistic form directly and by summarizing the system behaviour (sampled input-output pairs) of the operators. But designers find it difficult to obtain adequate fuzzy rules and membership functions because these are most likely to be influenced by the intuitiveness of the operators and the designers. Moreover, some information will be lost when human operators express their experience by linguistic rules. This results in a set of fuzzy rules which are usually not optimal. Thus a fuzzy system which is able to develop and improve its fuzzy rules and structure automatically on the basis of the monitoring of human controllers is highly desired.

During recent years, the fuzzy neural network approach has gained considerable interest for solving real world problems, including modelling and the control of highly complex systems, signal processing and pattern recognition [13]. Extensive experimentation has demonstrated that the class of feed-forward fuzzy neural networks exhibits a number of significant advantages compared to the neural network models [14]. First, the neural networks are global models where training is performed on the entire pattern range. On the contrary, owing to the partition of the input space, the fuzzy models perform a fuzzy blending of local models in space. As a result, faster convergence is achieved during learning for a specific task. Secondly, fuzzy neural networks are capable of incorporating both numerical data (quantitative information) and expert's knowledge (qualitative information) and describe them in the form of linguistic IF-THEN rules. In that respect, they provide a unified framework for integrating the computational parallelism and low-level learning of neural networks with the high-level reasoning of fuzzy systems. The above feature assists in the determination of the initial structure, also leading to models with fewer parameters compared to neural networks.

Neuro-fuzzy (NF) systems have been extensively used in pattern classification applications, including specialised medical ones [15]. Examples of NF systems as classifiers include schemes such as ANFIS [17, 18, 19], FSOM [20], Fuzzy-RBF [21], Fuzzy-ART [22]. The most famous example of Neuro-fuzzy network is the Adaptive Network based Fuzzy Inference System (ANFIS) developed by Jag [23] that implements a TS Fuzzy System [24, 25] in a multilayer architecture, and applies a mixture of both back-propagation and least mean squares procedure to train the system. In [26], a combination of fuzzy logic and neural network has been used to develop an adaptive control system for medical purpose.

1.1 EEG

An electroencephalogram (EEG) machine is a device used to create a picture of the electrical activity of the brain. It has been used for both medical diagnosis and neurobiological research. The essential components of an EEG machine include electrodes, amplifiers, a computer control module, and a display device.

EEG machines are used for a variety of purposes. In medicine, they are used to diagnose such things as seizure disorders, head injuries, and brain tumour.

EEGs are time-series signals with added noise. From such type of signals, many times a kind of information of the nature about these signals is required in real time in order to take crucial decisions [27, 28, 29, 30].

Electroencephalogram analysis is a very useful technique to investigate the activity of the central nervous system. It provides information related to the brain activity based on measurements of electrical recordings taken on the scalp of the subjects. Inference and studies about the subject's health and effective treatment of many diseases can be carried out by analysing the information obtained from the EEG.

EEG analysis has often been used to help doctors in Medicine and Information Technology to assist in their diagnostic procedures, especially whenever there are problems of different diagnosis in diseases. Methods from the domain of Intelligent Systems give the opportunity to formalise the medical knowledge and standardise various diagnostic procedures, in specific domains of Medicine and to store them in computer systems.

All EEG signals were recorded with the same 128-channel amplifier system, using an average common reference. The data were digitised at 173.61 samples per second using 12 bit resolution. Band-pass filter settings were 0.53–40 Hz (12dB/oct).

Typical EEGs are depicted in Fig. 1.

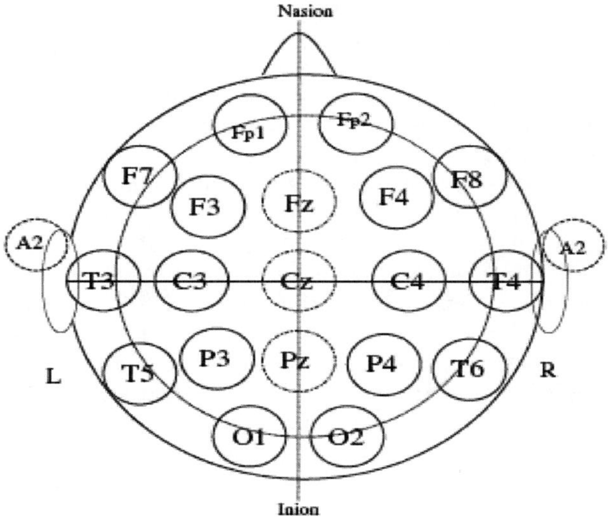

Fig. 1 The 10–20 international system of electrode placement c images of normal and abnormal cases.

The proposed EEG medical diagnostic system has been designed around the concept of a new neuro-fuzzy model.

The proposed diagnostic system, incorporating the proposed modelling scheme and also wavelet theory for the feature extraction section, achieved immensely satisfactory results on real test cases.

1.2 Stabilometry

Stabilometry is the branch of medicine responsible for examining balance in human beings. Balance and dizziness disorders are probably two of the most common illnesses that physicians have to deal with. Around 30% of population suffers from any kind of dizziness disorder before reaching the age of 65; for older people, this pathologic symptom occurs more frequently, and cause falling. The patient stands on a platform and completes a series of tests. These tests are designed to isolate the main sensorial, motor and biomechanical components that contribute to balance.

In order to examine balance, a device, called posturograph, is used to measure the balance-related functionalities. The patient stands on a platform and completes a series of tests, as shown in figure 2. These tests have been designed to isolate the main sensorial, motor and biomechanical components that contribute to balance. Emphasis has been given to the evaluation of the capacity for each individual components as well as the overall components capacity. The posturograph generates a time-series signal, where the main information normally is confined to events.

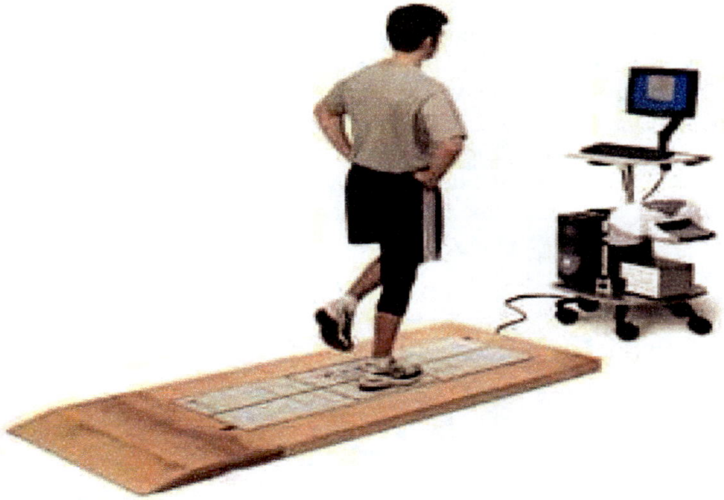

Fig. 2 Patient completing a test on a posturograph.

Initially, stabilometry was considered as a technique measuring only the balance of human beings under certain conditions [31]. Many researchers have studied the effect of closed eyes on balance [32] [33]. These works confirmed that the condition of having the eyes closed affects balance due to the fact that balance has a strong visual component.

Currently, stabilometry is also considered as a useful tool for diagnosing balance-related disorders like the Parkinson disease [34] or benign vertigo of childhood [35]. Regarding stabilometric data analysis, body sway parameters have been used for analysis balance-related functions [36], [37]. However, it appears that classic posturographic parameters, such as the measure of the sway of the centre of pressure [38] have failed in the detection of balance disorders [39]. The analysis of stabilometric time series using data mining techniques offers new possibilities. Recently, a new method has been developed for comparison of two stabilometric time-series [40]. This method calculates the level of similarity of two time-series and can be applied to compare either the balance of two patients or to study how the balance of one patient evolves with time. Stabilometry also plays an important role in the treatment of balance-related diseases. The NedSVE/IBV system has been utilised for the development of a new method that assists in the rehabilitation of patients who have lost their balance [41].

2 Objectives and Contributions

The proposed method builds reference models from a set of time series by means of the analysis of the events that they contain. This method is suitable for domains where the relevant information is concentrated in specific regions of the time series, known as events. The proposed method enables to define regions of interest according to the knowledge extracted from the domain experts, which is a plus compared with other methods addressing the time series as a whole without taking into account that certain regions can be irrelevant in the domain in question.

The technique developed throughout this article has been successfully applied to time series generated by Stabilometric devices that record the electrical activity of the brain. In this article, for classification of extracted features from Stabilometric events an Adaptive Fuzzy Inference Neural Network system (AFINN) has been applied.

3 Data Selection and Recording

3.1 EEG

We have used the publicly available data described in Andrzejak *et al.* [9]. The complete data set consists of five sets (denoted A–E) each containing 100 single-channel EEG segments. These segments were selected and cut out from continuous multi-channel EEG recordings after visual inspection for artefacts, e.g., due to muscle activity or eye movements. Sets A and B consisted of segments taken from surface EEG recordings that were carried out on five healthy volunteers. Volunteers were relaxed in an awake-state with eyes open (A) and eyes closed (B), respectively. Sets C, D, and E originated from EEG archive of pre-surgical diagnosis. : fig.3 shows examples of five different sets of EEG signals taken from different subjects.

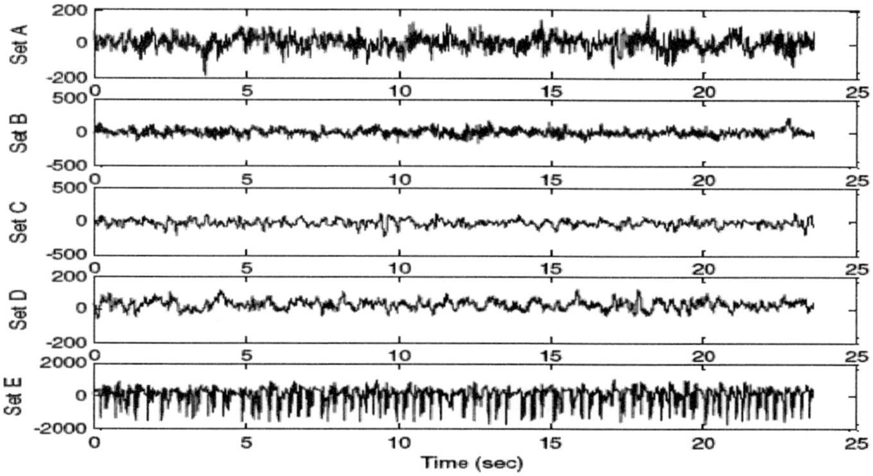

Fig. 3 Examples of five different sets of EEG signals taken from different subjects

3.1.1 Feature Extraction for EEG Signals

Numerous techniques from the theory of signal analysis have been used to obtain representations and extract the features of interest for classification purposes. Within this framework the signal is decomposed into sub-bands using fast wavelet transform algorithms.

The extracted wavelet coefficients provide a compact representation that shows the energy distribution of the EEG signal in time and frequency. Table 1 presents frequencies corresponding to different levels of decomposition for Daubechies order-2 wavelet with a sampling frequency of 173.6 Hz. In order to further decrease the dimensionality of the extracted feature vectors, statistics over the set of the wavelet coefficients was used. The following statistical features were used to represent the time frequency distribution of the EEG signals:

- Maximum of the wavelet coefficients in each sub-band.
- Minimum of the wavelet coefficients in each sub-band.
- Mean of the wavelet coefficients in each sub-band
- Standard deviation of the wavelet coefficients in each sub-band

Table 1 Frequencies corresponding to difference levels of decomposition

Decomposed signal	Frequency range (Hz)
D1	43.4–86.8
D2	21.7–43.4
D3	10.8–21.7
D4	5.4–10.8
D5	2.7–5.4
A5	0–2.7

To reduce the volume of data, the sample (time points) was partitioned into 16 windows of 256 times points each. From these sub-samples, we performed the DWT and derived measures of dispersion statistics from these windows (each corresponding to approximately 1.5 seconds). The DWT was performed at 4 levels, and resulted in five sub-bands: d1-d4 and a4 (detail and approximation coefficients respectively). For each of these sub-bands, we extracted four measures of dispersion, yielding a total of 20 attributes per sample window. Since our classifiers use supervised learning, we must also provide the outputs, which was simply a class label.

3.2 Stabilometry

Throughout this research, a static Balance Master posturograph has been used. In a static posturograph, the platform on which the patient stands is static, i.e. does not move. The platform has four sensors, one at each of the four corners: right-front (RF), left-front (LF), right-rear (RR) and left-rear (LR). Each sensor records a datum every 10 milliseconds during the test. This datum is sent to the computer connected to the posturograph. The datum is the intensity of the pressure that the patient is exerting on that sensor. Data are recorded as multidimensional time-series.

The posturograph Balance Master can be used to run a wide range of tests according to a predefined protocol. This chapter has focused on the Unilateral Stance (UNI) test that is the most useful for domain experts (physicians) in terms of output information. UNI test aims to measure how well the patient is able to keep his or her balance when standing on one leg with either both eyes open or both eyes closed for 10 seconds. The UNI test generates time-series signals containing events, that is, regions of special interest for experts in the domain. Next section describes the possible events appearing in the time series of UNI test and the features used to characterise these events. Both the events and their features were determined according to the physicians' criteria. The following cases are the four different conditions of UNI test:

- Left leg with Open Eyes: The patient is asked to hold still with his or her left leg on the platform while his or her right leg has to be lifted.
- Right leg with Open Eyes: The patient is asked to hold still with his or her right leg on the platform while his or her left leg has to be lifted.
- Left leg with Closed Eyes: The patient is asked to hold still with his or her left leg on the platform while his or her right leg has to be lifted.
- Right leg with Closed Eyes: The patient is asked to hold still with his or her right leg on the platform while his or her left leg has to be lifted.

The current research has been carried out on time series from a set of healthy sportspeople, including both genders.

3.2.1 Model Generation Method

The model generation method presented here is suited for domains where only particular regions of the time series, known as events, contain relevant information for that domain while the remaining of the time series hardly provides any information. In order to deal with events, each event is characterized by a set of attributes.

The method proposed in this article receives a set of time series $S = \{S_1, S_2, ..., S_n\}$, each containing a particular number of events, and generates a reference model M that represents this set of time series. The model M is built on the basis of the most characteristic events. The most characteristic events of S are those events that appear in the highest number of timer series of S.

To find out whether a particular event in a time series S_i also appears in another time series S_j ($j \neq i$), the event has to be characterized with an attribute vector and compared with the other events of the other series. To speed up this process, all the events present in the time series are clustered, so similar events belong to the same cluster. On the one hand, the clustering process is useful to know the different groups of events. On the other hand, it facilitates the extraction of the most characteristic events. Once we have a set of clusters, the objective is to find those clusters containing events that appear in the highest number of time series, that is, characteristic events. Having located those groups with similar events, an exhaustive cluster analysis is run in order to extract the event representative of each of these groups. This will be described later (steps 5 to 9 of the algorithm). These extracted representative events are the characteristic events of S and will be part of the final model.

Let $S = \{S_1, S_2, ..., S_n\}$ be a set of n time series and m the typical number of events that appear in the time series of S. The algorithm for generating a reference model M representing the set S is as detailed below (with the purpose of making the algorithm more legible key decisions are justified at the end of the algorithm):

1. **Initialize the model.**
   $M = \emptyset$.
2. **Identify events.**
   Extract all the events E_v from the series of S and use an attribute vector to characterize each event. This vector covers what the expert considers to be the key features for each type of domain event. This step is domain dependent, as the event characterization will depend on the time series type. To extract the events, the time series is examined in search of regions that meet the conditions identifying each event type defined according to the knowledge extracted from the expert. Section 3.2.2 describes how the events are identified and characterised according to those conditions.
3. **Determine the typical number of events *m*.**
   m is the typical number of events in each time series of S. At the end of the algorithm it will be discussed how to determine this value.

4. Cluster events.
Cluster all the events extracted in step 2. Bottom-up hierarchical clustering techniques have been used. Taking into account that the proposal described here should be a general-purpose method and there is no a priori information for specifying the optimum number of clusters in each domain, bottom-up hierarchical clustering is a good option, as it is not necessary to specify the number of clusters k beforehand.

Repeat steps 5 to 9 *m* times
5. Get the most significant cluster C_k.
Determine which cluster C_k of all the clusters output in step 4 is the most significant. Cluster significance is measured using Equation (1).

$$\text{SIGNF}(C_k) = \frac{\#\text{TS}(C_k)}{n} \tag{1}$$

That is, cluster significance is given by the number of time series that have events in that cluster over the total number of time series n. Events that have already been examined (step 8 and 9) are not taken into account to calculate the numerator.
6. Extract the event E_c that best represents the cluster.
Extract the event that is most representative of the cluster C_k, that is, the event E_c that minimizes the distance to the other events in the cluster. Let S_j be the time series in which the event E_c was found.
7. Add the event E_c to the model.
$M = M \cup E_c$.
8. Mark event E_c as examined.
9. Mark the most similar events to E_c as examined.
From the cluster C_k obtain, for each time series $S_i \neq S_j$, the event E_p from S_i that is the most similar to the representative event (E_c) output in step 6. Each E_p will be represented in the model by the event E_c and therefore these E_p events will also be discarded in order not to be considered in later iterations.
10. Return M as a model of the set S.

The most significant clusters, that is, those clusters that contain events present in the highest number of time series were analysed to output the events that are part of the model. To do this, the process of identifying the most significant cluster is repeated m times, outputting a representative and marking as examined both this representative and similar events in each time series. With regard to the algorithm, note that:

a) The identification of events is domain dependent because the criteria to define events in each domain are required. The rest of the algorithm is domain independent and it can be applied to any domain without any change. Figure 2 shows the overall structure of the proposed method that receives a set of time series S and generates a model M that represents it.

b) After the representative event of the most significant cluster has been output, it should not be taken into account again for the next iteration, and it is marked as an already examined event.
c) A cluster may contain not just one but several events from each time series. For this reason, even if a cluster is selected as the most significant, the cluster in question is not omitted in later iterations. The events already processed are marked as examined and will not be taken into account in future iterations.

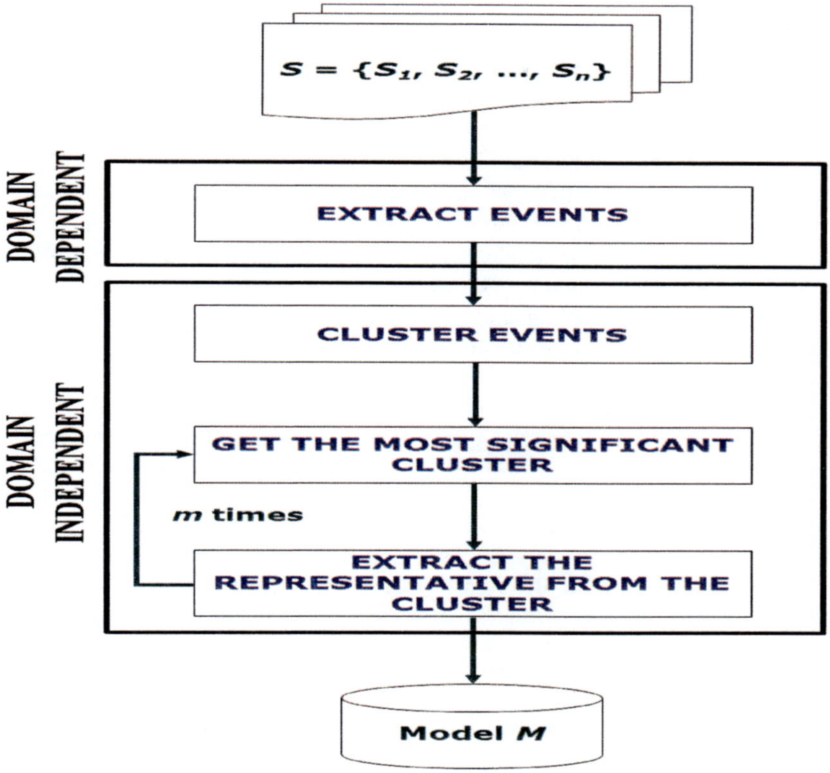

Fig. 4 Overall structure of the proposed method.

Another important issue is the number of events making up the model. In this case, we have chosen the mode (m) of the number of events of the time series of S. This decision is based on the fact that if the original time series have a typical number of events m, it makes sense for the model that represents them to have the same number of events m. The typical number of events in the time series of S

may not be unimodally distributed. This could happen especially if there are not many time series in the set *S*. For non-unimodal distributions, we have opted to take the integer value closest to the mean of the number of events.

A last point to be considered is the distance between events that has been used in the algorithm for clustering, representative event selection and discarding similar events. The *city block* distance is used. Given two vectors, the city block distance calculates the sum of the absolute value of the difference of each of the coordinates of the above vectors:

$$d(x, y) = \sum_{i=1}^{p} |x_i - y_i| \tag{2}$$

In Equation (2), x and y are the vectors (that is, the event descriptors) for comparison and p is the number of coordinates (dimension). Other distance measures have been studied, but the city block distance was finally chosen. The main reason for this choice is that the clustering algorithm uses the mean distance per attribute as the threshold for determining whether or not two elements are similar enough to belong to the same cluster. This mean distance per attribute is obtained simply by dividing the total city block distance $d(x,y)$ by the number of attributes p. The use of the city block distance then saves time as it obviates additional transformations that would make the clustering process more complex to develop and more computationally intensive.

Figure 5 shows an example of the application of the proposed method to a set $S = \{S_1, S_2, S_3, S_4\}$ of 4 time series (n=4). In this case, S_1 contains 2 events (E_{11} and E_{12}, S_2 contains 2 events (E_{21} and E_{22}), S_3 contains 3 events (E_{31}, E_{32} and E_{33}) and finally S_4 contains 2 events (E_{41} and E_{42}). Therefore, the typical number of events is 2 (m=2). Once the events are extracted, they are clustered into three different clusters (C_1, C_2 and C_3). Then, the most significant cluster is obtained. To do that, it is necessary to calculate the significance of each cluster according to Equation (EQ). In this case, cluster C_1 have events present in 3 out of the 4 time series, cluster C_2 have events that appear in 1 out of the 4 time series and cluster C_3 have events present in 4 out of the 4 time series of *S*. Then, the significance of C_1 is $SIGNF(C_1) = \frac{3}{4} = 0.75$, the significance of C_2 is $SIGNF(C_2) = \frac{1}{4} = 0.25$ and the significance of C_3 is $SIGNF(C_3) = \frac{4}{4} = 1$. Therefore, the most significant cluster is C_3. In the next step, the event E_{12} is extracted as the representative event of the cluster C_3 because E_{12} is the event in C_3 that minimizes the distance to the other events in that cluster. Thus, the event E_{12} is a characteristic event of *S* and will be part of the final model *M*. This process has to be repeated twice (because m=2) to build the final model that consists of the events E_{12} and E_{32}.

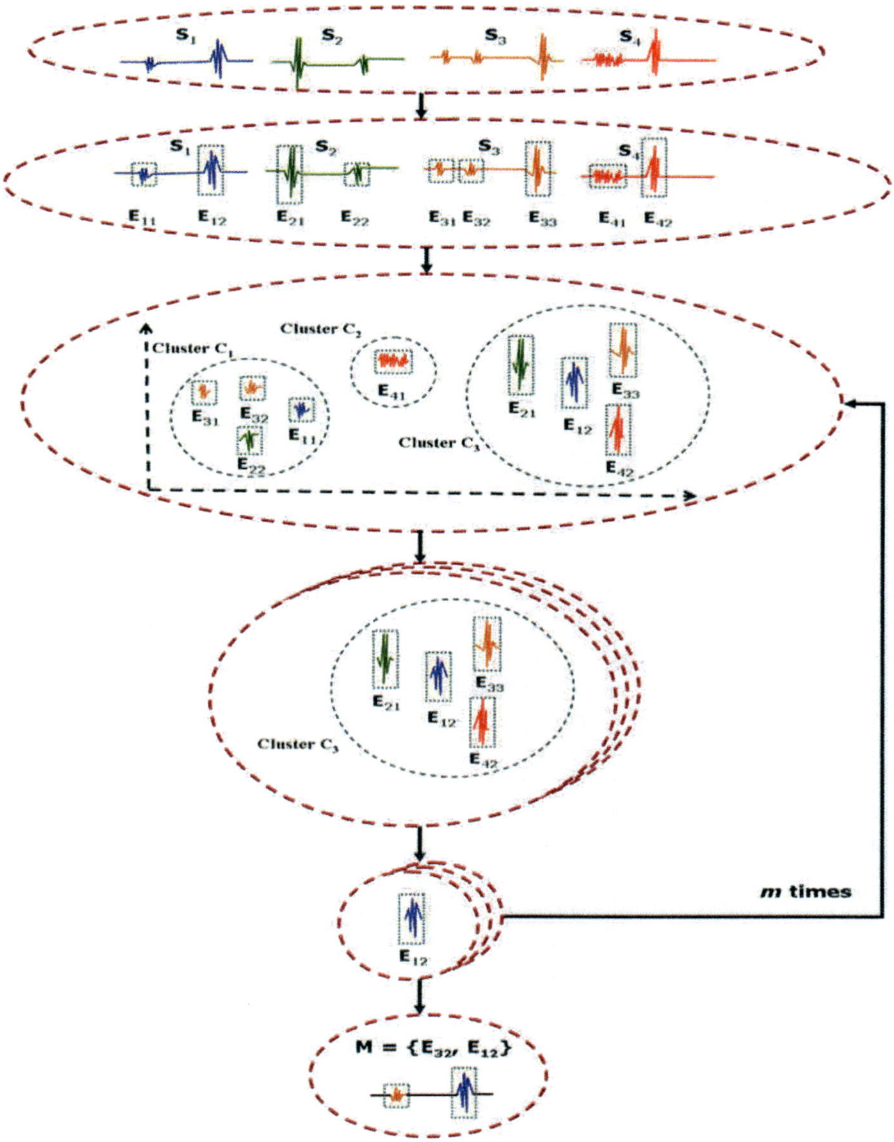

Fig. 5 Example of the application of the proposed method

3.2.2 Feature Extraction of Stabilometric Events

There are several stabilometric tests that can be carried out on the posturograph. In this chapter we have focused on the Unilateral Stance (UNI) test that is the most useful for the experts on the domain in terms of output information. The UNI test aims to measure how well the patient is able to keep his or her balance when standing on one leg with both eyes either open or shut for 10 seconds (see figure 5).

The ideal situation for the UNI test would be for the patient not to wobble at all but to keep a steady stance throughout the test. According to the knowledge extracted from the expert physicians, the interesting events in this test occur when the patient becomes unsteady, loses balance and puts the lifted leg down on the platform. This type of event is known in the domain as a fall. The features characterising the falls are as follows:

- **Duration:** It is the amount of time between the moment when the patient starts to lose balance and the moment when he or she is stable again, after falling.
- **Intensity:** It is the strength that the patient exerts on the platform when he or she falls down onto it.
- **Test time at which the event occurs:** It is the timestamp when the fall starts.

When there is a fall, the respective sensors for the lifted leg will register the pressure increase. Figure 6 shows the time-series of a patient who has taken the UNI test. The curves at the top of the figure are the values recorded by the RR and RF sensors, that is, the right leg sensors, the leg the patient was standing on. The curves at the bottom of the figure are the values recorded by sensors LR and LF, that is, the left-leg sensors, the leg that should be lifted. The pressure peaks generated when there is a fall event are highlighted. Figure 6 shows that the LR and LF time series are almost static (with a small variation margin) throughout, except when there is a fall. Therefore, it could be possible to define a stability value for the pressure exerted on the respective sensors for the above time series. This stability value, which in the case of this example would be around 20, could be a statistical measure like the mode. Every time there is a fall, there is a particular time instant where the LF and LR sensors record a local maximum and the RR and RF sensors record a local minimum. This point is more or less midway through the fall.

To identify the fall events and determine their features, the proposed method calculates the mode of the time-series related to the leg that must be lifted (bottom of figure 6). This value represents the balance value as shown in figure 7. The method identifies points where there is a local maximum whose distance to the balance value is higher than a certain threshold (∂). That distance is precisely the intensity of the fall. The duration of the fall is then calculated by analysing the two intersections between the time series and the balance value line. The timestamp at which the event starts is the first intersection between the time series and the balance value line.

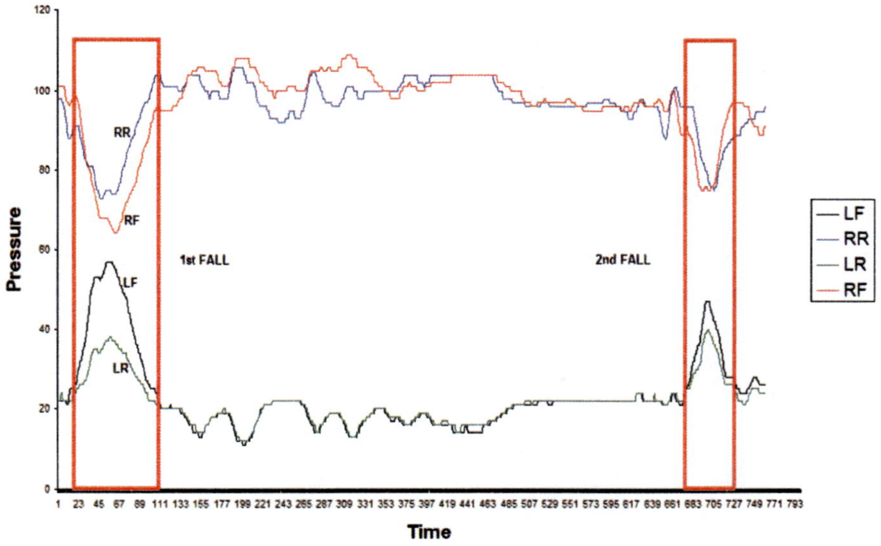

Fig. 6 UNI test time series, highlighting two events (falls)

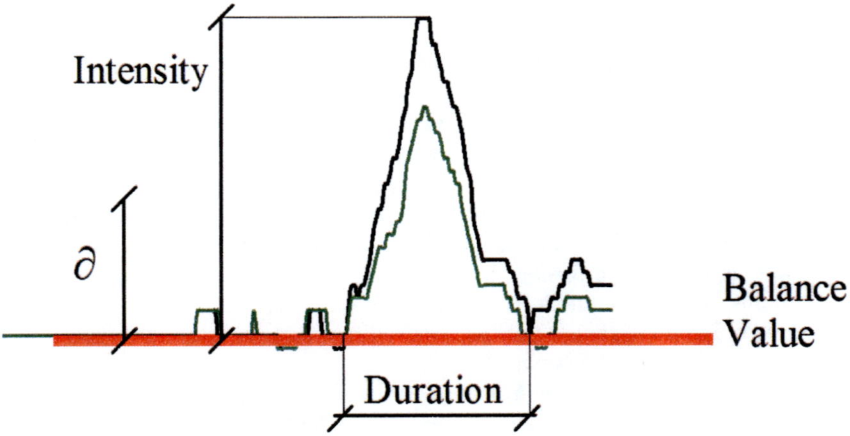

Fig. 7 Fall event taken from a stabilometric time series

4 Architecture of AFINN

There are many different combinations of fuzzy logic systems and neural networks.

This thesis proposes a connectionist model of fuzzy system in the form of feed-forward multi-layer networks, which can be trained using an iterative algorithm. This kind of neuro-fuzzy system employs a perceptron-like structure and a hybrid supervised learning procedure of neural networks for fuzzy inference system with rule base and fuzzy reasoning. The most important problem in fuzzy systems is to find fuzzy rules. Some methods can generate fuzzy rules from input-output pairs [42].

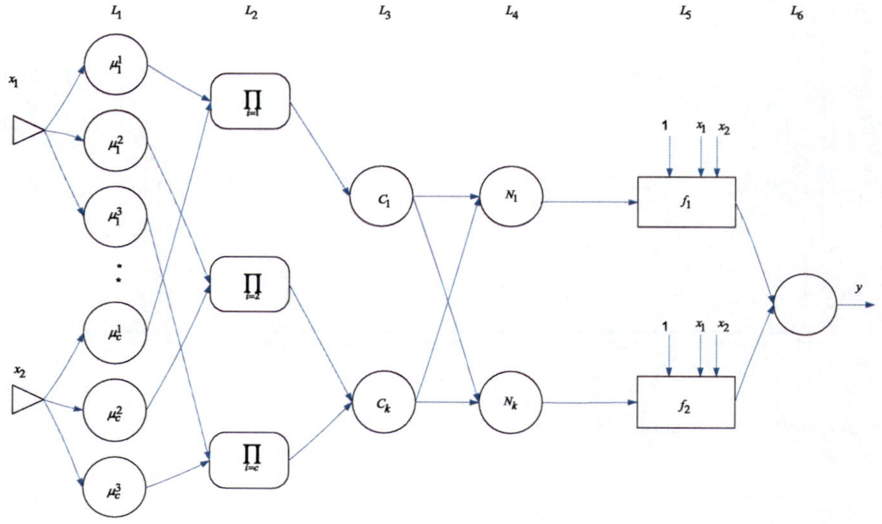

Fig. 8 Structure of AFINN system

The architecture of the proposed neuro-fuzzy network is shown in Figure 8 which consists of five layers. The first two layers L1 and L2, premise part correspond to the fuzzification section IF part of fuzzy rules whereas layers L4 and L5 the consequence part contain information about the THEN part of these rules and perform the defuzzification task. In Layer L3 a mapping between the rule layer and the output layer is performed by a competitive learning process. The local linear systems at L4 are associated with each term of layer L3 rather than that of rule base layer L2. Thus the size of the required matrices for least-squares estimation is considered to be much smaller.

The clustering algorithm which is applied in this framework at Layer L2 consists of two stages. In the first stage the method similar to the LVQ algorithm generates crisp c-partitions of the data set. The number of clusters c and the cluster centres $v_i, i = 1, \ldots\ldots, c$ obtained from this stage are used by the FCM algorithm in the second stage.

- The first stage clustering algorithm determines the number of clusters by dividing the learning data into these crisp clusters and calculates the cluster centres which are the initial values of the fuzzy cluster centres derived from the second stage algorithm. Let $X = [x_1, \ldots\ldots, x_n] \in R^{np}$ be learning data. The first cluster is created starting with the first data vector from X and the initial value of the cluster centre is taking as a value of this data vector. Then other data vectors are included in the cluster but only the ones that satisfy the following condition

$$\|x_k - v_i\| < D \tag{3}$$

Where $x_k \in X, k = 1,....,n$ and $v_i, i = 1,........,c$ are cluster centres, $[v_1,....,v_n] \in R^{cp}$, the constant value D is fixed at the beginning of the algorithm. Cluster centres v_i are modified for each cluster (i.e., $i = 1,.....,c$) according to the following equation

$$v_i(t+1) = v_i(t) + a_t(x_k - v_i(t)) \tag{4}$$

Where $t = 0,1,2,......$ denotes the number of iterations, $a_i \in [0,1]$ is the learning rate and decreases during the performance of the algorithm (depending on the number of elements in the cluster). Recursion of Eq. 4, originates from the LVQ algorithm. As a result of the performance of this algorithm we get the number of clusters c that we have divided the data set into, and thus we know the values of the cluster centres $v_i, i = 1,........,c$ which we can use as initial values for the second stage clustering algorithm.

- In the second stage the fuzzy c-means algorithm has been used. FCM is a constrained optimisation procedure which minimises the weighted within-groups sum of squared errors objective functions J_m with respect to the fuzzy membership's u_{ik} cluster centres v_i, given the training data $x_i,\ i = 1,....,c \quad k = 1,.....,n$

$$\min_{(U,V)}\{J_m(U,V;X) = \sum_{k=1}^{n}\sum_{i=1}^{c}(u_{ik})^m \|x_k - v_i\|_A^2\} \tag{5}$$

Where $U = [u_{ik}]_{c \times n}$ and u_{ik}'s satisfy the following conditions:

$$\begin{aligned} &0 \le u_{ik} \le 1 && \forall_{i,k} \\ &0 < \sum_{k=1}^{n} u_{ik} < n && \forall_i \\ &\sum_{i=1}^{c} u_{ik} = 1 && \forall_k \end{aligned} \tag{6}$$

The matrix $U \in R^{cn}$ is called a fuzzy c-partition of X, where $X = [x_{1,\ldots\ldots,x}$ X, where $X = [x_1, \ldots\ldots, x_n] \in R^{np}$ is a data set.

The distance between x_k and v_i is the Euclidean norm when the distance matrix A is chosen as the identity matrix I.

The centroid of a cluster is the mean of all points, weighted by their degree of belonging to the cluster:

$$c_j = \frac{\sum_{i=1}^{N} u_{ij}^m \cdot x_i}{\sum_{i=1}^{N} u_{ij}^m} \tag{7}$$

The optimal values of u_{ik} and v_i which create the optimal fuzzy c-partition of the data set into c clusters are calculated as follows:

$$u_{ik} = \left[\sum_{j=1}^{c} \left(\frac{\|x_k - v_i\|_A}{\|x_k - v_j\|_A} \right)^{2/(m-1)} \right]^{-1} \qquad \forall_{i,k} \tag{8}$$

Formulas (7) and (8) are repeated until J_m no longer decreases.

The number of clusters c and the initial values of cluster centres v_i come from the first stage clustering algorithm. The most important problem in fuzzy systems is to find fuzzy rules. In this chapter the fuzzy rule base is derived using results obtained from the clustering algorithm described in this section.

In our proposed system L3 (Layer 3) is an additional layer of output partitions, each of which is associated with a local number of rule nodes which will be reduced and each node in Layer 3 will only be connected to one node in Layer 2. Nodes in this layer represent the partitions of the output variables, link at this layer form as consequences of the rules confusing for me.

The main rational underlying the work described in this chapter, which represents the core of the thesis is the development of a new neuro-fuzzy network. We will consider an Adaptive Fuzzy Inference Neural Network system (AFINN) which is made up of Gaussian-membership functions associated with local linear systems. The proposed fuzzy logic system is based on the Sugeno-type modified with the introduction of an additional layer of output partitions. Unlike the ANFIS system, in which the number of local linear systems is the same as that of the number of rules, AFINN provides a means of controlling the growth of the number of local linear systems when the order of the system under consideration increases, so that least-squares estimation can be applied without performance degradation. A clustering algorithm is applied for the sample data in order to organise feature vectors into clusters such that points within a cluster are closer to each other than vectors belonging to different clusters. Then fuzzy rules will be created using results obtained from this algorithm. Unlike Sugeno's method [43], the fuzzy implication of the fuzzy system is based on fuzzy partitions of the input space directly rather than fuzzy partitions of each dimension of the input space. Thus the membership functions considered in the proposed system are multidimensional membership functions. In this sense, there is a similarity with the construction of Gaussian centres in Radial Basis Function networks (RBFN). Since the input space is considered to be partitioned instead of each dimension of the input space, the number of rules can be small and hence the number of local linear systems is also small. In addition, a competitive learning technique is applied to locate space partitions according to the clustering of the fuzzy rules at the beginning of training. The proposed methodology is implemented and its performance evaluated against Multilayer Perceptron (MLP).

Essentially, we can say that the interpretability of fuzzy model depends on two main aspects: the complexity of the fuzzy rule base (i.e. the number of rules) and the readability of the fuzzy sets used in the fuzzy rules.

The number of rules is a crucial condition to obtain linguistically interpretable fuzzy models. Therefore the problem of finding a balance between the complexity (rule base size) and the accuracy of the fuzzy model is of considerable importance.

4.1 Training Procedure for AFINN

In the tuning phase, emphasis is given to the nature of AFINN scheme itself. Two different sets of parameters exist and need to be tuned. These include the nonlinear premise parameters in the fuzzification part and the linear consequent parameters in the defuzzification part. A hybrid learning approach thus has been adopted for the AFINN scheme. The network can be considered as a cascade of nonlinear systems and linear systems. In this phase, the error back-propagation is applied to tune the premise parameters of the membership functions and recursive least squares estimation is applied to find the consequence parameters of local linear systems.

The hybrid algorithm is composed of a forward pass and a backward pass. The least squares method is used to optimize the consequent parameters with the premise parameters fixed. Once the optimal consequent parameters are found, the backward pass starts immediately.

5 Evaluation

5.1 Evaluation of the Model Generation Method

The chapter focused on the UNI test data. Thorough the evaluation process, we used time series taken from a total of 30 top-competition sportspeople, divided into two groups. The first group was composed of 15 professional basketball players, whereas the second was made up of 15 young elite skaters. Thirty is a reasonable number of patients, taking into account that top-competition athletes do not abound and the stabilometric tests are quite complex (a single patient check-up takes up 2-3 Mb).

The ultimate aim of the evaluation is to measure how good the model generation method is. Two models from each of the above groups of sportspeople have been created. The first model ($M_{basketball}$) was created from a training set composed of 10 of the 15 basketball players. The other 5 players constituted the test set. The second model ($M_{skating}$) was generated from a training set composed of 10 of the 15 skaters. The other 5 skaters were used as test set. The sportspeople in the test set were chosen at random from all the sportspeople in each group.

Once the models have been created, they have been evaluated by checking whether the $M_{basketball}$ model properly represents the group of professional basketball players and whether the $M_{skating}$ model is representative of the group of elite skaters. To do that, we have compared each of the ten individuals in the test group against each of the two created models, making use of the Stabilometric time series comparison method described in [37]. This process was repeated five times changing the training set and the test set. The final results show that the 90% of sportspeople across all the experiments were successfully classified. It demonstrates the ability of the proposed method to generate representative reference models in the field of Stabilometry.

5.2 *Evaluation of AFINN*

Classification problems, also referred to as pattern recognition tasks, arise in a variety of engineering and scientific disciplines such as biology, psychology, medicine and artificial intelligence.

To illustrate the applicability of the proposed methodology to classification tasks, four classification datasets have been considered: the EEG data set, Cancer dataset, the Iris dataset and the MONK's dataset.

To assess the prediction quality of the resulting model on the basis of the available data, we have to estimate the unknown prediction risk; the estimate of the true generalization error of a model based of a finite set of observations can be obtained via data re-sampling techniques such as holdout, cross-validation and bootstrap. The basic idea is to build a model, from one part of the available data, and then use that model to predict the rest of the data. A 5-fold cross validation was performed, which involves splitting the whole data set into 5 equally sized parts and using each part as a test set for the model extracted from the remaining data. The estimate of the generalisation error is computed as the average of classification errors provided by models extracted from the 5 partitions.

Split the dataset into two groups: The Training set is used to train the classifier and testing set to estimate the error of the trained classifier.

The classification methodology proposed in this thesis and described in this chapter has been tested on several benchmark examples for classifications.

The results of these tests can be found in many papers published on international journal and conference proceedings which contributed to the work of this chapter [44, 45, 46 and 47].

These benchmark examples cover a wide range of applications.

All experiments have been tested using the MATLAB software package.

The training data set was used to train the AFINN model, whereas the testing data set was used to verify the accuracy and the effectiveness of the trained AFINN model for classification

6 Discussion of Results

Automated diagnostic systems aim to enhance the ability to detect pathological structures in medical examinations and to support evaluation of pathological findings during the diagnostic procedure. Most techniques developed for automated Stabilometric data analysis have focused on the study of the centre of pressure of the patient. However, balance-related events (falls and imbalances) contain useful information for the physicians. In this research, the proposed AFINN network has been implemented for Stabilometric time-series classification, employing the most significant features of the events contained in UNI time series. As explained in section 2, the UNI test consists of four different conditions. This case study focused on the second trial of Left leg with Closed Eyes condition. The first and third trials have not been considered because the first trial contains noise and during the third trial the patient has already learnt how to be stable. In this chapter, 56 Stabilometric time series with 1000 timestamps have been used. The data set was divided into two classes according to the gender of patients: MALE and FEMALE. 38 out of the 56 time series belong to male patients while 18 belong to female patients. 5 balance-related features were extracted from each time series.

In our experiments, the training data set was used to train the AFINN model, whereas the testing data set was used to verify the accuracy and effectiveness of the trained AFINN model for classification for the 2 classes of Stabilometric time series. The proposed scheme has high classification accuracy with within 5 epochs. The results of the proposed classifier, using 10 different training sets for Stabilometry are illustrated at Table 2.

Table 2 AFINN performance for Stabilometric time series

System	Rules or number of Nodes	Epoch	Class 1 (Female)	Class 2 (Male)
AFFIN	7/4	5	95%	94.3%

The clustering Fuzzification part resulted in 7 rules, while after the competitive layer, the rules were reduced to 4, which resulted in fewer consequent parameters at the Defuzzification layer.

AFFIN performance for A, B, C, D and E classes illustrated in table 3.

Table 3 AFINN performance for A, B, C, D, and E classes

Data sets	No. rules	Class A	Class B	Class C	Class D	Class E
1	9/5	96.88	99.4	98.44	98.44	97.5
2	9/5	99.4	98.4	99.1	99.1	99.4
3	6/4	98.75	99.4	99.4	99.1	98.75
4	9/5	99.4	99.1	96.88	98.75	98.75
5	7/5	98.44	98.44	98.75	99.1	96.88
Average	7/4	98.6	99.1	98.5	98.9	98.26

The training data set was used to train the AFINN model, whereas the testing data set was used to verify the accuracy and the effectiveness of the trained AFINN model for classification of the original five classes of EEG signals.

7 Conclusions

The historical development of machine learning and its applications in medical diagnosis shows that from simple and straightforward to use algorithms, systems and methodologies have emerged that enable advanced and sophisticated data analysis.

Fuzzy set theory plays an important role in dealing with uncertainty when making decisions in medical applications. Using fuzzy logic has enabled us to use the uncertainty in the classifier design and consequently to increase the credibility of the system output.

With the recent advancement of computational technologies along with the technologies in health care and biomedical field, the research community in medical intelligent systems and in machine learning is facing new challenges and opportunities to contribute substantially to clinical medicine.

Several hybrid systems combining different soft computing paradigms such as neural networks, fuzzy systems and genetic algorithms, have been proposed as a technique ideal for predictive modelling or classification. In particular considerable work has been done to integrate the excellent learning capability of neural networks with the representation power of fuzzy inference systems. This results in neuro-fuzzy modelling approaches, which combine the benefits of these two powerful paradigms into a single case and provide a powerful framework to extract fuzzy rules from numerical data. The aim of using a neuro-fuzzy system is to find, through learning from data, a fuzzy model that represents the process underlying the data.

A neuro-fuzzy methodology for modelling is presented as a neural network implementation of the new fuzzy system. We have studied a two stage clustering algorithm to determine the rules, number of fuzzy sets, and the initial values of the parameters i.e. the centres and widths of the fuzzy membership functions. Unlike ANFIS in which the number of local linear systems is the same as that of the rules, the proposed system provides a means of controlling the growth of the number of local linear systems when the order of the system under consideration increases so that least square estimation can be applied. The proposed network was trained and tested with the extracted features using a discrete wavelet transform of the EEG signals, and then Principal Component Analysis (PCA) was used to reduce the data dimensionality.

The simulation results reveal an almost perfect performance compared to the classic MLP neural network.

The proposed network was trained and tested with the extracted features using a statistical method for the identifications and characterisation of events in Stabilometric time series.

We have developed a method to generate reference models from a set of time series by matching up the events that they contain. This method is suitable for domains where the key information is concentrated in specific regions of the series, called events, and the remaining regions are irrelevant. The proposed method enables to define regions of interest according to the knowledge extracted from the domain experts, which is a plus compared with other methods addressing the time series as a whole without taking into account that certain regions can be irrelevant in the domain in question.

The method was evaluated on stabilometric time series, obtaining very satisfactory results, especially as regards the representativeness of the reference models generated by the proposed method. The results confirm the generality of the model generation method.

The proposed AFINN scheme characterised by its high performance accuracy, high training speed provides an efficient solution to the "curse of dimensionality" problem inherited in the classical NF scheme.

References

[1] Baxt, W.G.: Application of Ariticial Neural Networks to Clinical Medicine. Lancet 346, 1135–1138 (1995)

[2] Jang, J.-S.R.: ANFIS: Adaptive Network Based Fuzzy Inference System. IEEE Transactions on Systems, Man, and Cybernetics (1993)

[3] Jang, J.-S.R.: Neuro-fuzzy Modelling and Control. The Proc. of the IEEE (1995)

[4] Sugeno, M., Kang, G.T.: Structure identification of fuzzy model. Fuzzy Sets and systems 28, 15–33 (1988)

[5] Takagi, T., Sugeno, M.: Fuzzy identification of systems and its applications to modelling and control. IEEE Trans. On systems, Man, and Cybernetics 15, 116–132 (1985)

[6] Kovalerchuk, B., Vityaev, E., Ruiz, J.F.: Consistent Knowledge Discovery in Medical; Diagnosis. IEEE Engineering in Medicine and Biology Magazine 19(4), 26–37 (2000)

[7] Wiegerinck, W., Kappen, H., ter Braak, E., Nijman, M., Neijt, J.: Approximate inference for medical diagnosis. Pattern Recognition Letters 20(11-13), 1231–1239 (1999)

[8] Kovalerchuk, B., Vityaev, E., Ruiz, J.F.: Consistent Knowledge Discovery in Medical; Diagnosis. IEEE Engineering in Medicine and Biology Magazine 19(4), 26–37 (2000)

[9] Walter, D., Mohan, C.: ClaDia: a fuzzy classifier system for disease diagnosis. In: Proceedings of the 2000 Congress on Evolutionary Computation, pp. 1429–1435 (2000)
[10] Zahan, S.: A fuzzy approach to computer-assisted myocardial Ischemia diagnosis. Artificial Intelligence in Medicine 21(1-2), 271–275 (2001)
[11] Pena-Reyes, C.A., Sipper, M.: Designing Breast Cancer Diagnostic via a Hybrid Fuzzy-Genetic Methodology. In: Proc. of the 1999 IEEE Int. Fuzzy Systems Conf., pp. 135–139 (1999)
[12] Pattichis, C., Schizas, C., Middleton, L.: Neural Network models in EMG diagnosis. IEEE Trans. On Biomedical Engineering 42(5), 486–496 (1995)
[13] Boulougoura, M., Wadge, E., Kodogiannis, V.S., Chowdrey, H.S.: Intelligent systems for computer-assisted clinical endoscopic image analysis. In: 2nd IASTED Int. Conf. on BIOMEDICAL ENGINEERING, Innsbruck, Austria, pp. 405–408 (2004)
[14] Czogala, E., Leski, J.: Fuzzy and Neuro-fuzzy Intelligent Systems. Springer, Heidelberg (2000)
[15] Rutkowska, D.: Fuzzy Neural Networks with an application to medical diagnosis. BioCybernetics and biomedical Engineering (1-2), 71–78 (1998)
[16] Szczepaniak, P., Lisboa, P., Kacprzyk, J.: Fuzzy Systems in Medicine. Springer, Heidelberg (2000)
[17] Jang, J.S.: ANFIS: Adaptive–network based fuzzy inference systems. IEEE Trans. On Systems, Man, & Cybernetics 23(3), 665–685 (1993)
[18] Sun, R.: Robust reasoning: Integrating rule-based and similarity-based reasoning. Artificial Intelligence 75(2), 214–295 (1995)
[19] Castro, J., Delgado, M.: Fuzzy Systems with Defuzzification are Universal Approximators. IEEE Trans. on systems, Man and Cybernetics 26, 149–152
[20] Vuorimaa, P., Jukarainen, Karpanoja, E.: A neuro-fuzzy system for chemical agent detection. IEEE Trans. On Fuzzy Systems 3(4), 415–424 (1995)
[21] Nanayakkara, T., Watanabe, K., Kiguchi, K., Izumi, K.: Fuzzy Self-Adaptive RBF Neural Network Based Control of a Seven-Link Industrial Robot Manipulator. Advanced Robotics 15(1), 17–43 (2001)
[22] Tontini, G., de Queiroz, A.: RBF Fuzzy-ARTMAP: a new fuzzy neural network for robust on-line learning and identification of patterns. In: Proc. IEEE Int. Conf. on Systems, Man & Cybernetics, vol. 2, pp. 1364–1369 (1996)
[23] Jang, J.-S.R.: Neuro-fuzzy Modelling and Control. The Proc. of the IEEE (1995)
[24] Sugeno, M., Kang, G.T.: Structure identification of fuzzy model. Fuzzy Sets and systems 28, 15–33 (1988)
[25] Takagi, T., Sugeno, M.: Fuzzy identification of systems and its applications to modelling and control. IEEE Trans. On systems, Man, and Cybernetics 15, 116–132 (1985)
[26] Castro, J., Delgado, M.: Fuzzy Systems with Defuzzification are Universal Approximators. IEEE Trans. On systems, Man and Cybernetics 26, 149–152
[27] Wolpaw, J.R., Birbaumer, N., Heetderks, W.J., McFarland, D.J., Peckham, P.H., Schalk, G., Donchin, E., Quatrano, L.A., Robinson, C.J., Vaughan, T.M.: Brain-computer interface technology: A review of the first international meeting. IEEE Transactions on Rehabilitation Engineering 8(2), 164–173 (2000)
[28] Tzallas, A.T., Tsipouras, M.G., Fotiadis, D.I.: Automatic seizure detection based on time-frequency analysis and artificial neural networks. Computational Intelligence and Neuroscience 7(3), 1–13 (2007)

[29] Barry, R.J., Clarke, A.R., Johnstone, S.J.: A review of electrophysiology in attention-deficit/hyperactivity disorder:1 Qualitative and quantitive electroencephalography 2. Event-related potentials. Clinical Neurophysiology 114, 171–183, 184–198 (2003)
[30] Kovalerchuk, B., Vityaev, E., Ruiz, J.F.: Consistent Knowledge Discovery in Medical; Diagnosis. IEEE Engineering in Medicine and Biology Magazine 19(4), 26–37 (2000)
[31] Romberg, M.H.: Manual of the Nervous Disease of Man, pp. 395–401. Syndenham Society, London (1853)
[32] Paulus, W.M., Straube, A., Brandt, T.: 'Visual stabilization of posture: physiological stimulus characteristics and clinical aspects'. Brain 107, 1143–1163 (1984)
[33] Gagey, P., Gentaz, R., Guillamon, J., Bizzo, G., Bodot-Braeard, C., Debruille, Baudry, C.: Normes 1985. Association Française de Posturologie, Paris (1988)
[34] Ronda, J.M., Galvañ, B., Monerris, E., Ballester, F.: Asociación entre Síntomas Clínicos y Resultados de la Posturografía Computerizada Dinámica. Acta Otorrinolaringol. Esp. 53, 252–255 (2002)
[35] Barona, R.: Interés clínico del sistema NedSVE/IBV en el diagnóstico y valoración de las alteraciones del equilibrio. Biomechanics Magazine of the Institute of Biomechanics of Valencia, IBV (February 2003)
[36] Rocchi, L., Chiari, L., Cappello, A.: Feature selection of stabilometric parameters based on principal component analysis. In: Medical & Biological Engineering & Computing 2004, vol. 42 (2004)
[37] Demura, S., Kitabayashi, T.: 'Power spectrum characteristics of body sway time series and velocity time series of the center of foot pressure during a static upright posture in preschool children'. Sport Sciences for Health 3(1), 27–32 (2008)
[38] Diener, H.C., Dichgans, J., Bacher, M., Gompf, B.: Quantification of postural sway in normals and patients with cerebellar diseases. Electroenc. and Clin. Neurophysiol. 57, 134–142 (1984)
[39] Corradini, M.L., Fioretti, S., Leo, T., Piperno, R.: Early Recognition of Postural Disorders in Multiple Sclerosis Through Movement Analysis: A Modeling study. IEEE Transactions on Biomedical Engineering 44(11) (1997)
[40] Lara, J.A., Moreno, G., Perez, A., Valente, J.P., López-Illescas, A.: Comparing posturographic time series through events detection. In: 21st IEEE International Symposium on Computer-Based Medical Systems, CBMS 2008, June 2008, pp. 293–295 (2008)
[41] Peydro, M.F., Vivas, M.J., Garrido, J.D., Barona, R.: Procedimiento de rehabilitación del control postural mediante el sistema NedSVE/IBV. Biomechanics Magazine of the Institute of Biomechanics of Valencia, IBV (2006)
[42] Dubes, R.C.: Cluster analysis and related issues. In: Chen, C.H., Pau, L.F., Wang, P.S.P. (eds.) Handbook of pattern Recognition & Computer Vision, pp. 3–32. World Scientific Publishing Co., Inc., River Edge
[43] Rutkowska, D.: Neuro-Fuzzy Architectures and Hybrid Learning. Springer, Heidelberg (2002)
[44] Jahankhani, P., Revett, K., Kodogiannis, V., Lygouras, J.: Classification Using Adaptive Fuzzy Inference Neural Network. In: Proceedings of the Twelfth IASTED International Conference Artificial Intelligence and Soft Computing (ASC 2008), Palma de Mallorca, Spain, September 1-3 (2008), ISBN 978-0-88986-756-7
[45] Jahankhani, P., Revett, K., Kodogiannis, V.: Data Mining an EEG Dataset With an Emphasis on Dimensionality Reduction. In: IEEE Symposium on Computational Intelligence and Data Mining (CIDM), April 1-5 (2007)

[46] Jahankhani, P., Revett, K., Kodogiannis, V.: EEG Signal Classification Using Wavelet Feature Extraction and Neural Networks. In: IEEE John Vincent Atanasoff 2006 International Symposium on Modern Computing, Sofia, Bulgaria, October 3-6, pp. 120–125 (2006)

[47] Jahankhani, P.K., Revett, V.: Automatic Detection of EEG Abnormalities Using Wavelet Transforms. WSEAS Transactions on Signal Processing 1(1), 55–61 (2005) ISSN 1790-5022

[48] Lara, J.A., Jahankhani, P., Pérez, A., Valente, J.P., Kodogianniz, V.: Classification of Stabilometric Time-Series using an Adaptive Fuzzy Inference Neural Network System. In: 10th International conference, ICAISC 2010 Zakopane, Poland, June 2010, pp. 635–643 (2010), ISBN 0302-9743

Glossary of Terms and Acronyms

Adaptive: A system that can be modified during operation to meet a specified criterion.

Artificial neural network: An artificial neural network consists of a number of a very simple and highly interconnected processors, called nenurons, which are analogous to the biological neurons in the brain.

Back-propagation: A supervised learning rule for multilayer perceptrons that operates by calculating the value of the error function for a known input, then back-propagating the error from one layer to the previous one. Each neuron has its weights adjusted so that it reduces the value of the error function until a stable state is reached.

Class: A group of objects with common attributes.

Clustering: The process of dividing a heterogeneous group of objects into homogeneous subgroups.

Consequent: Action in the If part of a rule.

Data mining: Data mining is the extraction of knowledge from data. It can also be defined at the exploration and analysis of large quantities of data in order to discover meaningful patterns and rules.

Defuzzification: Finding the best crisp representation of a given fuzzy set.

Expert: A person who has deep knowledge in the form of facts and rules and string practical experience in a narrow domain.

Fuzzification: The first step in fuzzy inference; the process of mapping crisp input into degrees to which these inputs belong to the respective fuzzy sets.

Fuzzy inference: is a process of mapping from a given input to an output by using the theory of fuzzy set.

Fuzzy rules: A conditional statement in the IF x is THEN y is B, where x and y are linguistic variables, and A and B linguistic values determined by fuzzy sets.

Fuzzy system: Fuzzy systems are well suited for modelling human decision making. Important decisions are often based on human intuition, common sense and experience, rather than on the availability and precision of data.

Membership function: The mapping that associates each element in a set with its degree of membership. It can be expressed as discrete values or as continues function.

Neural network: represent a class of general-purpose tools that are success fully applied to prediction, classification and clustering problems.

Tuning: Tuning is the most laborious and tedious part in building a fuzzy system. It often involves adjusting existing fuzzy sets and fuzzy rules.

Chapter 14
Demand on Computational Intelligence Paradigms Synergy

SOA and Mobility for Efficient Management of Resource-Intensive Applications on Constrained Devices

N. Kryvinska, C. Strauss, and L. Auer

Abstract. Enterprises are migrating towards SOA-based models in order to meet the greater than ever needs for integration and consolidation. Besides, driven by the dissemination of more refined mobile devices in the enterprise, and the rapid growth of wireless networks based on IEEE 802.11 WiFi Standards, mobile applications have been increasingly used in mission-critical business applications. The SOA-based next generation mobility management model analyzed here provides a baseline framework for the successful architecting, deployment and maintenance of mobile applications. We introduce and analyze the requirements to the architecture design needed to comply with new mobility management concept development. We also examine the architecture planning and design issues for the successful implementation of mobility management solutions. Furthermore, we provide a scenario example of the framework for SOA (Service-oriented Architecture) mobile appliances implementation, namely, a model that demonstrates "the customer search" mobile application. Finally, we present a practical case, e.g., a "mobile messaging" application, to show how applying a SOA approach can make the writing of mobile clients using remote services simple and intuitive, which in turn can increase the number of services available on the market, as well as their functionalities and features.

1 Introduction

"Synergy" - Benefits resulting from combining two different groups, people, objects or processes– Wiktionary (http://en.wiktionary.org/wiki/synergy).

N. Kryvinska · C. Strauss · L. Auer
Department of eBusiness, School of Business, Economics and Statistics,
University of Vienna, Vienna, Austria
e-mail: {natalia.kryvinska,christine.strauss}@univie.ac.at,
lukas.auer@univie.ac.at

N. Bessis, F. Xhafa (Eds.): Next Generation Data Technologies for CCI, SCI 352, pp. 357–382.
springerlink.com

To provide market-leading personalization, mobile service providers have to develop an agile IT infrastructure, one that can be reconfigured quickly for new offerings and enables the modification of existing products. Besides, these changes have to be made available immediately in dynamic websites where customers may purchase more phone time and add more services. To build an IT infrastructure that could respond with this kind of agility, a new architecture is required, one in which servers communicate and provide services across disparate domains in a highly reconfigurable SOA-based design. Such a design can enable providers to build functionality as Web services and then expose those services via multiple channels. These channels include: the wireless application protocol (WAP), the short message service (SMS), interactive voice response (IVR), call center, and, of course, the Web (Fig. 1).

By deploying such a SOA-WSD (web service delivery) configuration, service providers can easily scale the architecture at low cost, thus ensuring that its capabilities match the demands of the provider's fast-growing customer base. Moreover, by using SOA-WSD, they can reconfigure systems quickly and many components that have been developed can be reused. In addition, this platform can ease the integration with other systems from wireless carriers, CRM (Customer Relationships Management) packages, and billing solutions.

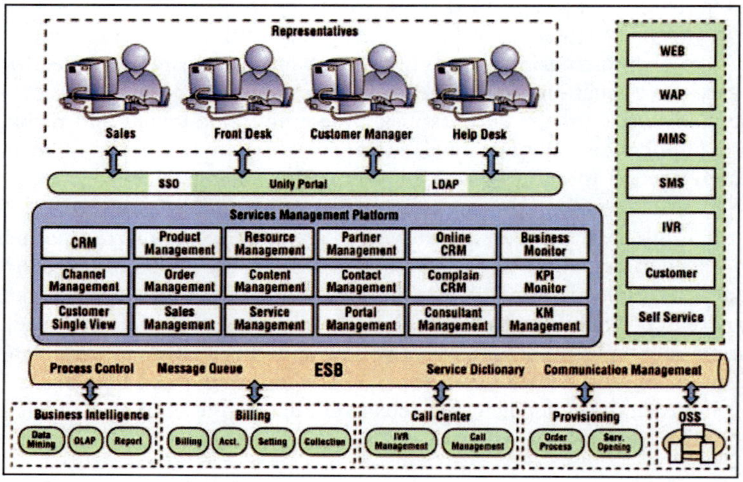

Fig. 1 Enterprise SOA Management Platform [7].

As a result, a new infrastructure model such as a SOI (Service-Oriented Infrastructure) can provide access to the computational resources automatically on demand and deliver the requested services at an appropriate quality level (Fig. 2). This infrastructure model allows mobile devices to use SOA services, and therefore enables the execution of complex, resource-intensive applications on the constrained devices [1 ÷ 6].

For that reason, we analyze the SOI model, including its interactions with local mobile devices while accessing the services on SOA-WSD. Furthermore, we attempt to build an efficient framework that integrates the mobility of resource-limited devices (e.g. handhelds or cellular phones) with the processing power of the SOA-WSD platform [1, 4 ÷ 6].

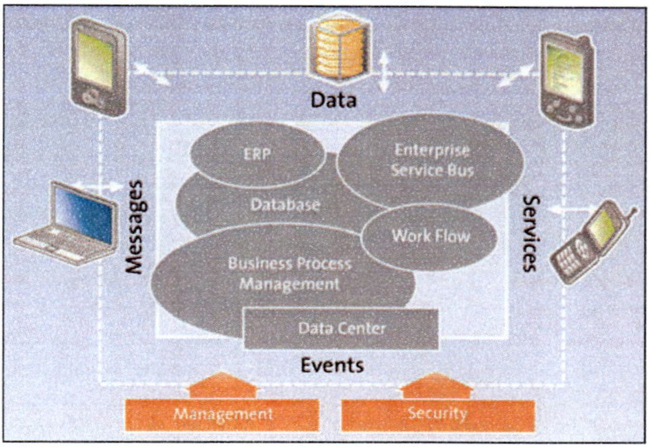

Fig. 2 Enterprise Mobile SOI [4].

The rest of the chapter is organized as follows: In the Introduction, we describe the current situation in the research field. In Section 2, we analyze existing challenges and the current position of SOA in mobile world. In Section 3, we discuss the requirements to the architecture design needed to comply with the new mobility management concept development. Next, in Section 4, we deal with the architecture planning and design issues, which represent the focal point for a successful mobility management solutions implementation. In Section 5, we provide scenario examples of SOA mobile appliances implementation, namely - two practical handling models that demonstrate "m-learning" and "the customer search" mobile applications. In the next step (Section 6), we compare how a SOA approach implemented to the connected-device model is better than using the "servlets" approach. In Section 7, we illustrate a "mobile messaging" application example in order to provide a realistic case that outlines how applying the SOA approach can make writing of mobile clients, using remote services, simple and intuitive. Subsequently, we address the development a mathematical model of Logical Architecture for mobile connectivity services, applying a mathematical model for the monitoring of services delivery in two interconnected systems in tandem. Finally, we present our conclusions.

2 Challenges and Current Position of SOA in Mobile World

The service providers implement Service-Oriented Architecture (SOA) in order to connect distinct applications, enabling interactions between the various components built on web-based standards that allow communications and data exchange (e.g., XML, HTTP, or SOAP). Built on middleware and a standards-based interface, data can be presented in a number of ways and via a number of easily constructed user interfaces. Standards such as XML, SOAP, WSDL, JavaBeans, AJAX, etc. allow purpose built applications to reuse and retrieve information from existing data bases, repositories and application data warehouses.

However, an efficient deployment of SOA to mobile devices, which are rapidly gaining an importance in enterprises and eventually the consumer space, is a primary challenge. SOA can potentially be used to create new clients and methodologies for deployment in the mobile domain. Whereas high-end smart devices employ fairly capable browsers to serve, for instance, as the key delivery platform for many SOA applications, they nevertheless generally do not have the same level of capability as a full featured PC-based browser. Also, it is particularly important for smart devices to deploy such capability, since significant power is spent in supporting the real-time back-and-forth communications of large amounts of data. While this is not a problem with mains-connected systems such as a PC, on wireless devices frequent radio transmissions consume a large amount of power, and hence dramatically reduce battery life between recharges.

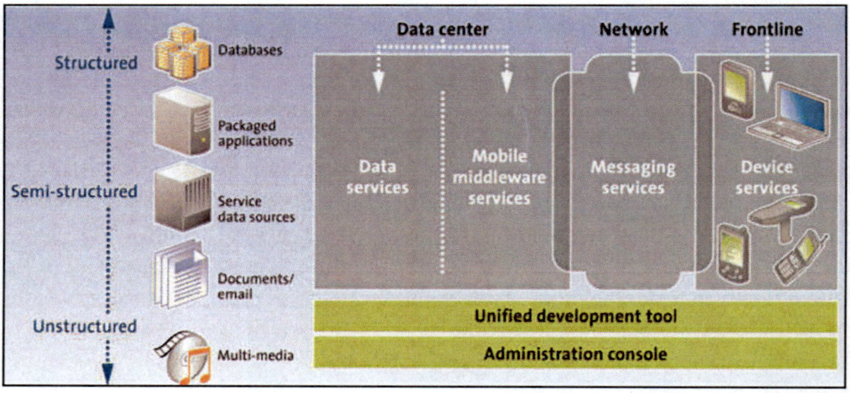

Fig. 3 A Model of Next Generation Mobility Architecture [4].

As a result, a new infrastructure management model is needed in order to provide access to the computational resources automatically on demand and deliver the requested services at an appropriate quality level. This model can enable mobile devices use of SOA services, and therefore facilitate the execution of complex, resource-intensive applications on the constrained devices [4, 8 ÷ 10].

For that reason, we analyze the next generation SOA-based mobility management model, including its interactions with local mobile devices while accessing the services. Furthermore, we attempt to build an efficient framework that integrates the mobility of resource-limited devices (e.g. handhelds, cellular phones) with the processing power of the SOA platform.

3 Components and Integrated Technology Stack for the Next Generation Mobility Architecture

In this section, we introduce and discuss the requirements to the components and system architecture design principles that enterprises need to adhere to as they develop a new mobility management concept. Our aim is to form an integrated technology stack, including components detailing, that is consistent and provides a comprehensive solution for mobile enterprise initiatives (Fig. 3):

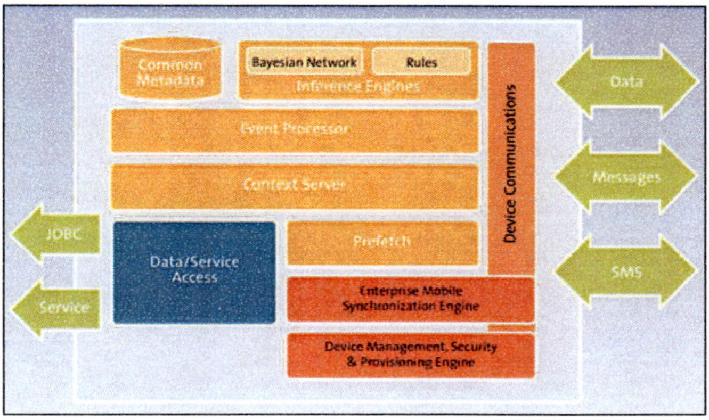

Fig. 4 Mobile Middleware Services Layer [4].

- *Data Services* - this layer provides bi-directional access to the virtualized data layer over diverse data sources ranging from structured to unstructured. These data sources could be packaged applications like SAP, databases, web services, etc.
- *Mobile Middleware Services* – this component provides the bridge between the data network and the device network. With its value added services like pre-fetching, inference, etc., this component makes it possible for data to be synchronized in an intelligent fashion between the data services layer and the applications running on various devices leveraging multiple communication channels, such as SMS, HTTP, etc. (Fig. 4).
- *Device Services* - this component provides a standardized interface for accessing various device functionalities such as persistence, synchronization, security, etc. Mobile developers can program to this interface for accessing device-specific functionality in a uniform way, thereby reducing the complexity in developing applications (Fig. 5).

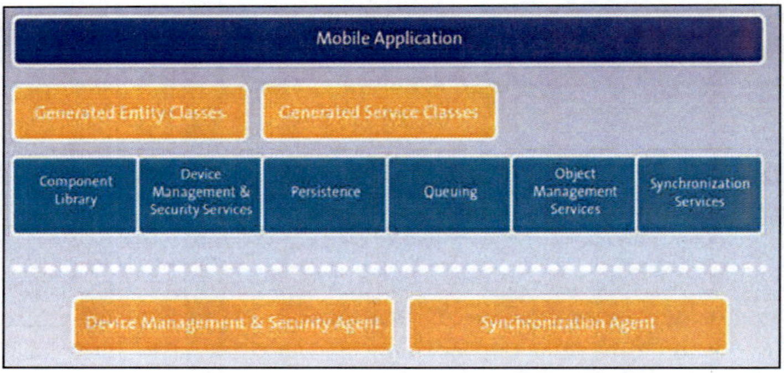

Fig. 5 Device Services Layer [4].

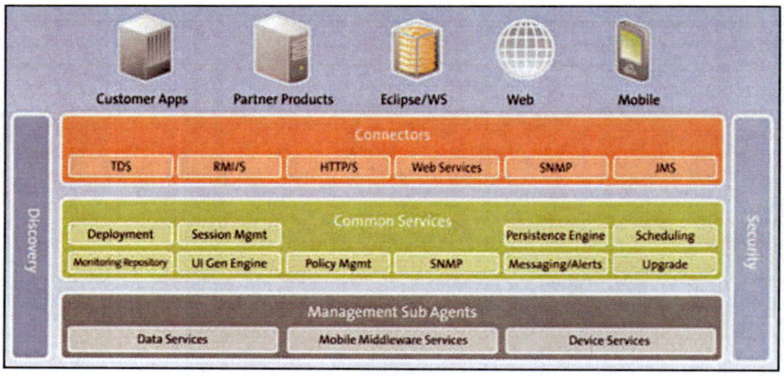

Fig. 6 Management Administration Console [4].

- *Administration Console* - by providing a standards-centric approach, the runtime can be monitored and administered through various third-party tools. With the feedback generated by the monitoring information, IT administrators can tune the operational systems to perform better under various load conditions (Fig. 6) [4, 11 ÷ 13].

4 The Architecture Planning and Design

The architecture planning and design is the focal point for the successful implementation of mobility management solutions. From a business perspective, the following architecture phases aid in achieving a successful implementation:

- *Contextual Architecture* - focuses on important characteristics and requirements and places the mobile solution in a business context.
- *Conceptual Architecture* - defines the functional requirements necessary to support the goals established in the contextual architecture.
- *Logical Architecture* - defines the architecture services that need to be in place, such as data storage, communication, security, integration, and so on.

– *Physical Architecture* - details components to be developed, sets development standards, and names products and tools to be used.

During the Contextual Architecture phase ("Why?"/Scope), important business aspects of the mobile solution are defined. The main deliverables, the Architecture Scope and the Architectural Principles, are defined by working through the following aspects:

– Specifications of important characteristics and requirements (business and IT principles);
– Future requirements;
– Requirements and impact from other initiatives;
– Corporate standards, policies, and guidelines.

The Conceptual Architecture ("What?") is functional by nature. The functions required to address the goals of the solution can be identified and described by answering the "what?" question. The activities to be performed in this phase are meant to define the following aspects:

– Business process overview or model;
– Business information flow;
– Use-case view or model;
– Functional requirements by: reliability/availability, supportability, volumes.

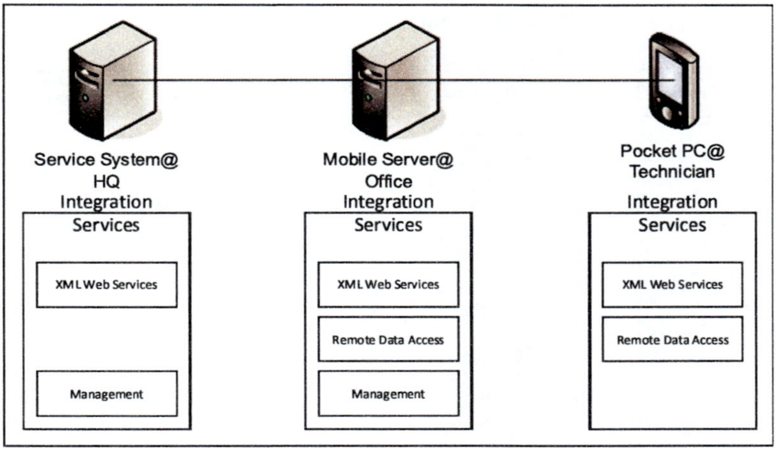

Fig. 7 Logical Integration View [14].

The Logical Architecture ("How?") phase deals with services and mechanisms that need to be in place to eventually support the physical implementation of the solution. Logically, mobile solutions often relate to services such as connectivity, communication, storage, security, integration, and distribution. The three most challenging software development aspects related to mobile solutions are network dependency, integration, and security. The Logical Architecture phase aims at addressing these challenges and describes how to solve them. The main deliverables out of the Logical Architecture phase include the following:

- Information services view;
- Component model view describing the structure of the application services and components;
- Integration views with description of logical interfaces;
- Data or object model (entity model);
- Technical infrastructure view with logical platform services.

A Logical Integration view, shown in Fig. 7, provides a picture of system integration services that need to be in place.

The Logical Architecture phase does not deal with physical implementations of services, such as "Pocket PC" or "XML Web Service", but rather with logical terms such as "mobile device" and "communication".

All the physical elements of the architecture are defined at the final architecture phase, the Physical Architecture ("With what?") phase. This includes platform, integration standards, tools, languages, and so on. The core deliverables of a physical architecture definition include the following:

- Detailed information systems components;
- Physical integration, interfaces, and protocols;
- Definition of which standards to comply with;
- Development, test, and production environment;
- Development and management tools.

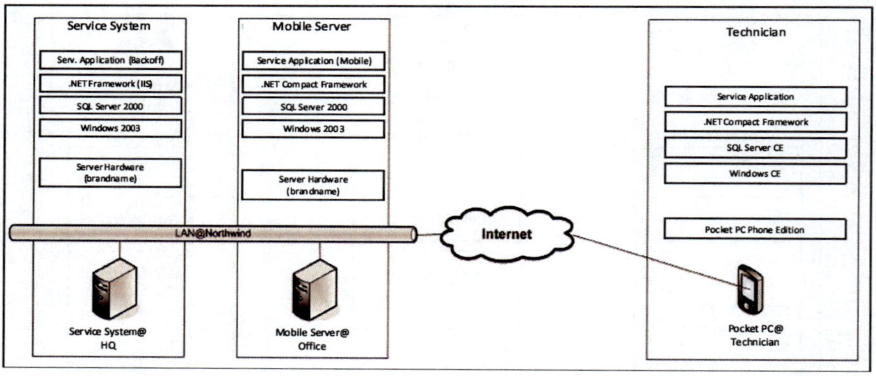

Fig. 8 Physical Deployment View [14].

Fig. 8 illustrates a physical deployment view. The purpose of this view is to show each physical element of the system architecture.

The components shown in the previous architecture phases can now be further detailed with physical entities and moved to the Physical Architecture deliverables [14 ÷ 16].

5 SOA for Mobile Applications – Practical Handling Use Cases

The main drivers of SOA models for the IT environment are increased integration and consolidation demands. These demands are particularly sensitive in heterogeneous computing environments with a variety of applications, operating systems, and hardware.

Besides, the use of PDAs and short-range wireless has grown progressively, particularly with regard to the 802.11 family of protocols that has paved the way for wide wireless mobile applications dissemination. The unique combination of characteristics associated with SOAs and mobile applications has presented new challenges for the software architects and application designers, as well as new opportunities, together with ensuring the integration across heterogeneous environments and partial wireless connectivity [17 ÷ 20]. Therefore, it is important to analyze practical examples of such applications performance in small resource-limited environment, in order to find some effective solution(s) for the large scale IT infrastructures.

5.1 Mobile SOA Implemented in University e-Learning Environment

Fig. 9 presents an example of Mobile SOA implemented in academic/university, e.g., learning environment [8, 16, 19].

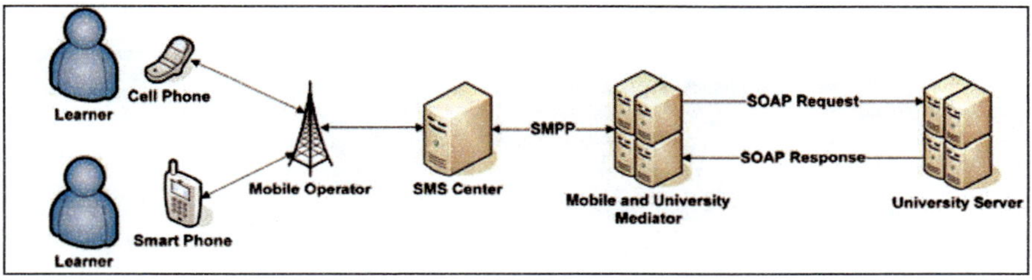

Fig. 9 An Example of Mobile Service Communication Architecture [8].

The learning management system (LMS) is created to enable m-learning. Assessment is one of the learning activities that can be achieved electronically and via mobile devices, with mobile assessment relying on services that are not part of the LMS. Integrating different external systems and services to be virtually part of LMS is a leading integration challenge, which this case study addresses by presenting the SOA integration into LMS. The proposed architecture consists of two layers: an Interface layer and a Service layer. The interface layer interacts with instructors and students via portals and with external organization services via Web services. The mobile assessment utilizes mobile services architecture to deliver interactive messaging automatically, with short messaging services (SMS) used to both send assessment questions and receive multiple responses.

SMS responses are integrated within LMS to enable m-learning. The mobile services architecture presented in Fig. 9 enables students to interact via mobile SMS with the university's LMS server. The students are connected to a mobile service provider by cell/smart phones. With mobile operators implementing one or more SMS centers, the university mediator is connected directly to different SMS centers using Short-Message Peer-to-Peer (SMPP) protocol over the Internet. SMPP is often used to allow third parties to submit messages en masse and has been designed to support services over diverse cellular networks. A university instructor/examiner can receive and send SMSs by using SOAP requests/responds, with the university's LMS managing sessions with different users, using data extracted from SMSs [8, 16, 19].

5.2 SOA-Based Service "Customer Search"

The model depicted in Fig. 10 demonstrates a "customer search" example of mobile application in SOA. The static view of the architecture presents the actors, services, and components - that is, the "what?" aspect of the architecture.

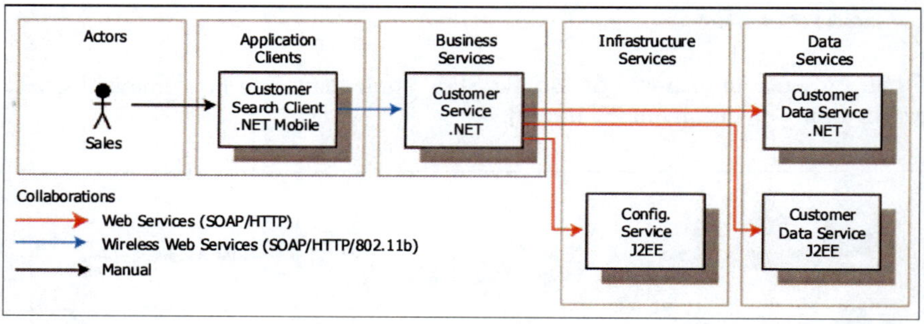

Fig. 10 "Customer Search" Application SOA Service [17].

An actor is the sales agent using the customer search mobile application to retrieve information about customers on the basis of the category of products they have purchased in their order history, their rank in terms of total sales, and their geographical region. The services are the coarse-grained "objects" in the SOA and may be broadly classified as business, infrastructure, and data services:

- Business services are vertical, application-oriented, and closely related to the business domain.
- Infrastructure services are horizontal foundation services that support business services.
- Data services are similar to infrastructure in that they are horizontal and supporting, but differ in that they are oriented specifically toward data access.

The customer service is the business service for the application in which the business logic is implemented. This includes interaction with the configuration infrastructure service on startup in order to retrieve configuration information, and interaction with customer data services at runtime to retrieve, and then process, customer information requested by the client. Responsibilities also include aggregation, sorting, and filtering of customer data. This service is implemented in C#, hosted on .NET, and published using Web services.

The configuration service is an infrastructure service responsible for providing configuration information used to initialize other services, including (for example) properties defining the availability of customer data services. This service is implemented in Java, hosted on J2EE, and published using Web services with the Axis framework.

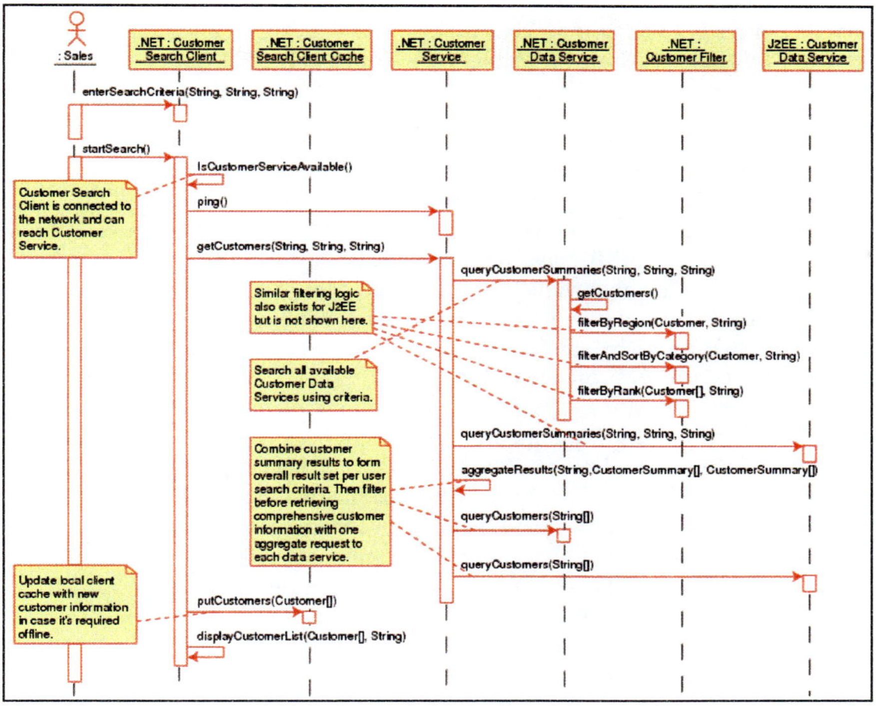

Fig. 11 Connected Use Sequence Diagram [17].

The "customer search" application client is implemented in C# on the .NET Compact Framework to run on a Microsoft Pocket PC mobile device with IEEE 802.11b WiFi short-range wireless connectivity with the SOA. One of the challenges in wireless connectivity is accommodating partial connectivity, situations in which the network is not always available. Also, mobile clients have relatively limited screen and other input/output capabilities and resources (such as memory and CPU). Therefore, care must be paid to UI (User Interface) design for mobile clients to maximize use of the device capabilities, while staying within its bounds to make the client usable. In general, trying to apply desktop UI design to devices does not work well, even if it is technically possible.

Subsequently, as soon as wireless connectivity is established, the mobile client reaches the services in the SOA and the use case named "Search Customers in Central Databases" applies, for which the normal end-to-end scenario is shown in the sequence diagram in Fig. 11.

So, what happens when the "customer search" client tests connectivity by calling a ping method on the customer service in the SOA? In this case, the method call succeeds since connectivity is established, and the client proceeds to use the services to retrieve the most up-to-date customer information.

When no wireless connectivity is available, the use case "Search Customers in Local Cache" is applied. In this case, the ping method call from the client to the customer service fails, so the client uses its local cache, filtering and displaying the latest information from the memory cache of the mobile device that is available to the sales agent [17 ÷ 20].

6 "Applications – End-Devices" Connection Models

Different computing technologies such as CORBA (Common Object Request Broker Architecture), RMI (Java Remote Method Invocation), and Web services have been used as highly efficient "connection" tools for large-scale SOA-based systems. Here we would like to analyze the design patterns applicable to SOA and how these patterns can be applied to Java devices both on the move and for fixed clients that need to access different services.

Java mobile technology is available on millions of mobile devices, with the majority of Java phones now supporting at least a basic J2ME implementation of the Mobile Information Device Profile (MIDP 1.0) and the Connected Limited Device Configuration (CLDC 1.0). The combination of CLDC underlying the MIDP libraries provides sufficient functionality to write interactive clients that can connect to back-end business systems. These interactive clients take the form of MIDlets (named after the applet model). As an applet runs in a browser, a MIDlet runs on a mobile phone or device.

Besides, most of mobile devices also have Web browsers and provide interactive content via WAP. However, in certain situations, Java is much more useful and effective than simple systems based on the Hyper Text Transfer Protocol (HTTP). Sometimes, the WAP protocol can be limiting and inappropriate. So, on one hand, accessing the internet via WAP could be effective. On the other hand, this technology can be limiting when compared with the range of abilities available by Java devices.

Thus, as a next step, we would like to compare how a SOA approach implemented to the connected-device model is better than using the Servlets approach [21 ÷ 24].

6.1 First Scenario – Connecting via Servlets

The mobile Java applications come together with the basic packages to allow systems to connect to remote hosts via HTTP. However, these devices have generally not been considered powerful enough to support Java Remote Method Invocation (RMI). In its current standards, RMI is ruled out of the set of possible technologies, so the first approach most developers take when starting to build a mobile business application is to connect via a Java servlet (Fig. 12).

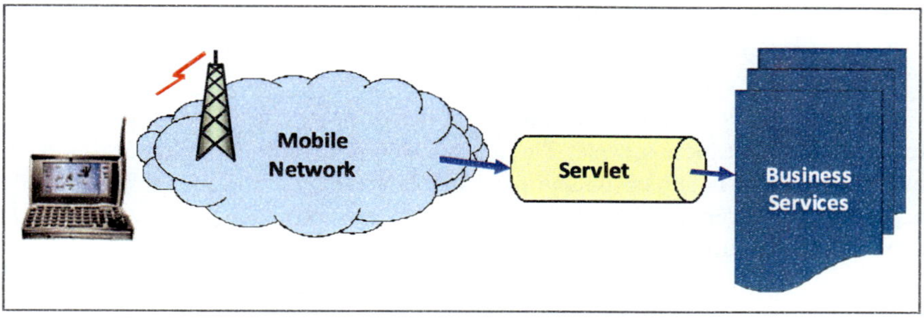

Fig. 12 Connecting via a Servlet.

This approach is a rather easy method to start implementing a simple prototype or demo system. Nevertheless, it becomes quickly difficult to maintain, and it offers poor scalability, because each time some functionality is added, the following is necessary:

- write new mapping code in the servlet;
- write new decoding code in your MIDlet;
- test if the code works together on all target platforms (server and device).

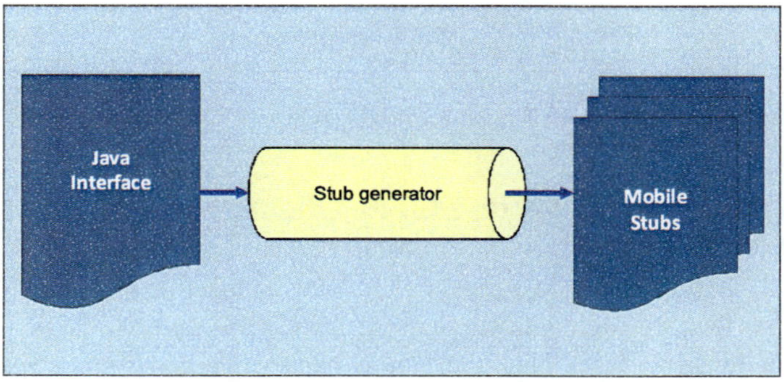

Fig. 13 Mobile Stub Generation.

6.2 Second Scenario – SOA-Based

Comparing the SOA-based approach with the first one is relatively easy, but it does not scale well and is expensive to maintain. The SOA model provides a "transparent" transport layer and connectors to the remote business systems that could be generated from within an IDE in a manner similar to the compiling of an RMI stub.

The stub generator builds a small set of classes capable of providing the transport layer the assembling and disassembling of method calls over the air Fig. 13.

The UML-sequence diagram displayed in Fig. 14 below shows the interactions between the generated proxy, which uses the stubs (not shown), the gateway, and a remote business service.

The stubs contain special encoding that the gateway uses to identify and route requests to the business service. There is considerably more going on during these interactions than is shown on the diagram; however, for the moment, it is only necessary to understand these basic concepts and principles in order to build applications that use the generated code [21 ÷ 24].

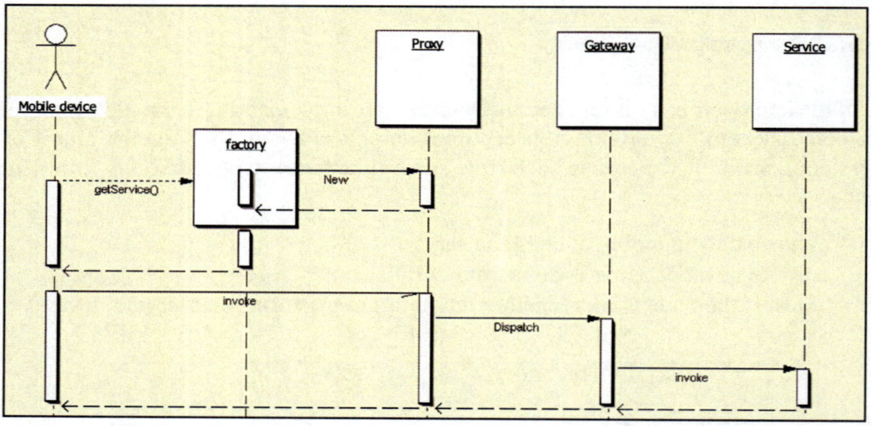

Fig. 14 Interactions Between Generated Proxy [21].

7 SOA Approach to Write Mobile Clients Using Remote Services

Since the unified messaging service is implemented as an RMI service, it essentially has a Java interface. This interface is named `UnifiedMessenger`. It is ready to run within fixed "server-side" network. The interface to the service is shown below:

```
public interface UnifiedMessenger  {
  String getMessage() throws RemoteException;
  void sendMessage(String recipient, String Message)
       throws RemoteException;
}
```

In order to produce the stubs, we need to communicate with the RMI service. The stub generator (Fig. 13) is then used. In addition to generating the stubs, the generator produces a factory. The mobile client uses the service-specific factory to get an interface to the `UnifiedMessenger` service, as is shown in the following code.

```
//MIDP code
UnifiedMessenger msgr = UnifiedMessengerFactory
.getService();
String msg = msgr.getMessage();
```

The `UnifiedMessengerFactory` is also created by the stub generator, and returns a proxy that knows how to connect to the running `UnifiedMessenger` service. The MIDP client now has access to the `UnifiedMessenger` service running on the server via the "shadow" interface running within the MIDP client. The MIDP code has a very direct and simple mapping to the RMI equivalent code, which can be found on a standard RMI system.

```
  //RMI client code
  UnifiedMessenger msgr = (UnifiedMessen-
ger)Naming.lookup( url );
  String msg = msgr.getMessage();
```

Considering the above example, it is evident that applying the SOA approach makes writing mobile clients using remote services truly simple and intuitive, which in turn decreases the time-to-market of every single service and can significantly increase the number of services, as well as their functionalities/features [21 ÷ 24].

8 Mathematical Model of Logical Architecture for the Mobile Connectivity Services

As mentioned Section 4, Logical Architecture provides services and mechanisms to support the solution's physical implementation, e.g., in our case, for the mobile solutions such as connectivity, communication, and distribution. The most challenging development aspects related to mobile solutions are network dependency and integration [14 ÷ 16]. To that end, this Section addresses these challenges and describes how to solve them by means of mathematical methods. The Logical Architecture is shown in Fig. 7 and one of its practical realizations in Fig. 9.

The key objective of the Logical Architecture is to deliver/support services at anytime, anywhere, to any device [25]. Besides, regardless of many similarities between the service delivery models used by different providers, customization is highly desired to enable communications within diverse operations and business support systems (OSS/BSS) [26]. We also take into consideration that an analytical performance evaluation is crucial for the justification of the effectiveness of the modeling of different operational conditions in delivering of high-quality services [27]. Consequently, we apply a mathematical model for the monitoring of services delivery in two interconnected systems in tandem. We consider a queuing network model used to represent a series of two single-server queues, each with unlimited waiting space and the FIFO service discipline [28]. We also develop our model in order to obtain feasible values of main performance features [29, 30].

8.1 Structural Design and Implementation

Designing networking architectures for hundreds or thousands of users, transactions, and diverse content types requires an approach that differs from standard infrastructure designs. Such infrastructures must support services, with high levels of reliability and performance, while maintaining both manageability and security. Also, computing architectures and applications must be designed to support ubiquitous access. In addition, flexibility is essential to support new or evolving requirements, and "always-on" access has become a condition, regardless if services are accessed over the Internet, intranet or extranet [25, 26, 29].

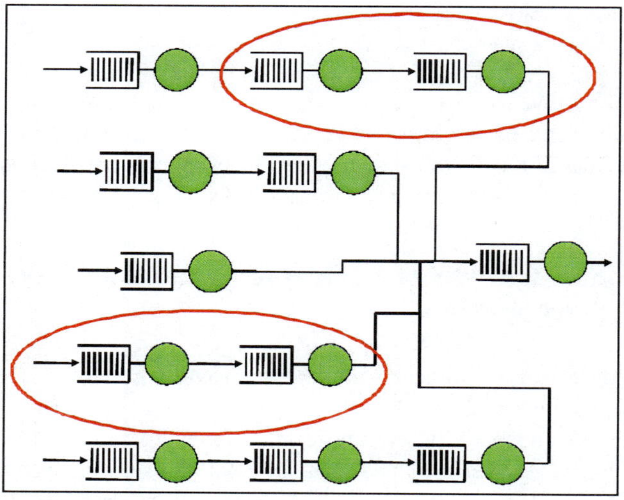

Fig. 15 Interconnected Systems in Tandem.

The most critical factor in the real-time services delivery is the response time or delay. Therefore, we examine a mathematical model in order to apply it to monitoring end-to-end delay in the services delivery over two interconnected systems in tandem.

Interconnected systems in tandem (Fig. 15) have received significant attention in literature on account of their pervasiveness in real life implementations. For example, Avi-Itzhak [31] studied the system with arbitrary input and regular service times [32]. Other related work, in the sense that it focuses on the response time as opposed to the joint queue length, has been done by Knessl and Tier [33]. Knessl and Tier have studied the first two moments of the response time in an open two-node queuing network with feedback for the case with an exponential processor sharing (PS) node and a FIFO node, while the arrivals at the PS node are Poisson. Chao and Pinedo [34] examined the case of two tandem queues with batch Poisson arrivals and no buffer space in the second queue. They allowed the service times to be general and obtained the expected time in system [32].

The behavior of two systems in tandem analyzed here is as follows: when q_1 is M/M/1 or M/G/1 then q_2, is not. The arrival time (hence waiting time) is highly correlated, for example, a long packet is more likely to have a smaller waiting time at q_2 [35]. The first system in tandem is M/G/1. The second system is G/M/1.

8.2 *General Definitions of Queuing Systems Used*

For arrival processes other than Poisson, it is rarely possible to find an exact expression for the mean waiting time except in the case where the holding times are exponentially distributed. In general, it is assumed that either the arrival process or the service process should be Markovian. For GI/G/1 queuing system, it is possible to give the theoretical boundary for the mean waiting time. By denoting the variance of the inter-arrival times as σ_a^2 and the variance of the holding time distribution as σ_h^2, we can find a good realistic estimation for the actual mean waiting time from Marchall's approximation:

$$T_W \approx \frac{aT_S}{2(1-a)}\left[\left(\frac{\sigma_A^2+\sigma_S^2}{T_S^2}\right)\left(\frac{T_S^2+\sigma_S^2}{T_A^2+\sigma_S^2}\right)\right] \tag{1}$$

where, a is offered traffic, T_a is the mean inter-arrival time (T_a = $T_{s/a}$). The approximation seems to be a downward scaling of Kingman's inequality so it just agrees with the Pollaczek-Khintchine's formula in the case M/G/1.

The example for a non-Poisson arrival process is the queuing system GI/M/1, where the distribution of the inter-arrival times is a general distribution given by the density function *f(t)* [36]. When we consider the system at an arbitrary point of time, the state probabilities will not be described by a Markov process only, because the probability that the occurrence of an arrival will depend on how much time has passed since the occurrence of the last arrival. When the system is considered immediately before (or after) an inter-arrival time, there will be independence in the traffic process since the inter-arrival times are stochastic independent and the holding times are exponentially distributed. The inter-arrival times are balance points, and it is taken into consider the so-called embedded Markov chain.

The probability that immediate before an inter-arrival time to observe the system in state i is $p(i)$, and α is the positive real root, that satisfies the equation:

$$\alpha = \int_0^{\infty} e^{-\mu(1-\alpha)t} f(t)dt \tag{2}$$

In statistic equilibrium we will have the following result:

$$p(i) = (1-\alpha)\alpha^i \qquad i = 0, 1, 2, \ldots \tag{3}$$

The steady state probabilities can be obtained by considering two for each of the following inter-arrival times t_1 and t_2. When the departure process is a Poisson process with the constant intensity i, with j customers in the system, the probability $q(j)$ that there are j customers who have completed service between two inter-arrival times can be expressed by details in the Poisson process. The normalization constant is as usual:

$$\sum_{i=0}^{\infty} p_{t_1}(i) = \sum_{j=0}^{\infty} p_{t_2}(i) = 1 \tag{4}$$

The $p(i)$ is not the probability to find the system in state i at an arbitrary point of time (e.g., time mean value), but to find the system in state i immediately before an arrival (e.g., call mean value) [36].

In G/M/1 system, the probability that arriving customer finds the server busy θ is not the same as ρ - the server utilization, because of the general pattern of arrivals. Only the random arrivals have $\theta = \rho$. The value of θ can be obtained from the Eq. 5.

$$\theta = f^*(s)\left(\frac{1-\theta}{T_S}\right) \qquad 0 \le \theta < 1 \tag{5}$$

where $f^*(s)$ is the Laplace-Stieltjes transform of the pdf of inter-arrival times. In some cases Eq. 5 can be solved analytically, but in general a numerical procedure is required. In Fig. 16 we present the probability that arriving customer finds the server busy for different distributions of inter-arrival time [37]. The average waiting time we calculate using Eq. 6.

$$T_W = \frac{\theta T_S}{1-\theta} \tag{6}$$

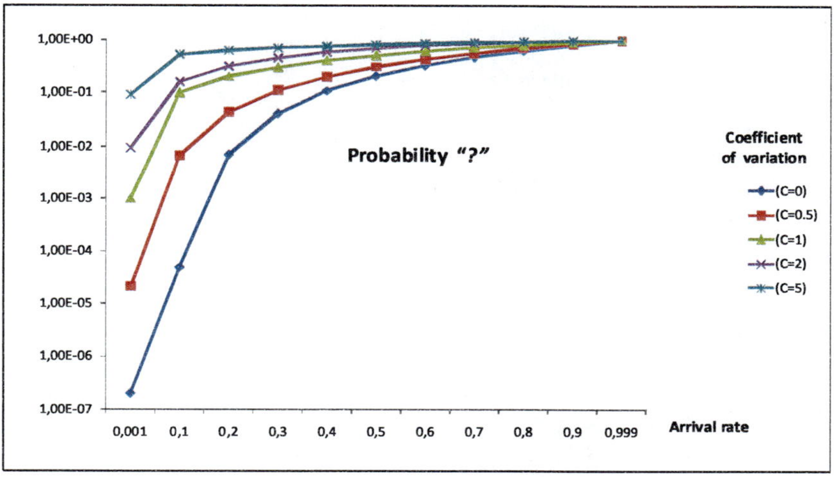

Fig. 16 Arriving Customer Finds Server Busy with Probability θ.

Furthermore, when we know the exact form of the distribution for waiting time, we can obtain standard deviation (e.g., variance) of time that user spends in queue (Table 1, Fig. 18):

$$\sigma_{T_W} = \sqrt{\frac{\theta(2-\theta)}{(1-\theta)^2} T_S^2} \tag{7}$$

Besides, we can observe that the average waiting time increases when the arrival pattern becomes more irregular. Fig. 17 shows average waiting time for different values of C_A^2. The effect of increased variance in the inter-arrival time is apparent, and is very marked at high utilizations.

Table 1 The average waiting time.

λ	Tw_D	Tw_{E2}	Tw_M	Tw_{H2}	Tw_{Ga}
0.001	$2*10^{-7}$	0.000021	0.001001	0.0094892	0.101322
0.1	0.00005	0.00683642	0.111111	0.186634	1.11297
0.2	0.007029	0.0449102	0.25	0.444189	1.68029
0.3	0.0426224	0.121491	0.428571	0.811135	2.31221
0.4	0.12027245	0.244292	0.666667	1.34973	3.09766
0.5	0.2550043	0.432521	1	2.16226	4.15413
0.6	0.479815319	0.728997	1.5	3.43302	5.70093
0.7	0.8761726	1.23669	2.33333	5.57289	8.24078
0.8	1.6927323	2.26733	4	9.819	13.2753
0.9	4.1786639	5.38325	9	22.4082	28.3057
0.999	37.6652747	67.7711	999	4165.67	18866.9

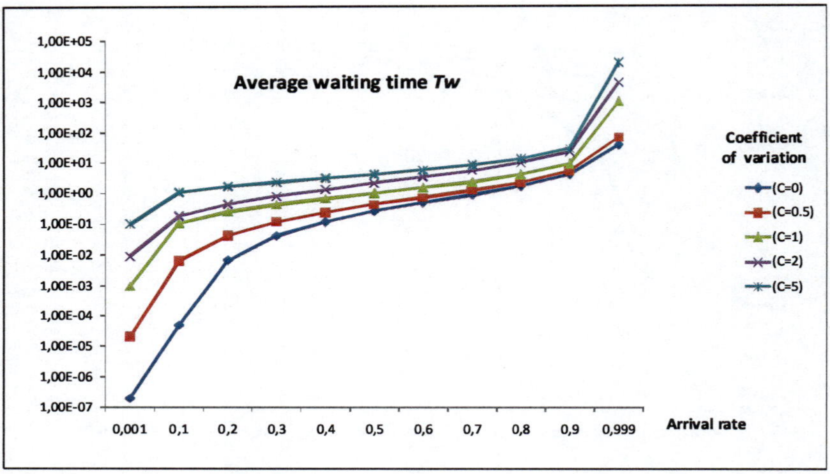

Fig. 17 Average Waiting Time

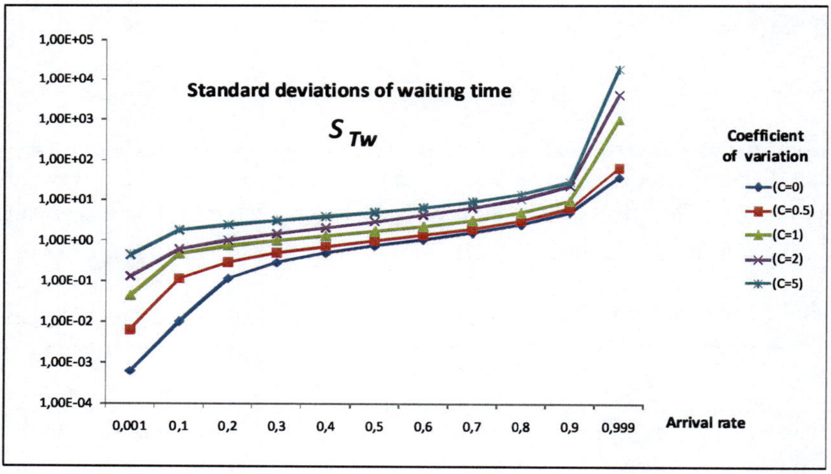

Fig. 18 Standard Deviations of Waiting Time

Table 2 Standard Deviations of Waiting Time

λ	$\sigma_{T_{WD}}$	$\sigma_{T_{WE2}}$	$\sigma_{T_{WM}}$	$\sigma_{T_{WH2}}$	$\sigma_{T_{WGa}}$
0.001	0.000632456	0.00648084	0.0447549	0.138089	0.461421
0.1	0.0100004	0.117131	0.484322	0.638827	1.86135
0.2	0.118775	0.303047	0.75	1.04196	2.48676
0.3	0.295062	0.507683	1.0202	1.51004	3.15765
0.4	0.504986	0.740447	1.33333	2.12632	3.97376
0.5	0.758311	1.02573	1.73205	2.99998	5.05619
0.6	1.0908	1.41047	2.29129	4.31875	6.62589
0.7	1.58746	2.00069	3.1798	6.49638	9.18651
0.8	2.50016	3.11054	4.89898	10.7727	14.2402
0.9	5.0812	6.30443	9.94987	23.3869	29.2887
0.999	38.6523	68.7638	999.999	4166.67	18867.9

The average time in system for the G/M/1 queuing system has the following pattern (Table 3, Fig. 19):

$$T = \frac{T_S}{1-\theta} > \frac{T_S}{1-\rho} \tag{8}$$

In comparison with the M/M/1 system, we can see that for T value G/M/1 is bigger because $\theta > \rho$, where θ is the probability that an arriving customer will have to wait.

Table 3 The average time in system, equal standard deviation

λ	$T_D = \sigma_{T_{WD}}$	$T_{E2} = \sigma_{T_{E2}}$	$T_M = \sigma_{T_M}$	$T_{H2} = \sigma_{T_{H2}}$	$T_{Ga} = \sigma_{T_{Ga}}$
0.001	1	1.00002	1.001	1.00949	1.10132
0.1	1.00005	1.00684	1.11111	1.18663	2.11297
0.2	1.00703	1.04491	1.25	1.44419	2.68029
0.3	1.04262	1.12149	1.42857	1.81113	3.31221
0.4	1.12027	1.24429	1.66667	2.34973	4.09766
0.5	1.255	1.43252	2	3.16226	5.15413
0.6	1.47982	1.729	2.5	4.43302	6.70093
0.7	1.87617	2.23669	3.33333	6.57289	9.24078
0.8	2.69273	3.26733	5	10.819	14.2753
0.9	5.17866	6.38325	10	23.4082	29.3057
0.999	38.6653	68.7711	1000	4166.67	18867.9

Then, the standard deviation (e.g., variance) time in system is (Fig. 19):

$$\sigma_T \cong T \tag{9}$$

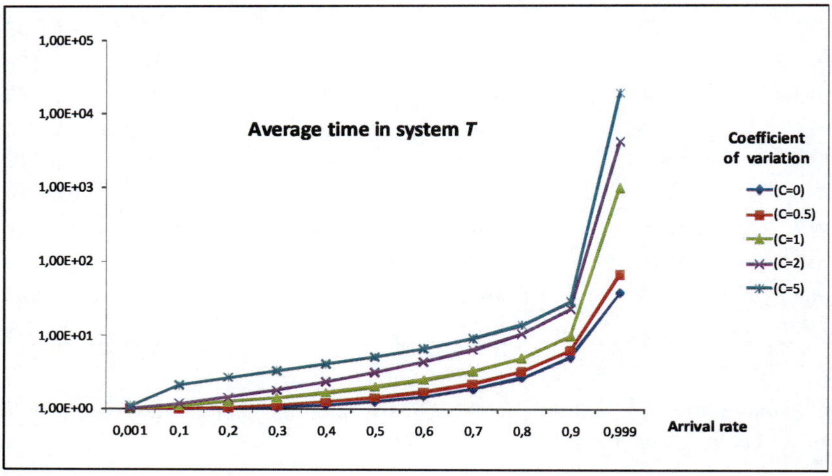

Fig. 19 Average Time in System vs. Standard Deviation

The two stations tandem network with a MAP (Markovian Arrival Process) external input process is studied in this chapter. This is a generalization of the Jackson network and cannot be easily handled by the standard modeling and solution method. To illustrate the performance of our method, we have analyzed some numerical examples with a range of parameter settings. Both the M/G/1 and G/M/1 systems show greater variability in longer times in system, with the M/G/1 system demonstrating this for service times, while the G/M/1 model demonstrates doing so for inter-arrival times [10, 38 ÷ 42].

9 Conclusions

Contemporary mobile virtual network operators (MVNOs) are the fastest-growing local mobile/wireless service providers. They have been achieving such significant growth by providing highly adaptable and personalized services. The key to MVNOs' continued success is their ability to adapt their offerings to best reflect the needs of individual end-users. Consequently, MVNOs require a highly-scalable hardware and software infrastructure to drive their customer portals. At the same time, MVNOs seek to keep costs at an exposed minimum, as they operate in a very competitive market. The challenge they face is improving delivery of numerous new services that are extremely customizable, while decreasing IT costs.

This chapter, therefore, has addressed an open issue: what kind of infrastructure is needed to succeed in such a competitive service provisioning marketplace. To that end, the chapter has also examined and evaluated some examples of an architectural model that allows mobile appliances to use composite services and subsequently enables the execution of complex and resource-intensive applications on constrained devices.

References

1. Technical Paper, Virgin Mobile USA: Moving to a service-oriented architecture on BEA WebLogic Platform 8.1 and Intel® Xeon™ processor-based servers. BEA part number: PSS0826E1104-1A (2004)
2. White Paper, Going Mobile: Developing an application mobilization plan for your business. Research In Motion Limited, BlackBerry®, RIM®, Research In Motion® (2008)
3. Guan, T., Zaluska, E., Roure, D.: A Grid Service Infrastructure for Mobile Devices. In: Proceedings of 1st Semantic Knowledge and Grid Conference, Beijing, China, November 27-29 (2005)
4. White Paper, Next Generation Mobility Architecture. ANYWHERE Solution, a subsidiary of SYBASE (2007)
5. Auer, L., Kryvinska, N., Strauss, C.: Service-oriented Mobility Architecture Provides Highly-configurable Wireless Services. In: Proceedings of the IEEE Wireless Telecommunications Symposium (WTS 2009), Prague, Czech Republic, April 22-24 (2009) (short paper)
6. Auer, L., Kryvinska, N., Strauss, C., Zinterhof, P.: SOA as an Effective Tool for the Flexible Management of Increased Service Heterogeneity in Converged Enterprise Networks. In: The IEEE Second Workshop on Engineering Complex Distributed Systems (ECDS 2008), Barcelona, Spain, March 4-7 (2009)
7. Case study, SOA and the value received when business drives IT decisions. Shanxi Mobile supports rapid business growth with SOA. IBM SOA DeepView, IBM (November 2007)
8. Riad, M., El-Ghareeb, H.A.: A Service Oriented Architecture to Integrate Mobile Assessment in Learning Management Systems. Turkish Online Journal of Distance Education-TOJDE 9(2), 1302–6488 (2007) ISSN 1302-6488
9. Technology Brief, The Case for MOA: Mobile Oriented Architecture, and Moving Towards Mobile Intelligence (MI). J. Gold Associates, December 26 (2006), `http://www.jgoldassociates.com`
10. Samimi, F.A., McKinley, P.K., Sadjadi, S.M.: Mobile Service Clouds: A Self-Managing Infrastructure for Autonomic Mobile Computing services? In: Keller, A., Martin-Flatin, J.-P. (eds.) SelfMan 2006. LNCS, vol. 3996, pp. 130–141. Springer, Heidelberg (2006)
11. Lepaja, S., Lila, A., Kryvinska, N., Nguyen, H.M.: A Framework for End-to-End QoS Provisioning in Mobile Internet Environment. In: Proceedings of the IFIP Fifth IEEE Int. Conference Mobile and Wireless Communications Networks, MWCN 2003, Singapore, October 27-29 (2003)
12. Auer, L., Kryvinska, N., Strauss, C.: Managing an Increased Service Heterogeneity in Converged Enterprise Infrastructure with SOA. Int. Journal of Web and Grid Services (IJWGS) 4(4) (2008)

13. White Paper, OSS/BSS reference architecture and its implementation scenario for fulfillment. Nokia Corporation Networks, Tietoenator Corporation (2005)
14. Sjostrom, A., Forsberg, C.: Northwind Pocket Service: Field Service for Windows Mobile-based Pocket PCs. Pocket PC (General) Technical Articles, Business anyplace and Odyssey Software (July 2004), `http://www.odysseysoftware.com/default.aspx`
15. Kryvinska, N., Strauss, C., Collini-Nocker, B., Zinterhof, P.: A Scenario of Voice Services Delivery over Enterprise W/LAN Networked Platform. In: Proceedings of the Third International Workshop on Broadband and Wireless Computing, Communication and Applications BWCCA 2008, ACM SIGMM, Linz, Austria, November 24-26 (2008)
16. Sanchez-Nielsen, E., Martin-Ruiz, S., Rodriguez-Pedrianes, J.: An Open and Dynamic Service Oriented Architecture for Supporting Mobile Services. In: Proceedings of the 6th International Conference on Web Engineering, ICWE 2006, Palo Alto, California, USA, July 11-14 (2006)
17. Houlding, D.: A Service-Oriented Architecture for Mobile Applications - A Framework for Developing Mobile Apps. (Juli 01, 2004), `http://www.ddj.com/mobile/184405730?pgno=1`
18. Valiente, P., van der Heijden, H.: A method to identify opportunities for mobile business processes. Stockholm School of Economics, SSE/EFI Working Paper Series in Business Administration, vol. 10 (2002)
19. Natchetoi, Y., Kaufman, V., Shapiro, A.: Service-Oriented Architecture for Mobile Applications. In: Proceedings of the ACM SAM 2008, Leipzig, Germany, May 10 (2008)
20. Ritz, T., Stender, M.: Modeling of B2B Mobile Commerce Processes. In: Proceedings of the 17th International Conference on Production Research ICPR-17, Virginia Tech, Blacksburg (2003)
21. Warren, N., Bishop, P.: Taking Service-Oriented Architectures Mobile, Part 1: Thinking Mobile, June 21 (2005), `http://today.java.net/pub/a/today/2005/06/21/mobile1.html`
22. Pajunen, L., Ruokonen, A.: Modeling and Generating Mobile Business Processes. In: Proceedings of the IEEE International Conference on Web Services (ICWS), Salt Lake City, Utah, USA (July 2007)
23. McKinley, P., Samimi, F., Shapiro, J., Tang, C.: Service Clouds: A distributed infrastructure for composing autonomic communication services. Techn. Rep. MSUCSE-05-31, Michigan University (2005)
24. Anckar, B., D'Incau, D.: Value-added services in mobile commerce: an analytical framework and empirical findings from a national consumer survey. In: Proceedings of the 35th Hawaii International Conference on System Sciences, Hawaii, IEEE (2002)
25. Lofstrand, M., Carolan, J.: Sun's Pattern-Based Design Framework: the Service Delivery Network. Sun BluePrints™ OnLine, Sun Microsystems (September 2005)
26. White Paper, Enabling Service Delivery Using the Microsoft Connected Services Framework. Microsoft Corporation (January 2005)
27. Mun, Y.: Performance Analysis of Banyan-Type Multistage Interconnection Networks Under Nonuniform Traffic Pattern. The Journal of Supercomputing 33(1), 33–52 (2005)
28. Glynn, P.W., Whitt, W.: Departures from Many Queues in Series. The Annals of Applied Probability 1(4), 546–572 (1991)

29. Kryvinska, N., Strauss, C., Zinterhof, P.: Minimizing Queuing Constrains in Service Delivery on Enterprise Communication & Computing Platform. In: Proceedings of the 6th Vienna International Conference on Mathematical Modelling (MATHMOD 2009), Vienna, Austria, ARGESIM Report no. 35, February 11-13, pp. 2560-2563 (2009)
30. Waluyo, A.B., Taniar, D., Rahayu, W., Srinivasan, B.: Mobile service oriented architectures for NN-queries. Journal of Network and Computer Applications 32(2), 434–447 (2009)
31. Avi-Itzhak, B.: A sequence of service stations with arbitrary input and regular service times. Management Science 11(5), 565–571 (1965)
32. Van Houdt, B., Alfa, A.S.: Response time in a tandem queue with blocking, Markovian arrivals and phase-type services. Operations Research Letters 33, 373–381 (2002)
33. Knessl, C., Tier, C.: Approximation to the moments of the sojourn time in a tandem queue with overtaking. Stochastic Models 6(3), 499–524 (1990)
34. Chao, X., Pinedo, M.: Batch arrivals to a tandem queue without an intermediate buffer. Stochastic Models 6(4), 735–748 (1990)
35. Vastola, K.S.: Performance Modeling and Analysis of Computer Communication Networks. Electrical Computer and Systems Engineering Dept. Rensselaer Polytechnic Institute Troy, NY, `http://poisson.ecse.rpi.edu/~vastola/pslinks/perf/perf.html`
36. Iversen, V.B.: Fundamentals of Teletraffic Engineering (2001), `http://www.tele.dtu.dk/teletraffic`
37. Hashida, O., Ueda, T., Yoshida, M., Murao, Y.: Queueing Tables. The Electrical Communication Laboratories Nippon Telegraph and Telephone Public Corporation, Tokyo, Japan (1980)
38. Tanner, M.: Practical Queueing Analysis. IBM McGraw-Hill Series (1995)
39. Kleinrock, L.: Queueing Systems. Volume II: Computer Applications, vol. 2. Wiley-Interscience publication, Hoboken (1986)
40. Liu, B., Alfa, A.S.: Performance Analysis of a Mobile Communication Network: Unidirectional Tandem Case with Phase Type Service. Kluwer Telecommunication Systems 20(3,4), 241–254 (2002)
41. Lian, Z., Zhao, N., Liu, L.: Departure Processes of a Tandem Network. In: Proceedings of the 7th International Symposium on Operations Research and Its Applications (ISORA 2008), Lijiang, China, October 31 - November 3, pp. 98–103 (2008)
42. Subercaze, J., Maret, P., Calmet, J., Pawar, P.: A service oriented framework for mobile business virtual communities. In: Pervasive Collaborative Networks. IFIP, vol. 283, pp. 493–500. Springer, Boston (2008)

Glossary of Terms and Acronyms

BSS	Business Support Systems
CLDC	Connected Limited Device Configuration
CORBA	Common Object Request Broker Architecture
CRM	Customer Relationships Management
HTTP	Hyper Text Transfer Protocol
IVR	Interactive Voice Response
MAP	Markovian Arrival Process
MIDP	Mobile Information Device Profile

MVNO	Mobile Virtual Network Operator
OSS	Operations Support Systems
RMI	Java Remote Method Invocation
SMPP	Short-Message Peer-to-Peer Protocol
SMS	Short Message Service
SOA	Service-oriented Architecture
SOI	Service-Oriented Infrastructure
UI	User Interface
WAP	Wireless Application Protocol
WSD	Web Service Delivery

Chapter 15
Evolutionary Algorithms towards Generating Entertaining Games

Zahid Halim and A. Raif Baig

Abstract. Computer games are gaining popularity by every passing day. This has increased the number of choices in computer games for the users. At the same time the quality of entertainment provided by these games has also decreased due to abundance of games in the market for personal computers. On the other hand the task of game development for the developers is becoming tiresome, which requires scripting the game, modeling its contents and other such activities. Still it cannot be known how much the developed game is entertaining for the end users. As entertainment is a subjective term. What might be entertaining for one user may not be entertaining for others. Another issue from the point of view of game developers is the constant need of writing new games, requiring investment both in terms of time and resources. In this work we create a set of metrics for measuring entertainment in computer games. The genres we address are board based games and predator/prey type of games. The metrics devised are based on different theories of entertainment specifically related to computer games, taken from literature. Further we use Evolutionary Algorithm (EA) to generate new and entertaining games using the proposed entertainment metrics as the fitness function. The EA starts with a randomly initialized set of population and using genetic operators (guided by the proposed entertainment metrics) we reach a final set of population that is optimized against entertainment. For the purpose of verifying the entertainment value of the evolved games with that of the human we conduct a human user survey and experiment using the controller learning ability.

1 Introduction

Nowadays computer games have became a major source of entertainment for all age groups, especially children. The reason for computer games being the primary source of entertainment could be many including these being highly interactive, high resolution graphics, diverse level of choice and challenge. According to

---

Zahid Halim · A. Raif Baig
National University of Computer and Emerging Science, Islamabad, Pakistan
e-mail: {zahid.halim,rauf.baig}@nu.edu.pk

N. Bessis, F. Xhafa (Eds.): Next Generation Data Technologies for CCI, SCI 352, pp. 383–413.
springerlink.com

a survey conducted in [1] on 1254 subjects, only 80 were found playing no electronic games in the last 6 months. The results in [1] show the popularity of computer and video games in young generation.

Thinking from the point of view of game developers it has always been a challenge to measure the entertainment value of the human player. This is due to the fact that entertainment is very subjective. It also depends upon the genre of game and contents of the game in addition to the subject (user) playing it. Keeping this fact in mind it would be very convenient for the game developers to develop entertaining games if they could somehow measure entertainment, the way other things like temperature, weight and many such things are measured. This would give a quantitative representation of the entertainment a game has as against the subjective one nowadays. Still based upon the measurable entertainment, the responsibility of producing a game like nowadays will be on the shoulders of game developer. Game developer will have to define the complete game from start till end along with each stage with its components and complexities. It would be very convenient that we could also produce the game automatically based upon the measurable entertainment. This would lead to interesting application in the area of computer game development.

1.1 Our Contribution

In this chapter we address the two issues in game development: (a) measuring entertainment value of a game and (b) automatic generation of entertaining games. We propose an entertainment metrics to quantitatively measure the entertainment value of the game. At present we address two genres of games and a separate set of entertainment metrics is proposed for each. The genres of games addressed include board based games and predator/prey games. There might be other sources of entertainment, other than the one we have considered for devising our entertainment metrics, like graphics and sound effects but these factors are not in the scope of the basic ingredients of a game and that is why they have not been considered. We have further shown the utility of the proposed entertainment metrics by generating new and entertaining games through computational intelligence techniques like genetic algorithms and evolutionary strategy which uses the proposed entertainment metrics as fitness function, guiding the evolution towards entertaining set of games. In order to counter check the entertainment value of the evolved games we have conducted a human user survey to verify that the results correlate with those produced by the system. In contrast to the user survey, the entertainment value of the evolved games is also verified using the Schmidhuber's theory of artificial curiosity [2].

Remaining of the chapter is organized as follows: section 2 covers background work, section 3 covers entertainment theories, section 4 lists search space, section 5 and 6 cover fitness function and chromosome encoding, respectively, section 7 explains the software agents, section 8 lists experiments and section 9 concludes the chapter.

2 Background Work

The concept of measuring entertainment and automatic generation of games and/or its contents is fresh and quite a limited amount of literature is available on the topic. This section is dedicated to the work done in the domain of measuring entertainment in computer games and their automatic generation. We have studied the work done in this regard by different researchers and listed it here.

2.1 *Board Based Games*

Iida [3], in 2003, has proposed a measure of entertainment for games and used it to analyze the evolution of game of chess over the centuries. This measure is considered to be the pioneer in quantification of entertainment. Even though Iida's work is limited to chess variants, the measure of entertainment can be easily applied to other board games. According to this measure, the entertainment value of a game is equal to the length of the game divided by the average number of moves considered by a player on his turn. The game is more entertaining if the value of this measure is low. The main idea is that the player should have many choices (moves) on the average and the length of the game should not be large. Long games with few choices per move are boring. The authors differentiate between possible moves and the moves considered by a player. The set of considered moves is smaller than the set of possible moves and the metric is based on the moves considered by a player.

In [4] the authors introduce the uncertainty of game outcome as a metric of entertainment. If the outcome is known at an early stage then there is not much interest in playing it. Similarly if it is found at the last move then it is probably probabilistic. The outcome should be unknown for a large duration of the game and should become known in the last few moves of the game. Authors state that it is easy to create new board games and variants of classical games but to make a game attractive to the human user, is challenging. In [4] a simple technique based on synchronism and stochastic elements is used to refine the game of Hex. Authors proof that the game's attraction has increased by conducting experiments to show an increment of the outcome uncertainty.

In [5] Symeon uses board games for e-learning. He proposes an e-learning board game that adopts the basic elements of a racing board game and cultivates in students skills like creativity, problem-solving, and imagination, as students try to reach the end by improving their performance in a variety of learning activities.

2.2 *Video Games*

Togelius [6] has presented an approach to evolve entertaining car racing tracks for a video game. Tracks were represented as b-splines and the fitness of a track depended on how an evolved neural network based controller (modeled after a player) performed on the track. The objectives were for the car to have made maximum progress in a limited number of time steps (high average speed), high maximum speed (so that at least one section of the track is such that high speeds can be achieved), and high variability in performance (as measured by the final progression made) between trials (so that the track is challenging: neither too easy nor too hard). The game model used for experimentations in [6] is simple both graphically and physically (being 2D).

In [7] three metrics (which are combined into one) have been proposed for measuring the entertainment value of predator/prey games. The first metric is called appropriate level of challenge (T). It is calculated as the difference between the maximum of a player's lifetime and his average lifetime over N games. This metric has a higher value if the game is neither too hard nor too easy and the opponents are able to kill the player in some of the games but not always. The second metric is behaviour diversity metric (S). It is standard deviation of a player's lifetime over N games. It has a high value if there is diversity in opponent's behaviour. The third metric is spatial diversity metric E{Hn}. It is the average entropy of grid-cell visits by the opponents over N games. Its value is high if the opponents move all the time and cover the cells uniformly. This movement portrays aggressive opponent behaviour and gives an impression of intelligent game play. The three metrics are combined into one single metric $I = [\gamma T + \delta S + \varepsilon E\{Hn\}]/[\gamma + \delta + \varepsilon]$ where I is the interest value of the predator/prey game; γ, δ and ε are weight parameters. The work in [8] is some sort of extension of [7].

In [9], the authors have developed a computer game called "Glove" with three levels of incongruity: hard, easy and balanced. Their assumption is that the player would get frustrated or bored respectively, with the first two settings and would enjoy with the third one. The verification of this assumption has not been actually done in their paper. They argue that the actual complexity of a game can be defined as its difficulty level and the incongruity, i.e. the difference between the actual complexity and a player's mental complexity of a game can be measured indirectly by observing the player's behavior in the game.

In [10] an effort has been made to evolve rules of the game. The evolution of games in [10] is guided by a fitness function based on "learning ability". It gives low scores to games that do not require any skill to play and also to those which are hard and impossible whereas it assigns high fitness to games which can be learnt quickly. Although there are games being created automatically but they are not being measured against their entertainment value present in the game due to its rules and contents. They employ theory of artificial curiosity based fitness function introduced in [1] which focuses on the predictability of the game environment.

Chris in [11] modified the level generator to create new level on the basis of four parameters, three of which deals with the performing different operations with holes and the last parameter deals with the direction of the movement of Mario. If one or more direction switch is present, the level will suddenly be mirrored at random points, forcing the player to turn around and go the other way, until reaching the end of the level or the next direction switch. The aforementioned parameters are then categorized into two sub parts that are high or low thus making 16 possible combinations in total. In [11] several statistical features are noted during the playing of the game. These include completion time, time spent on various tasks (e.g jumping), killed enemies (e.g way of killing) and information on how the player dies. Chris used neuroevolutionary preference learning of simple non linear perceptron to predict certain player emotions from game play features.

In [12] Nicola presents the concept of fun in the game of Pac-man based on the concept of flow. He argues that the fun factor in a game depends upon the psychological flow concept. Work in [12] deal with the question whether flow is a more reliable measure than asking human players directly for the fun experienced during the game. For the purpose of detecting flow a measure based on interaction time fraction between the human-controlled Pac-Man and the ghosts is introduced. The outcome of the measure is compared with the work done in this regard by Yannakakis and Hallam [13].

3 Theories on Entertainment

According to Csikszentmihalyi's theory of flow [14,15] the optimal experience for a person is when he is in a state of flow. In this state the person is fully concentrated on the task that he is performing and has a sense of full control. The state of flow can only be reached if the task is neither too easy nor too hard. In other words the task should pose the right amount of challenge.

In addition to the right amount of challenge, Malone [16] proposes two more factors that make games engaging: fantasy and curiosity. If a game has the capability of evoking the player's fantasy and makes him feel that he is somewhere else or doing something exotic then that game is more enjoyable than a game which does not do so. Curiosity refers to the game environment. The game environment should have the right amount of informational complexity: novel but not incomprehensible. Koster's theory of fun [17] states that the main source of enjoyment while playing a game is the act of mastering it. If a game is such that it is mastered easily and the player does not learn anything new while playing then the enjoyment value of that game is low.

Rauterberg [18, 19] has introduced the concept of incongruity as a measure of interest in a task. Given a task, humans make an internal mental model about its complexity. Incongruity refers to the difference between the actual complexity of the task and the mental model of that complexity. We have positive congruity if this difference is positive and negative congruity otherwise. In case of negative incongruity a person would be able to accomplish the task easily. Interest in a task is highest when the incongruity is neither too positive nor negative. In case of large positive incongruity the humans have a tendency to avoid the task and in situations of large negative incongruity they get bored. This requirement of right amount of incongruity is similar to the right amount of challenge in the concept of flow mentioned above. It has been further proposed that in case of reasonable positive incongruity the humans have a tendency to learn more about the task so that their mental model comes at par with the actual complexity of the task.

Several other related and derivative works are available on this topic. Many of them are covered in Yananakakis's recent survey [20].

4 Search Space

For the purpose of generating new games we need to define a search space that will be used by the evolutionary algorithm for this purpose, for these games to be entertaining the evolutionary algorithm will be guided by a fitness function, which will be our proposed entertainment metrics. As we are addressing two different genres of games we need to have separate search space and fitness functions for both.

4.1 Board Based Games

For defining the search space for board based games we use the search space of the popular board games of chess and checkers as a super set. Figure 1 summarizes the search space.

Search Space Dimension	Values
Play Area	Both white & black squares are used
Types of Pieces	6
Number of pieces/type	variable but at maximum 24
Initial position	First 3 rows & Both white & black
Movement direction	All directions, straight forward, straight forward and backward, L shaped, diagonal forward
Step Size	One Step, Multiple Steps
Capturing Logic	Step over, step into
Game ending logic	No moves, no king
Conversion Logic	Depends upon rules of the game
Mandatory killed	Depends upon rules of the game
Turn passing allowed	No

Fig. 1 Board based game search space dimensions.

The size of play area in our search space is a grid of 8x8 squares, alternating white and black, all squares can be used. Combining the rule space of chess and checkers we have total six types of pieces in our search space. Each type can have a minimum of 0 and a maximum of 16 pieces. However, total pieces should not be zero nor exceed 16. The initial positions of the pieces are the nearest three rows of a player. A cell can have one piece of type 0 to 6, type 0 means no piece is present in that cell. The search space to evolve new games consists of only those six movement logics as in both chess and checkers. Which include diagonal forward, diagonal forward and backward, movement in all directions, L shaped movement, straight forward and straight forward and backward. The step size for a piece can either be one or up to an occupied cell.

Capturing is done by jumping over or moving into the opponent's cell. The result of capturing is death of captured piece. The game ends if there are no more pieces left of a specific type. We call this type of piece the "piece of honour".

There can be zero or one piece type declared to be a piece of honour. A game ends if a piece of honour of any player is dead or the player without moves is the loser. A game can have a maximum of 100 moves. A piece may or may not convert to another type after reaching last row. Evolution decides which type is convertible. Each piece has a conversion logic which decides which type it will convert to when last row is reached. Turn passing is not allowed.

4.2 Predator/Prey Games

The predator/prey genre consists of one or more predators, predators may be homogeneous or heterogeneous in their behaviour, obstacles (which may or may not be for both predator and prey) and some objective for the prey to achieve. Pac-Man is a very popular game of this genre. Keeping the above constrained in mind and inspired by its closeness to the rule space defined by J. Togelius in his work [10]. We have defined our rule space as follows.

Play area consists of 20 X 20 cells. There are *N* predators of type *M*; each type is represented by a different colour. We have selected *N* to be 0-20 and *M* in range 0-3. Where colours are red, green and blue. Each type of predator moves around the play area according to any of the following three schemes: still, turn clockwise upon encountering an obstacle and turn counter clockwise upon encountering an obstacle. The predators may collide with each other and the prey. As the different predators may have different behaviour so the response to collision needs to be different of each type. The possible types of responses to a collision are as: death of the prey and/or predator, random change in current location of the prey and/or predator, no effect on the prey and/or predator. The score is calculated for the prey only, which is one of +1, 0, or -1 upon collision with a predator. For the predators for which upon collision with the prey, prey's score increases prey is predator for them and the predators are prey. This shows the uncertain nature of the game which adds some level of entertainment in the game as well. The time for which the game will be played vary from 1-100 time steps and the maximum score that a prey can achieve vary from 1-2000. The game will stop if any of the following is true: the time exceeds its maximum limit, prey has died, or the prey score exceeds the maximum score. Figure 2 displays a typical environment of one such game created based on the rule space defined. The yellow is the prey and remaining are the predators of different types.

Fig. 2 The play area of the predator-prey game.

5 Structure of the Chromosome

For the purpose of evolving games we have used Evolutionary Algorithms. Each individual chromosome of the EA population represents one complete set of rules for the game; whereas each gene of the chromosome represents one rule of the game.

5.1 Board Based Games

Based upon the above search space, the structure of the chromosome used is listed in figure 3. The chromosome consists of a total 50 genes. First 24 genes may contain values from 0 to 6 were 1 represents a piece of type 1, 2 for piece of type 2 and so on. Zero is interpreted as no piece. The piece type represented by gene 1 is placed in the cell 1 of the game board; piece type represented by gene 2 is placed in the cell 2 of the game board and so on.

Gene	Title	Va lue
1-24	Placement of gene of each type	0-6
25-30	Movement logic of each type	1-6
31-36	Step Size	0/1
37-42	Capturing logic move into cell or jump over 0/1	0/1
43	Piece of honor	0-6
44-49	Conversion Logic 0-6	0-6
50	Mandatory to capture or not	0/1

Fig. 3 Structure of the chromosome

Gene 25 to 30 represents movement logic for each piece type respectively, where 1 is for diagonal forward, 2 for diagonal forward and backward, 3 for all directions, 4 for L shaped movement, 5 for straight forward and backward and 6 for straight forward. Genes 31 to 36 are used for step size of each type, where 0 is used to indicate single step size and 1 for multiple step sizes. Genes 37 to 42 are used for step size of each type, where 0 is used to indicate step into and 1 for step over. Gene number 43 indicates the type of piece that will be the piece of honour, possible values include 0-6, where 1-6 indicate the piece type and 0 represents that there is no piece of honour in the game. Genes 44-49 represents the conversion logic, of piece type 1 to 6 respectively, when they reach the last row of the game board. Where 0 represents the piece will not be converted to any type and 1-6 represents the type of piece. The last gene represent whether it is mandatory in the game to capture the opponent piece in case it could be, 0 represents no and 1 represents yes.

5.2 Predator/Prey Games

There are a total of 30 genes in a chromosome; the chromosome encoding is shown in Figure 4. The rules of the game they represent and their possible values are as follows:

- First three genes each for representing the number of red, green and blue type of pieces. The values they can have range form 0-20.
- Next three genes each representing the movement logic of red, green and blue type of pieces. The possible values for these genes are 0 to 4 representing: still, clockwise, counter clockwise, random short and random long movement respectively.
- A total of next 15 genes for representing collision logic between two pieces or between any piece and an agent. Since there are a total of 4 entities (three pieces and an agent), hence the possible effects of collision between any two entities can be represented by a 4x4 -1 matrix. A collision between an agent and another agent is not possible because there is only one agent. Hence one element of the 4x4 matrix is empty and needs not be represented as a gene in the chromosome. The possible values of the collision logic genes are 0, 1 and 2 representing no effect on the colliding entity, the entity dies and are removed from the game and the piece is moved to some randomly chosen location respectively.
- Next 9 genes for representing the score addition or depletion on the collision of any two entities. The possible values of these genes are -1, 0, and 1.
- Next nine genes represent score effect for collision between: agent and red piece, agent and green piece, agent and blue piece, red and red piece, red and green piece, red and blue piece, green and green piece, green and blue piece, and blue and blue piece.

	Red	0-20		Blue-Green	0-2
	Green	0-20		Blue-Blue	0-2
Number of predators	Blue	0-20		Blue-Agent	0-2
	Red	0-4		Agent-Red	0-2
	Green	0-4		Agent-Green	0-2
Movement logic	Blue	0-4	Collision logic	Agent-Blue	0-2
	Red- Red	0-2		Red- Red	-1,0,+1
	Red- Green	0-2		Green-Green	-1,0,+1
	Red-Blue	0-2		Blue-Blue	-1,0,+1
	Red- Agent	0-2		Agent-Red	-1,0,+1
	Green-Red	0-2		Agent Green	-1,0,+1
	Green-Green	0-2		Agent-Blue	-1,0,+1
	Green-Blue	0-2		Green-Red	-1,0,+1
	Green-Agent	0-2		Blue-Red	-1,0,+1
Collision logic	Blue-Red	0-2	Score logic	Blue-Green	-1,0,+1

Fig. 4 Chromosome encoding along with possible values a gene can have.

6 Fitness Function

Each chromosome encodes the rules of a game. In other words, it is a complete game. The aim of the evolutionary process is to evolve a population of games and find a best one which is entertaining for the player. For this purpose we have assumed that better entertainment is based on four different aspects described below for each of the genre discussed. In three of these aspects (for board based games) we assume that both the players play each game with the same strategy (random controller). Hence both have the same chances of winning.

6.1 Board Based Games

6.1.1 Duration of the Game

In general, a game should not be too short or too long, as both are uninteresting. For example, if a game is such that it usually ends after a few moves (like Tic-Tac-Toe) then it would not appeal to adults. On the other hand, if a game usually continues for several hundred moves then the players may choose not to play it due to lack of enough time.

The duration of play (D) of a game is calculated by playing the game n times and taking the average number of moves over these n games. For the games evolved in this chapter, the maximum moves are fixed at 100 (50 for each player). If a game does not end in 100 moves then it is declared a draw. The average value of D is taken because if the game is played multiple times with a different strategy (or even by the same strategy which has probabilistic components) then we do not get the same value of D every time. For averaging, the game is played n = 20 times in our experiments. Equation (1) shows the mathematical representation of D.

$$D = \frac{\sum_{k=0}^{n} L_K}{n} \quad (1)$$

Where, L_K is the life of the game playing agent in game K. In order to reward games neither too short nor too long raw value of D is scaled in range 0-1. The boundaries for scaled value of D are shown in figure 5.

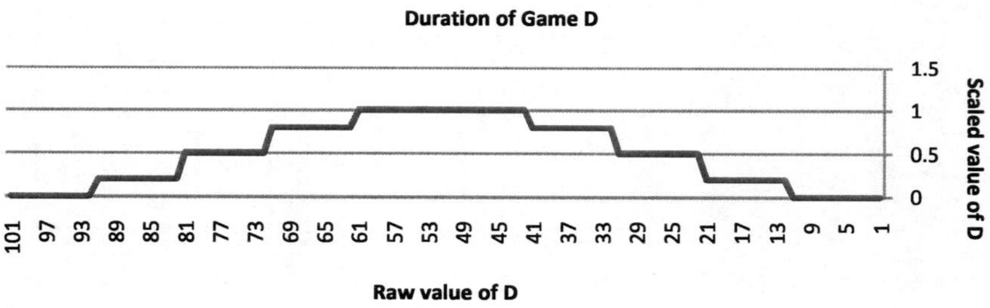

Fig. 5 Scaling ranges for raw value for duration of game.

For the raw duration of games 0 to 10 and 100 to 90 a scaled value of 0 is assigned, for ranges 11-20 and 81-90 a value of 0.2 is assigned, for 21-30 and 71-80 a scaled value of 0.5 is used, 31-40 and 61-70 are converted to 0.8 and a range of 41-60 is assigned the highest value i.e. 1.

6.1.2 Intelligence for Playing the Game

A game is interesting if the rules of the game are such that the player having more intelligence should be able to win. The intelligence (I) is defined as the number of wins of an intelligent controller over a controller making random (but legal) moves. For this purpose the game is played *n* times (*n* = 20 times in our experiments). Higher number of wins against the random controller means that the game requires intelligence to be played and does not have too many frustrating dead ends. Intelligence I is calculated using equation (2).

$$I = \frac{\sum_{k=0}^{n} I_K}{n} \tag{2}$$

Where, I_K is 1 if intelligent controller wins the game otherwise its 0.

6.1.3 Dynamism Exhibited by the Pieces

This aspect assumes that a game whose rules encourage greater dynamism of movement in its pieces would be more entertaining than a game in which many pieces remain stuck in their cells for the entire duration of the game. The dynamism is captured by the following fitness function given in equation (3).

$$Dyn = \frac{\sum_{i=1}^{n}\left(\frac{\sum_{i=1}^{m}(C_i)/Li}{m}\right)}{n} \tag{3}$$

Where,
C_i, is the Number of cell changes made by piece i during a game
L_i, is life of the piece i
And m is the total number of pieces specified in a chromosome.

The dynamism is averaged by calculating it for 20 games for the same chromosome. This fitness function has a higher value if the pieces show a more dynamic behaviour.

6.1.4 Usability of the Play Area

It is interesting to have the play area maximally utilized during the game. If most of the moving pieces remain in a certain region of the play area then the resulting game may seem strange. The usability is captured using equation (4).

$$U = \frac{\sum_{i=1}^{n}\left(\frac{\sum_{k=0}^{m}(C_k)}{|C_u|}\right)}{n} \tag{4}$$

Where,
C_k, is usability counter value for a cell k.
$|C_u|$, is the total number of usable cells.
n, is 20 as explained previously.

A usability counter is set up for each cell which increments when a piece arrives in the cell. The usability U is averaged by playing twenty different games for a chromosome. A cell which is never visited during a game will have a counter value of zero, thus contributing nothing to the usability formula. Furthermore, a cell which has a few visits would contribute less than a cell having large number of visits.

6.1.5 Combined Fitness Function

The above four metrics are combined in the following manner. All chromosomes in a population are evaluated separately according to each of the four fitness functions. Then the population is sorted on "duration of game" and a rank based fitness is assigned to each chromosome. The best chromosome of the sorted population is assigned the highest fitness (in our case it is 20 because we have 10 parents and 10 offsprings), the second best chromosome is assigned the second best fitness (in our case 19), and so on. The population is again sorted on the basis of "intelligence" and a rank based fitness is assigned to each chromosome. Similarly, rank based fitness is assigned after sorting on "diversity" and "usability". The four rank based fitness values obtained for each chromosome are multiplied by corresponding weights and then added to get its final fitness.

$$FF = aD + bI + cDyn + dU \tag{5}$$

Where, a, b, c, and d are constants. In our experiments we keep the values of these constants fixed at 1. The multiplication with a corresponding weight allows us to control the relative influence of an aspect. The calculation of rank based fitness gets rid of the problem of one factor having higher possible values than another factor.

6.2 *Predator/Prey Games*

6.2.1 Duration of the Game

The fitness function should be such that it discourages such a possibility. The duration of play, D, is calculated as in equation 6.

$$D = \frac{\sum_{k=0}^{n} L_K}{n} \tag{6}$$

Where, L_K, is the life of the game playing agent in game K ,

And n is the total number of times the agent plays a game represented by a single chromosome, in this case n is fixed to 20.

Since there are many probabilistic factors in the game, we will not get the same value of D if the game is played multiple times. For this purpose the game is played n times (n= 20 in our experiments) for a chromosome and an average is taken.

6.2.2 Appropriate Level of Challenge

The level of challenge of the game can be directly measured from the score of the player in the game. Higher a player score more is his interest and motivation to play the game again. Too high a score, achieved easily is not challenging enough and similarly too low a score even after an intelligent game play is discouraging. There has to be an appropriate level of challenge provided by the rules. The factor of uncertainty in the rules of the game where entities other than the agent can have positive and/or negative effect on the player's score, introduces a factor of good or bad luck in the game, as in snake and ladders game, and can enhance the enjoyment level, provided that this uncertainty factor is not too high. The challenge C is converted into a fitness function using equation 6:

$$c = e^{\left(\frac{-|S_m - S_a|}{S_m}\right)} \tag{6}$$

Where

$$S_a = \frac{\sum_{k=0}^{n} S_K}{n}$$

S_K, is score of the agent when it plays it K^{th} time.
n, is 20 as explained previously.

Since the value of S_a can also be negative, hence we use the following processing:

$$S_a = \begin{cases} S_a, & \text{if } S_a \geq 0 \\ |S_a| + 20, & \text{otherwise.} \end{cases} \tag{7}$$

6.2.3 Diversity

The diversity of the game is based upon the diversity of the pieces in the game. The behavior of the moving pieces of the game should be sufficiently diverse so that it cannot be easily predicted. Pieces with complex movement logic including random reallocation (teleport) would be more entertaining than static or simply moving ones. The diversity is captured by equation 8.

$$\text{Div} = \frac{\sum_{i=1}^{n}\left(\sum_{k=0}^{m}(\partial_k)\right)}{n} \tag{8}$$

Where,
m, is the total number of pieces (all three types) specified in a chromosome.
∂_k, Number of cell changes made by piece k during a game.
n, is 20 as explained previously.

6.2.4 Usability

Usability is the fourth and last factor we have considered for our metrics of entertainment. It is interesting to have the play area maximally utilized during the game. If most of the moving pieces remain in a certain region of the play area then the resulting game may seem strange. The usability is captured by equation 9:

$$U = \frac{\sum_{i=1}^{n}\left(\frac{\sum_{k=0}^{m}(C_k)}{|C_u|}\right)}{n} \tag{9}$$

Where,
C_k, is usability counter value for a cell k.
$|C_u|$, is the total number of usable cells.
n, is 20 as explained previously.

A usability counter is set up for each cell which increments when a piece arrives in the cell. The usability U is averaged by playing ten different games for a chromosome. The total cells for our current experimentation are 14x14 minus the 7+7=14 cells used by two walls.

6.2.5 Combined Fitness Function

To assign a chromosome a single fitness value we use rank based fitness using the above mentioned four metrics, as explained in section 6.1.5. The four rank based fitness values obtained for each chromosome are multiplied by corresponding weights and then added to get its final fitness, as in equation 10.

$$FF = \alpha D + \beta C + \chi Div + \delta U \tag{10}$$

Where, α, β, χ, and δ are constants. In our experiments we keep the value of these constants fixed at 1. The multiplication with a corresponding weight allows us to control the relative influence of an aspect.

7 Software Agents

Evolutionary algorithm evolves a population of games and the fitness of each game has to be determined in each generation we may have a total of several thousands of such fitness evaluations. Since a fitness evaluation means playing the game several times, it is not possible to do so manually. We need software game playing agents. The more intelligent the agent the better will be the accuracy of fitness evaluation. We have developed two types of such agents for board based games (as required by the fitness function).

- Random agent.
- Agent using Min-Max with rule based evaluation function.

7.1 Random Agent

As the name suggests the random game playing agent plays the game by randomly selecting a legal move at each step. The agent follows the following algorithm listed in figure 6:

Input: Game Board current state
1. *Generate all legal moves*
2. *Store the moves in a queue*
3. *Shuffle the queue*
4. *If Not mandatory to kill*
5. *Randomly select a move from the queue.*
6. *Else*
7. *Select a move that captures an opponent's piece, if such move exists*
8. *Otherwise, randomly select a move from the queue.*

Output: Next move to take

Fig. 6 Algorithm for the random playing agent

The agent initially generates all the legal moves and stores them in a queue. The queue is shuffled once all the moves are saved in it. Shuffling is important, as we take an average of 20 games to calculate the individual metrics values, if the queue is not shuffled then each time the game is played it will use the same sequence of moves to play the game and fitness values will remain the same in each iteration of the game play. If the mandatory to capture bit is "on" in a chromosome which is being evaluated then the agent first tries to find a move that will capture an opponent's piece. If no such move is found it randomly selects a move from the queue.

7.2 Agent Using Min-Max with Rule Based Evaluation Function

This type of agent is intelligent as compared to the random one. It generates all the possible one ply depth game boards using a min-max algorithm. Each of the resulting game board is evaluated using a rule based evaluation function and the one with the highest evaluation is selected as a next move.

Evaluation function for this type of agent assigns priorities (weights) to piece-type according to whether its disappearance would cause the game to end, flexibility of movement (more directions and multiple step sizes are better), and capturing logic (capturing by moving into opponent's cell is better). Once the priority of a piece is calculated we multiply each piece with its corresponding weight and calculate weighted summation for self and opponent. The board evaluation is the self weighted summation minus opponents weighted summation. Figure 7 lists the algorithm for the evaluation function we use.

```
Input: Game Board current state
For each piece
    priority=0
For each piece
    if is piece of honor
        priority = priority +1 000
    if movement logic all directions
        priority = priority + 8
    if movement logic diagonal Forward and Backward
        priority = priority + 7
    if movement logic Straight Forward and Backward
        priority = priority + 7
    if movement logic diagonal Forward
        priority = priority + 6
    if movement logic Straight Forward
        priority = priority + 6
    if movement logic L shaped
        priority = priority + 5
    if capturing logic step into
        priority = priority + 4
    if capturing logic step over
        priority = priority + 3
Count the number of pieces of Player A
Multiply the number of pieces of a type with its relevant priority
Count the number of pieces of Player B
Multiply the number of pieces of a type with its relevant priority
Calculate boardValue = WeightSumofA-WeightSumofB
Check if the Piece of Honour is dead add -1000 to boardValue
Check if the Piece of Honour is NOT dead add +1000 to boardValue
Output: boardValue
```

Fig. 7 Algorithm for evaluation of board positions

Since we are using mini-max of single ply hence we had to incorporate a mechanism in the evaluation function to overcome the randomness effect near the end of the game when pieces are few and may be far apart. In such cases the evaluation function listed in figure 7 gives same evaluation for all the board positions thus increasing the duration of game. To avoid such situation we restrict the agent to select the move which decreases distance between its own piece and one of an opponent's pieces, provided all next game board position have equal evaluations.

7.3 *Agents for Predator/Prey Games*

For predator/prey games the controller is implemented using an (Artificial Neural Network) ANN and a rule based controller, used to evolve two separate populations to study the effect of controller on evolution. For ANN based controller we have used a multi-layer fully feed forward connected neural network, architecture is shown in Figure 8. There are a total of 6 neurons in the input layer, 5 neurons in the hidden layer and 4 output layer neurons. Sigmoid activation function is used at each neuron. The weights on the edges range between -5 to +5. The input vector to the neural network is (Δxr, Δyr, Δxg, Δyg, Δxb, Δyb) where Δxr, Δyr, Δxg, Δyg, Δxb and Δyb represent agent's distance from nearest red type predator in x-coordinate, agent's distance from nearest red type predator in y-coordinate, agent's distance from nearest green type predator in x-coordinate, agent's distance from nearest green type predator in y-coordinate, agent's distance from nearest blue type predator in x-coordinate and agent's distance from nearest blue type predator in y-coordinate. The ANN outputs a 4 dimensional vector (Nu,Nd,Nl,Nr) where Nu,Nd,Nl and Nr represents agent's next position for up, agent's next position for down, agent's next position for left and agent's next position for right. Agent moves in the direction having the maximum value.

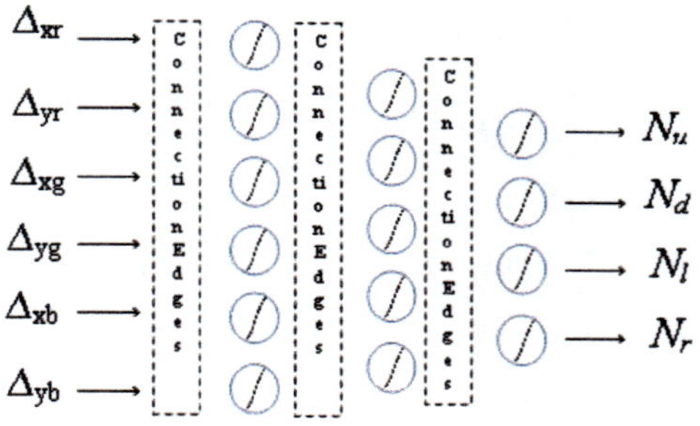

Fig. 8 ANN architecture used to control agent

For the purpose of training of ANN weights we have employed GA where each chromosome of the population represents the set of weights for the entire ANN. In this case the chromosome length is of 97 genes. Mutation is used only as a genetic operator. For each game a random population of GA representing the weights of the edges is created. The game is played 10 times using these weights and a score is assigned which is the average score achieved by the controller. The GA is run for 10 iterations and the best chromosome of each is saved. As the GA finishes we select the best chromosome out of 10, based upon the highest average score achieved.

The rule based controller is implemented as a human supplied rule set. The same controller is used for playing all games (chromosomes) during the entire evolutionary process. Our rule based agent controller is composed of rules formulated to implement the following policy. According to the game rules, at each simulation step the agent must take exactly one step. The agent looks up, down, left and right. It notes the nearest piece (if any) in each of the four directions, and then it simply moves one step towards the nearest score increasing piece. If there are no score increasing piece present it determines its step according to the following priority list:

- Move in the direction which is completely empty (there is only the wall at the end). If more than one directions are empty move towards the farthest wall (in the hope that subsequent position changes would show it a score increasing piece)
- Move in the direction which contains a score neutral piece. The farthest, the better.

- Move in the direction which contains a score decreasing piece. The farthest, the better.
- Move in the direction which contains a death causing piece. The farthest, the better.

Going into walls is not allowed, and if there is a wall present in the adjoining cell, the possibility of going in that direction is automatically curtailed. The above mentioned controller rules encourage the agent to maximize its score by trying to collide with the piece which increase its score and at the same time try to avoid collision with the rest.

8 Experiments

Different experiments carried out are explained in the sub-sections below for each of the game type.

8.1 Board Based Games

The methodology of the experimentations is such as we evolve new board based games using 1+1 Evolutionary Strategy (ES). Once we get a set of evolved games we first select minimum number of games, using equation (12), for analysis. In order to analyze and study the entertainment value contained in the evolved games we follow a twofold strategy i.e. by conducting a controller learnability experiment and a human user survey.

8.1.1 Evolution of New Games

In order to generate new and entertaining board based games we use 1+1 Evolutionary Strategy (ES). Initially a population of 10 chromosomes is randomly initialized with permissible values. The evolutionary algorithm is run for 100 iterations. Mutation is the only genetic operator used with a mutation probability of 30 percent. In each iteration of the ES one parent produce one child and a fitness difference is calculated between them. If it is greater than 4 (i.e. the child is at least half times better than its parent) child is promoted to the next population. We use the formula given in equation (11) to calculate the fitness difference.

$$Fitness\ Difference = \sum_{for\ all\ metrics} \left(1 - \frac{fitness_p - fitness_c}{fitness_p}\right) \tag{11}$$

Where,

$fitness_p$ is the fitness value of parent for current metrics
$fitness_c$ is the fitness value of child for current metrics

We keep an archive of 8 slots and in each iteration update it with the best 2 chromosomes based on each of the fitness metrics. Figure 9 shows the metrics values of one family of chromosome in shape of a graph (figure 9), over a period of 100 iterations.

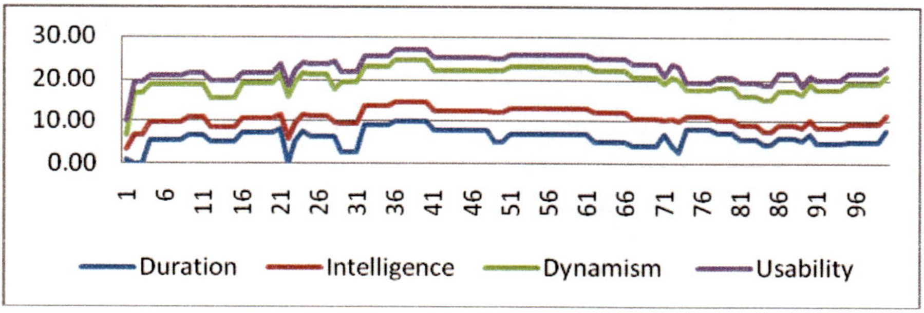

Fig. 9 Metrics values of a typical family of chromosome.

As we use 1+1 ES the best chromosome found in an iteration may get lost in iterations to come. For this purpose as mentioned previously we keep an archive of 8 for best 2 chromosomes against each of the metrics. As the evolutionary process progresses the number of changes, child beating its parent using fitness difference formula (equation (11)) decreases.

8.1.2 Games for Analysis

The evolutionary process gives 8 games evolved against the entertainment based on duration, intelligence, dynamism and usability. For further analysis of these games we select the set of most diverse games. Diversity (from each other) of these games is calculated using their fitness values and is listed in table 1 for one experiment.

Table 1 Fitness values of chromosomes in archive

		Game No.	Duration	Dynamism	Intelligence	Usability
Archive	Duration 1	1	0.885	0.081	1.000	21.051
	Duration 2	2	0.850	0.068	0.700	16.775
	Dynamism 1	3	0.021	0.181	1.000	22.065
	Dynamism 2	4	0.221	0.172	1.000	25.866
	Intelligence 1	5	0	0.086	1.000	23.085
	Intelligence 2	6	0	0.068	1.000	21.028
	Usability 1	7	0.400	0.066	0.850	84.927
	Usability 2	8	0.216	0.039	0.700	80.997

We calculate the diversity based on each of the four metrics for all pair of games using equation (12).

$$Game\ diversity = \left| \frac{Game\ Column\ Fitness - Game\ Row\ Fitness}{Selected\ Metrics\ Maximum\ Value} \right| \qquad (12)$$

Selecting a threshold value of 0.6 table 2 shows the diversity count of evolved games. Diversity count indicates that a game is different from how many other games based on all the four metrics of entertainment.

Table 2 Diversity count of evolved games

Game No.	Different from number of games (for threshold ≥ 0.6)
1	5
2	5
3	3
4	1
5	0
6	1
7	6
8	3

Based upon the above statistics game number 1, 2 and 7 seems to be the most diverse. We select these three for further analysis. From this point onwards we will refer to these as game 1, 2 and 3 respectively. Rules of these games are listed in figure 22-24 (appendix I). For the sake of simplicity we do not name the pieces rather we identify them by their type number which range from 1-6.

We also create a randomly initialized game that has not been passed through the evolutionary process for optimization against entertainment. This game is used to analyze the learnability of the game experiment covered in next section and also in the user survey, to compare with the evolved games. Rules of the random game are shown in figure 25 (appendix I).

8.1.3 Learnability of Evolved Games

The entertainment value of the evolved games needs to be verified against some criteria other than the proposed entertainment metrics. For this purpose we use the Schmidhuber's theory of artificial curiosity [2]. We need to see how quickly a player learns an evolved game. Games learned very quickly will be trivial for the player and thus not contributing anything towards entertainment. Those taking large amount of time to learn will be too difficult. Games between these two boundaries will fall in the range of entertaining games. To observe the learnability of the evolved games there are two options first to ask a human to play a game multiple times and see how fast she/he learns and second is to do the same task using a software based controller.

We have used an ANN based controller. The architecture of the controller is the same architecture used by Chellapilla [21] for evolving an expert checkers player. There are total 5 layers in the ANN, input with 64 neurons, first hidden layer with

91 neurons; second hidden layer with 40 neurons third with 10 neurons and the output layer with 1 neuron. A hyperbolic tangent function is used in each neuron. The connection weights range from [-2, 2]. The training of the ANN is done using co-evolution. A set of genetic algorithm (GA) population is initialized that represent the weight of the ANN. Each individual of the population is played against randomly selected 5 others. Mutation is the only genetic operator used; we have kept the ANN and its training as close to Chellapilla [21] work as possible, except for the number of iteration for which the ANN is trained. We train the set of ANN until we get a set of weights that beats all others. Such individual will have its fitness equal to 1. The number of iterations that take to get such individual in the archive is called the learning duration or learnability of the game. Figure 10 shows the learnability of all the 4 games including the random game (referred to as game 4).

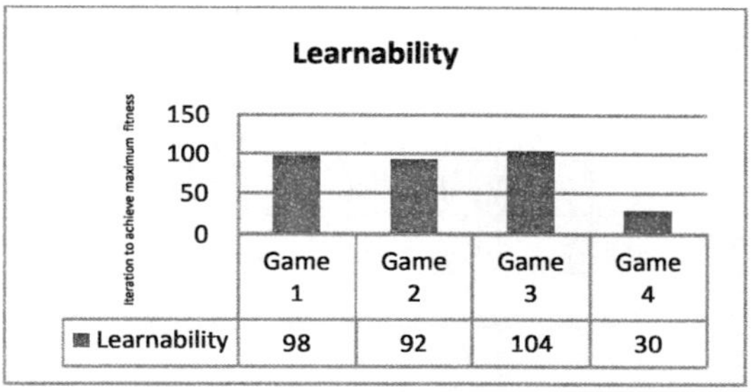

Fig. 10 Learnability of evolved and random game

It takes about 80 to 110 iterations to get a chromosome representing ANN weights that achieve a fitness of 1 during co-evolution, for the evolved games. In case of the randomly initialized game (game 4) it takes only 30 iterations, thus showing game 4 is trivial and uninteresting. Thus we can conclude that game 1, 2 and 3 proof to be entertaining as an ANN based controller neither takes too short a time nor too long to learn these games.

8.1.4 User Survey on Entertainment Value of Evolved Games

To validate the results produced against human entertainment value we have to perform a human user survey. For this purpose we select a set of 10 subjects. Subjects are chosen such that they have at least some level of interest towards computer games. Each individual was given 4 games while s/he was supposed to play each game 3 times, the rules of the game were already explained to the subjects and also displayed on the software they used. This makes a total of 12 games to be played by each subject.

The four games given to the user are marked as Game 1, Game 2, Game 3 and Game 4. Game 4 is the randomly initialized one (but will legal values) whereas remaining three games are evolved for entertainment. Each subject was asked to rank the game they play as 1- liked, 2-disliked and 3-neutral. The results of the human user survey are shown in table 3. For visual purposes we use for √ liked, × for disliked and ~ for neutral.

Table 5 Human user survey results

	Game 1			Game 2			Game 3			Game 4		
Subject	Run 1	Run 2	Run 3	Run 1	Run 2	Run 3	Run 1	Run 2	Run 3	Run 1	Run 2	Run 3
1	~	√	√	~	√	√	~	√	√	~	√	×
2	~	~	√	~	√	√	~	√	√	~	×	×
3	×	√	√	~	×	×	√	×	×	×	×	×
4	~	√	√	√	√	√	√	√	√	√	√	√
5	~	√	√	√	√	√	~	√	√	~	√	×
6	~	√	√	~	√	√	√	√	√	~	√	×
7	~	~	√	~	√	√	√	√	√	~	×	×
8	×	√	√	√	√	√	~	√	√	×	×	×
9	√	√	√	×	√	√	~	√	√	×	×	×
10	√	√	√	×	√	√	~	√	√	~	~	~

For the purpose of collecting statistics from the human user survey we mark a game liked by the subject if during any run he has liked the game and for rest he has either liked or been neutral. If in any run he has disliked the game we mark the game to be disliked. Figure 11 shows the statistics based upon this scheme. About 70- 90 percent subjects have liked the evolved games and found these entertaining, whereas only 10 percent say that the random game (game 4) was entertaining.

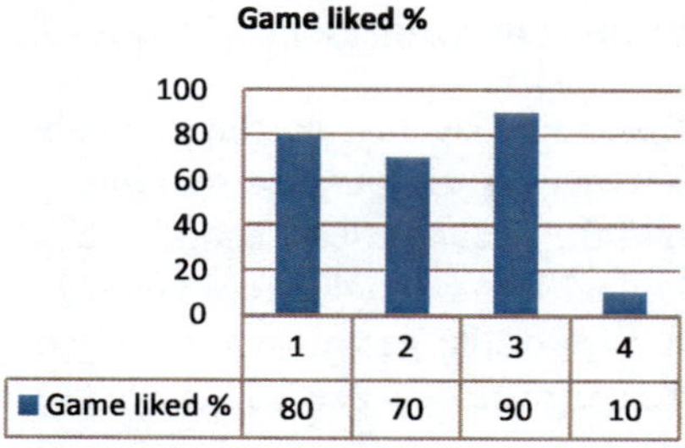

Fig. 11 Statistics of the human user survey

8.2 Predator Prey Games

A population of 10 chromosomes is randomly initialized by the GA. In each generation one offspring is created for each chromosome by duplicating it and then mutating any one of its gene, where all genes have equal probability to be selected. The mutation is done by replacing the existing value with some other permissible value, where each permissible value has equal probability to be selected. This result in a parent and offspring pool of total 20 chromosomes from which 10 best, based on their fitness rank, are selected for the next generation. This evolutionary process is continued for 100 generations.

The fitness of a chromosome, in our case, is based on data obtained by playing the game according to the rules encoded in the chromosome. As there are a number of probabilistic components in the game, hence the data obtained from playing the same game multiple times is never the same. To minimize the effect of this noise in the fitness, we take the average of ten games for every fitness evaluation. To analyze the effect of each of the four proposed factors of entertainment individually we evolve four different populations. Each of these populations is guided by one of the proposed fitness function. We also evolve a population using the combined fitness function.

8.2.1 Duration of the Game

Considering the duration of the game play as determined by the average life span of the agent we rapidly evolve a population of chromosome in which there are very less possibilities for the agent to die in the allotted 100 steps. Figure 12 shows one such evolved chromosome. It is worth noticing in figure 12 (a) that none of the collision logic column 19, 20 and 21 contain 1 which represents death of the agent. Rules of the games in figure 12 (b) shows death of the agent only in case it collides with red type predator but as shown in column 1 there are 0 instances of red predator in the game. Thus these games will be played for a longer duration of time.

	Number of Predators			Movement Logic			Collision Logic															Score Logic								
	R	G	B	R	G	B	R-R	R-G	R-B	R-A	G-R	G-G	G-B	G-A	B-R	B-G	B-B	B-A	A-R	A-G	A-B	R-R	G-G	B-B	A-R	A-G	A-B	G-R	B-R	B-G
(a)	3	17	13	3	0	3	2	1	1	0	0	1	0	2	1	2	2	2	0	0	2	-1	-1	0	-1	0	0	-1	-1	0
(b)	0	18	10	0	1	3	2	2	1	1	2	0	0	1	1	1	1	1	1	0	2	0	1	0	1	-1	-1	1	-1	-1

Fig. 12 The best chromosome evolved by optimizing duration of game. a) Evolved using rule based controller b) Evolved using ANN based controller

8.2.2 Appropriate Level of Challenge

Considering this metric alone, we were able to evolve chromosomes which encouraged a score of 10 to 20. An example is shown in Figure 13. The game rules represented by figure 13 (a) shows that agent looses point only when it collides with blue predators, although agents score also decreases when green/red predator collides or if there is collision between blue type predators. The score remains neutral when collision occurs between blue/green and agent/red. In all other cases score is increased. Agent does not die out in any of the collision case. The rules represented by figure 13 (b) shows a decrease of agents score in case of red/red collision only. Here agent does not die out in any of the collision case.

	Number of Predators			Movement Logic			Collision Logic															Score Logic								
	R	G	B	R	G	B	R-R	R-G	R-B	R-A	G-R	G-G	G-B	G-A	B-R	B-G	B-B	B-A	A-R	A-G	A-B	R-R	G-G	B-B	A-R	A-G	A-B	G-R	B-R	B-G
(a)	10	0	11	0	0	3	1	2	1	1	0	0	2	0	2	2	2	2	2	0	2	1	1	-1	0	1	-1	-1	1	0
(b)	7	7	8	2	4	3	2	0	2	2	0	0	2	1	2	0	0	0	2	0	2	-1	0	0	0	1	1	1	1	0

Fig. 13 The best chromosome evolved by optimizing appropriate level of challenge. a) Evolved using rule based controller b) Evolved using ANN based controller

8.2.3 Diversity

When the diversity is considered in isolation, the solutions evolved tend to have somewhat higher number of predators of each type. Another tendency is to have one of the dynamic movement logics. Yet another observation was that the death of agent was avoided in most of the better chromosomes so that the game may continue for maximum number of steps. A sample chromosome is shown in Figure 14.

	Number of Predators			Movement Logic			Collision Logic															Score Logic								
	R	G	B	R	G	B	R-R	R-G	R-B	R-A	G-R	G-G	G-B	G-A	B-R	B-G	B-B	B-A	A-R	A-G	A-B	R-R	G-G	B-B	A-R	A-G	A-B	G-R	B-R	B-G
(a)	0	10	0	2	4	2	0	1	2	1	2	0	2	0	1	1	0	2	1	0	2	-1	0	-1	1	0	-1	0	-1	-1
(b)	0	17	0	4	4	4	1	2	0	0	0	1	2	2	1	1	1	2	1	0	2	-1	-1	0	0	-1	-1	-1	1	0

Fig. 14 The best chromosome evolved by optimizing diversity. a) Evolved using rule based controller b) Evolved using ANN based controller

8.2.4 Usability

The chromosomes evolved by using usability of the play area metric seem similar to that for diversity metric. One such chromosome is shown in figure 15. The game rules shown in figure 15 show that usability as an entertainment metrics alone encourages rules with maximum number of predators of each type none having movement logic 0 which means predator does not move.

	Number of Predators			Movement Logic			Collision Logic															Score Logic								
	R	G	B	R	G	B	R-R	R-G	R-B	R-A	G-R	G-G	G-B	G-A	B-R	B-G	B-B	B-A	A-R	A-G	A-B	R-R	G-G	B-B	A-R	A-G	A-B	G-R	B-R	B-G
(a)	20	18	19	1	2	2	2	0	0	2	2	1	2	0	2	2	2	0	2	0	2	1	-1	0	1	0	0	-1	0	0
(b)	19	20	20	4	4	2	2	0	2	2	2	2	0	2	2	2	0	2	2	0	2	-1	-1	1	-1	1	1	0	-1	-1

Fig. 15 The best chromosome evolved by optimizing usability. a) Evolved using rule based controller b) Evolved using ANN based controller

8.2.5 The Combined Fitness Function

The above mentioned combined fitness function was used to evolve a population of chromosome. Most of the resulting evolved chromosomes are playable and seem interesting. They were much better than random games. Also they seem better than some games evolved using individual components of the fitness function. However, an extensive user survey is needed to verify and quantify all these observations, which is covered in next sub section of this chapter. A chromosome showing the best evolved chromosome is shown in figure 16.

	Number of Predators			Movement Logic			Collision Logic															Score Logic								
	R	G	B	R	G	B	R-R	R-G	R-B	R-A	G-R	G-G	G-B	G-A	B-R	B-G	B-B	B-A	A-R	A-G	A-B	R-R	G-G	B-B	A-R	A-G	A-B	G-R	B-R	B-G
(a)	15	8	3	1	1	0	1	2	0	0	1	0	1	2	0	2	2	0	0	0	0	1	1	-1	-1	-1	1	-1	-1	-1
(b)	11	20	3	2	4	2	2	2	0	2	2	2	0	1	1	1	1	2	0	0	2	0	1	1	-1	1	0	-1	0	1

Fig. 16 The best chromosome evolved by optimizing the combined fitness function. a) Evolved using rule based controller b) Evolved using ANN based controller

8.2.6 Human User Survey

To validate the results produced against human entertainment value we performed a human user survey on 10 subjects, chosen such that they have at least some aptitude towards playing computer games. The experiment was conducted in two different sets on different days using the same subjects, one for the games evolved using rule based controller and second for the games evolved by ANN based controller. Each individual was given 6 games while s/he was supposed to play 2 times, the rules of the game were already explained to the subjects and also displayed on the software they used. This makes a total of 12 games to be played by each subject in each set of experiment. The six games given to the user were marked as A, B, C, D, E and F. Where A was a randomly initialized game, B was the game evolved by using duration of game as the fitness function in isolation, C was the game evolved by using appropriate level of challenge as entertainment metrics, D was evolved against diversity, E for usability and F was the game which was evolved based on all the four entertainment matrices combined. The subjects were unaware of criteria of evolution for each game. Randomly selected game rules are shown in figure 17. Each subject was asked to rank the game they play as 1- liked, 2-disliked and 3-neutral.

The randomly generated game, game code A, rules consists of 0 red predators, 12 green predators collision with them cause 0 effect on agents score but causes it

Number of Predators			Movement Logic			Collision Logic															Score Logic								
R	G	B	R	G	B	R-R	R-G	R-B	R-A	G-R	G-G	G-B	G-A	B-R	B-G	B-B	B-A	A-R	A-G	A-B	R-R	G-G	B-B	A-R	A-G	A-B	G-R	B-R	B-G
0	12	0	3	0	2	2	1	0	0	0	1	1	2	2	1	1	1	1	0	1	0	1	1	1	0	1	0	0	-1

Fig. 17 Randomly select game rules for user survey

to die, 20 blue predators and they decreases the agent's score by 1 and also kills the agent when collision occurs. The game rules has the tendency to kill the agent very early and always with 0 or negative scores. The survey suggests that every subject disliked this game.

A pie chart representation of the survey is given in figure 18 and 19. As all the games are played by each subject twice we have marked a game to be liked by a subject if for once s/he has liked it and second time been neutral. If in any of the play subject disliked the game we mark it as disliked. If a subject has been neutral in both iterations of game play it is not considered. In all other cases the game is marked as liked by the subject.

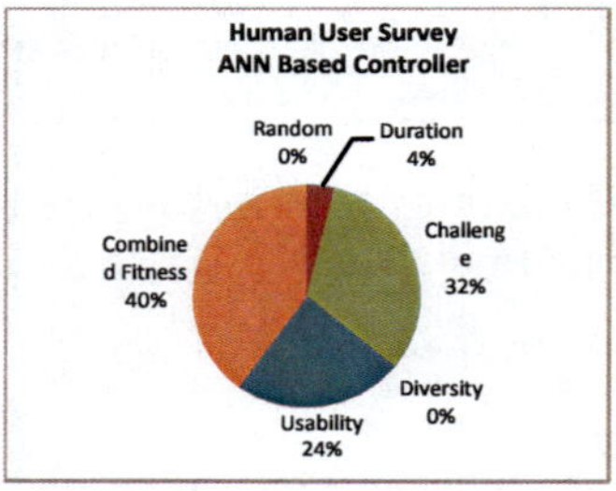

Fig. 18 Pie chart representation of the human user survey using rule based controller.

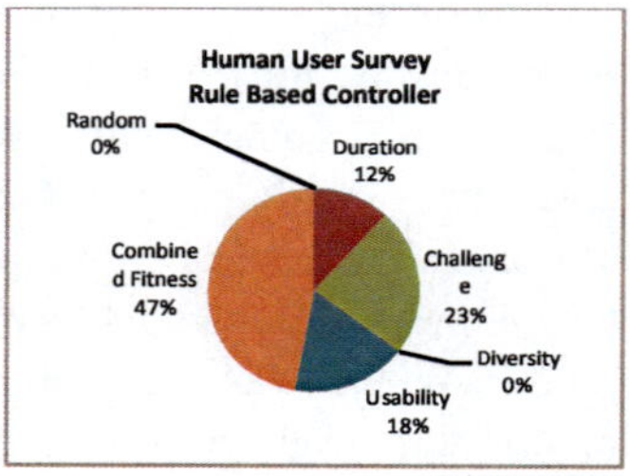

Fig. 19 Pie chart representation of the human user survey using ANN based controller.

The purpose of evolving two sets of population was to see the effect of controller type on the evolved games. Theoretically the rule based controller will evolve the games which are entertaining only while the game is played using the rules encoded in the rule based controller. Thus the idea of using a relatively intelligent and generic controller using an ANN was implemented. Although the games evolved using both the controllers seems somewhat same but in comparison the games evolved using the ANN based controller is more liked by the subjects in the user survey. Figure 20 shows a comparison of the two controllers rating, in general except for the duration metrics games evolved by the ANN based controller are rated higher.

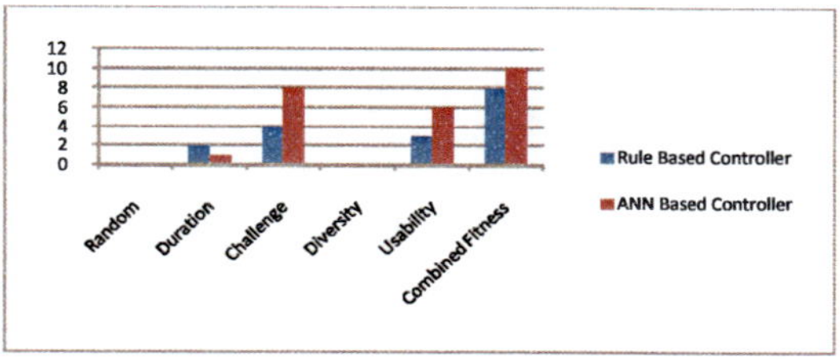

Fig. 20 comparison of the rule based and ANN based controller

8.2.7 Controller Learning Ability

The controller we use is the same ANN based controller we used to evolve the games as shown in figure 8. First of all the weights of the neural network are trained using GA against each of the evolved games and a randomly initialized game, separately.

The learning ability of the controller is calculated as follows:

- The controller plays the game for N times, where N is 10.
- Calculate average score of N games.
- Repeat step 1 and 2 until the standard deviation of the last M runs is minimized. Where M is 3 and the minimized standard deviation is set to 5. We also round the average score to an integer value.

If for 1000 runs condition in step 3 is not satisfied learning ability of the controller is set to 1000. Figure 21 lists the results of the controller learning experiment.

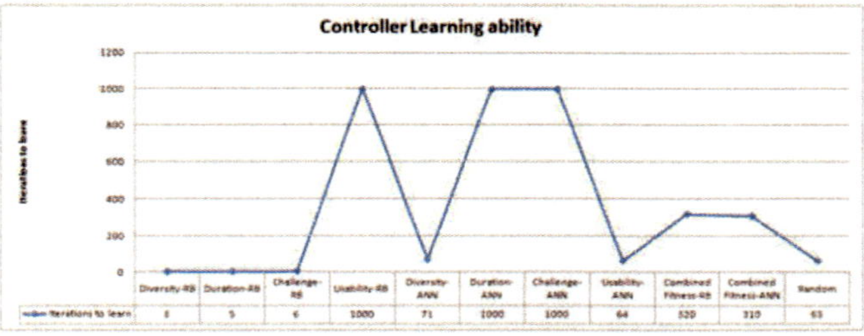

Fig. 21 Learning ability of a controller

The results of the experiments shows that the games evolved using the individual entertainment metrics in isolation were either too easy for the controller to learn, as in case of diversity, duration, challenge evolved using rule based controller and usability and diversity evolved using ANN based controller, or too hard as in case of usability for rule based controller and duration and challenge for ANN based controller. Same goes for the randomly initialized game rules which were learned by the controller in only 63 iterations. In contrast the game rules evolved using combined fitness function (both based on rule based and ANN based controller) were neither too hard nor too easy and lied between the two extremes.

9 Conclusion

The work presented in this chapter is focused on two primary questions: how we can measure the entertainment value of a computer game and how we can automatically generate entertaining games using computational intelligence techniques. For this purpose we used two different genres of games as our test bed which are, board based games and predator/prey games. Based on different theories of entertainment in computer games we identify different factors that contribute towards entertainment in each genre of game. In the process we have presented some metrics for entertainment which are based on duration of game, level of challenge, diversity, intelligence, and usability. These metrics are combined in a fitness function to guide the search for evolving the rules of the game.

Further work needs to be done to make the proposed entertainment metrics more generic so they can automatically cover more types of computer games. The automatic game creation approach can be applied to other genre of games like platform games, real time strategy games and many others. Some other direction could be to use co-evolution where one population tries to evolve the rules and other tries to evolve the strategy to play on those rules. In the context of entertainment metrics in our case the four fitness criterion are linearly combined, an alternate would be use of multi objective genetic algorithm for this purpose. There may be a study conducted on effect of the type of controller being used for evolution process. It will be interesting to see if different types of controllers producing entertaining yet totally different types of games. The idea of measuring entertainment and automatic generation needs to be extended to other real world applications. One such application could be developing strategies for wars.

References

[1] Olson, C.K., Kutner, L.A., Warner, D.E., Almerigi, J.B., Baer, L., Nicholi, A.M., Beresin, E.V.: Factors correlated with violent video game use by adolescent boys and girls. Journal of Adolescent Health, 77–83 (2007)

[2] Schmidhuber, J.: Developmental robotics, optimal artificial curiosity, creativity, music, and the fine arts. Connection Science 18, 173–187 (2006)

[3] Iida, H., Takeshita, N., Yoshimura, J.: A Metric for Entertainment of Board Games: Its Application for Evolution of Chess Variants. In: Nakatsu, R., Hoshino, J. (eds.) Entertainment Computing: Technologies and Applications Proceedings of IWEC 2002 , pp. 65–72. Kluwer Academic Publishers, Boston (2003)

[4] Cincotti, A., Iida, H.: Outcome Uncertainty And Interestedness In Game-Playing: A Case Study Using Synchronized Hex. In: New Mathematics and Natural Computation (NMNC), vol. 02(02), pp. 173–181 (2006)

[5] Retalis, S.: Creating Adaptive e-Learning Board Games for School Settings Using the ELG Environment. Journal of Universal Computer Science, 2897–2908 (2008)

[6] Togelius, J., Nardi, R.D., Lucas, S.M.: Towards automatic personalised content creation for racing games. In: Proceedings of the IEEE Symposium on Computational Intelligence and Games, Piscataway, NJ, April 1-5 (2007)
[7] Yannakakis, G.N., Hallam, J.: Towards Optimizing Entertainment In Computer Games. Applied Artificial Intelligence 21(10), 933–971 (2007)
[8] Gallagher, M., Ryan, A.: Learning to Play Pac-Man: An Evolutionary Rule-based Approach. In: Proceedings of IEEE Congress on Evolutionary Computation, Canberra, Australia, December 8-12 (2003)
[9] Lankveld, G., Spronck, P., Rauterberg, M.: Difficulty Scaling through Incongruity. In: Proceedings of the Fourth Artificial Intelligence and Interactive Digital Entertainment Conference, Stanford, California, October 22-24 (2008)
[10] Togelius, J., Schmidhuber, J.: An Experiment in Automatic Game Design. In: Proceedings of IEEE Computational Intelligence and Games, Perth, Australia, December 15-18 (2008)
[11] Pedersen, C., Togelius, J., Yannakakis, G.N.: Optimization of platform game levels for player experience. In: Proceedings of Artificial Intelligence and Interactive Digital Entertainment, Stanford, California, October 14-16 (2009)
[12] Compton, K., Mateas, M.: Procedural Level Design for Platform Games. In: Proceedings of 2nd Artificial Intelligence and Interactive Digital Entertainment Conference, Stanford, California, June 20-23 (2006)
[13] Yannakakis, G.N., Hallam, J.: Evolving Opponents for Interesting Interactive Computer Games. In: Proceedings of the 8th International Conference on the Simulation of Adaptive Behavior, Los Angeles, USA, July 13-17 (2004)
[14] Csikszentmihalyi, M.: Flow.: The Psychology of Optimal Experience. Harper & Row, New York (1990)
[15] Csikszentmihalyi, M., Csikszentmihalyi, I.: Introduction to Part IV in Optimal Experience: Psychological Studies of Flow in Consciousness. Cambridge University Press, Cambridge (1988)
[16] Malone, T.W.: What makes computer games fun? Byte 6, 258–277 (1981)
[17] Koster, R.: A Theory of Fun for Game Design. Paraglyph Press (2005)
[18] Rauterberg, M.: Amme: An Automatic Mental Model Evaluation to Analyze User Behavior Traced in a Finite, Discrete State Space. In: Proceedings of the Annual Conference of the European Association of Cognitive Ergonomics, EACE 2005, Chania, Crete, September 29-October 1 (2005)
[19] Rauterberg, M.: About a Framework for Information and Information Processing of Learning Systems. In: Falkenberg, E., Hesse, W., Olibve, A. (eds.) Information System Concepts, pp. 54–69. IFIP Chapman & Hall, Boca Raton (1995)
[20] Yannakakis, G.N.: How to Model and Augment Player Satisfaction: A Review. In: Proceedings of the 1st Workshop on Child, Computer and Interaction, ICMI 2008, Chania, Crete (October 2008)
[21] Chellapilla, K., Fogel, D.B.: Evolving an expert checkers playing program without using human expertise. IEEE Trans. Evolutionary Computation 5(4), 422–428 (2001)

Glossary of Terms and Acronyms

ANN: Artificial Neural Network
Chromosomes: One individual of the genetic algorithm population representing one complete game
ES: Evolutionary Strategy
Fitness function: the objective of the evolution process
GA: Genetic Algorithm
Learnability: The duration in number of steps an artificial intelligence based controller takes to learn a game
Piece of honour: A piece in a board based game having the highest priority and whose absence will make relevant player loose the game
Search space: The total scope of the game rules which will be used to evolve new games.
Software agents: the artificial intelligence based controllers to automatically play the game

Appendix I

Piece No	Movement Logic	Step Size	Capturing Logic	Conversion Logic
1	L	Multiple	Step Into	6
2	Diagonal Forward & Backward	Single	Step Over	5
3	All Directions	Multiple	Step Into	Nil
4	Straight Forward	Multiple	Step Into	1
5	Straight Forward	Multiple	Step Over	2
6	All Directions	Multiple	Step Over	3
	Piece of Honour	5		
	Mandatory to Capture	No		

Fig. 22 Rules of game 1 and pieces board positions

Piece No	Movement Logic	Step Size	Capturing Logic	Conversion Logic
1	L	Single	Step Over	1
2	Diagonal Forward	Single	Step Into	3
3	Diagonal Forward	Single	Step Over	Nil
4	All Directions	Multiple	Step Over	Nil
5	Straight Forward	Multiple	Step Into	2
6	All Directions	Single	Step Into	1
	Piece of Honour	5		
	Mandatory to Capture	Yes		

Fig. 23 Rules of game 2 and pieces board positions

Piece No	Movement Logic	Step Size	Capturing Logic	Conversion Logic
1	L	Multiple	Step Into	5
2	L	Single	Step Into	2
3	Straight Forward	Multiple	Step Over	3
4	Straight Forward	Single	Step Over	5
5	Diagonal Forward & Backward	Single	Step Over	3
6	Diagonal Forward	Multiple	Step Over	3
	Piece of Honour	Nil		
	Mandatory to Capture	Yes		

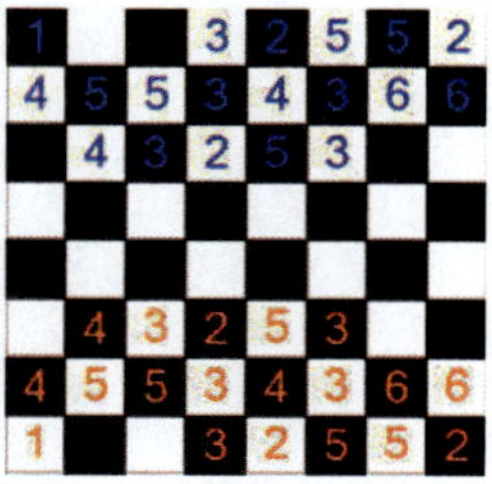

Fig. 24 Rules of game 3 and pieces board positions

Piece No	Movement Logic	Step Size	Capturing Logic	Conversion Logic
1	All Directions	Multiple	Step Into	Nil
2	All Directions	Single	Step Over	1
3	Straight Forward & Backward	Multiple	Step Over	5
4	Diagonal Forward & Backward	Multiple	Step Over	2
5	Straight Forward	Multiple	Step Into	6
6	Straight Forward & Backward	Single	Step Into	6
	Piece of Honour	5		
	Mandatory to Capture	No		

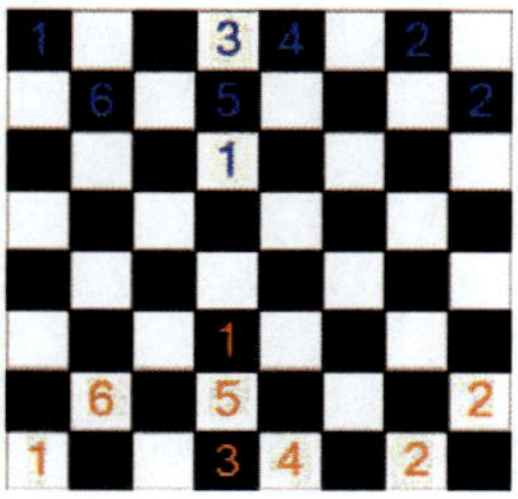

Fig. 25 Rules of random game and pieces board positions

Chapter 16
Leveraging Massive User Contributions for Knowledge Extraction

Spiros Nikolopoulos, Elisavet Chatzilari, Eirini Giannakidou,
Symeon Papadopoulos, Ioannis Kompatsiaris, and Athena Vakali

Abstract. The collective intelligence that emerges from the collaboration, competition, and co-ordination among individuals in social networks has opened up new opportunities for knowledge extraction. Valuable knowledge is stored and often "hidden" in massive user contributions, challenging researchers to find methods for leveraging these contributions and unfold this knowledge. In this chapter we investigate the problem of knowledge extraction from social media. We provide background information for knowledge extraction methods that operate on social media, and present three methods that use Flickr data to extract different types of knowledge namely, the community structure of tag-networks, the emerging trends and events in users tag activity, and the associations between image regions and

Spiros Nikolopoulos
Informatics & Telematics Institute, Thermi, Thessaloniki, Greece and School of Electronic Engineering and Computer Science - Queen Mary University of London
e-mail: `nikolopo@iti.gr`

Elisavet Chatzilari
Informatics & Telematics Institute, Thermi, Thessaloniki, Greece and Centre for Vision, Speech and Signal Processing University of Surrey Guildford, GU2 7XH, UK
e-mail: `ehatzi@iti.gr`

Eirini Giannakidou · Symeon Papadopoulos
Informatics & Telematics Institute, Thermi, Thessaloniki, Greece and Department of Computer Science, Aristotle University of Thessaloniki, Greece
e-mail: `{igiannak,papadop}@iti.gr`

Ioannis Kompatsiaris
Informatics & Telematics Institute, Thermi, Thessaloniki, Greece
e-mail: `ikom@iti.gr`

Athena Vakali
Department of Computer Science, Aristotle University of Thessaloniki, Greece
e-mail: `avakali@csd.auth.gr`

N. Bessis, F. Xhafa (Eds.): Next Generation Data Technologies for CCI, SCI 352, pp. 415–443.
springerlink.com

tags in user tagged images. Our evaluation results show that despite the noise existing in massive user contributions, efficient methods can be developed to mine the semantics emerging from these data and facilitate knowledge extraction.

1 Introduction

Content sharing through the Internet has become a common practice for the vast majority of web users. Due to the rapidly growing new communication technologies, a large number of people all over the planet can now work together in ways that were never before possible in the history of humanity. This user-driven approach is characterized by the fact that its structure and dynamics are similar to those of a complex system, yielding stable and knowledge-rich patterns after a specific usage period [9]. Combining the behavior preferences and ideas of massive users that are imprinted in collaborative data can result into novel insights and knowledge [47], often called Collective Intelligence. Analyzing such data will enable us to acquire a deep understanding of their inner structure, unfold the hidden knowledge and reveal new opportunities for the exploitation of collaborative data.

Collective Intelligence is mainly addressed in Web 2.0 applications, that have experienced an unprecedented information explosion. Social networks, like Facebook, Flickr, and Twitter, enable users to easily create and share information-rich and visually appealing content. Content is also often integrated or reused from other web pages, at the ease of a mouse click. The main vehicles for generating collaborative data in Social Networks are the Social Tagging Systems (STS), which are systems that enable their users to upload digital resources (e.g., bookmarks, pictures, blogs, etc) and annotate them with tags (i.e., freely chosen keywords). An established means of modeling the structure of Collaborative Tagging Systems is the folksonomy model [36] which encodes the associations among the different types of entities (i.e., users, resources and tags) in the form of a network. Based on this model a wide range of techniques have been developed for performing knowledge extraction from social networks.

Extracting the knowledge hidden in Social Networks can help tackle a variety of issues in different disciplines, such as content consumption (e.g., poor recall and precision), knowledge management (e.g., obsolescence, expertise), etc. Several analysis and extraction approaches are being developed towards extracting knowlege from social media. Community detection involves the analysis of a folksonomy with the goal of identifying communities, i.e., groups of objects (which are represented as nodes in the network) that are more densely connected to each other than with the rest of the objects on the network. Similarly, the incorporation of a temporal dimension into the analysis process reveals the macroscopic and microscopic views of tagging, highlights links between objects for specific time periods and, in general, lets us observe how the user tagging activity changes over time. Facilitating the learning process of image analysis models is another use of the knowledge extracted from leveraged user contributed content. All these approaches are motivated by the fact that the intelligence emerging from the collaboration, competition and coordination among individuals is greater than the sum of the individuals' intelligence.

Numerous applications can be envisaged for exploiting the knowledge extracted from massive user contributions. It is common to derive community-based views of networks, i.e. networks of which the nodes correspond to the identified communities of the original networks and the edges to the relations between the communities. Such views are more succinct and informative than the original networks. It is for this reason that community detection has found applications in the field of recommendation systems [37, 51, 15, 45] , as well as for representing user profiles [1, 20]. Other applications that make use of the knowledge extracted from tag communities include sense disambiguation [2] and ontology evolution/population [51]. Despite the great potential of user contributed content as a source for knowledge extraction, there is a series of challenges involved in such an endeavor. First, the unprecendented growth of user content and associated metadata presents extreme scalability and efficiency challenges to knowledge discovery methods, which so far have been applicable in medium-to-large scale. In addition, the unconstrained nature of uploading and sharing such content has resulted in large amounts of spammy and noisy content and metadata, thus considerably compromising the quality of data to be analyzed. A related challenge stems from the fact that there is currently a variety of metadata associated with online content items; for instance, a photo can be described by a title, a set of tags, and GPS coordinates. However, not all photos consistently contain all of these metadata. Therefore, it is hard to devise sufficiently resilient knowledge discovery and content retrieval methods given that metadata is incomplete or of dubious quality.

Our main objective in this chapter is to demonstrate how the massive user contributions can be leveraged to facilitate the extraction of valuable knowledge. In order to extract the knowledge that is stored and often "hidden" in social data, various approaches have been employed. However, despite the active research efforts in this area, the full potential of Web 2.0 data has not been exploited yet, mainly due to the limitations mentioned earlier. In this chapter we contribute towards overcoming the aforementioned limitations and present three methods for extracting knowledge from Flickr data. A technique for detecting communities in folksonomy-derived tag networks, a time-aware user/tag co-clustering approach which groups together similar users and tags that are very "active" during the same time periods, and a technique that relies on user contributed content to guide the learning process of an object recognition detector. In all cases we use massive amounts of social data and exploit the semantics emerging from their collaborative nature to facilitate knowledge-related tasks.

The remaining of the chapter is organized as follows. In Section 2 we review the related literature with a special focus on the fields related with the presented methods. Sections 3, 4 and 5 are devoted in presenting our methods for extracting knowledge from flickr data and evaluating their results. Concluding remarks and avenues for future research are described in Section 6.

2 Related Work

There is a growing number of research efforts that attempt to exploit the dynamics of social tagging systems, exploit the Collective Intelligence that is fostered by this type of content and facilitate different types of applications. Here, we focus on three research directions that concern the methods to be presented in Sections 3, 4, and 5, respectively. That is, in the following we review the related literature in the fields of tag clustering, temporal tag analysis and using social media to facilitate image analysis. Specifically, emphasis is placed on: *i)* studying the tag clustering problem using *community detection* methods, *ii)* applying temporal analysis on social media for *event identification*, and, *iii)* combining tag and visual information from social media to assist image analysis algorithms.

2.1 Tag Clustering and Community Detection

The problem of tag clustering has recently attracted considerable research interest since it is a challenging task from a data mining perspective, but at the same time it also holds the potential for benefiting a variety of Information Retrieval (IR) applications due to the fact that tag clusters typically correspond to semantically related concepts or topics. For instance, tag clustering is considered important for extracting a hierarchical topic structure from a tagging system in order to improve content retrieval and browsing [7]. Similar conclusions are reached by [5] who point that the use of tags in their raw form limits the potential for content exploration and discovery in an information system; thus, there is a need for an additional level of organization through tag clustering. In [20], tag clusters are used as proxies for the interests of users. Using tag clusters instead of plain tags for profiling user interests proved beneficial for personalized content ranking. An additional application of tag clustering is presented in [2]. There, the clusters were used as a means of identifying the different contexts of use for a given tag, i.e., for sense disambiguation. It was shown that using the tag clusters results in improved performance compared to the use of external resources such as WordNet.

The methods used for performing tag clustering mainly adopt one of two approaches: (a) conventional clustering techniques, such as Hierarchical Agglomerative Clustering (HAC) [7, 20] and (b) Community detection methods [5, 49, 2]. HAC suffers from high complexity (quadratic to the number of tags to be clustered) and the need to set ad-hoc parameters (e.g. three parameters need to be set in the clustering scheme used in [20]). Community detection methods largely address the shortcomings of HAC since efficient implementations exist with a complexity of $O(Nlog(N))$ for finding the optimal grouping of N tags into communities. Furthermore, community detection methods rely on the measure of modularity [38] as a means to assess the quality of the derived cluster structure. Thus, modularity maximization methods do not require any user-defined parameters. However, a problem of modularity maximization methods, also pointed in [49] and confirmed by our experiments, is their tendency to produce clusters with a highly skewed size distribution. This makes them unsuitable for the problem of tag clustering in the context of IR applications.

2.2 Temporal Tag Analysis

Temporal analysis has been an active topic of research in many disciplines [12, 29]. In Social Tagging environments, where activities are performed in specific temporal contexts, such analysis can be used for extracting knowledge, such as dominant topics over specific time periods, emerging trends, and events that attract users' interest. More specifically, a number of researchers performed temporal tag analysis to locate coherent topics out of unstructured sets of tags in social media and identify "hot" topics that signify emerging trends. In [52] the authors use a statistical model [54], to discover tags that constitute "topics of interest" at particular timeframes. A trend detection measure is introduced in [26], which captures topic-specific trends at each timeframe and is based on the weight-spreading ranking of the PageRank algorithm [6]. The association of tags signified as topics or trends with specific users may be used for extracting user interests in personalized applications [30, 24].

A subdomain of topic detection research involves *event recognition*, that is the analysis of tags/time usage patterns along with geo-related information available in social media, to infer the event semantics of tags. The authors of [42] search for tags in Flickr that can be mapped to events by examining the tags' distribution over time and space. The intuition behind their method is that a tag describing an event usually occurs at a specific segment of time and is assigned on photos geo-tagged around a specific place (e.g., "olympics2008"). In order to capture events of different time scales, they introduce an approach that does not rely on a-priori defined timeframes, but searches for low-entropy clusters in the time usage distribution of a tag that are robust at many time scales. A set of similarity metrics for tracking social media content that is related to events and enable event-based browsing is presented in [4].

Furthermore, the potential of knowledge extraction from social media has been investigated by analyzing the dynamics of these systems and monitoring the activity over time. More specifically, Halpin et al. were the first that introduced the temporal dimension in tag distributions' analysis and presented results for tag dynamics over a dataset from del.icio.us, considering 35 distinct timeframes, [25]. The authors of [61] studied tag recurrence dynamics, by modeling a social media environment as a time-ordered series of posts. The analysis of dynamics of social media shows resemblance with those of complex systems, i.e., a consensus is built incrementally in a decentralized manner, proving, thus, that there is value in analyzing data and extracting knowledge from social media, since these environments are characterized by some kind of stability over time and use. Such techniques may be applied on tag prediction/suggestion approaches.

Finally, temporal analysis can also be used in many applications to illustrate tagging activity in social media with an explicit temporal dimension. In [16] the authors developed a browser-based application in which the user may navigate through interesting tags of various timeframes in Flickr, at varying timescales. They grasp a tag's interestingness on a particular timeframe by counting its frequency in this timeframe over other timeframes. In order to achieve efficiency, they employ backend algorithms that pre-compute tag interestingness scores for varying sized timeframes. Russell presented a tool that visualizes the collective tagging activity on a resource over time, highlighting periods of stable and changing tagging patterns, [43]. The latter denote a change in users' awareness of the described resource.

2.3 Image Analysis Using Collaborative Data

The works combining user contributed tags with visual features are used to facilitate various tasks, such as image collection browsing and retrieval [3], tag-oriented clustering of photos [22], ranking the results of a video retrieval system [21], or even identifying photos that depict a certain object, location or event [28, 41]. Lately, considerable interest has also been placed on the potential of collaborative data to serve as the training samples for various image analysis tasks. The common objective of these approaches is to compensate for the loss in learning from weakly annotated and noisy training data, by exploiting the massive amount of available samples. Web 2.0 and collaborative tagging environments have further boosted this idea by making available plentiful user tagged data. From the perspective of exploring the trade-offs between analysis efficiency and the characteristics of the dataset, we can mention the works of [27, 13]. In [27] the authors explore the trade-offs in acquiring training data for image classification models through automated web search as opposed to human annotation. The authors set out to determine when and why search-based models manage to perform satisfactory and design a system for predicting the performance trade-off between annotation- and search-based models. In [13] the authors investigate both theoretically and empirically when effective learning is possible from ambiguously labeled images. They formulate the learning problem as partially-supervised multiclass classification and provide intuitive assumptions under which they expect learning to succeed.

Some indicative works that rely on the assumption that due to the common background that most users share, the majority of them tend to contribute similar tags when faced with similar type of visual content include [58, 53, 56]. In [58] the authors are based on social data to introduce the concept of Flickr distance. Flickr distance is a measure of the semantic relation between two concepts using their visual characteristics. The authors rely on the assumption that images about the same concept share similar appearance features and use images obtained from Flickr to represent a concept. The authors present some very interesting results demonstrating that collaborative tagging environments can serve as a valuable source for various computer vision tasks. In [53] the authors make the assumption that semantically related images usually include one or several common regions (objects) with similar visual features. Based on this assumption they build classifiers using as positive examples the regions assigned to a cluster that is decided to be representative of the concept. They use multiple region-clusters per concept and eventually they construct an ensemble of classifiers. Similarly in [56] the authors investigate non-expensive ways to generate annotated training samples for building concept classifiers using supervised learning. The authors utilize clickthrough data logged by retrieval systems that consist of the queries submitted by the users, together with the images in the retrieval results, that these users selected to click on in response to their queries. The method is evaluated using global concept detectors and the conclusion that can be drawn from the experimental study is that although the automatically generated data cannot surpass the performance of the manually produced ones, combining both automatically and manually generated data consistently gives the best results.

The employment of clustering for mining images of objects has been also explored [28, 41]. In [28] the authors make use of user contributed photo collections and demonstrate a location-tag-vision-based approach for retrieving images of geography-related landmarks. They use clustering for detecting representative tags for landmarks, based on their location and time information. Subsequently, they combine this information with a vision-assisted process for presenting the user with a representative set of images. Eventually, the goal is to sample the formulated clusters with the most represenative images for the selected landmark. In [41] the authors are concerned with images that are found in user contributed collections and depict objects (such as touristic sights). The presented approach is based on geo-tagged photos and the task is to mine images containing objects in a fully unsupervised manner. The retrieved photos are clustered according to different modalities (including visual content and text labels) and Frequent Itemset Mining is applied on the tags associated with each cluster in order to assign cluster labels.

3 Tag Clustering through Community Detection in Tag Networks

The free nature of tagging (no constraints on the tags used, no requirement for expert users) has been responsible for the wide uptake of tagging in numerous web applications. At the same time, such lack of constraints with respect to tagging is the source of numerous annotation quality problems, such as spam, misspellings, and ambiguity of semantics. Coupled with the huge volume of tagging data, these problems compromise the performance (in terms of accuracy) of tag-based information retrieval applications. Given the above observation, tag clustering, i.e. the process of organizing tags in groups, such that tags of the same group are topically related to each other, can provide a powerful tool for addressing the annotation quality and large volume problems that are inherent in real-world tagging applications. Since tag clustering results in a form of semantic organization for the tags of the system, it can be seen as a knowledge organization process. Furthermore, since the extracted tag clusters correspond to meaningful concepts and topics, which are often non-obvious, tag clustering can also be seen as a knowledge extraction process.

There are several approaches for tackling tag clustering. Several works have made use of classic clustering schemes, such as k-means [22] and hierarchical agglomerative clustering [7, 20] to discover clusters of tags in folksonomies. According to them, tags are represented as vectors and the employed clustering scheme makes use of some similarity function (e.g. cosine similarity, inverse of Euclidean distance) in order to group tags into clusters. Such approaches suffer from two important limitations: (a) they are hard to scale, since they rely on all pairwise similarities/distances to be computed, (b) they need the number of clusters to be set a priori, which is usually not possible to estimate in real-world tagging systems.

Lately, tag clustering schemes have appeared [5, 49, 2] that are based on community detection in tag networks. Tag networks are very fast to build by use of tag co-occurrence analysis in the context of the tagged resources. Then, community detection methods identify sets of vertices in the networks that are more densely connected to each other than to the rest of the network. The majority of the aforementioned works rely on some modularity maximization scheme [38] in order to perform tag clustering. Modularity maximization methods are reasonably fast (given a network of size m there are methods with a complexity of $O(m \cdot \log m)$) and they do not require the number of clusters to be provided as a parameter. However, such methods suffer from the "gigantic" community problem, i.e. they tend to produce community structures consisting of one or few huge communities and numerous very small ones. In addition, they result in a tag partition, thus assigning every tag to some cluster even in the case that the tag is spam or of low quality.

To this end, we describe MultiSCAN, a new tag clustering scheme that is based on the notion of (μ,ε)-cores [60]. The proposed scheme results in a partial clustering of tags, and distinguishes between tags that belong to a cluster, tags that are associated with many clusters (hubs) and tags that should not be assigned to any cluster (outliers). Furthermore, the proposed scheme addresses an important issue present in the original SCAN scheme [60] that it extends. It does not require setting parameters μ and ε by conducting an efficient parameter space exploration process.

3.1 Description of MultiSCAN

The proposed scheme builds upon the notion of (μ,ε)-cores introduced in [60] and recapped in subsection 3.1.1. Subsequently, it conducts an efficient iterative search over the parameter space (μ,ε) in order to discover cores for different parameter values (subsection 3.1.2). In that way, it alleviates the user from the need of setting parameters μ and ε. An extended variant of this scheme is presented in [39]. The extended version contains an additional cluster expansion step that aims at attaching relevant tags to the extracted tag clusters. Here, we focus solely on the parameter exploration step to study in isolation its effect on the extracted cluster structure.

3.1.1 Core Set Discovery

The definition of (μ,ε)-cores is based on the concepts of *structural similarity*, *ε-neighborhood* and *direct structure reachability*.

Definition 1. *The* ***structural similarity*** σ *between two nodes v and w of a graph* $G = \{V, E\}$ *is defined as:*

$$\sigma(v,w) = \frac{|\Gamma(v) \cap \Gamma(w)|}{\sqrt{|\Gamma(v)| \cdot |\Gamma(w)|}} \tag{1}$$

where $\Gamma(v)$ *is the* structure *of node v, i.e., the set of nodes comprising the node itself and its neighbors:* $\Gamma(v) = \{w \in V | (v,w) \in E\} \cup \{v\}$.

Definition 2. *The* ***ε-neighborhood*** *of a node is the subset of its structure containing only the nodes that are at least ε-similar with the node; in math notation:*

$$N_\varepsilon(v) = \{w \in \Gamma(v) | \sigma(v,w) \geq \varepsilon\} \tag{2}$$

Definition 3. *A vertex v is called a* (μ, ε)***-core*** *if its ε-neighborhood contains at least* μ *vertices:* $CORE_{\mu,\varepsilon}(v) \Leftrightarrow |N_\varepsilon(v)| \geq \mu$.

Definition 4. *A node is directly* ***structure reachable*** *from a* (μ, ε)*-core if it is at least ε-similar to it:* $DirReach_{\mu,\varepsilon}(v,w) \Leftrightarrow CORE_{\mu,\varepsilon}(v) \wedge w \in N_\varepsilon(v)$.

For each (μ, ε)-core identified in the network, a new community is built and adjacent nodes are attached to it provided they are directly reachable to it or reachable through a chain of nodes which are directly structure reachable to each other. The rest of the nodes are considered to be *hubs* or *outliers* depending on whether they are adjacent to more than one communities or not. An example of computing structural similarity values for the edges of a network and then identifying the underlying (μ, ε)-cores, hubs and outliers of the network is illustrated in Figure 1. This

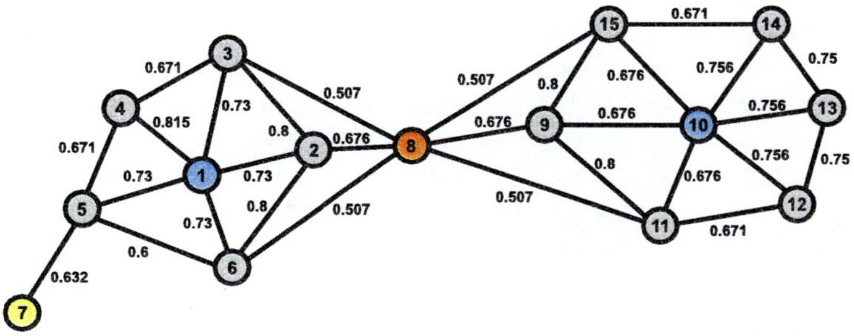

Fig. 1 Example of community structure in an artificial network. Nodes are labeled with successive numbers and edges are labeled with the structural similarity value between the nodes that they connect. Nodes 1 and 10 are (μ, ε)-cores with $\mu = 5$ and $\varepsilon = 0.65$. Nodes 2-6 are structure reachable from node 1 and nodes 9, 11-15 are structure reachable from node 10. Thus, two community seed sets have been identified: the first consisting of nodes 1-6 and the second consisting of nodes 9-15.

technique for identifying communities is computationally efficient since its complexity is $O(\overline{k} \cdot n)$ for a network of n nodes and average degree $\overline{k}$. Computing the structural similarity values of the m network edges introduces an additional $O(\overline{k} \cdot m)$ complexity in the community detection

3.1.2 Parameter Space Exploration

One issue that is not addressed in [60] pertains to the selection of parameters μ and ε. Setting a high value for ε (to a maximum value of 1.0) will render the core detection step very eclectic, i.e. few (μ, ε)-cores will be detected. Higher values for μ will also result in the detection of fewer cores (for instance, all nodes with degree lower than μ will be excluded from the core selection process). For that reason, we employ an iterative scheme, in which the community detection operation is carried out multiple times with different values of μ and ε so that a meaningful subspace of these two parameters is thoroughly explored and the respective (μ, ε)-cores are detected.

The exploration of the (μ, ε) parameter space is carried out as follows. We start by a very high value for both parameters. Since the maximum possible values for μ and ε are k_{max} (maximum degree on the graph) and 1.0 respectively, we start the parameter exploration by two values dependent on them (for instance, we may select $\mu_0 = 0.5 \cdot k_{max}$ and $\varepsilon_0 = 0.9$; the results of the algorithm are not very sensitive to this choice). We identify the respective (μ, ε) cores and associated communities and then relax the parameters in the following way. First, we reduce μ; if it falls below a certain threshold (e.g. $\mu_{min} = 4$), we then reduce ε by a small step (e.g. 0.05) and we reset $\mu = \mu_0$. When both μ and ε reach a small value ($\mu = \mu_{min}$ and $\varepsilon = \varepsilon_{min}$), we terminate the community detection process. This exploration path ensures that communities with very high internal density will be discovered first and subsequently less profound ones will also be detected. In order to speed up the parameter exploration process, we employ a logarithmic sampling strategy when moving along the μ parameter axis. The computational complexity of the proposed parameter scheme is a multiple of the original SCAN (excluding the structural similarity computation which is performed only once). The multiplicative factor is $C = s_\varepsilon \cdot s_\mu$, where s_ε is the number of samples along the ε axis ($\simeq 10$) and s_μ is the number of samples along the μ axis ($\simeq \log k_{max}$). This improves over the original proposal in [40], which requires k_{max} samples along the μ axis.

3.2 *Evaluation of Tag Clustering*

In order to evaluate the behavior of community detection in real-world tagging systems, we conduct a study comparing the performance of our method (MultiSCAN) against two competing community detection methods on two datasets coming from different tagging applications, namely BibSonomy and Flickr. The first of the two community detection methods under study is the well-known greedy modularity

Table 1 Folksonomy datasets used for evaluation

Dataset	#triplets	U	R	T	$\|V\|$	$\|E\|$	$\bar{k}$	$\overline{cc}$
BIBS-200K	234,403	1,185	64,119	12,216	11,949	236,791	39.63	0.6689
FLICKR-1M	927,473	5,463	123,585	27,969	27,521	693,412	50.39	0.8512

maximization scheme presented by Clauset, Newman and Moore (CNM) [11][1] and the second is the SCAN algorithm of [60], which constitutes the basis for MultiScan. The two datasets used for our study are described below.

BIBS-200K: BibSonomy is a social publication bookmarking application. The BibSonomy dataset was made available through the ECML PKDD Discovery Challenge 2009[2]. We used the "Post-Core" version of the dataset, which consists of a little more than 200,000 tag assignments (triplets) and hence the label "200K" was used as part of the dataset name.

FLICKR-1M: Flickr is a popular online photo sharing application. For our experiments, we used a focused subset of Flickr comprising approximately 120,000 images that were located within the city of Barcelona (by use of a geo-query). The number of tag assignments for this dataset approaches one million.

Starting from each dataset, we built a tag graph, considering an edge between any two tags that co-occur in the context of some resource. The raw graph contained a large component and several very small components and isolated nodes. For the experiments we used only the large component of each graph. Some basic statistics of the analyzed large components are presented in the right part of Table 1. The nodes of the three tag graphs appear to have a high clustering coefficient on average, which indicates the existence of community structure in them. We applied the three competing clustering schemes, CNM, SCAN and MultiSCAN, on the tag graphs and proceeded with the analysis of the derived communities. Since SCAN is parameter-dependent, we performed the clustering multiple times for many (μ, ε) combinations and selected the best solution.

We used the derived tag clusters for tag recommendation in order to quantify their effect on the IR performance of a cluster-based tag recommendation system. More specifically, we created a simple recommendation scheme, which, based on an input tag, uses the most frequent tags of its containing cluster to form the recommendation set. In case more than one tags are provided as input, the system produces one tag recommendation list (ranked by tag frequency) for each tag and then aggregates the ranked list by summing the tag frequencies of the tags belonging to more than one list. Although this recommendation implementation is very simple, it is suitable for benchmarking the utility of cluster structure since it is directly based on it.

[1] We used the publicly available implementation of this algorithm, which we downloaded from `http://www.cs.unm.edu/~aaron/research/fastmodularity.htm`

[2] `http://www.kde.cs.uni-kassel.de/ws/dc09`

The evaluation process was conducted as follows: Each tag dataset was divided into two sets, one used for training and the other used for testing. Based on the training set, the corresponding tag graph was built and the tag clusters based on the three competing methods were extracted. Then, by using the tag assignments of the test set, we quantified the extent to which the cluster structure found by use of the training set could help predict the tagging activities of users on the test set. For each test resource tagged with L tags, one tag was used as input to the tag recommendation algorithm and the rest $L-1$ were predicted. In that way, both the number of correctly predicted tags and the one of missed tags is known. In addition, a filtering step was applied on the tag assignments of the test set. Out of the test tag assignments, we removed the tags that (a) did not appear in the training set, since it would be impossible to recommend them and (b) were among the top 5% of the most frequent tags, since in that case recommending trivial tags (i.e., the most frequent within the dataset) would be enough to achieve high performance.

Table 2 presents a comparison between the Information Retrieval (IR) performance of tag recommendation when using the CNM, SCAN and MultiSCAN tag clusters. According to it, using the SCAN and MultiSCAN tag clusters results in significantly better tag recommendations than by use of CNM across both datasets. For instance, in the FLICKR-1M dataset, the MultiSCAN-based recommendation achieves five times more correct recommendations (R_{TP}) than the CNM-based one ($9,909$ compared to $2,074$). A large part of the CNM-based recommendation failure can be attributed to the few gigantic communities that dominate its community structure. Compared to the best run of SCAN, MultiSCAN performs better in terms of number of unique correct suggestions (U_{TP}) and $P@1$, but worse in terms of precision. In terms of F-measure, SCAN performs somewhat better in both datasets. Given the fact that SCAN requires parameter tuning to achieve this performance and that MultiSCAN provides more correct unique suggestions, we may conclude that the MultiSCAN tag cluster structure is more suitable for the task of tag recommendation.

There are several pertinent issues on the topic that have not been addressed here. First, the tag network creation step can be performed in different ways. Here, we used plain cooccurrence of tags in the context of some resource. There are other tag network creation approaches, such as vector-based tag similarities or tag-focused networks [2]. Depending on the employed tag network creation approach, the produced network will present different characteristics (e.g., edge density) that may affect the subsequent community detection process. An additional issue that we did not address pertains to the existence of multiple scales of community structure in a folksonomy. For instance, a division of a tag network into few large clusters would correspond to a high-level topic description (e.g. "sports", "politics", etc.), while a more fine-grained division would discover specific *micro-topics* (e.g. "firefox plugins", "brownies recipe"). Instead, most community detection methods (including the one presented here) discover a single configuration of nodes into communities that is more "natural" given the properties of the tag network. The optimal scale of community structure depends on the information retrieval problem at hand. For

Table 2 IR performance of CNM, SCAN and MultiSCAN community structures in tag recommendation. The following notation is used: R_T denotes the number of correct tags according to the ground truth, R_{out} the number of tag suggestions made by the recommender, R_{TP} the number of correct suggestions, U_{TP} the number of unique correct suggestions, P, R, and F stand for precision, recall and F-measure respectively, and $P@1$, $P@5$ denote precision at one and five recommendations respectively.

	BIBS-200K			FLICKR-1M		
	CNM	SCAN	MultiSCAN	CNM	SCAN	MultiSCAN
R_T		15,344			57,206	
R_{out}	15,271	4,762	7,346	57,021	19,063	33,714
R_{TP}	377	**979**	2,545	2,074	9,781	**9,909**
U_{TP}	196	588	**705**	263	1,103	**1,437**
P (%)	2.47	**20.56**	13.10	4.46	**51.31**	29.39
R (%)	2.46	**6.38**	6.27	4.45	17.10	**17.32**
F (%)	2.46	**9.74**	8.48	4.46	**25.65**	21.80
$P@1$ (%)	2.54	2.97	**5.03**	1.89	5.03	**10.09**
$P@5$ (%)	2.39	**26.36**	19.94	3.04	**46.30**	34.09

instance, as was observed in our experimental study, the existence of large communities harms the performance of a cluster-based tag recommender.

4 Time-Aware Tag/User Co-clustering

The ability to capture emerging trends and dominant topics over time is another form of challenge that could be addressed by data mining approaches in social media content. Indeed, as more and more people tend to express themselves through tagging in social media environments on a daily basis, it can be drawn that monitoring these systems over time allows us to watch the evolution of community focus. Therefore, analysis of such content within its temporal context enables knowledge extraction regarding real world events, long-term or short term topics of interest, and trends. Difficulties arise, though, from the fact that the knowledge extracted from this kind of analysis is particularly sensitive to the time-scale used. For example, the tag `Olympics2008` does not appear to be an event at the hour or single day scale, but does exhibit distinctive temporal patterns at larger time scales. The approach presented in this section overcomes this concern by defining a time-aware co-clustering method that can be applied at multiple time-scales, τ.

4.1 The Proposed Framework

The knowledge extraction approach we propose here is based on the analysis of both users' and tags' activity patterns. The patterns are extracted from two sources of information: i) the meaning of the tags used, and, ii) the time period each activity occurs. The intuition behind this decision is as follows. The social media associated with an event mainly exhibit similarity in terms of their tags and their time locality. Likewise, the users that are attracted by an event or trend tend to use tags related to this incident during the time it is happening. In this context, we follow the following assumption:

> *An **event** or **trend** can be tracked in a social media environment as a dense cluster that consists of two types of objects: related tags with frequent patterns of usage in a given period, and, many users that use these tags around the same period.*

To materialize this observation, a co-clustering method is utilized that employs the time locality similarity and yields a series of clusters, each of which contains a set of users together with a set of tags. Co-clustering is proposed as an approach which may be applied in grouping together elements from different datasets [14]. In our case, co-clustering is used to relate tags and users. In an effort for the clusters to better reflect user choices at particular time intervals, our approach examines tag-based similarity, as well. To examine tag-based similarity, we use the *Wu & Palmer* metric [59], which is based on WordNet to evaluate the similarity in meaning between two terms [18], as follows:

$$TagSim(u_x, t_y) = \max_{t_z} \frac{2 \times depth(LCS)}{[depth(t_z) + depth(t_y)]}, \tag{3}$$

$\forall t_z$ assigned by u_x, where $depth(t_x)$ is the maximum path length from the root to t_x and LCS is the least common subsumer of t_x and t_y.

To quantify the locality in the temporal patterns between a user and a tag at a given timescale τ, we divide the entire time period into I sequential timeframes of size τ and represent: i) each user as $u_x = [u_{x1}, u_{x2}, \ldots, u_{xI}]$, where u_{xj} is the number of tags user u_x has assigned during the timeframe j, and ii) each tag as $t_y = [t_{y1}, t_{y2}, \ldots, t_{yI}]$, where t_{yj} is the number of times the tag t_y has been used during the timeframe j. Then, we calculate the similarity between any two user or tag vectors, by taking their inner product:

$$TimSim(u_x, t_y) = \frac{\sum_{k=1}^{I} u_{ik} \cdot t_{jk}}{\sqrt{\sum_{k=1}^{I} u_{ik}^2 \cdot \sum_{k=1}^{I} t_{jk}^2}}, \tag{4}$$

Having calculated temporal and tag-based similarities between users and tags, we compute the dot product of *TagSim* and *TimSim* between any two objects, in order to get a matrix that embeds both kinds of similarities between users and tags:

$$Sim = TagSim \bullet TimSim, \tag{5}$$

Given *Sim*, we may proceed with the application of the co-clustering algorithm [14], in order to get clusters containing users and tags with similar patterns over time. The applied algorithm is based on the spectral clustering theory, as discussed in [23, 29], and relies on the eigenstructure of the similarity matrix, *Sim*, to partition users and tags into K disjoint clusters. The steps of the applied spectral clustering

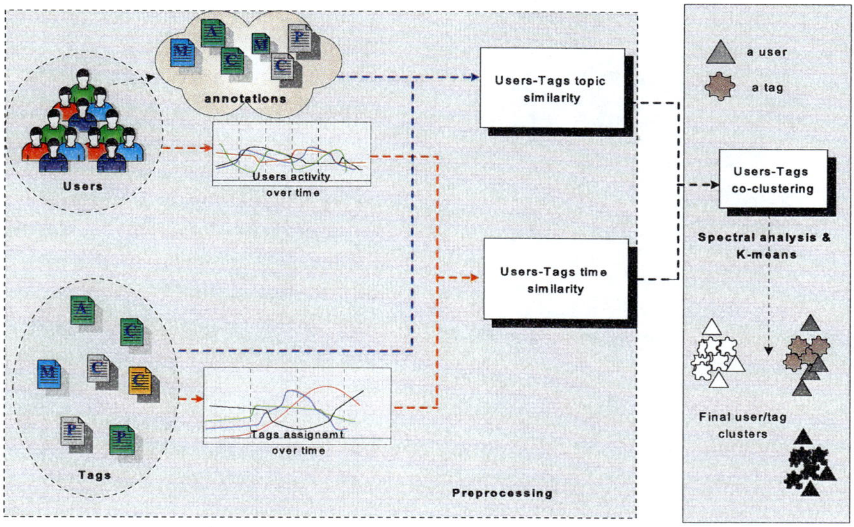

Fig. 2 The proposed time-aware co-clustering algorithm overview

algorithm, which are illustrated in Figure 2, are: i) normalization, ii) computation of eigenvalues, and iii) eigenvector clustering, using K-means.

4.2 Evaluation of Time-Aware User/Tag Co-clustering

We tested our method on a Flickr dataset of 1218 users, 6764 photos, and 2496 unique tags that span the time period from Sep. 2007 to Sep. 2008. To examine the method's applicability in tracking time-related happenings, e.g., events or trends, we used the following four seed tags that are associated with many real-world events, to create the dataset: `Olympics`, `earthquake`, `marriage`, `ancient greece`. The input parameters used are the cluster number K and the time scale τ.

First of all, we aim at studying the impact of the proposed similarity metric on capturing trends or events, in comparison with other similarity metrics. As suggested by the assumption presented in Section 4.1, we examine the compactness of the extracted clusters in terms of gathering together objects that have tag-based and temporal similarity. In order to check this, we performed a rough annotation of our dataset as follows. We assumed four thematic ranges in the dataset, each one associated with one seed tag. Then, we record the activity in time of both tags and users. We divide the time period in timeframes of duration τ. If the activity of an object in a timeframe is above a certain threshold ϑ, we assume that an event or trend is associated with this activity. Thus, a number of events or trends are generated. The value of the threshold ϑ at each time scale is defined empirically. Then, each object (i.e. user or tag) is assigned to the event or trend in which it was more active and had the closest proximity in time. Thus, a ground truth of our dataset is created.

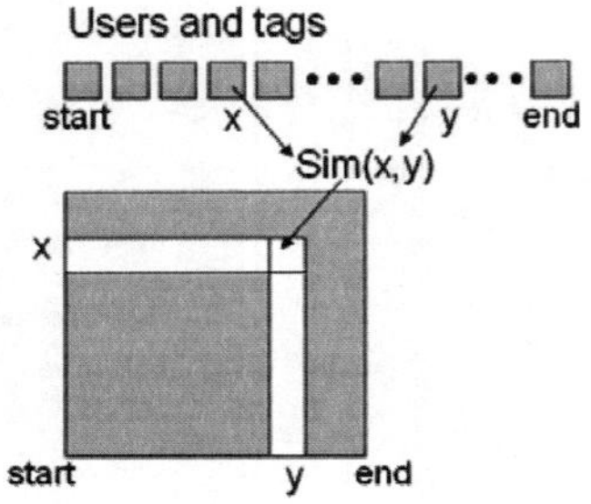

Fig. 3 Tag/user similarity matrix

Then, to evaluate the performance of the proposed similarity metric, we compute a similarity matrix for the 1218 users and 2496 tags using both tag-based and temporal similarity as described in Section 4. The matrix is filled by calculating the similarity between every pair $< user, tag >$. Specifically the (i, j) element of the matrix quantifies the similarity between the ith and the jth object, as depicted in Figure 3. Then, the matrix is reordered, so that objects that have been assigned to the same event or trend during the ground truth generation are contiguous (in rows and in columns). The darker the coloring of a cell (i, j) where $1 \leq i, j \leq |U| + |T|$ the more similar the objects at position (i, j) are. Thus, clusters appear as symmetrical dark squares across the main diagonal. A checkerboard pattern of the described similarity matrix across the main diagonal indicates good clustering, whereas grey rectangles across the diagonal and dark or grey areas outside the diagonal imply that the similarity metric used in the clustering process does not capture the objects assigned to the same trend or event in the same cluster.

In the same way we created a similarity matrix solely based on the temporal locality of objects and a similarity matrix solely based on the tag-based similarity. We conducted experiments for various values of τ. For each τ, we selected the value of K based on the ground truth generation. In Figure 4 we indicatively present the clustering outline for $K = 7$ and $\tau = 10$, in these three different cases. Particularly, the plot shown in (a) was extracted from the proposed similarity metric, while the plots in (b) and (c) were derived using the temporal and the tag-based similarity metric, respectively. It is obvious that the combination of both temporal and tag-based features in the similarity metric succeeds in finding more coherent clusters that according to our original assumption can be mapped to events or trends. The

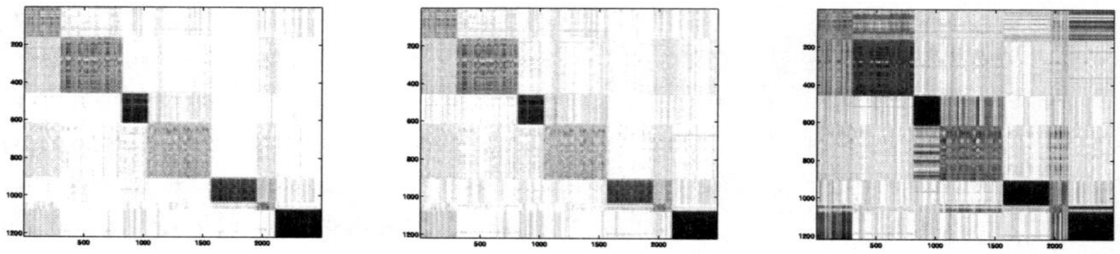

(a) clustering based on temporal and tag-based features

(b) clustering based on tag-based features

(c) clustering based on temporal features

Fig. 4 Events or trends capturing(darker blocks indicate better capturing). All similarity matrices are ordered according to ground truth

coherence deteriorates in case we use only tag-based or temporal similarity between objects.

Next, we want to show that the proposed method is sensitive to various values of τ and performs knowledge extraction at various time scales. While the overall analysis on the entire dataset facilitates the extraction of massive activities, such as events or trends, the analysis at a user level allows the extraction of long-term or short-term user interests and the inclination of that user to follow certain trends or events. Figure 5 illustrates the tagging activity of three users during a yearly period (solid curves). The macroscopic tag clouds indicate each user's most frequent tags during the entire time period, while the microscopic tag clouds reflect each user's most frequent tags in specific timeframes. Given the little overlap in the users' macroscopic tag clouds and their differentiated tagging curves, one would expect that these three users would not be grouped together, if their similarity is evaluated

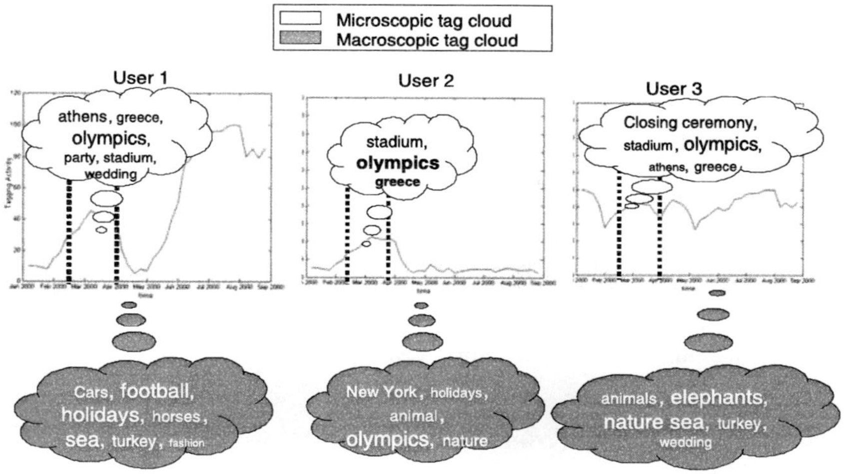

Fig. 5 Tag clouds of three users in an STS

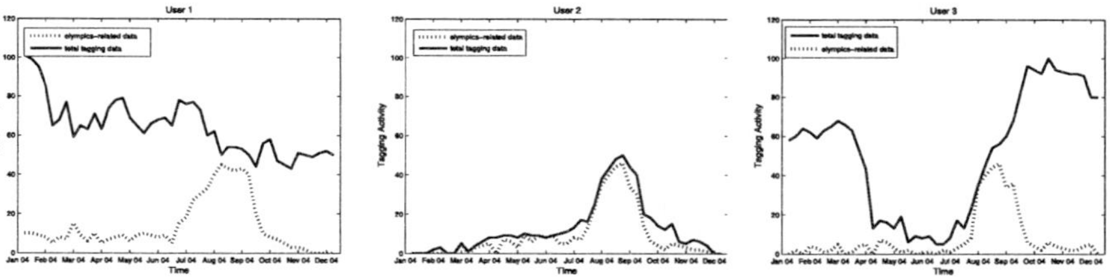

(User1) (User2) (User3)

Fig. 6 Tagging activity of users over time

solely on the similarity of their associated tags. However, focusing on short-term views of the users' tagging activity and examining their microscopic tag clouds at monthly timescale, we observe that for the time interval highlighted with dotted lines in Figure 5 the similarity in these users' tags is very high, as they all use many "Olympics" related tags. This indicates a simultaneous preference by these users for this particular topic and at this specific time period. At the same time, this massive tagging preference may imply that a related event occurred at that period and attracted Olympics-friends to comment on it, through tags.

The users' similar Olympics-related tagging activity is highlighted by the dotted curves in Figure 6, which displays the usage of "Olympics" related tags of each user over time. We claim that such user groups, exhibiting both semantic and temporal cohesion, can only be extracted via a time-aware clustering method, which will examine user behavior at varying time scales. Each time scale selection reveals a different micro-view of users' interests that affects the current clustering, since the microscopic tag cloud of each user is likely to change as the selected time interval's length τ slides across the timeline.

To summarize, in this section, a technique was presented that performs temporal analysis on social media and is based on co-clustering tags and users by considering jointly their temporal and tag-based similarity. The extracted clusters may be used for event or trend recognition and for capturing users' interests at different timescales. An evaluation based on generated ground truth from a Flickr dataset demonstrates that the proposed framework performs better in tracking events or trends than other methods that consider solely tag-based or temporal locality. A number of applications can benefit from such a technique. For example, Olympics-related clusters can be exploited by a sports commercial advertising campaign or be embedded in an application, so that users receive personalized Olympics-related news (e.g., announcement of upcoming events).

5 Enhancing Image Analysis Using Collaborative Data

Semantic object detection is considered one of the most useful operations performed by the human visual system and constitutes an exciting problem for computer vision scientists. Due to its fundamental role in the detection process, many researchers have focused their efforts on trying to understand the mechanisms of learning and particularly the way that humans learn to recognize material, objects, and scenes from very few examples and without much effort. In this direction the authors of [31] make the hypothesis that, once a few categories have been learned with significant cost, some information may be abstracted from the process to make learning further categories more efficient. Based on this hypothesis, when learning new categories, they take advantage of the "general knowledge" extracted from previously learned categories by using it in the form of a prior probability density function in the space of model parameters. Similarly in [32] when images of new concepts are added to the visual analysis model, the computer only needs to learn from the new images.

What has been learned about previous concepts is stored in the form of profiling models, and the computer needs no re-training.

On the other hand in [55] the authors claim that with the availability of overwhelming amounts of data, many problems can be solved without resorting to sophisticated algorithms. The authors mention the example of Google's "Did you mean" tool, which corrects errors in search queries by memorizing billions of query-answer pairs and suggesting the one closest to the user query. In their paper the authors present a visual analog to this tool using a large dataset of 79 million images and a non-parametric approach for image annotation that is based on nearest neighbor matching. Additionally, the authors of [8] employ multiple instance learning to learn models from images labeled as containing the semantic concept of interest, but without indication of which image regions are observations of that concept. Similarly in [17] object recognition is viewed as machine translation that uses expectation maximization in order to learn how to map visual objects (blobs) to concept labels. In all cases, the authors are trying to discover a scalable (in terms of the number of concepts) and effortless (in terms of the necessary annotation) way to teach the machine how to recognize visual objects the way a human does. Motivated by the same objective, in this work we investigate whether the knowledge aggregated in social tagging systems by the collaboration of web users can help in this direction.

While model parameters can be estimated more efficiently from strongly annotated samples, such samples are very expensive to obtain. On the contrary, weakly annotated samples can be found in large quantities especially from social media sources. Social Tagging systems such as Flickr accommodate image corpora populated with hundreds of user tagged images on a daily basis. Motivated by this fact, our work aims at combining the advantages of both strongly supervised (learn model parameters more efficiently) and weakly supervised (learn from samples obtained at low cost) methods, by allowing the strongly supervised methods to learn from training samples that are found in collaborative tagging environments. Specifically, drawing from a large pool of weakly annotated images, our goal is to benefit from the knowledge aggregated in social tagging systems, in order to automatically determine a set of image regions that can be associated with a certain tag.

5.1 Framework Description

The proposed framework for leveraging social media to train object detection models is depicted in Figure. 7. The analysis components of the framework are: tag-based image selection, image segmentation, extraction of visual features from image regions, region-based clustering using their visual features and supervised learning of object detection models using strongly annotated samples.

More specifically, given an object c that we wish to train a detector for, our method starts from a large collection of user tagged images and performs the following actions. Images are selected based on their tag information in order to formulate image group(s) that correspond to thematic entities. Given the tendency of social tagging systems to formulate knowledge patterns that reflect the way content is perceived by the web users [34], tag-based image selection is expected to identify these patterns and create image group(s) emphasizing on a certain object. By emphasizing we refer to the case where the majority of the images within a group depict a certain object and that the linguistic description of that object can be obtained from the most frequently appearing tag (see Section 5.2 for more details). Subsequently, region-based clustering is performed on all images belonging to the image group that emphasizes on object *c*, that have been pre-segmented by an automatic segmentation algorithm. During region-based clustering the image regions are represented by their visual features and each of the generated clusters contains visually similar regions. Since the majority of the images within the selected group depicts instances of the desired object *c*, we anticipate that the majority of regions representing the object of interest will be gathered in the most populated cluster, pushing all irrelevant regions to the other clusters. Eventually, we use as positive samples the visual features extracted from the regions belonging to the most populated cluster, to train in a supervised manner a model detecting the object *c*.

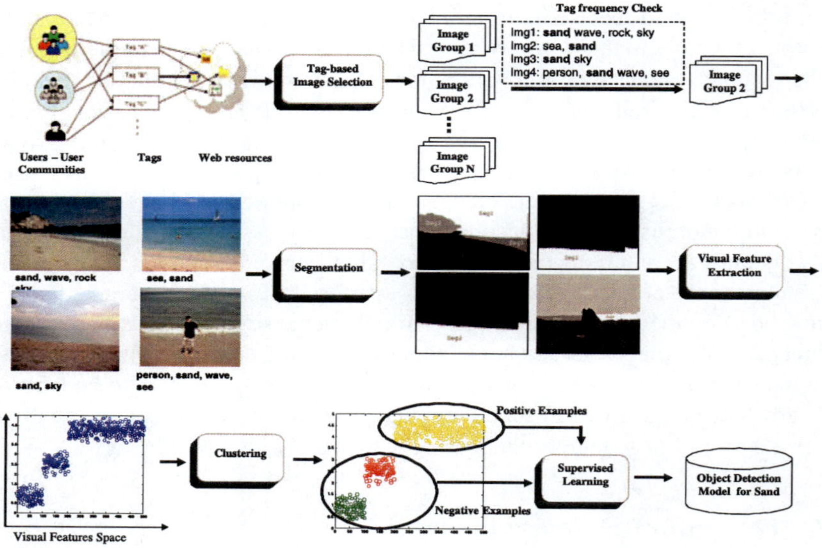

Fig. 7 Leveraging a set of user tagged images to train a model for detecting the object *sand*.

5.2 Analysis Components

Tag-based image selection: Refers to the techniques used to select images from a large dataset (S) of arbitrary content, based on their tag information. We employ one of the following three approaches based on the associated annotations:

1. **Keyword-based selection:** This approach is used for selecting images from strongly annotated datasets. In order to create $S^c \subset S$ we need only to select the images that are labeled with the name of the object *c*.

2. **Flickr groups:** Flickr groups are virtual places hosted in collaborative tagging environments that allow social users to share content on a certain topic. In this case, S^c is created by taking all images contained in a Flickr group titled with the name of the object c. From here on we will refer to those images as roughly-annotated images.
3. **SEMSOC:** SEMSOC [22, 23] is applied by our framework on weakly annotated images (i.e., images that have been tagged by humans in the context of a collaborative tagging environment, but no rigid annotations have been provided) in order to create semantically consistent groups of images. In order to obtain the image group S^c that emphasizes on object c, we select the SEMSOC-generated group S^{c_i} where its most frequent tag relates with c.

Segmentation: Segmentation is applied on all images in S^c with the aim to extract the spatial masks of visually meaningful regions. In our work we have used a K-means with connectivity constraint algorithm as described in [35]. The output of this algorithm for an image I_q is a set of segments $R_{I_q} = \{r_i^{I_q}, i = 1, \dots, m\}$, which roughly correspond to meaningful objects.
Visual Descriptors: In order to describe visually the segmented regions we have employed the following: a) the Harris-Laplace detector and a dense sampling approach for determining the interest points, b) the SIFT descriptor as proposed by Lowe [33] in order to describe each interest point using a 128-dimensional feature vector, and c) the bag-of-words model initially proposed in [50] in order to obtain a fixed-length feature vector for each region. The feature extraction process is similar to the one described in [44] with the important difference that in our case descriptors are extracted to represent each of the identified image segments, rather than the whole image. Thus, $\forall r_i^{I_q} \in R_{I_q}$ and $\forall I_q \in S^c$ a 300-dimensional feature vector $f(r_i^{I_q})$ is extracted.
Clustering: For performing feature-based region clustering we applied the affinity propagation clustering algorithm [19] on all extracted feature vectors $f(r_i^{I_q})$, $\forall r_i^{I_q} \in R_{Iq}$ and $\forall I_q \in S^c$. This is an algorithm that takes as input the measures of similarity between pairs of data points and exchanges messages between data points, until a high-quality set of centers and corresponding clusters is found.
Learning Model Parameters: Support Vector Machines (SVMs) [46] were chosen for generating the object detection models due to their ability in smoothly generalizing and coping efficiently with high-dimensionality pattern recognition problems. All feature vectors assigned to the most populated of the created clusters were used as positive examples for training a binary classifier. Negative examples were chosen arbitrarily from the remaining dataset. For training the object detection models we have used the libSVM library [10]. The radial basis function(RBF) kernel was used to map the samples into a higher dimensional space. In order to find the optimal parameters for the RBF kernel (C and γ) we performed 10-fold cross validation (i.e., divide the training set into 10 subsets of equal size and evaluate the performance using each time one of the subsets for testing and the remaining 9 for training). A "grid-search" on the exhaustive range of C and γ parameters provides us with

various pairs of (C,γ) values. These pairs are evaluated using cross-validation and the one with the best cross-validation accuracy is selected.

5.3 Evaluation of Object Detection Models

The goal of our experimental study is to compare the quality of object models trained using samples leveraged by the proposed framework, against the models trained using manually provided, strongly annotated samples. To carry out our experiments we have relied on three different types of datasets. The first type includes the strongly annotated datasets constructed by asking people to provide region detail annotations of images pre-segmented with the automatic segmentation algorithm of Section 5.2. For this case we have used a collection of 536 images S^B from the *Seaside* domain annotated in our lab. The second type refers to the roughly-annotated datasets like the ones formed in Flickr groups. In order to create a dataset of this type S^G, for each object of interest, we have downloaded 500 member images from a Flickr group that is titled with a name related to the name of the object. The third type refers to the weakly annotated datasets like the ones that can be collected freely from the collaborative tagging environments. For this case, we have crawled 3000 images S^{F3K} from Flickr using the wget[3] utility and Flickr API facilities. Moreover, in order to investigate the impact of the dataset size on the robustness of the generated models we have also crawled from Flickr another dataset consisting 10000 images S^{F10K}. Depending on the annotation type we use the tag-based selection approaches presented in Section 5.2 to construct the necessary image groups S^c.

In order to compare the efficiency of the models generated using training samples with different annotation type (i.e., strongly, roughly, weakly), we need a set of objects that are common in all three types of datasets. For this reason after examining the contents of S^B, reviewing the availability of groups in Flickr and applying SEMSOC on S^{F3K} and S^{F10K}, we determined four object categories C^{bench}={**sky, sea, vegetation, person**}. These objects exhibited significant presence in all different datasets and served as benchmarks for comparing the quality of the different models. The factor limiting the number of benchmarking objects is on the one hand the need to have strongly annotated images for these objects and on the other hand the un-supervised nature of SEMSOC that restricts the eligible objects to the ones emphasized by the generated image groups. C^{bench} is the maximum set of objects shared between all different dataset types. For each object $c_i \in C^{bench}$ one model was trained using the strong annotations of S^B, one model was trained using the roughly-annotated images contained in S^G, and two models were trained using the weak annotations of S^{F3K} and S^{F10K}, respectively. In order to evaluate the performance of these models, we test them using a subset (i.e., 268 images) of the strongly annotated dataset $S^B_{test} \subset S^B$, not used during training. F-Measure was used for measuring the efficiency of the models.

By looking at the bar diagram of Figure 8, we derive the following conclusions: a) Model parameters are estimated more efficiently when trained with strongly

[3] wget: http://www.gnu.org/software/wget

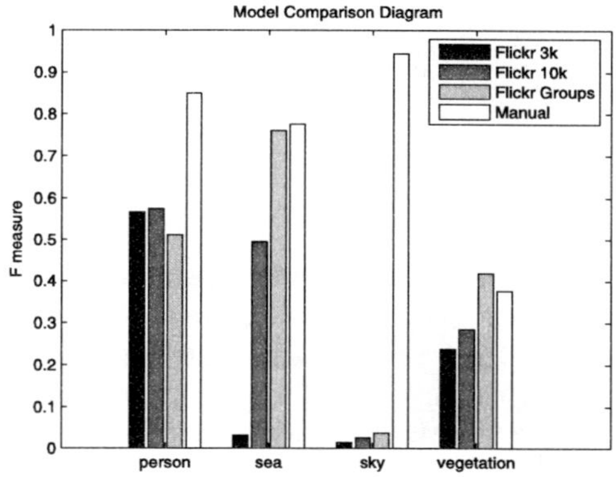

Fig. 8 Performance comparison between four object recognition models that are learned using samples of different annotation quality (i.e., strongly, roughly and weakly)

annotated samples, since in three out of four cases they outperform the other models and sometimes by a significant amount (e.g., sky, person). b) Flickr groups can serve as a less costly alternative for learning the model parameters, since using the roughly-annotated samples we get comparable and sometimes even better (e.g., vegetation) performance than manually trained models, while requiring considerable less effort to collect the training samples. c) The models learned from weakly annotated samples are usually inferior to the other cases, especially in cases where the proposed approach for leveraging the data has failed in selecting the appropriate cluster (e.g., *sea* and *sky*).

One drawback of Flickr groups derives from the fact that since they are essentially virtual places they are not guaranteed to constantly increase their size and therefore provide larger datasets that could potentially increase the efficiency of the developed models. This is why we also employ SEMSOC for constructing the necessary images sets. SEMSOC is an unsupervised selection procedure that operates directly on image tags and its goal is to provide a set of images the majority of which depict the object of interest. Naturally the image sets generated by SEMSOC are not of the same quality as those obtained from Flickr groups. However, the motivation for using SEMSOC is that it can potentially produce considerably larger image sets. Given that in Flickr groups the user needs to classify an image in one of the existing groups (or create a new group), the total number of positive samples that can be extracted from the images of a Flickr group, has an upper limit on the total number of images that have been included in this group by the users. On the other hand, the total number of positive samples that can be obtained by SEMSOC in principle, is only limited by the total number of images that are uploaded on the entire Flickr repository and depict the object of interest. However, given that collaborative tagging environments like Flickr are growing rapidly, we can accept that SEMSOC will manage to produce arbitrary large image sets. In this respect, in our experiment

Table 3 Comparing with existing methods in object recognition

	Building	Grass	Tree	Cow	Sheep	Sky	Aeroplane	Water	Face	Car	Bicycle	Flower	Sign	Bird	Book	Chair	Road	Cat	Dog	Body	Boat	Average
Textonboost [48]	62	**98**	**86**	58	50	83	60	**53**	**74**	63	**75**	63	**35**	19	**92**	15	**86**	54	19	**62**	7	**58**
PLSA-MRF/I [57]	45	64	71	**75**	**74**	**86**	**81**	47	1	**73**	55	**88**	6	6	63	18	80	27	26	55	8	50
Prop. Framework	**87**	9	65	45	45	14	29	**53**	56	12	**75**	**88**	27	**30**	25	**50**	44	**59**	**71**	29	**41**	45

we also examine how the efficiency of the developed models is affected by the size of the image set that has been used to obtain their training samples.

From the bar diagram of Figure 8 it is clear that when using the S^{F10K} the incorporation of more indicative examples into the training set improves the generalization ability of the generated models in all four cases. However, in the case of object *sea* we note also a drastic improvement of the model's efficiency. This is attributed to the fact that the increment of the dataset size alleviates the error introduced by the employed algorithms (i.e., segmentation, feature extraction, clustering) and allows the proposed method to select the appropriate cluster for training the model. On the other hand, in the case of object *sky* it seems that the correct cluster is still missed despite the use of a larger dataset. In this case the size of the dataset should grow even larger in order to compensate for the aforementioned error and select the appropriate cluster.

In order to compare our framework with existing methods we used the publicly available MSRC dataset[4] consisting of 591 images. In order to train the models, for each of the 21 objects, we have downloaded 500 member images from a Flickr group that is titled with a name related to the name of the object. We compare the region label annotations that were automatically acquired by our framework using Flickr groups with the patch level annotations of the approach proposed Verbeek and Triggs [57] and the ones obtained from Textonboost [48]. The the classification rates per object for each method are shown in Table 3. Looking at the individual objects we can see that despite the low cost for annotation our method yields the best performance in 9 out of 21 cases, compared to 7 out of 21 for the PLSA-MRF/I and 8 out of 21 for the Textonboost (note that in three cases Water, Flower, Bicycle the classification rates are identical for two different methods). On average, the accuracy obtained from our approach (45%) is inferior to the one obtained from PLSA-MRF/I (50%) which is again inferior to the accuracy obtained from Textonboost (58%). This is in accordance with our expectation since the performance scores obtained by the three methods are ranked proportionally to the amount of annotation effort required to train their models. Based on the above we can claim that the significant gain in effort that we achieve by leveraging social media to obtain the necessary training samples, compensates for the limited loss in performance that we suffer when compared with state of the art object recognition systems.

In this Section we have shown that the collective knowledge encoded in the user contributed content can be successfully used to remove the need for close human supervision when training object detectors. The experimental results have

[4] http://research.microsoft.com/vision/cambridge/recognition

demonstrated that although the performance of the detectors trained using leveraged social media is inferior to the one achieved by manually trained detectors, there are cases where the gain in effort compensates for the small loss in performance. In addition we have seen that by increasing the number of utilized images we manage to improve the performance of the generated detectors, advocating the potential of social media to facilitate the creation of reliable and effective object detectors. Finally, despite the fact that there will always be strong dependence between the discriminative power of the employed feature space and the efficiency of the proposed approach in selecting the appropriate set of training samples, our experimental study has shown that we can maximize the probability of success by using large volumes of user contributed content.

6 Conclusions

In this chapter we have demonstrated how massive user contributions can be leveraged to extract valuable knowledge. The community structure of tag networks, the emerging trends and events in users tag activity, as well as the associations between image regions and tags in user tagged images, all form different types of knowledge that was made possible to extract due to the collaborative and massive nature of the data. It is true that with the abundant availability of social data on the Web, analysis can now use the information coming both from the content itself, the social context and the emergent social dynamics. Although noisy and of questionable reliability, user contributions exhibit noise reduction properties when considered massively, given that they encode the collective knowledge of multiple users. Thus, the common objective among all methods performing knowledge extraction on social media, is to exploit those noise reduction properties and capture the knowledge provided by multiple users.

Our review on the methods performing knowledge extraction from massive user contributions has resulted in the following observations. Unsupervised approaches constitute the main vehicle for extracting the statistical patterns of the data. Either through the use of clustering, co-clustering or community detection techniques, we have noticed the tendency of keeping human intervention to a minimum and favoring algorithms that are able to extract all necessary parameters (e.g., the number of clusters) by pre-processing the available data. This tendency is basically motivated by the need to process a huge amount of data, which renders impractical schemes that require supervision. This tendency is further explained by the fact that the effectiveness of the methods extracting knowledge from social media is tightly bound to the amount of data that need to be processed. Given that the knowledge-rich patterns encoded in the data become stable and thus "visible" only after a specific usage period, many are the cases where the proposed approaches are unable to produce meaningful results, unless applied on large scale datasets. This is the reason why scalability constitutes an important requirement for such methods.

As avenues for future research we can identify the tendency of scientific efforts to optimally combine the information carried by the different modalities hosted by social networks (i.e., images, tags, friendship links, etc). Being different in nature and heterogeneous in representation, this information should be analyzed by appropriately designed methods in order to become exploitable under a certain task. Finally, as a particularly challenging objective we also identify the potential of employing all those knowledge extraction methods, for automating the process of making the content contributed by users part of the Linked Open Data (LOD) cloud.

Acknowledgements. This work was sponsored by the European Commission as part of the Information Society Technologies (IST) programme under grant agreement n215453 - WeKnowIt and the contract FP7-248984 GLOCAL.

References

1. Ching-man, Gibbins, N., Yeung, N.S.A.: A study of user profile generation from folksonomies. In: SWKM (2008)
2. Au Yeung, C.m., Gibbins, N., Shadbolt, N.: Contextualising tags in collaborative tagging systems. In: HT 2009: Proceedings of the 20th ACM Conference on Hypertext and Hypermedia, pp. 251–260. ACM, New York (2009)
3. Aurnhammer, M., Hanappe, P., Steels, L.: Augmenting navigation for collaborative tagging with emergent semantics. In: Cruz, I., Decker, S., Allemang, D., Preist, C., Schwabe, D., Mika, P., Uschold, M., Aroyo, L.M. (eds.) ISWC 2006. LNCS, vol. 4273, pp. 58–71. Springer, Heidelberg (2006)
4. Becker, H., Naaman, M., Gravano, L.: Learning similarity metrics for event identification in social media. In: WSDM 2010, pp. 291–300. ACM, New York (2010)
5. Begelman, G.: Automated tag clustering: Improving search and exploration in the tag space. In: Proc. of the Collaborative Web Tagging Workshop at WWW 2006 (2006)
6. Brin, S., Page, L.: The anatomy of a large-scale hypertextual web search engine. Computer Networks and ISDN Systems 30, 107–117 (1998)
7. Brooks, C.H., Montanez, N.: Improved annotation of the blogosphere via autotagging and hierarchical clustering. In: WWW 2006, pp. 625–632. ACM, New York (2006)
8. Carneiro, G., Chan, A.B., Moreno, P.J., Vasconcelos, N.: Supervised learning of semantic classes for image annotation and retrieval. IEEE Trans. Pattern Anal. Mach. Intell. 29(3), 394–410 (2007)
9. Cattuto, C.: Collaborative tagging as a complex system. talk given at internationl school on semiotic dynamics. In: Language and Complexity, Erice (2005)
10. Chang, C.C., Lin, C.J.: LIBSVM: a library for support vector machines (2001), Software, available at `http://www.csie.ntu.edu.tw/cjlin/libsvm`
11. Clauset, A., Newman, M.E.J., Moore, C.: Finding community structure in very large networks. Phys. Rev. E 70(6), 066,111 (2004)
12. Cooper, M., Foote, J., Girgensohn, A., Wilcox, L.: Temporal event clustering for digital photo collections. ACM Trans. Multimedia Comput. Commun. Appl. 1(3), 269–288 (2005)
13. Cour, T., Sapp, B., Jordan, C., Taskar, B.: Learning from ambiguously labeled images. In: IEEE Computer Society Conference on Computer Vision and Pattern Recognition, CVPR 2009 (2009)
14. Dhillon, I.S.: Co-clustering documents and words using bipartite spectral graph partitioning. In: Proceedings of KDD 2001, San Francisco, California, pp. 269–274 (2001)

15. Diederich, J., Iofciu, T.: Finding communities of practice from user profiles based on folksonomies. In: Proceedings of the 1st International Workshop on Building Technology Enhanced Learning Solutions for Communities of Practice, TEL-CoPs 2006 (2006)
16. Dubinko, M., Kumar, R., Magnani, J., Novak, J., Raghavan, P., Tomkins, A.: Visualizing tags over time. In: Proceedings of WWW 2006, pp. 193–202. ACM, Edinburgh (2006)
17. Duygulu, P., Barnard, K., de Freitas, J.F.G., Forsyth, D.: Object recognition as machine translation: Learning a lexicon for a fixed image vocabulary. In: Heyden, A., Sparr, G., Nielsen, M., Johansen, P. (eds.) ECCV 2002. LNCS, vol. 2353, pp. 97–112. Springer, Heidelberg (2002)
18. Fellbaum, C. (ed.): WordNet: An Electronic Lexical Database (Language, Speech, and Communication). The MIT Press, Cambridge (1998)
19. Frey, B.J., Dueck, D.: Clustering by passing messages between data points. Science 315, 972–976 (2007), http://www.psi.toronto.edu/affinitypropagation
20. Gemmell, J., Shepitsen, A., Mobasher, B., Burke, R.: Personalizing navigation in folksonomies using hierarchical tag clustering. In: Song, I.-Y., Eder, J., Nguyen, T.M. (eds.) DaWaK 2008. LNCS, vol. 5182, pp. 196–205. Springer, Heidelberg (2008)
21. Ghosh, H., Poornachander, P., Mallik, A., Chaudhury, S.: Learning ontology for personalized video retrieval. In: MS 2007: Workshop on Multimedia Information Retrieval on The Many Faces of Multimedia Semantics, pp. 39–46. ACM, New York (2007)
22. Giannakidou, E., Kompatsiaris, I., Vakali, A.: Semsoc: Semantic, social and content-based clustering in multimedia collaborative tagging systems. In: ICSC, pp. 128–135 (2008)
23. Giannakidou, E., Koutsonikola, V.A., Vakali, A., Kompatsiaris, Y.: Co-clustering tags and social data sources. In: WAIM, pp. 317–324 (2008)
24. Giannakidou, E., Koutsonikola, V.A., Vakali, A., Kompatsiaris, Y.: Exploring temporal aspects in user-tag co-clustering. In: Special session: Interactive Multimedia in Social Networks, WIAMIS (2010)
25. Halpin, H., Robu, V., Shepherd, H.: The complex dynamics of collaborative tagging. In: Proceedings of WWW 2007, pp. 211–220. ACM, New York (2007)
26. Hotho, A., Ja"schke, R., Schmitz, C., Stumme, G.: Trend detection in folksonomies. In: Avrithis, Y., Kompatsiaris, Y., Staab, S., O'Connor, N.E. (eds.) SAMT 2006. LNCS, vol. 4306, pp. 56–70. Springer, Heidelberg (2006)
27. Kennedy, L.S., Chang, S.F., Kozintsev, I.: To search or to label?: predicting the performance of search-based automatic image classifiers. In: Multimedia Information Retrieval, pp. 249–258 (2006)
28. Kennedy, L.S., Naaman, M., Ahern, S., Nair, R., Rattenbury, T.: How flickr helps us make sense of the world: context and content in community-contributed media collections. ACM Multimedia, 631–640 (2007)
29. Koutsonikola, V.A., Petridou, S., Vakali, A., Hacid, H., Benatallah, B.: Correlating time-related data sources with co-clustering. In: Bailey, J., Maier, D., Schewe, K.-D., Thalheim, B., Wang, X.S. (eds.) WISE 2008. LNCS, vol. 5175, pp. 264–279. Springer, Heidelberg (2008)
30. Koutsonikola, V., Vakali, A., Giannakidou, E., Kompatsiaris, I.: Clustering of social tagging system users: A topic and time based approach. In: Vossen, G., Long, D.D.E., Yu, J.X. (eds.) WISE 2009. LNCS, vol. 5802, pp. 75–86. Springer, Heidelberg (2009)
31. Li, F.F., Fergus, R., Perona, P.: One-shot learning of object categories. IEEE Trans. Pattern Anal. Mach. Intell. 28(4), 594–611 (2006)
32. Li, J., Wang, J.Z.: Real-time computerized annotation of pictures. IEEE Trans. Pattern Anal. Mach. Intell. 30(6), 985–1002 (2008), http://dx.doi.org/10.1109/TPAMI.2007.70847

33. Lowe, D.G.: Distinctive image features from scale-invariant keypoints. Int. J. Comput. Vision 60(2), 91–110 (2004)
34. Marlow, C., Naaman, M., Boyd, D., Davis, M.: Ht06, tagging paper, taxonomy, flickr, academic article, to read. In: Hypertext, pp. 31–40 (2006)
35. Mezaris, V., Kompatsiaris, I., Strintzis, M.G.: Still image segmentation tools for object-based multimedia applications. IJPRAI 18(4), 701–725 (2004)
36. Mika, P.: Ontologies are us: A unified model of social networks and semantics. Web Semant 5(1), 5–15 (2007), http://dx.doi.org/10.1016/j.websem.2006.11.002
37. Nanopoulos, A., Gabriel, H.H., Spiliopoulou, M.: Spectral clustering in social-tagging systems. In: Vossen, G., Long, D.D.E., Yu, J.X. (eds.) WISE 2009. LNCS, vol. 5802, pp. 87–100. Springer, Heidelberg (2009)
38. Newman, M.E.J., Girvan, M.: Finding and evaluating community structure in networks. Phys. Rev. E 69(2), 026,113 (2004)
39. Papadopoulos, S., Kompatsiaris, Y., Vakali, A.: A graph-based clustering scheme for identifying related tags in folksonomies. In: Bach Pedersen, T., Mohania, M.K., Tjoa, A.M. (eds.) DAWAK 2010. LNCS, vol. 6263, pp. 65–76. Springer, Heidelberg (2010)
40. Papadopoulos, S., Vakali, A., Kompatsiaris, Y.: Community detection in collaborative tagging systems. In: Pardede, E. (ed.) Community-Built Database: Research and Development. Springer, Heidelberg (2010)
41. Quack, T., Leibe, B., Gool, L.J.V.: World-scale mining of objects and events from community photo collections. In: CIVR, pp. 47–56 (2008)
42. Rattenbury, T., Good, N., Naaman, M.: Towards automatic extraction of event and place semantics from flickr tags. In: SIGIR 2007: Proceedings of the 30th Annual International ACM SIGIR Conference on Research and Development in Information Retrieval, pp. 103–110. ACM, New York (2007)
43. Russell, T.: Cloudalicious: Folksonomy over time. In: Proceedings of the 6th ACM/IEEE-CS Joint Conference on Digital Libraries, pp. 364–364. ACM, Chapel Hill, NC, USA (2006)
44. van de Sande, K., Gevers, T., Snoek, C.: Evaluating color descriptors for object and scene recognition. IEEE Transactions on Pattern Analysis and Machine Intelligence 99(1) (5555)
45. Schifanella, R., Barrat, A., Cattuto, C., Markines, B., Menczer, F.: Folks in folksonomies: social link prediction from shared metadata. In: WSDM 2010: Proceedings of the Third ACM International Conference on Web Search and Data Mining, pp. 271–280. ACM, New York (2010)
46. Scholkopf, B., Smola, A., Williamson, R., Bartlett, P.: New support vector algorithms. Neural Networks 22, 1083–1121 (2000)
47. Segaran, T.: Programming Collective Intelligence. O'Reilly Media Inc., Sebastopol (2007)
48. Shotton, J., Winn, J.M., Rother, C., Criminisi, A.: *textonBoost*: Joint appearance, shape and context modeling for multi-class object recognition and segmentation. In: Leonardis, A., Bischof, H., Pinz, A. (eds.) ECCV 2006. LNCS, vol. 3951, pp. 1–15. Springer, Heidelberg (2006)
49. Simpson, E.: Clustering tags in enterprise and web folksonomies. HP Labs Techincal Reports (2008), http://www.hpl.hp.com/techreports/2008/HPL-2008-18.html
50. Sivic, J., Zisserman, A.: Video google: A text retrieval approach to object matching in videos. In: ICCV 2003: Proceedings of the Ninth IEEE International Conference on Computer Vision, p. 1470. IEEE Computer Society, Washington, DC, USA (2003)

51. Specia, L., Motta, E.: Integrating folksonomies with the semantic web. In: Franconi, E., Kifer, M., May, W. (eds.) ESWC 2007. LNCS, vol. 4519, pp. 624–639. Springer, Heidelberg (2007)
52. Sun, A., Zeng, D., Li, H., Zheng, X.: Discovering trends in collaborative tagging systems. In: Yang, C.C., Chen, H., Chau, M., Chang, K., Lang, S.-D., Chen, P.S., Hsieh, R., Zeng, D., Wang, F.-Y., Carley, K.M., Mao, W., Zhan, J. (eds.) ISI Workshops 2008. LNCS, vol. 5075, pp. 377–383. Springer, Heidelberg (2008)
53. Sun, Y., Shimada, S., Taniguchi, Y., Kojima, A.: A novel region-based approach to visual concept modeling using web images. In: ACM Multimedia, pp. 635–638 (2008)
54. Swan, R., Allan, J.: Extracting significant time varying features from text. In: Proceedings of the Eighth International Conference on Information and Knowledge Management, pp. 38–45 (1999)
55. Torralba, A., Fergus, R., Freeman, W.T.: 80 million tiny images: A large data set for nonparametric object and scene recognition. IEEE Transactions on Pattern Analysis and Machine Intelligence 30, 1958–1970 (2008), `http://doi.ieeecomputersociety.org/10.1109/TPAMI.2008.128`
56. Tsikrika, T., Diou, C., de Vries, A.P., Delopoulos, A.: Image annotation using click-through data. In: 8th ACM International Conference on Image and Video Retrieval, Santorini, Greece (2009)
57. Verbeek, J.J., Triggs, B.: Region classification with markov field aspect models. In: CVPR (2007)
58. Wu, L., Hua, X.S., Yu, N., Ma, W.Y., Li, S.: Flickr distance. ACM Multimedia, 31–40 (2008)
59. Wu, Z., Palmer, M.: Verm semantics and lexical selection. In: Proceedings of the 32nd Annual Meeting of the Association for Computational Linguistics, New Mexiko, USA, pp. 133–138 (1994)
60. Xu, X., Yuruk, N., Feng, Z., Schweiger, T.A.J.: Scan: a structural clustering algorithm for networks. In: KDD 2007: Proceedings of the 13th ACM SIGKDD International Conference on Knowledge Discovery and Data Mining, pp. 824–833. ACM, New York (2007)
61. Zhang, J., Marszalek, M., Lazebnik, S., Schmid, C.: Local features and kernels for classification of texture and object categories: A comprehensive study. Int. J. Comput. Vision 73(2), 213–238 (2007), `http://dx.doi.org/10.1007/s11263-006-9794-4`

Chapter 17
Validating Privacy Requirements in Large Survey Rating Data

Xiaoxun Sun, Hua Wang, and Jiuyong Li

Abstract. Recent study shows that supposedly anonymous movie rating records are de-identified by using a little auxiliary information. In this chapter, we study a problem of protecting privacy of individuals in large public survey rating data. Such rating data usually contains both ratings of sensitive and non-sensitive issues, and the ratings of sensitive issues belong to personal privacy. Even when survey participants do not reveal any of their ratings, their survey records are potentially identifiable by using information from other public sources. To amend this, in this chapter, we propose a novel (k, ε, l)-anonymity model to protect privacy in large survey rating data, in which each survey record is required to be "similar" with at least $k-1$ others based on the non-sensitive ratings, where the similarity is controlled by ε, and the standard deviation of sensitive ratings is at least l. We study an interesting yet nontrivial satisfaction problem of the proposed model, which is to decide whether a survey rating data set satisfies the privacy requirements given by the user. For this problem, we investigate its inherent properties theoretically, and devise a novel slice technique to solve it. We discuss the idea of how to anonymize data by using the result of satisfaction problem. Finally, we conduct extensive experiments on two real-life data sets, and the results show that the slicing technique is fast and scalable with data size and much more efficient in terms of execution time and space overhead than the heuristic pairwise method.

1 Introduction

The problem of privacy-preserving data publishing has received a lot of attention in recent years. Privacy preservation on relational data has been studied extensively. A

Xiaoxun Sun
Australian Council for Educational Research, Australia
e-mail: sun@acer.edu.au

Hua Wang
University of Southern Queensland, Australia
e-mail: wang@usq.edu.au

Jiuyong Li
University of South Australia, Australia
e-mail: jiuyong.li@unisa.edu.au

N. Bessis, F. Xhafa (Eds.): Next Generation Data Technologies for CCI, SCI 352, pp. 445–469.
springerlink.com

major category of privacy attacks on relational data is to re-identify individuals by joining a published table containing sensitive information with some external tables. Most of existing work can be formulated in the following context: several organizations, such as hospitals, publish detailed data (called microdata) about individuals (e.g. medical records) for research or statistical purposes [32, 23, 22, 30].

Privacy risks of publishing microdata are well-known. Famous attacks include de-anonymisation of the Massachusetts hospital discharge database by joining it with a public voter database [32] and privacy breaches caused by AOL search data [16]. Even if identifiers such as names and social security numbers have been removed, the adversary can use linking [32], homogeneity and background attacks [23] to re-identify individual data records or sensitive information of individuals. To overcome the re-identification attacks, k-anonymity was proposed [25, 26, 27, 32]. Specifically, a data set is said to be k-anonymous ($k \geq 1$) if, on the quasi-identifier (QID) attributes (that is, the maximal set of join attributes to re-identify individual records), each record is identical with at least $k-1$ other records. The larger the value of k, the better the privacy is protected. Several algorithms are proposed to enforce this principle [1, 7, 12, 19, 20, 21, 18]. Machanavajjhala et al. [23] showed that a k-anonymous table may lack of diversity in the sensitive attributes. To overcome this weakness, they propose the l-diversity [23]. However, even l-diversity is insufficient to prevent attribute disclosure due to the skewness and the similarity attack. To amend this problem, t-closeness [22] was proposed to solve the attribute disclosure vulnerabilities inherent to previous models.

Recently, a new privacy concern has emerged in privacy preservation research: how to protect individuals' privacy in large survey rating data. Though several models and many algorithms have been proposed to preserve privacy in relational data (e.g., k-anonymity [32], l-diversity [23], t-closeness [22], etc.), most of the existing studies are incapable of handling rating data, since the survey rating data normally does not have a fixed set of personal identifiable attributes as relational data, and it is characterized by high dimensionality and sparseness. The survey rating data shares the similar format with transactional data. The privacy preserving research of transactional data has recently been acknowledged as an important problem in the data mining literature [14, 36]. To our best knowledge, there is no current research addressing the issue of how to efficiently determine whether the survey rating data satisfies the privacy requirement. In this chapter, we propose a (k, ε, l)-anonymity model to protect privacy in the large survey rating data and study the *Satisfaction Problem* (Section 5) of the proposed model, which is to decide whether a survey rating data set satisfies the given privacy requirements. By utilizing the largeness and sparseness properties, we develop a novel slicing technique solving the satisfaction problem. Our extensive experiments confirm that our new slicing algorithm is fast and scalable in practical compared with the heuristic pairwise algorithm. The main contributions of the chapter are summarized as follows:

(1) Propose a novel (k, ε, l)-anonymity model to protect individual's privacy in large survey rating data. The principle demands that each transaction be similar with $k-1$ others, where the similarity is measured by ε metric, and it further requires the standard deviation of the sensitive ratings be at least l. ε captures

the protection range of each individual, whereas k is to lower an adversary's chance of beating that protection, and l reflects diversity of the sensitive ratings.

(2) Investigate the theoretical properties of (k, ε, l)-anonymity model. Specifically, we prove a sufficient condition of the existence of at least one (k, ε, l)-anonymity solution in large survey rating data, and we prove the lower and upper bound of the parameter l.

(3) Apply the flag matrix to index the rating data and devise a novel slicing technique by searching closest neighbors in large, sparse and high dimensional rating data to determine the satisfaction problem, which is to decide if the given rating data satisfies privacy requirements.

(4) Conduct extensive experiments to show that the slicing approach is scalable, time efficient and space efficient compared with the heuristic pairwise method.

The focus of this chapter is about the methods and technologies that use novel data protection technologies to analyze and anonymize distributed data, which are available from various social network users. The developed techniques in this chapter are also beneficial for the use of collaborative decision and management support systems. The rest of the chapter is organized as follows. The motivation of the chapter and its rationality are introduced in Section 2. We survey the related work in Section 3. We formally defined the (k, ε, l)-anonymity model and investigate its theoretical properties in Section 4. The novel slicing algorithm is presented in Section 5. The extensive experiments are included in Section 6. Finally, we conclude the chapter in Section 7.

2 Motivation

On October 2, 2006, Netflix, the world's largest online DVD rental service, announced the \$1-million Netflix Prize to improve their movie recommendation service [15]. To aid contestants, Netflix publicly released a data set containing 100,480,507 movie ratings, created by 480,189 Netflix subscribers between December 1999 and December 2005. Narayanan and Shmatikov shown in their recent work [24] that an attacker only needs a little information to identify the anonymized movie rating transaction of the individual. They re-identified Netflix movie ratings using the Internet Movie Database (IMDb) as a source of auxiliary information and successfully identified the Netflix records of known users, uncovering their political preferences and other potentially sensitive information.

We consider the privacy risk in publishing anonymous survey rating data. For example, in a life style survey, ratings to some issues are non-sensitive, such as the likeness of book "Harry Potter", movie "Star Wars" and food "Sushi". Ratings to some issues are sensitive, such as the income level and sexuality frequency. Assume that each survey participant is cautious about his/her privacy and does not reveal his/her ratings. However, it is easy to find his/her preferences on non-sensitive issues from publicly available information sources, such as personal weblog or social networks. An attacker can use these preferences to re-identify an individual in the anonymous survey rating data and consequently find sensitive ratings of a victim.

Table 1 (a) A published survey rating data set containing ratings of survey participants on both sensitive and non-sensitive issues. (b) Public comments on some non-sensitive issues of some participants of the survey.

	non-sensitive			sensitive
ID	issue 1	issue 2	issue 3	issue 4
t_1	6	1	*null*	6
t_2	1	6	*null*	1
t_3	2	5	*null*	1
t_4	1	*null*	5	1
t_5	2	*null*	6	5

(a)

	non-sensitive issues		
name	issue 1	issue 2	issue 3
Alice	excellent	so bad	-
Bob	awful	top	-
Jack	bad	-	good

(b)

Based on the public preferences, person's ratings on sensitive issues may be revealed in a supposedly anonymized survey rating data set. An example is given in the Table 1. In a social network, people make comments on various issues, which are not considered sensitive. Some comments can be summarized as in Table 1(b). People rate many issues in a survey. Some issues are non-sensitive while some are sensitive. We assume that people are aware of their privacy and do not reveal their ratings, either non-sensitive or sensitive ones. However, individuals in the anonymoized survey rating data are potentially identifiable based on their public comments from other sources. For example, Alice is at risk of being identified, since the attacker knows Alice's preference on issue 1 is 'excellent', by cross-checking Table 1(a) and (b), s/he will deduce that t_1 in Table 1(a) is linked to Alice, the sensitive rating on issue 4 of Alice will be disclosed. This example motivates us the following research question:

(Satisfaction Problem): Given a large survey rating data set T with the privacy requirements, how to efficiently determine whether T satisfies the given privacy requirements?

Although the satisfaction problem is easy and straightforward to be determined in the relational databases, it is nontrivial in the large survey rating data set. The research of the privacy protection initiated in the relational databases, in which several state-of-art privacy paradigms [32, 23, 22] are proposed and many greedy or heuristic algorithms [12, 19, 20, 30] are developed to enforce the privacy principles. In the relational database, taking k-anonymity as an example [26, 32], it requires each record be identical with at least $k-1$ others with respect to a set of quasi-identifier attributes. Given an integer k and a relational data set T, it is easy to determine if T satisfies k-anonymity requirement since the equality has the transitive property, whenever a transaction a is identical with b, and b is in turn indistinguishable with c, then a is the same as c. With this property, each transaction in T only needs to be check once and the time complexity is at most $O(n^2d)$, where n is the number of transactions in T and d is the size of the quasi-identifier attributes. So the satisfaction problem is trivial in relational data sets. While, the situation is different for the large rating data. First of all, the survey rating data normally does not have a fixed set of personal identifiable attributes as relational data. In addition, the survey rating data is characterized by high dimensionality and sparseness. The lack of a clear set of personal identifiable attributes together with its high dimensionality and sparseness make the determination of satisfaction problem challenging. Second, the defined dissimilarity distance between two transactions (ε-proximate) does not possess the transitive property. When a transaction a is ε-proximate with b, and b is ε-proximate with c, then usually a is not ε-proximate with c. Each transaction in T has to be checked for as many as n times in the extreme case, which makes it highly inefficient to determine the satisfaction problem. It calls for smarter technique to efficiently determine the satisfaction problem before anonymizaing the survey rating data. To our best knowledge, this research is the first touch of the satisfaction of privacy requirements in the survey rating data.

3 Related Work

Privacy preserving data publishing has received considerable attention in recent years, especially in the context of relational data [1, 7, 12, 18, 19, 20, 23, 25, 35, 29]. All these works assume a given set of attributes QID on which an individual is identified, and anonymize data records on the QID. Their main difference consist in the selected privacy model and in various approaches employed to anonymize the data. The author of [1] presents a study on the relationship between the dimensionality of QID and information loss, and concludes that, as the dimensionality of QID increases, information loss increases quickly. Transactional databases present exactly the worst case scenario for existing anonymisation approaches because of high dimension of QID. To our best knowledge, all existing solutions in the context of k-anonymity [26, 27], l-diversity [23] and t-closeness [22] assume a relational table, which typically has a low dimensional QID.

There are few previous work considering the privacy of large rating data. In collaboration with MovieLens recommendation service, [11] correlated public mentions of movies in the MovieLens discussion forum with the users' movie rating histories in the internal MovieLens data set. Recent study reveals a new type of attack on anonymized data for transactional data [24]. Movie rating data supposedly to be anonymized is re-identified by linking non-anonymized data from other source. No solution exists for high dimensional large survey rating databases.

Though we consider data publishing for data mining purposes, we assume that the data publisher has no capability or interests in data mining. Therefore, it is not realistic to expect such data publishers to perform privacy-preserving data mining on behalf of the recipient. In fact, the data may be published on the Internet without a specific recipient. For this reason, techniques for privacy-preserving data mining [2, 3, 9] cannot be applied to data publishing.

Privacy-preservation of transactional data has been acknowledged as an important problem in the data mining literature. There is a family of literature [5, 6] addressing the privacy threats caused by publishing data mining results such as frequent item sets and association rules. Existing works on topic [4, 34] focus on publishing patterns. The patterns are mined from the original data, and the resulting set of rules is sanitized to present privacy breaches. In contrast, our work addresses the privacy threats caused by publishing data for data mining. As discussed above, we do not assume that the data publisher can perform data mining tasks, and we assume that the data must be made available to the recipient. The two scenarios have different assumptions on the capability of the data publisher and the information requirement of the data recipient. The recent work on topic [14, 36] focus on high dimensional transaction data, while our focus is preventing linking individuals to their ratings. Our recent work [28] discusses how to anonymize survey rating data with balanced data utility and privacy.

This chapter is loosely related to the work on anonymizing social networks [8]. A social network is a graph in which a node represents a social entity (e.g., a person) and an edge represents a relationship between the social entities. Although the data is very different from transaction data, the model of attacks is similar to ours: an attacker constructs a small subgraph connected to a target individual and then matches the subgraph to the whole social network, attempting to re-identify the target individual's node, and therefore, other unknown connection to the node. [8] demonstrates the severity of privacy threats in nowadays social networks, but does not provide a solution to prevent such attacks.

4 Problem Formalization

We assume that a survey rating data set publishes people's ratings on a range of issues. In a lifestyle survey, some issues are sensitive, such as income level and sexuality frequency, while some are non-sensitive, such as the likeness of a book, a movie or a kind of food. Each survey participant is cautious about his/her privacy and does not reveal his/her ratings. However, an attacker can use the public available information to identify an individual's sensitive ratings in the supposedly anonymous survey rating data. Our objective is to design effective models to protect privacy of people's sensitive ratings in the published survey rating data.

Given a survey rating data set T, each transaction contains a set of numbers indicating the ratings on some issues. Let $(o_1, o_2, \cdots, o_p, s_1, s_2, \cdots, s_q)$ be a transaction, $o_i \in \{1 : r, null\}$, $i = 1, 2, \cdots, p$ and $s_j \in \{1 : r, null\}$, $j = 1, 2, \cdots, q$, where r is the maximum rating and *null* indicates that a survey participant did not rate. $o_1, \cdots, o_p$ stand for non-sensitive ratings and $s_1, \cdots, s_q$ denote sensitive ratings. Each transaction belongs to a survey participant.

Although each survey participant is wary about their privacy and does not disclose his/her ratings, an attacker may find a victim's preference (not exact rating scores) by personal familiarity or by reading the victim's comments on some issues from personal Weblog or social networks. We consider that attackers know preferences of non-sensitive issues of a victim but do not know exact ratings and want to find out the victim's ratings on some sensitive issues.

4.1 Background Knowledge

The auxiliary information of an attacker includes: (i) the knowledge that a victim is in the survey rating data; (ii) preferences of the victims on some non-sensitive issues. The attacker wants to find ratings on sensitive issues of the victim.

In practice, knowledge of Types (i) and (ii) can be gleaned from an external database [24]. For example, in the context of Table 1(b), an external database may be the IMDb. By examining the anonymous data set in Table 1(a), the adversary can identify a small number of candidate groups that contain the record of the victim. It will be the unfortunate scenario where there is only one record in the candidate group. For example, since t_1 is unique in Table 1(a), Alice is at risk of being identified. If the candidate group contains not only the victim but other records, an adversary may use this group to infer the sensitive value of the victim individual. For example, although it is difficult to identify whether t_2 or t_3 in Table 1(a) belongs to Bob, since both records have the same sensitive value, Bob's private information is identified.

In order to avoid such attack, we propose a two-step protection model. Our first step is to protect individual's identity. In the released data set, every transaction should be "similar" to at least to $(k-1)$ other records based on the non-sensitive ratings so that no survey participants are identifiable. For example, t_1 in Table 1(a) is unique, and based on the preference of Alice in Table 1(b), her sensitive issues can be re-identified in the supposed anonymized data set. Jack's sensitive issues, on the other hand, is much safer. Since t_4 and t_5 in Table 1(a) form a similar group based on their non-sensitive rating.

The second step is to prevent the sensitive rating from being inferred in an anonymized data set. The idea is to require that the sensitive ratings in a similar group should be diverse. For example, although t_2 and t_3 in Table 1(a) form a similar group based on their non-sensitive rating, their sensitive ratings are identical. Therefore, an attacker can immediately infer Bob's preference on the sensitive issue without identifying which transaction belongs to Bob. In contrast, Jack's preference on the sensitive issue is much safer than both Alice and Bob.

4.2 (k,ε,l)-Anonymity

Let $T_A = \{o_{A_1}, o_{A_2}, \cdots, o_{A_p}, s_{A_1}, s_{A_2}, \cdots, s_{A_q}\}$ be the ratings for a survey participant A and $T_B = \{o_{B_1}, o_{B_2}, \cdots, o_{B_p}, s_{B_1}, s_{B_2}, \cdots, s_{B_q}\}$ be the ratings for a participant B. We define the dissimilarity between two non-sensitive ratings as follows.

$$Dis(o_{A_i}, o_{B_i}) = \begin{cases} |o_{A_i} - o_{B_i}| & \text{if } o_{A_i}, o_{B_i} \in \{1:r\} \\ 0 & \text{if } o_{A_i} = o_{B_i} = null \\ r & \text{otherwise} \end{cases} \tag{1}$$

Definition 1 (ε-proximate). *Given a survey rating data set T with a small positive number ε, two transactions T_A, $T_B \in T$, where $T_A = \{o_{A_1}, \cdots, o_{A_p}, s_{A_1}, s_{A_2}, \cdots, s_{A_q}\}$ and $T_B = \{o_{B_1}, o_{B_2}, \cdots, o_{B_p}, s_{B_1}, s_{B_2}, \cdots, s_{B_q}\}$. We say T_A and T_B are ε-proximate, if $\forall\, 1 \le i \le p$, $Dis(o_{A_i}, o_{B_i}) \le \varepsilon$. We say T is ε-proximate, if every two transactions in T are ε-proximate.*

If two transactions are ε-proximate, the dissimilarity between their non-sensitive ratings is bounded by ε. In our running example, suppose $\varepsilon = 1$, ratings 5 and 6 may have no difference in interpretation, so t_4 and t_5 in Table 1(a) are 1-proximate based on their non-sensitive rating. If a group of transactions are in ε-proximate, then the dissimilarity between each pair of their non-sensitive ratings is bounded by ε. For example, if $T = \{t_1, t_2, t_3\}$, then it is easy to verify that T is 5-proximate.

Definition 2 ((k,ε)-anonymity). *A survey rating data set T is said to be (k,ε)-anonymous if every transaction is ε-proximate with at least $(k-1)$ other transactions. The transaction $t \in T$ with all the other transactions that ε-proximate with t form a (k,ε)-anonymous group.*

For instance, there are two (2,5)-anonymous groups in Table 1(a). The first one is formed by $\{t_1, t_2, t_3\}$ and the second one is formed by $\{t_4, t_5\}$. The idea behind this privacy principle is to make each transaction contains non-sensitive attributes are similar with other transactions in order to avoid linking to personal identity. (k,ε)-anonymity well preserves identity privacy. It guarantees that no individual is identifiable with the probability greater than the probability of $1/k$. Both parameters k and ε are intuitive and operable in real-world applications. The parameter ε captures the protection range of each identity, whereas the parameter k is to lower an adversary's chance of beating that protection. The larger the k and the smaller the ε are, the better protection it will provide.

Although (k,ε)-anonymity privacy principle can protect people's identity, it fails to protect individuals' private information. Let us consider one (k,ε)-anonymous group. If the transactions of the group have the same rating on a number of sensitive issues, an attacker can know the preference on the sensitive issues of each individual without knowing which transaction belongs to whom. For example, in Table 1(a), t_2 and t_3 are in a $(2,1)$-anonymous group, but they have the same rating on the sensitive issue, and thus Bob's private information is breaching.

This example illustrates the limitation of the (k,ε)-anonymity model. To mitigate the limitation, we require more diversity of sensitive ratings in the anonymous groups. In the following, we define the distance between two sensitive ratings, which leads to the metric for measuring the diversity of sensitive ratings in the anonymous groups.

First, we define dissimilarity between two sensitive rating scores as follows.

$$Dis(s_{A_i}, s_{B_i}) = \begin{cases} |s_{A_i} - s_{B_i}| & \text{if } s_{A_i}, s_{B_i} \in \{1:r\} \\ r & \text{if } s_{A_i} = s_{B_i} = null \\ r & \text{otherwise} \end{cases} \tag{2}$$

Note that there is only one difference between dissimilarities of sensitive ratings $Dis(s_{A_i}, s_{B_j})$ and dissimilarities of non-sensitive ratings $Dis(o_{A_i}, o_{B_j})$, that is, in the definition of $Dis(o_{A_i}, o_{B_j})$, $null - null = 0$, and for the definition of $Dis(s_{A_i}, s_{B_j})$, $null - null = r$. This is because for sensitive issues, two *null* ratings mean that an attacker will not get information from two survey participants, and hence are good for the diversity of the group.

Next, we introduce the metric to measure the diversity of sensitive ratings. For a sensitive issue s, let the vector of ratings of the group be $[s_1, s_2, \cdots, s_g]$, where $s_i \in \{1 : r, null\}$. The means of the ratings is defined as follows:

$$\bar{s} = \frac{1}{Q}\sum_{i=1}^{g} s_i$$

where Q is the number of non-*null* values, and $s_i \pm null = s_i$. The standard deviation of the rating is then defined as:

$$SD(s) = \sqrt{\frac{1}{g}\sum_{i=1}^{g}(s_i - \bar{s})^2} \tag{3}$$

For instance in Table 1(a), for the sensitive issue 4, the means of the ratings is $(6+1+1+1+5)/5 = 2.8$ and the standard deviation of the rating is 2.23 according to Equation (3).

Definition 3 **((k,ε,l)-anonymity).** *A survey rating data set is said to be (k,ε,l)-anonymous if and only if the standard deviation of ratings for each sensitive issue is at least l in each (k,ε)-anonymous group.*

Still consider Table 1(a) as an example. t_4 and t_5 is 1-proximate with the standard deviation of 2. If we set $k = 2, l = 2$, then this group satisfies (2,1,2)-anonymity requirement. The (k,ε,l)-anonymity requirement allows sufficient diversity of sensitive issues in T, therefore it could prevent the inference from the (k,ε)-anonymous groups to a sensitive issue with a high probability.

4.3 Characteristics of (k,ε,l)-Anonymity

In this section, we investigate the properties of (k,ε,l)-anonymity model.

Definition 4. *Given a subset G of T, $neighbor(t,G)$ is the set of tuples whose non-sensitive values are ε-proximate with t and $|neighbor(t,G)|$ indicates its cardinality. $maxsize(G)$ is the largest size $neighbor(t,G)$ of every $t \in G$. Formally, $maxsize(G) = \max_{\forall t \in G} |neighbor(t,G)|$.*

For example, let T be the data in Table 1(a), consisting of $t_1, \cdots, t_5$, and $G = T$. Assume $\varepsilon = 1$, then $|neighbor(t_1,G)| = \{t_1\}$ since no other transaction in G is 1-proximate with t_1 and $|neighbor(t_1,G)| = 1$. Similarly, $neighbor(t_2,G) = \{t_2,t_3\}$ with $|neighbor(t_2,G)| = 2$ because t_2 and t_3 are 1-proximate with t_1. $maxsize(G) = 2$, because no other transaction $t \in G$ has a $neighbor(t,G)$ higher than 2. $maxsize(G)$ has the following property:

Lemma 1. *Let G_1, G_2 be two partition of G and $G_1 \cup G_2 = G$. Then,*

$$\frac{maxsize(G)}{|G|} \leq max\{\frac{maxsize(G_1)}{|G_1|}, \frac{maxsize(G_2)}{|G_2|}\}$$

Proof: We first show $maxsize(G) \leq maxsize(G_1) + maxsize(G_2)$. Due to symmetry, assume $t \in G_1$, and that $maxsize(G)$ is the size of the neighbor covering set $neighbor(t,G)$ of a tuple $t \in G$. Use S_1 (S_2) to denote the set of tuples in $neighbor(t,G)$ that also belong to G_1 (G_2). Obviously, $neighbor(t,G) = S_1 \cup S_2$ and $S_1 \cap S_2 = \emptyset$. Let t' be the tuple in S_2 with the largest range. Notice that $S_1 \subseteq neighbor(t,G_1)$ and $S_2 \subseteq neighbor(t',G_2)$. Therefore, $maxsize(G) = |S_1| + |S_2| \leq |neighbor(t,G_1)| + |neighbor(t',G_2)| \leq maxsize(G_1) + maxsize(G_2)$.

Given any subset G of T, we define $\alpha(G) = maxsize(G)/|G|$, and $\alpha(G_1)$, $\alpha(G2)$ in the same manner. As $maxsize(G) \leq maxsize(G_1) + maxsize(G_2)$, we have $(|G_1| + |G_2|) \cdot \alpha(G) = |G_1| \cdot \alpha(G_1) + |G2| \cdot \alpha(G2)$, leading to $\frac{|G_1|}{|G_2|} \cdot (\alpha(G) - \alpha(G_1)) + \alpha(G) \leq \alpha(G_2)$. If $\alpha(G) \leq \alpha(G_1)$, lemma holds. If $\alpha(G) \geq \alpha(G_1)$, the term $\frac{|G_1|}{|G_2|} \cdot (\alpha(G) - \alpha(G_1)) > 0$; hence, $\alpha(G) \leq \alpha(G_2)$. No matter in which case, lemma holds. ■

Note that if $G = \cup_{i=1}^{n} G_i$, the result of the lemma can be extended to $\frac{maxsize(G)}{|G|} \leq \max_{i=1}^{n}\{\frac{maxsize(G_i)}{|G_i|}\}$. In our example with $\varepsilon = 5$, $G_1 = \{t_1,t_2,t_3\}$ and $G_2 = \{t_4,t_5\}$. Clearly, $G_1 \cup G_2 = T$. It is easy to verify that $maxsize(G_1) = neighbor(t_2,G_1) = 2$ and $maxsize(G_2) = neighbor(t_4,G_2) = 2$. Hence, $\frac{2}{5} < \max\{\frac{2}{3}, \frac{2}{2}\} = 1$, the inequality in Lemma holds.

Theorem 1. *Given ε and a partition of $T = \cup_{i=1}^{n} G_i$, if T has at least one (k,ε)-anonymity solution, then $k \leq \lceil \frac{maxsize(T) \cdot |G_j|}{|T|} \rceil$, where $\frac{maxsize(G_j)}{|G_j|} = max_{i=1}^{n}\{\frac{maxsize(G_i)}{|G_i|}\}$.*

Proof: Suppose $|neighbor(t,G_j)| = maxsizeG_j$ and $k > \lceil \frac{maxsize(G) \cdot |G_j|}{|T|} \rceil$. If T has a (k,ε)-anonymous solution, then the possibility of t being identified is at least $\frac{1}{neighbor(t,G_j)}$, which is greater than $\frac{|T|}{maxsize(T) \cdot |G_j|}$ due to the fact that $\frac{maxsize(T)}{|T|} \leq \frac{maxsize(G_j)}{|G_j|}$. With our assumption, we get that the possibility of t being identified is greater than $\frac{1}{k}$, which contradicts with the fact that T has a (k,ε)-anonymous solution. ■

Theorem 1 provides a sufficient condition for the existence of a (k,ε)-anonymity solution. In our running example with $\varepsilon = 1$, we already know that $maxsize(G) = 2$, then according to Theorem 1, if a (k,ε)-anonymity exists, then $k \leq \lceil \frac{2 \times 3}{5} \rceil = 2$.

Lemma 2. *Given $S = \{s_1, s_2, \cdots, s_n\}$ as the sensitive ratings of T. Let S_1 and S_2 be two partitions of S and $S_1 \cup S_2 = S$. Then,*

$$SD(S) \geq min\{SD(S_1), SD(S_2)\}$$

Proof: Without loss of generality, suppose $S_1 = \{s_1, s_2, \cdots, s_k\}$ and $S_2 = \{s_{k+1}, \cdots, s_n\}$ and $SD(S_1) \leq SD(S_2)$. $\bar{s} = \frac{\sum_{i=1}^{n} s_i}{n}$, $\bar{s_1} = \frac{\sum_{i=1}^{k} s_i}{n}$ and $\bar{s_2} = \frac{\sum_{i=k+1}^{n} s_i}{n}$. Next, we show that $SD(S) > SD(S_1)$.

$$SD^2(S) - SD^2(S_1) = \frac{\sum_{i=1}^{n}(x_i - \bar{x})^2}{n} - \frac{\sum_{i=1}^{k}(x_i - \bar{x_1})^2}{k}$$

$$= \frac{1}{nk}(k\sum_{i=1}^{n}(x_i - \bar{x})^2 - n\sum_{i=1}^{k}(x_i - \bar{x_1})^2)$$

$$= \frac{1}{nk}(k\sum_{i=1}^{n}(x_i - \bar{x})^2 - k\sum_{i=1}^{k}(x_i - \bar{x_1})^2 - (n-k)\sum_{i=1}^{k}(x_i - \bar{x_1})^2)$$

Since $SD(S_1) \le SD(S_2)$, $\frac{\sum_{i=1}^{k}(x_i - \bar{x_1})^2}{k} \le \frac{\sum_{i=1}^{n-k}(x_i - \bar{x_2})^2}{n-k}$

$$\ge \frac{1}{nk}(k\sum_{i=1}^{n}(x_i - \bar{x})^2 - k\sum_{i=1}^{k}(x_i - \bar{x_1})^2 - k\sum_{i=k+1}^{n}(x_i - \bar{x_2})^2) \tag{4}$$

$$= \frac{1}{n}(\sum_{i=1}^{n}(x_i - \bar{x})^2 - \sum_{i=1}^{k}(x_i - \bar{x_1})^2 - \sum_{i=k+1}^{n}(x_i - \bar{x_2})^2)$$

$$= \frac{1}{n}(\sum_{i=1}^{k}((x_i - \bar{x})^2 - (x_i - \bar{x_1})^2) + \sum_{i=k+1}^{n}((x_i - \bar{x})^2 - (x_i - \bar{x_2})^2)$$

Since $k\bar{x_1} = \sum_{i=1}^{k} x_i$ and $(n-k)\bar{x_2} = \sum_{i=k+1}^{n} x_i$, then

$$= \frac{1}{n}(k(\bar{x_1} - \bar{x})^2 + (n-k)(\bar{x_2} - \bar{x})^2) \ge 0$$

Therefore, the lemma holds. ∎

Note that if $S = \cup_{i=1}^{n} S_i$, the result of the lemma can be extended to $SD(S) \ge \min_{i=1}^{n}\{SD(S_i)\}$. In our example with $\varepsilon = 5$, the ratings of the sensitive issue 4 $S = \{6,1,1,1,5\}$ are divided into two groups $S_1 = \{6,1,1\}$ and $S_2 = \{1,5\}$. It is easy to verify that $SD(S) = 2.23$, $SD(S_1) = 2.35$ and $SD(S_2) = 2$. Therefore, $SD(S) > \min\{SD(S_1), SD(S_2)\}$, the inequality in Lemma holds.

Corollary 1. *Let S be the ratings of the sensitive issue of T, and be divided into n groups, $S_1, \cdots, S_n$. If $\forall i$, $SD(S_i) \ge l_0$. Then, $SD(S) \ge l_0$.*

The following theorem gives the upper bound of the parameter l in the (k, ε, l)-anonymity model.

Theorem 2. *Let S be the set of ratings of the sensitive issue of T. Suppose S_min and S_max be the minimum and maximum ratings in S, then the maximum standard deviation of S is $\frac{(S\_max - S\_min)}{2}$.*

Proof: For the ease of description, we write S_min as a and S_max as b, we only need to prove the following inequality holds with $(a \le c \le b)$:

$$\sqrt{\frac{(a - \frac{a+b+c}{3})^2 + (b - \frac{a+b+c}{3})^2 + (c - \frac{a+b+c}{3})^2}{3}} \le \frac{(b-a)}{2} \tag{5}$$

Let $f(c)$ be written as:

$$f(c) = \frac{(a - \frac{a+b+c}{3})^2 + (b - \frac{a+b+c}{3})^2 + (c - \frac{a+b+c}{3})^2}{3}$$

The graph of $f(c)$ is a parabola, and after simplifying the function, the axis of symmetry is $c = \frac{a+b}{2}$, and since $f'(x) = 6 > 0$ and $a \leq \frac{a+b}{2} \leq b$, the function has the minimum value $\frac{(b-a)^2}{6}$, then

$$\frac{(b-a)^2}{6} \leq f(c) \leq min\{f(a), f(b)\}$$

because $f(a) = f(b) = \frac{6(b-a)^2}{27}$, then

$$\frac{(b-a)^2}{6} \leq f(c) \leq \frac{6(b-a)^2}{27}$$

Due to the fact that $\frac{6(b-a)^2}{27} < \frac{(b-a)^2}{4}$, then Equation (5) holds. The proof of Theorem 2 completes. ■

5 Satisfying Privacy Requirements

In this section, we formulate the satisfaction problem and develop a slicing technique based on the properties discussed in Section 4.3 to determine the following *Satisfaction Problem.*

Problem 1 (Satisfaction Problem). Given a survey rating data set T and privacy requirements k, ε, l, the satisfaction problem of (k, ε, l)-anonymity is to decide whether T satisfies the k, ε, l privacy requirements.

The satisfaction problem is to determine whether the user's given privacy requirement is satisfied by the given data set. It is a very important step before anonymizing the survey rating data. If the data set has already met the requirements, it is not necessary to make any modifications before publishing. As follows, we propose a novel slice technique to solve the satisfaction problem.

5.1 Satisfaction Algorithms

Recall that we are given a survey rating data set consisting of a set of transactions $T = \{t_1, t_2, \cdots, t_n\}$, $|T| = n$. Each transaction $t_i \in T$ contains issues from an issue set $I = \{i_1, i_2, \cdots, i_m\}$, $|I| = m$. Consider that both n (the number of survey participants) and m (the number of issues) may be very large. For example, a million of users rate thousands of movies. The efficient identification of the violation to privacy requirement is nontrivial. Firstly, the dissimilarity matrix is very big if we try to compute all pairwise distances. The time complexity is $O(n^2m)$. Secondly, the data matrix may not fit in the memory. An algorithm needs to read data from disk frequently.

We plan to utilize the sparseness of the survey rating data set to speed up the algorithm. The data set is very spare if we consider *null* values as empty. Here, we define a binary flag matrix F to record if there is a rating or not for each issue (column).

$$F_{ij} = \begin{cases} 1 \text{ if } i_j \in t_i \\ 0 \text{ if } i_j \notin t_i \end{cases}$$

For instance, the flag matrix associated with Table 1(a) is:

$$\mathbf{F} = \begin{pmatrix} 1\ 1\ 0 \\ 1\ 1\ 0 \\ 1\ 1\ 0 \\ 1\ 0\ 1 \\ 1\ 0\ 1 \end{pmatrix} \tag{6}$$

in which, each row corresponds to survey participants and each column corresponds to non-sensitive issues. If we want to find the transactions that are ε-proximate with t_1, intuitively, we need not to compute the dissimilarity between t_1 and t_4, and between t_1 and t_5 since both t_4 and t_5 do not rate issue 2. Based on the sparseness property, it could significant reduce the amount of the pairwise dissimilarity computation.

Definition 5 (Hamming Distance). *[17] Hamming distance between two vectors in the flag matrix of equal length is the number of positions for which the corresponding symbols are different. We denote the Hamming distance between two vectors v_1 and v_2 as $H(v_1,v_2)$.*

In other words, Hamming distance measures the minimum number of substitutions required to change one into the other, or the number of errors that transformed one vector into the other. For example, if $v_1 = (1,1,0)$ and $v_2 = (1,0,1)$, then $H(v_1,v_2) = 2$. If the Hamming distance between two vectors is zero, then these two vectors are identical.

Definition 6 (Hamming Group). *Hamming group is the set of vectors, in which the Hamming distance between any two vectors of the flag matrix is zero. The maximal Hamming group is a Hamming group that is not a subset of any other Hamming group.*

For example, there are two maximal Hamming groups in the flag matrix (6), which are made of vectors $\{(1,1,0),(1,1,0),(1,1,0)\}$ and $\{(1,0,1),(1,0,1)\}$ and they are actually groups of $\{t_1,t_2,t_3\}$ and $\{t_4,t_5\}$ in T.

Now we focus on the how to group T in order to fulfill the privacy requirement. As we has explained in the previous example that the first three transactions form a maximal Hamming group and the last two transactions form the other one, which inspires us for the idea of the first step of the algorithm. It works as follows: firstly, we

Table 2 Sample rating data

	non-sensitive			sensitive
ID	issue 1	issue 2	issue 3	issue 4
t_1	3	6	*null*	6
t_2	2	5	*null*	1
t_3	4	7	*null*	4
t_4	5	6	*null*	1
t_5	1	*null*	5	1
t_6	2	*null*	6	5

find out all the maximal Hamming groups, namely $H_1, \cdots, H_k$. For each Hamming group H_i, $1 \leq i \leq k$, we test for the privacy requirement. In our running example, if given $\varepsilon = 5$, the two maximal Hamming groups made of $\{t_1, t_2, t_3\}$ and $\{t_4, t_5\}$ are already satisfying with the privacy requirement. However, if having a look at Table 2, the flag matrix of which is

$$\mathbf{F}' = \begin{pmatrix} 1 & 1 & 0 \\ 1 & 1 & 0 \\ 1 & 1 & 0 \\ 1 & 1 & 0 \\ 1 & 0 & 1 \\ 1 & 0 & 1 \end{pmatrix} \tag{7}$$

The maximal Hamming groups are $H_1 = \{t_1, t_2, t_3, t_4\}$ and $H_2 = \{t_5, t_6\}$. If given $\varepsilon = 1$, H_2 has already met the requirement, but H_1 does not. In this case, smarter technique is required to further process the group H_1. Here, we adopt a greedy slicing technique to solve challenge.

5.2 Search by Slicing

Our slicing algorithm is based on the projection search paradigm first used by Friedman [10]. Friedman's simple technique works as follows. In the preprocessing step, d dimensional training points are ordered in d different ways by individually sorting each of their coordinates. Each of the d sorted coordinates arrays can be thought of as a 1-D axis with the entire d dimensional space projected onto it. Given a point q, the nearest neighbor is found as follows. A small ε is subtracted from and added to each of q's coordinates to obtain two values. Two binary search searches are performed on each of the sorted arrays to locate the positions of both values. An axis with the minimum number of points in between the position is chosen. Finally, points in between the positions on the chosen axis are exhaustively searched to obtain the closest point. The complexity is $O(nd\varepsilon)$ and is clearly inefficient in high d.

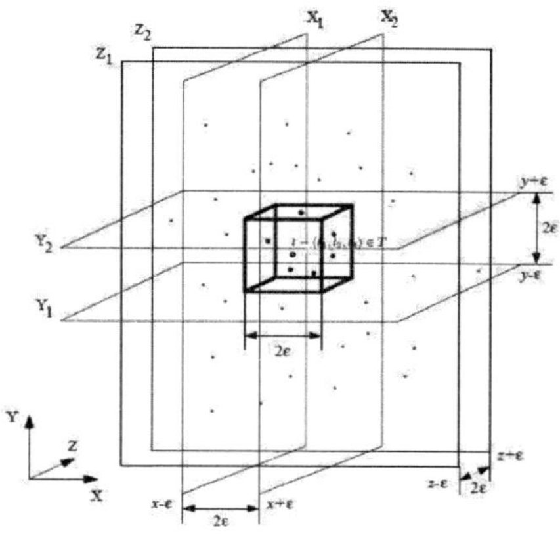

Fig. 1 The slicing technique finds a set of transactions C_t inside a cube of size 2ε within the ε-proximate of t. The ε-proximate of the set C_t can then be found by an exhaustive search in the cube.

5.2.1 To Determine k and l When Given ε

Our slicing technique is proposed to efficiently search for the neighbor within distance ε in high dimension. As we shall see, the complexity of the proposed algorithm grows very slowly with dimension for small ε. We illustrate the proposed slicing technique using a simple example in 3-D space, as shown in Figure 1. Given $t = (t_1, t_2, t_3) \in T$, our goal is to slice out a set of transactions T $(t \in T)$ that are ε-proximate. Our approach is first to find the ε-proximate of t, which is the set of transactions that lie inside a cube C_t of side 2ε centered at t. Since ε is typically small, the number of points inside the cube is also small. The ε-proximate of C_t' can then be found by an exhaustive comparison within the ε-proximate of t. If there are no transactions inside the cube C_t, we know that the ε-proximate of t is empty, so as the ε-proximate of the set C_t'.

The transactions within the cube can be found as follows. First we find the transactions that are sandwiched between a pair of parallel planes X_1, X_2 (See Figure 1) and add them to a *candidate set*. The planes are perpendicular to the first axis of coordinate frame and are located on either side of the transaction t at a distance of ε. Next, we trim the candidate set by disregarding transactions that are not also sandwiched between the parallel pair of Y_1 and Y_2, that are perpendicular to X_1 and X_2, again located on either side of t at a distance of ε. This procedure is repeated for Z_1 and Z_2 at the end of which, the candidate set contains only transactions within the cube of size 2ε centered at t. $Slicing(\varepsilon, T, t_0)$ (**Algorithm 1**) describes how to find the ε-proximate of the set C_{t_0} with $t_0 \in C_{t_0}$.

Since the number of transactions in the final ε-proximate is typically small, the cost of the exhaustive comparison is negligible. The major computational cost in the slicing process occurs therefore in constructing and trimming the candidate set.

Suppose the set C_t' ($t \in C_t'$) is finally ε-proximate. We repeat the process for another transaction on the set $T \setminus C_t'$. Finally, there comes to two situations. One is that all transactions are grouped into anonymous groups with each group having at least two transactions. The other situation is that for some $t' \in T$ there is no ε-proximate for it, in this case, we let t' form an (k, ε)-anonymous group by itself.

```
ALGORITHM 1: Slicing(ε, T, t0)(P)
Candidate ← {t0}; S ← ∅
/* To slice out the cube, ε-proximate of t0 */
for j ← 1 to n
do if |tj − t0| < ε
      then Candidate ← Candidate ∪ {tj}
           S ← S ∪ {j}
/* To trim the ε-proximate of t0 */
PCandidate ← Candidate
for i ← 1 to |S|
do for j ← 1 to |S|
    do if |tS(i) − tS(j)| > ε
          then PCandidate ← PCandidate \ {tS(i)}
return PCandidate
```

We use the sample rating data in Table 2 to illustrate how the slicing algorithm works. If we want to find a (k, ε)-anonymity solution with $\varepsilon = 1$. The first step is to slice out the transactions that are ε-proximate with the first transaction t_1, and we use C_t to denote the set of transactions, where $C_t = \{t_1, t_2, t_3\}$. The next step is to trim C_t to make it ε-proximate, and the method is to verify if the distance between any two elements in C_t is bounded by ε. In this example, dissimilarity between t_2 and t_3 is greater than ε, then we take one out of C_t (we choose t_3 here), and after that, we could obtain the new set $C_t' = C_t \setminus \{t_3\} = \{t_1, t_2\}$, which is already ε-proximate. Repeat this process on $T' = T \setminus C_t'$, and finally we can find one $(2, 1)$-anonymity solution consisting of three anonymous groups $\{\{t_1, t_2\}, \{t_3, t_4\}, \{t_5, t_6\}\}$. Further, if we consider sensitive issues, actually, there is enough diversity in each (k, ε)-anonymous group with $l = 1.5$. So for this example, it satisfies $(2, 1, 1.5)$-anonymity requirement.

Further, if we partition T into $\{G_1, G_2\}$, where $G_1 = \{t_1, t_2, t_3, t_4\}$ and $G_2 = \{t_5, t_6\}$, we get $maxsize(T) = 3$ and $maxsize(G_1) = 3$ with $\varepsilon = 1$. So according to Theorem 1, $k \leq \lceil \frac{maxsize(T) \cdot |G_1|}{|T|} \rceil$, which is $\frac{3 \times 4}{6} = 2$. This example also verifies Theorem 1.

5.2.2 To Determine ε and l When Given k

In this section, we discuss the situation when k is known, and how to find out a solution that satisfies (k, ε, l)-anonymity principle with ε as smaller as possible.

To solve this problem, we combine the slicing technique and binary search in our algorithm.

Binary search is a technique for locating a particular value in a sorted list of values. It makes progressively better guesses, and closes in on the sought value by selecting the middle element in the span (which, because the list is in sorted order, is the median value), comparing its value to the target value, and determining if the selected value is greater than, less than, or equal to the target value. A guess that turns out to be too high becomes the new upper bound of the span, and a guess that is too low becomes the new lower bound. Pursuing this strategy iteratively, it narrows the search by a factor of two each time, and finds the target value or else determines that it is not in the list at all.

Our algorithm starts from the upper bound $\varepsilon = r$ (r is the maximum rating in T) and begins with transaction $t_1 \in T$, at the initial stage, all transactions fall into one (k,ε)-anonymous group. We further our search by setting ε to $\frac{r}{2}$, which is a middle element between 0 and r. For this new ε, we need to find out all transactions that are $\frac{r}{2}$-proximate by running slicing technique discussed before. Our objective is to determine whether or not the set of transactions that is $\frac{r}{2}$-proximate neighborhood has the capacity greater than the given k. If yes, we set new upper bound to $\frac{r}{2}$ and search among the interval $[0, \frac{r}{2}]$. Continue this process for interval $[0, \frac{r}{2}]$ with middle element $\frac{r}{4}$. Else, we set the new lower bound to $\frac{r}{2}$ and continue searching in $[\frac{r}{2}, r]$ with middle element $\frac{3r}{4}$. Repeat this until reaching the *termination condition*. We terminate searching if for the interval [upper bound, lower bound], $|\text{upper bound} - \text{lower bound}| < 1$. Finally, ε returns to the unique integer in the interval [upper bound, lower bound].

Consider our running example with $k = 2$. We begin with $\varepsilon = 6$ and return to an anonymous solution with all transactions in one group. Next we try $\varepsilon = 3$ and the interval [0,6] is partitioned into [0,3] and [3,6]. By using the slicing algorithm, it returns that there is a set of transactions which is 3-proximate, and its capacity is less than 2. Then, we move to the interval [3,6] and try $\varepsilon = 4.5$, the ε is still not large enough. We finish the search until we get that ε is in the interval [4.5, 5.25], and since $|5.25 - 4.5| < 1$, the search terminates and ε returns to 5. Finally we can find one $(2,5,2)$-anonymous solution consisting of two anonymous groups $\{\{t_1,t_2,t_3\},\{t_4,t_5\}\}$.

5.2.3 To Determine k and ε When Given l

In this section, we discuss the situation when l is given, and how to find a solution satisfying (k,ε,l)-anonymity principle with ε as small as possible. Let S be the ratings of the sensitive issue of T, and $SD(S) = l_0$ be the standard deviation computed by Equation (3).

Case 1: When $l > l_0$. In this case, suppose there exists one solution that satisfies both principles. We let T be divided into n groups, and in each group, the similarity of any two transactions are bounded by ε, and the number of transactions in each group is at least k, and the standard deviation of the sensitive ratings in each group is at least l. According to Corollary 1, the standard deviation of the sensitive ratings of T $SD(S)$ is at least l as well, which makes $SD(S) > l_0$, and this is a contradiction. Hence, if $l > l_0$, there is no required solution.

Case 2: When $l \leq l_0$. The algorithm starts from $\varepsilon = r$, and at this initial stage, all transactions fall into one (k,ε,l)-anonymous group. Next, we continue our search by setting ε to $\frac{r}{2}$, which is a middle element between 0 and r. For this new ε, we need to verify if the standard deviation of the sensitive ratings in each group formed by this new ε is at least l. If yes, we set new upper bound to $\frac{r}{2}$ and search among the interval [0, $\frac{r}{2}$] and continue to test for the middle element $\frac{r}{4}$. Else, we set the new lower bound to $\frac{r}{2}$ and continue searching in [$\frac{r}{2}$, r] by testing the middle element $\frac{3r}{4}$. Repeat this until reaching the *termination condition*. We terminate searching if there exists an ε in the interval [upper bound, lower bound] with |upper bound − lower bound| < 1 and the sensitive ratings in each group formed by this ε is at least l. Finally, ε returns to the unique integer in the interval [upper bound, lower bound].

Consider the example in Table 2 with $l = 2$. The standard deviation of the sensitive ratings of T is 2.1. Since $l < 2.1$, then there exists a solution that meets the privacy principle. We begin with $\varepsilon = 6$, which returns to a solution containing all transactions in one group. Obviously, it meets both principles. Next we try $\varepsilon = 3$ and the interval [0,6] is partitioned into [0,3] and [3,6]. The (k,ε)-anonymous groups formed when $\varepsilon = 3$ are $\{t_1,t_2,t_3,t_4\}$ and $\{t_5,t_6\}$. We further verify the standard deviation of sensitive ratings in both group, and both are greater than 2. It means when $\varepsilon = 3$, there exists a solution that satisfies (2,3,2)-anonymity. In order to find the solution with smallest ε, we continue our search in the interval [0,3] and try the middle value $\varepsilon = 1.5$. It returns to three groups $\{t_1,t_2\}$, $\{t_3,t_4\}$ and $\{t_5,t_6\}$, however, the standard deviation of the sensitive ratings of the second group is $1.5 < l$. Next, we continue for search in [1.5, 3] and still could not meet the (k,ε,l)-anonymity requirement. We finish the search until we get that ε is in the interval [2.375, 3], and since $|3-2.375| < 1$, the search terminates and ε returns to 3. Finally we can find one solution that meets (2,3,2)-anonymity principle, and it consists of two anonymous groups $\{t_1,t_2,t_3,t_4\}$ and $\{t_5,t_6\}$.

6 Experimental Study

In this section, we experimentally evaluate the effectiveness and efficiency of the proposed algorithms for the satisfaction problem.

6.1 Data Sets

Our experimentation deploys two real-world databases. Netflix and MovieLens data sets. MovieLens data set `http://www.grouplens.org/taxonomy/term/14` was made available by the GroupLens Research Project at the University of Minnesota. The data set contains 100,000 ratings (5-star scale), 943 users and 1682 movies. Each user has rated at least 20 movies. Netflix data set `http://`

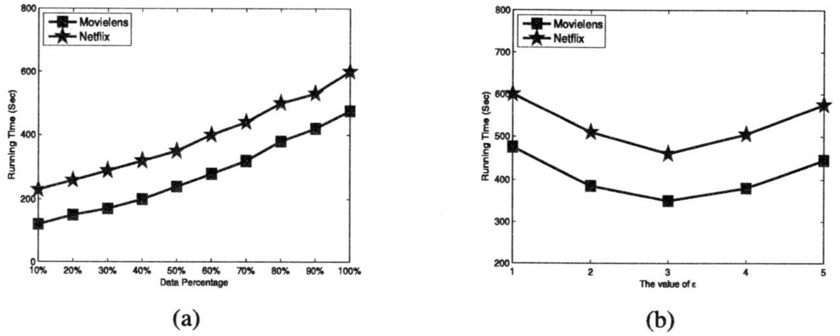

Fig. 2 Running time comparison on Movielens and Netflix data sets vs. (a) Data percentage varies (b) ε varies

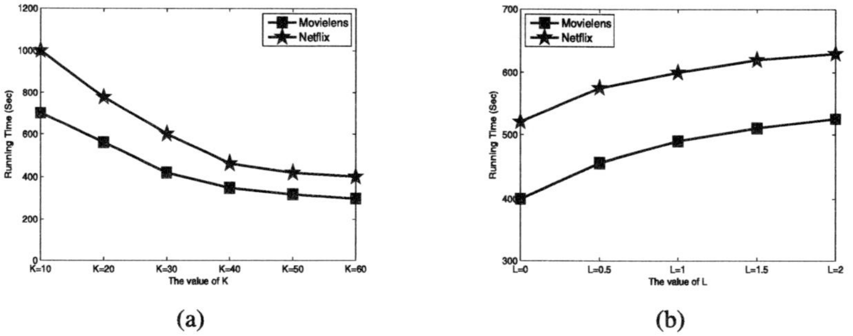

Fig. 3 Running time comparison on Movielens and Netflix data sets vs. (c) k varies (d) L varies

`www.netflixprize.com` was released by Netflix for competition. The movie rating files contain over 100,480,507 ratings from 480,189 randomly-chosen, anonymous Netflix customers over 17 thousand movie titles. The data were collected between October, 1998 and December, 2005 and reflect the distribution of all ratings received during this period. The ratings are on a scale from 1 to 5 (integral) stars. In both data sets, a user is considered as an object while a movie is regarded as an attribute and many entries are empty since a user only rated a small number of movies. Except for rating movies, users' ratings some simple demographic information (e.g., age range) are also included. In our experiments, we treat the users' ratings on movies as non-sensitive issues and ratings on others as sensitive ones.

6.2 Efficiency

Data used for Figure 2(a) is generated by re-sampling the Movielens and Netflix data sets while varying the percentage of data from 10% to 100%. For both data sets, we evaluate the running time for the (k, ε, l)-anonymity model with default setting $k =$

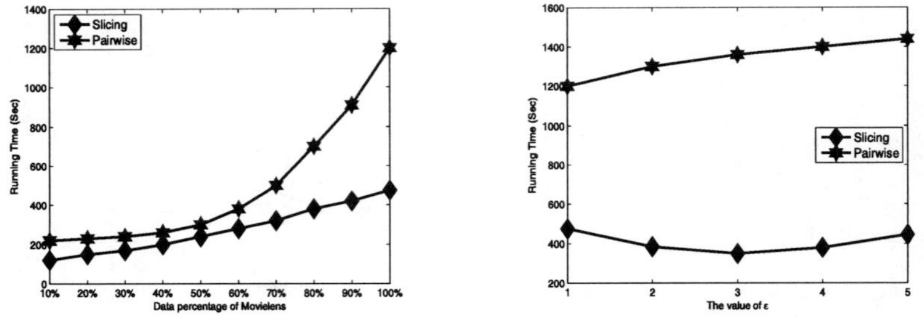

(a) (b)

Fig. 4 Running time comparison of Slicing and Pairwise methods on Movielens data set vs. (a) Data percentage varies (b) ε varies

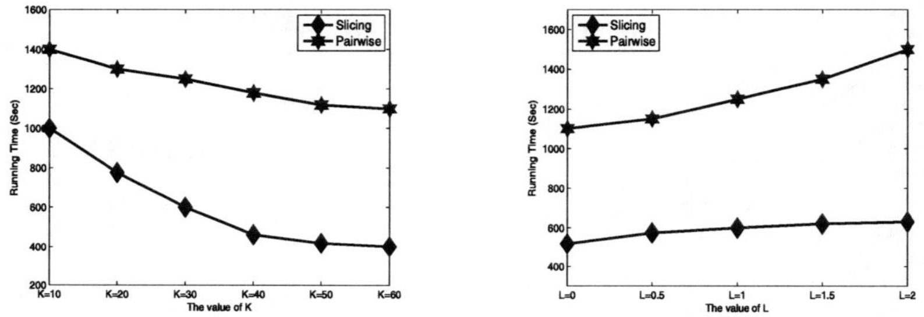

(c) (d)

Fig. 5 Running time comparison of Slicing and Pairwise methods on Netflix data set vs. (c) k varies (d) L varies

$20, \varepsilon = 1, l = 2$. For both testing data sets, the execution time for (k, ε, l)-anonymity is increasing with the increased data percentage. This is because as the percentage of data increases, the computation cost increases too. The result is expected since the overhead is increased with the more dimensions.

Next, we evaluate how the parameters affect the cost of computing. Data set used for this sets of experiments are the whole sets of MovieLens and Netflix data and we evaluate by varying the value of ε, k and l. With $k = 20, l = 2$, Figure 2(b) shows the computational cost as a function of ε, in determining (k, ε, l)-anonymity requirement of both data sets. Interestingly, in both data sets, as ε increases, the cost initially becomes lower but then increases monotonically. This phenomenon is due to a pair of contradicting factors that push up and down the running time, respectively. At the initial stage, when ε is small, more computation efforts are put into finding ε-proximate of the transaction, but less used in exhaustive search for proper ε-proximate neighborhood, and this explains the initial decent of overall cost. On the other hand, as ε grows, there are fewer possible ε-proximate neighborhoods, thus reducing the searching time for this part, but the number of transactions in the

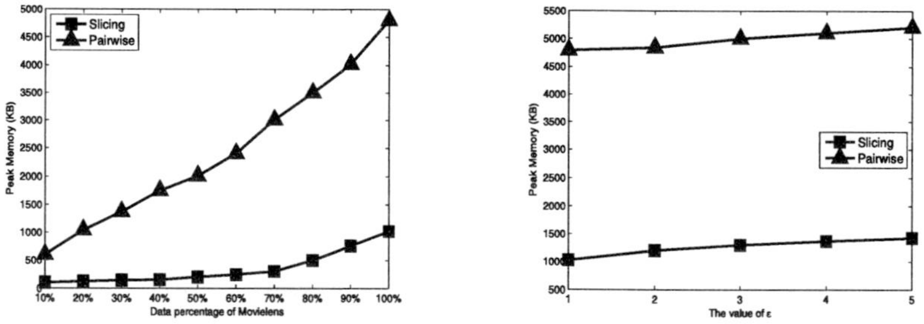

(a) (b)

Fig. 6 Space Complexity comparison of Slicing and Pairwise methods on Movielens data set vs. (a) Data percentage varies (b) ε varies

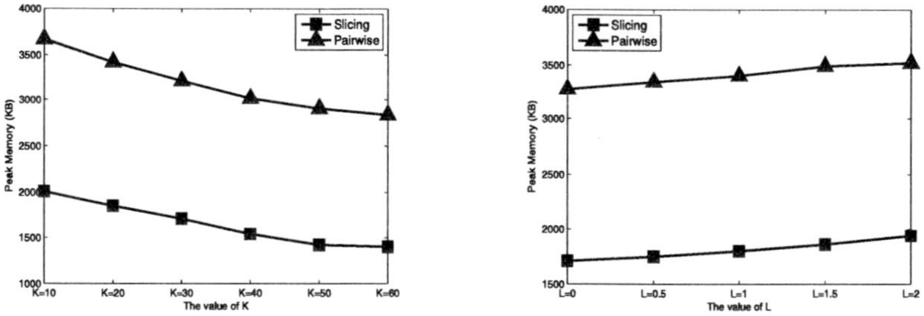

(c) (d)

Fig. 7 Space Complexity comparison of Slicing and Pairwise methods on Netflix data set vs. (c) k varies (d) L varies

ε-proximate neighborhood is increased, which results in huge exhaustive search for proper ε-proximate neighborhood and this causes the eventual cost increase. Setting $\varepsilon = 2$, Figure 3(a) displays the results of running time by varying k from 10 to 60 for both data sets. The cost drops as k grows. This is expected, because fewer search efforts for proper ε-proximate neighborhoods needed for a greater k, allowing our algorithm to terminate earlier. We also run the experiment by varying the parameter l and the results are shown in Figure 3(b). Since the rating of both data sets are between 1 and 5, then according to Theorem 2, 2 is already the largest possible l. When $l = 0$, there is no diversity requirement among the sensitive issues, and the (k, ε, l)-anonymity model is reduced to (k, ε)-anonymity model. As we can see, the running time increases with l, because more computation is needed in order to enforce stronger privacy control.

In addition to show the scalability and efficiency of the slicing algorithm itself, we also experimented the comparison between the slicing algorithm (Slicing) and the heuristic pairwise algorithm (Pairwise), which works by computing all the pairwise distance to construct the dissimilarity matrix and identify the violation of the privacy requirements. We implemented both algorithms and studied the impact of the execution time on the data percentage, the value of ε, the value of K and the value of L.

Figure 4 plots the running time of both slicing and pairwise algorithms on the Movielens data set. Figure 4(a) describe the trend of the algorithms by varying the percentage of the data set. From the graph we can see, the slicing algorithm is far more efficient than the heuristic pairwise algorithm especially when the volume of the data becomes larger. This is because, when the dimension of the data increases, the disadvantage of the heuristic pairwise algorithm, which is to compute all the dissimilarity distance, dominates the most of the execution time. On the other hand, the smarter grouping technique used in the slicing process makes less computation cost for the slicing algorithm. The similar trend is shown in Figure 4(b) by varying the value of ε, in which the slicing algorithm is almost 3 times faster than the the heuristic pairwise algorithm. The running time comparisons of both algorithms in Netflix data set by varying the value of K and L are shown in Figure 5(a) and (b). Even on a larger data set, the slicing algorithm outperformed the pairwise algorithm, and the running time of Slicing is quick enough to be used in practical.

6.3 Space Complexity

In addition to evaluate the efficiency of the proposed slicing technique, we also investigate the storage overheads of the algorithms. We adopt the peak memory to measure the storage overheads, which indicates the maximum memory used during the implementation.

Figure 6 shows the space complexity comparison of the slicing method and the pairwise approach on the Movielens data set by varying the percentage of the data and the value of ε. In both cases, the slicing algorithm takes less peak memory than the pairwise method, this is expected, since the pairwise approach computes all the possible distances and use them for identifying the validation of the privacy requirement, which takes much more space to store the dissimilarity matrix. We conduct the experiments by varying the value of K and L on a larger Netflix data set, and plot the storage overheads in Figure 7. The graph shows that the slicing algorithm need almost two times less memory than the heuristic pairwise approach.

6.4 Summary

Overall, in the experiential studies, we have adopted two real-life data sets to demonstrate the effectiveness and efficiency of our proposed approach. We have shown that our proposed slicing technique is fast, scalable and space efficient compared with traditional pair-wise approach. The slicing approach developed in this chapter is fast enough to be applied in practical.

7 Conclusion and Future Work

We proposed a novel (k, ε, l)-anonymity privacy principle for protecting privacy in such survey rating data. We studied the satisfaction problem, which is to decide whether a survey rating data set satisfies the privacy requirements given by the user. A fast slicing technique was proposed to solve the satisfaction problem by searching closest neighbors in large, sparse and high dimensional survey rating data. The experimental results show that the slicing technique is fast and scalable in practical.

This work also initiates the future investigations of approaches on anonymizing the survey rating data. Traditional approaches on anonymizing no matter relational data sets or transactional data set are by generalization or suppression, and the published data set has the same number of data but with some fields being modified to meet the privacy requirements. As shown in the literatures, this kind of anonymization problem is normally NP-hard, and several algorithms are devised along this framework to minimize the certain pre-defined cost metrics. Inspired by the research in this chapter, the satisfaction problem can be further used to develop a different method to anonymizing the data set. The idea is straightforward with the result of the satisfaction problem. If the rating data set has already satisfies the privacy requirement, it is not necessary to do any anonymization to publish it. Otherwise, we anonymize the data set by deleting some of the records to make it meet the privacy requirement. The criteria during the deletion can be various (for example, to minimize the number of deleted records) to make it as much as useful in the data mining or other research purposes. We believe that this new anonymization method is flexible in the choice of privacy parameters and efficient in the execution with the practical usage.

References

1. Aggarwal, C.: On *k*-Anonymity and the curse of dimensionality. In: Proceedings of the 31st International Conference on Very Large Data Bases, pp. 901–909 (2005)
2. Agrawal, R., Srikant, R.: Privacy-Preserving Data Mining. In: Proceedings of the 2000 ACM SIGMOD Conference on Management of Data, pp. 439–450 (2000)
3. Agrawal, D., Aggarwal, C.C.: On The Design and Qualification of Privacy Preserving Data Mining Algorithm. In: Proc. Symosium on Principles of Database Systems (PODS), pp. 247–255 (2001)
4. Atzori, M., Bonchi, F., Giannotti, F., Pedreschi, D.: Anonymity preserving pattern discovery. The International Journal on Very Large Data Bases 17(4), 703–727 (2008)
5. Atzori, M., Bonchi, F., Giannotti, F., Pedreschi, D.: Blocking anonymity threats raised by frequent itemset mining. In: Fifth IEEE International Conference on Data Mining, pp. 27–30 (2005)
6. Atzori, M., Bonchi, F., Giannotti, F., Pedreschi, D.: *k*-anonymous patterns. In: 9th European Conference on Principles and Practice of Knowledge Discovery in Databases, pp. 10–21 (2005)
7. Bayardo, R.J., Agrawal, R.: Data privacy through optimal *k*-anonymisation. In: Proceedings of 21st International Conference on Data Engineering, pp. 217–228 (2005)

8. Backstrom, L., Dwork, C., Kleinberg, J.: Wherefore Art Thou R3579x?: Anonymized Social Networks, Hidden Patterns, and Structural Steganography. In: International World Wide Web Conference, pp. 181–190 (2007)
9. Evfimievski, R., Srikant, R., Agrawal, R., Gehrke, J.: Privacy preserving mining of association rules. In: Proceedings of the eighth ACM SIGKDD international conference on Knowledge discovery and data mining, pp. 217–228 (2002)
10. Friedman, J.K., Bentley, J.L., Finkel, R.A.: An algorithm for finding best matches in logarithmic expected time, ACM Trans. on Math. Software 3, 209–226 (1977)
11. Frankowski, D., Cosley, D., Sen, S., Terveen, L.G., Riedl, J.: You are what you say: privacy risks of public mentions. In: Proceedings of the 29th Annual International ACM SIGIR Conference on Research and Development in Information Retrieval, pp. 565–572 (2006)
12. Fung, B.C., Wang, K., Yu, P.S.: Top-down specialization for information and privacy preservation. In: Proceedings of the 21st International Conference on Data Engineering, pp. 205–216 (2005)
13. Garey, M.R., Johnson, D.S.: Computers and Intractability: A Guide to the Theory of $\mathcal{NP}$-Completeness. Freeman, New York (1979)
14. Ghinita, G., Tao, Y., Kalnis, P.: On the Anonymisation of Sparse High-Dimensional Data. In: Proceedings of International Conference on Data Engineering (ICDE), pp. 715–724 (2008)
15. Hafner, K.: And if you liked the movie, a Netflix contest may reward you handsomely. New York Times, October 2 (2006)
16. Hansell, S.: AOL removes search data on vast group of web users. New York Times, August 8 (2006)
17. Hamming, R.W.: Coding and Information Theory. Prentice Hall, Englewood Cliffs (1980)
18. Iyengar, V.: Transforming data to satisfy privacy constraints. In: Proceedings of the Eighth ACM SIGKDD International Conference on Knowledge Discovery and Data Mining, pp. 279–288 (2002)
19. LeFevre, K., DeWitt, D., Ramakrishnan, R.: Incognito: efficient full-domain k-anonymity. In: Proceedings of the 2005 ACM SIGMOD International Conference on Management of Data, pp. 49–60 (2005)
20. LeFevre, K., DeWitt, D., Ramakrishnan, R.: Mondrian multidimensional k-anonymity. In: Proceedings of the 22nd International Conference on Data Engineering, p. 25 (2006)
21. Li, J., Tao, Y., Xiao, X.: Preservation of Proximity Privacy in Publishing Numerical Sensitive Data. In: ACM Conference on Management of Data (SIGMOD), pp. 473–486 (2008)
22. Li, N., Li, T., Venkatasubramanian, S.: t-Closeness: Privacy Beyond k-anonymity and l-diversity. In: Proceedings of International Conference on Data Engineering (ICDE), pp. 106–115 (2007)
23. Machanavajjhala, A., Gehrke, J., Kifer, D., Venkitasubramaniam, M.: l-Diversity: Privacy beyond k-anonymity. In: 22nd International Conference on Data Engineering, p. 22 (2006)
24. Narayanan, A., Shmatikov, V.: Robust De-anonymisation of Large Sparse Datasets. In: IEEE Symposium on In Security and Privacy, pp. 111–125 (2008)
25. Samarati, P., Sweeney, L.: Generalizing data to provide anonymity when disclosing Information. In: Proceedings of the Seventeenth ACM SIGACT-SIGMOD-SIGART Symposium on Principles of Database Systems, p. 188 (1998)
26. Samarati, P., Sweeney, L.: Protecting privacy when disclosing information: k-anonymity and its enforcement through generalization and suppression. Technical Report SRI-CSL-98-04, SRI Computer Science Laboratory (1998)

27. Samarati, P.: Protecting respondents' identities in microdata release. IEEE Transactions on Knowledge and Data Engineering 13(6), 1010–1027 (2001)
28. Sun, X., Wang, H., Li, J., Pei, J.: Publishing Anonymous Survey Rating Data. Data Mining and Knowledge Discovery. Springer, Heidelberg (2010) (accepted for publication)
29. Sun, X., Wang, H., Sun, L.: Extended k-Anonymity Models Against Sensitive Attribute Disclosure. Computer Communication. Elsevier, Amsterdam (2010) (accepted for publication)
30. Sun, X., Wang, H., Li, J.: Injecting purposes and trust into data anonymization. In: Proceeding of the 18th ACM Conference on Information and knowledge Management, pp. 1541–1544 (2009)
31. Sweeney, L.: Weaving technology and policy together to maintain confidentiality. J. of Law, Medicine and Ethics 25(2-3) (1997)
32. Sweeney, L.: *k*-Anonymity: A Model for Protecting Privacy. International Journal on Uncertainty Fuzziness Knowledge-based Systems 10(5), 557–570 (2002)
33. Traian, T.M., Bindu, V.: Privacy Protection: *p*-sensitive *k*-anonymity Property. In: Proceedings of the 22nd International Conference on Data Engineering Workshops, p. 94 (2006)
34. Verykios, V.S., Elmagarmid, A.K., Bertino, E., Dasseni, E., Saygin, Y.: Association Rule Hiding. IEEE Transactions on Knowledge and Data Engineering 16(4), 434–447 (2004)
35. Xiao, X., Tao, Y.: Anatomy: simple and effective privacy preservation. In: Proceedings of the 32nd International Conference on Very Large Data Bases, pp. 139–150 (2006)
36. Xu, Y., Wang, K., Fu, A.W.-C., Yu, P.S.: Anonymizing Transaction Databases for Publication. In: Proceeding of the 14th ACM SIGKDD International Conference on Knowledge Discovery and Data Mining, pp. 767–775 (2008)

Part IV
Future Trends and Concepts

Chapter 18
Next Generation Service Delivery Network as Enabler of Applicable Intelligence in Decision and Management Support Systems

Migration Strategies, Planning Methodologies, Architectural Design Principles

N. Kryvinska, C. Strauss, and P. Zinterhof

Abstract. The crucial challenge that the communication industry has been wrestling with is the issue of how its underlying technologies should evolve and be used to help service providers remain competitive for many years in an environment marked by increased rivalry and deregulation. The Next Generation Network (NGN), with its decomposed architecture, can take full advantage of sophisticated technologies both to offer new services that will increase service providers' revenues and reduce their operating costs. A strategy for evolving smoothly from modern networks to this new network structure is essential in order to minimize the required investment during the transition phase. However, any steps that are taken during this transition must make it easier for the networks to evolve ultimately to the NGN. Thus, the migration strategy proposed and explored in this chapter enables the development of a capable concept of how the structuring of networks must be changed, and in doing so takes into consideration the business needs of diverse service providers and network operators.

1 Introduction

Next Generation Networks (NGN) is a concept that brings together the collection of changes taking place in the way networks are structured. It represents a direction

N. Kryvinska · C. Strauss
Department of eBusiness, School of Business, Economics and Statistics,
University of Vienna, Vienna, Austria
e-mail: {natalia.kryvinska,christine.strauss}@univie.ac.at

P. Zinterhof
Department of Computer Sciences, University of Salzburg, Salzburg, Austria
e-mail: peter.zinterhof@sbg.ac.at

N. Bessis, F. Xhafa (Eds.): Next Generation Data Technologies for CCI, SCI 352, pp. 473–502.
springerlink.com

for the industry to follow, with the speed of its deployment depending very much on the business needs of different organizations.

Flexibility is the most important principle of NGN: the flexibility needed by established operators to adapt their networks to the changing marketplace, the flexibility new operators need to set up viable and profitable businesses, and the flexibility to provide business users with fixed and mobile services that will enhance the way they work and residential users with a whole raft of leisure services.

The next important principle of NGN is that it should be cost-effective for both established operators to migrate to it and for emerging operators to deploy it from scratch. In both cases, day-to-day operating costs should be lower than they currently are.

A further important aspect of NGN is its recognition of the pronounced need among end users for an ever greater variety of new services and applications (including multimedia), the majority of which were not even envisaged when modern networks were established. From the operators' viewpoint, transport no longer provides sufficient profits; in the future, operators will need to offer end users an extensive range of useful and easy-to-use services in order to remain competitive. Consequently, the NGN must be service-driven, providing all the means needed to offer new services and customize existing ones in order to generate future revenue.

The evolution towards NGN is now increasingly possible, because the principles of service creation platforms and the separation of service logic are ready to be extended to NGN. The cost-effective enabling technologies that can make NGN a reality are now commercially available. In addition, the dynamics of the market are putting pressure on operators to react to flat or declining revenues and margins in voice services. Consequently, the operators are seeking new opportunities to adapt their networks so that they can find new sources of profit. Most operators want an evolutionary strategy that builds on the strengths of their existing networks. However, some are considering making a complete break with the past and moving rapidly to NGN architecture. New, competitive operators trying to break into the market want to deploy an NGN structure from the outset.
Admittedly, NGN is based on complexity. However, in modern global networks, numerous generations of switches coexist, circuit and packet-switched networks operate alongside one another, and fixed and mobile networks interwork together uneasily. So, from this viewpoint, NGN not only looks less complex than modern networks, but also offers considerable savings in operating costs [1 ÷ 3].

The chapter structure is as follows: First, we evaluate migration schemes to the Next Generation Network. The specific guidelines for evolving smoothly from modern networks to the new network structure are essential in order to minimize the required investment during the transition phase, while taking early advantage of the NGN architecture's qualities. Next, we explore and develop an evolutional framework and modeling process for the NGN. Being able to compare competing technologies and understand the interrelationships and dependencies between complementary technologies is extremely important for designing and modeling the framework. For this reason, the process of advanced technology modeling is broken down into three parts (e.g., applicability of the technology, abstractions, and perspectives). Then, we discuss such challenges as management and control in NGN. The call processing and services development environments define the intelligence, flexibility, and interoperability that drive services, reliability, and thus also the revenues. Afterward, we examine NGN domains. NGN has both similarities to and differences with PSTN as well as with the Internet. But, NGN requires a clear separation of functions and domains, with a maximum degree of reuse built into the architecture and its components. The chapter continues by characterizing NGN services and addressing NGN's application model. We finalize our work by developing a business model for service engineering in NGN.

2 Migration Schemes to the Next Generation Network

Developing specific guidelines for evolving smoothly from modern networks to the new network structure is essential in order to minimize the required investment during the transition phase, while taking early advantage of the qualities of NGN architecture [4].

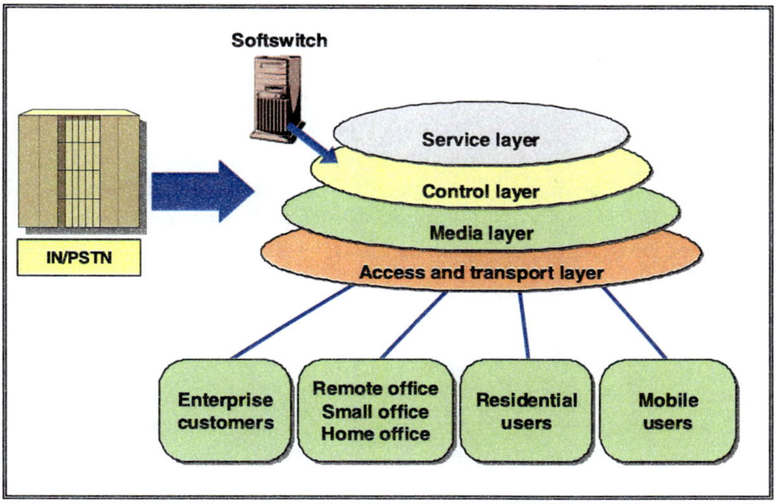

Fig. 1 Decomposed NGN architecture.

2.1 Decomposed NGN Architecture

The NGN architecture, as shown in Fig. 1, decomposes the monolithic blocks of traditional switches into individual network layers, which interwork via standard open interfaces. The basic call processing intelligence is essentially decoupled from the switching matrix hardware and now resides in a separate device, called a softswitch. This device, which is also known as a media gateway controller or call agent, acts as the controlling element in the new architecture. Open interfaces towards new application servers facilitate rapid service provisioning and ensure short time-to-market.

At the media layer, special media servers implement a variety of functions, such as the provision of dialing tone or announcements. The media servers' more advanced functions include interactive voice response and text-to-speech or speech-to-text conversion [5, 6].

2.2 Advantages of New Technologies

Open interfaces at each network layer enable a network operator to select the best vendor for each layer. Packet-based transport allows flexible bandwidth dimensioning, thus simplifying the management of network structures. Having fewer but more powerful call control entities in the network makes the upgrading of the software in the nodes that control the network more efficient, thereby reducing operating expenses.

Going beyond the technological issues, deregulation also has a considerable influence on an operator's mode of functioning. Through a process known as "local-loop unbundling", government regulators around the world are forcing incumbent operators to open their doors to rival companies. Once inside the exchange, these alternative carriers should be able to compete for local customers by taking direct control over the "last mile" of copper. This development is leading to increased competition between incumbent operators working outside their traditional regions and new network operators, all of whom want to win the most valuable customers, namely those with the highest volumes of spending on telecommunication services. NGNs are well suited to support the network architectures and business models enabled by deregulation [4, 6, 7].

3 Evolutional Framework and Modeling Process towards NGN

An architectural evolution involves both technical and technology management issues. At a technical level, complex problems are apparent in interfacing equipment, interworking between protocols and adaptation between architectures. At a technological level, network operators face increasingly complex issues when assessing emerging technologies for providing advanced services. Being able to compare competing technologies and understand the interrelationships and dependencies between complementary technologies is extremely important for designing and modeling the framework.

For this reason, the process of advanced technology modeling is broken down into three parts, which may be applied sequentially or in isolation. The first part considers the applicability of the technology, the second considers different abstractions of the technologies, while the third part evaluates the technology using perspectives [8, 9].

3.1 Applicability

This part of the modeling process considers a set of factors that determine whether the technology should be deployed. The applicability constraints of a particular model organize the status of a technology and specific factors to facilitate comparison and representation. Some of the identified applicability constraints include:

- *timeliness* - the time taken for technologies to mature and stabilize needs to be evaluated and considered in migration and evolution strategies;
- *installed network base* - the applicability of a technology is dependent on the existing network infrastructure and its capabilities. The installed network places constraints on future technologies and introduces interoperability requirements;
- *financial* - the cost of technologies in relation to possible returns will determine whether or not they acceptable;
- *marketing* - technologies are only applicable if there is a demand for the services that are offered [8, 10].

3.2 The Abstraction

This aspect of modeling involves different conceptual abstractions of the same technology or network. Abstractions provide a reference to compare architectures:

- *functional* - represents an abstract view of the functional relationships and interfaces without regard for physical implementation;
- *physical* - represents an abstract view of the elements or building blocks and their connections and protocols;
- *implementation view* - views the complete application of the technology with physical, connection, dimensioning, and geographic information [8, 11].

3.3 Perspectives

This element of the modeling process organizes the characteristics of a technology for consideration on an abstraction. The perspectives are related to the information represented in an abstraction. For example, the signaling perspective isolates technologies responsible for signaling from technologies performing other functions. With regard to the specific case of advanced service provision, some of the identified perspectives include:

- *management* - represents management requirements and is sub-divided into network and service management concerns;
- *service control* - contains entities that perform intelligent processing and database access;

- *call control* - represents all entities responsible for end-to-end network provision;
- *protocols* – is concerned with representing the protocol stack. The entities are mapped to reasonable representations of the OSI software model;
- *signaling* - has a relationship to the management, service control and call perspectives; between each of these perspectives, reference points are established across which protocol and signaling consistency must be maintained;
- *network* - represents all entities responsible for end-to-end network provision [8, 12].

3.4 Methodology for Evaluation and Planning

A methodology for planning the evolution and strategy of a network or technology is a requirement when evaluating high technology. This methodology is captured and represented as a process, which is depicted in Fig. 2.

Using this methodology, tone can perform an evaluation of specific circumstances, as there are a number of considerations that may significantly influence a technology's feasibility. The initial steps involve the identification and definition of the technical capabilities that a technological implementation will deliver. Separately, but in parallel, a set of generic business benefits are generated. Carefully selected technical application cases can then be developed and analyzed to determine how a particular scenario may further accentuate a potential business benefit. As the technical scenarios take place in an existing environment, an investigation must be undertaken of both market conditions/considerations and of the

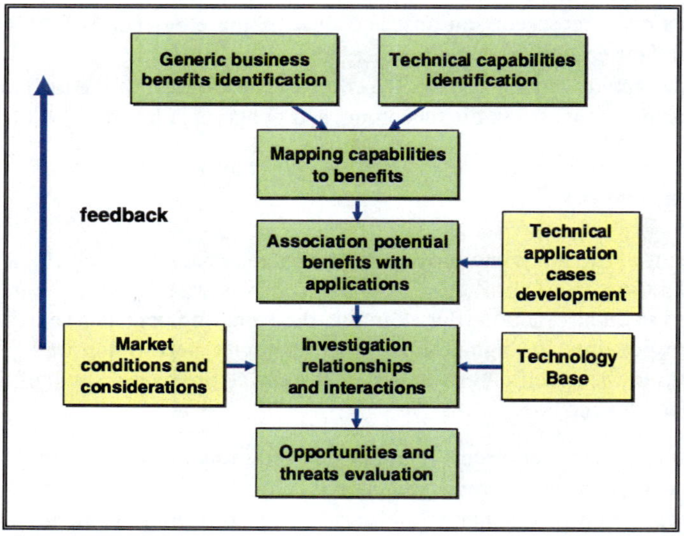

Fig. 2 Evaluation process methodology.

available technology base. These stages provide information with which relationships and interactions can be assessed. Having completed this assessment, the final stage of the process evaluates the opportunities and threats associated with each scenario. The feedback arrow indicates that the process is iterative. For example, further development of technical scenarios could provide more information on technical capabilities [8, 11, 12].

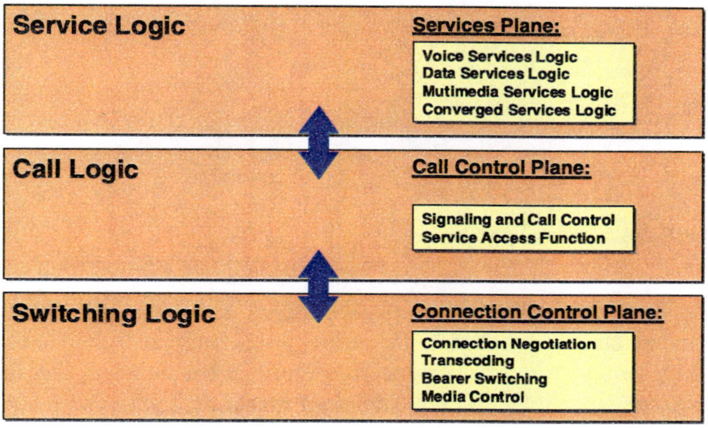

Fig. 3 Generic Open Call Control Architecture.

4 Management/Control in NGN and Network Intelligence for Features Functionality

The call processing and services development environments define the intelligence, flexibility, and interoperability that drive services, reliability, and thus the revenues.

The NGN underlying packet-switching hardware is independent of the call control logic. Likewise, the call control logic is highly flexible and provides open interfaces that enable the development of services (Fig. 3).

Call control logic and its application programming interfaces (APIs) have now become flexible enough to support services that transcend voice telephony and encompass data, unified messaging, and multi-media services [13, 14].

4.1 Network Intelligence Features

In circuit-switched telephone networks, the intelligence for feature functionality was provided by network switches and other servers. The evolution of functionality in such networks has been slow. On the other hand, IP (Internet Protocol) is based on intelligent endpoints that can participate in application and network layer protocols. Intelligent endpoints have the potential to enable tremendous innovation in the types of features and functionality available to the user. This potential is especially compelling when the endpoint integrates many different services (Fig. 4).

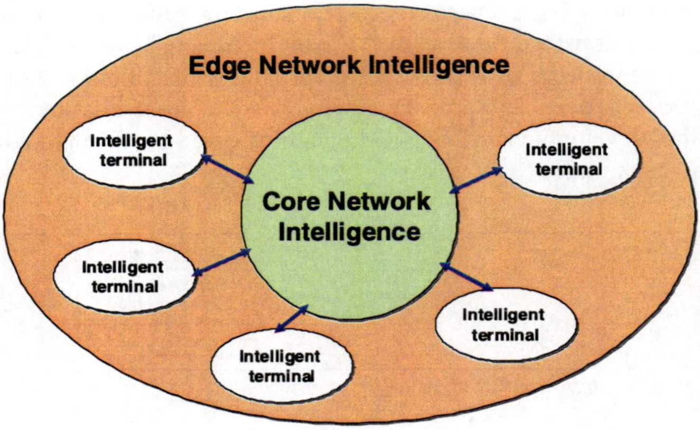

Fig. 4 Distributed Network Intelligence.

The move to push intelligence to the boundaries of the network, up to the level of the terminal, started even before the arrival of the Internet. Telephone sets were enhanced by adding memory to store frequently dialed numbers and to support features such as last number redial. Advanced functions, such as an integrated answering machine or fax, were added later and the sets of functionality developed further to even include such features as a Web browser, email, etc. However, this evolution does not necessarily imply that all intelligence has to be removed from the network core. To the contrary: many new features can only be realized through a combination of intelligence in the terminal and in the network.

A similar tendency can be observed in the IT world. Although PCs were originally designed to operate in a standalone mode, now most of them are connected to a network in which intelligence is distributed between the PCs and different servers. In this case, intelligence has not been removed entirely from the network: many functions still reside on the network servers. The following serve as examples: security servers (firewalls, admission control, etc), fileservers, version management of application software on the clients (e.g. automatic and remotely controlled upgrading of applications such as browsers, virus scanners, etc), and so on. The key to this evolution lies in minimum or zero administration using the power of the clients, under the control of the network operator, while minimizing configuration and administration costs.

The examples presented above show that service intelligence in the terminal should not be thought of as separate applications, but rather as an integral part of a distributed service platform that is largely driven by functions in the network.

With the rapid evolution of service and network technologies, new services may also require upgrades and extensions to the underlying communication protocols. The core logic of the intelligent terminal has to take care of the adaptation to the underlying platform and network environment, so that the service

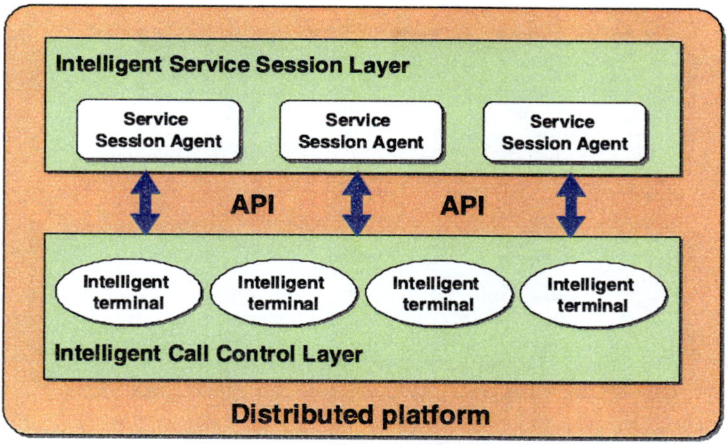

Fig. 5 Next Generation Network Intelligence Architecture.

provider perceives a uniform service platform on which new service logic can be deployed. The intelligent terminal, being a part of a distributed platform (Fig. 5), has to offer clear application programming interfaces (API) to add new units.

The requirements for extending service intelligence from the network to the terminal are:

- control of communication functions on the terminal by service provider;
- intelligent terminal adaptation to the characteristics of the terminal on which it runs, as well as to its network environment;
- flexibility to plug in or upgrade application components and communication protocols;
- intelligent terminal placed in a platform for value-added services and applications;
- definition of APIs for communications between intelligent terminal and the network [13, 15, 16].

4.2 Signaling Architecture

The signaling architecture for a communication system is essential for providing service mechanisms and primitives, as well as for achieving system scalability and robustness. The signaling architecture includes a signaling protocol that sets up, releases, and controls communication sessions, as well as the set of necessary system components on the control path (Fig. 6).

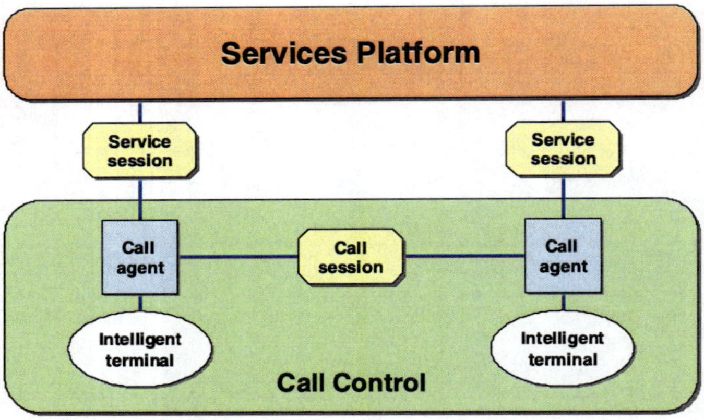

Fig. 6 Distributed Signaling Architecture for Next Generation Network.

The system components and their functions and properties constitute a signaling architecture. The signaling architecture's configuration and capabilities directly affects the services the communication system can support. And, it is driven by the following types of services:

- *communication services of any configuration and any terminal* - any-to-any communication refers to the ability to support communication between all types of devices effectively.
- *customized communication services* - they allow end users to customize their communication service (for example, when they want to be called, on what device, under what condition, and by whom).
- *communication services based on user activity* - this type of service generalizes the location-based services that have appeared in many other systems. Instead of customizing the communication service based on the current user location, the current user behavior can be tracked and used for customization.
- *personal mobility services* - personal mobility means treating people, rather than devices, as communication endpoints [13, 17, 18].

5 Separation of Functions and Domains in NGN

As implied by Table 1, NGN has both similarities with and differences to PSTN/IN, as well as with the Internet. But, NGN requires a clear separation of functions and domains, with a maximum degree of reuse built into the architecture and its components. The three major domains constitute the NGN:

- *service domain* - encompasses the service-related aspects of data and logic, and provides coherent end-to-end functionality to the customer;

- *transport domain* - provides the connectivity requested by the service domain with the required QoS and within specified policy constraints;
- *distributed processing environment (DPE) domain* - provides a ubiquitous middleware infrastructure for the distributed components of the service and transport domains to communicate with each other.

These domains are illustrated in Fig. 7. The two types of customers (private and commercial) are each depicted as having their own internal networks, with gateways connecting them to the service provider's infrastructure. Business networks could also contain servers, which can be thought of as third-party servers. All four types of relationships (e.g., residence-residence, residence-business, business-residence, and business- business) need to be supported, with the service provider directly or indirectly supplying and supporting components of all three domains, as well as the application content servers, to enable NGN services on a ubiquitous unified infrastructure [19, 20].

Table 1 The Attributes of PSTN/IN, Internet, and NGN.

	PSTN/IN	Internet	NGN
Multimedia services	No	Yes	Yes
QoS-enabled	Yes (voice)	No	Yes
Network intelligence	Yes	No	Yes
Intelligent CPE	No	Yes	Yes
Underlying transport network	TDM	Packet	Packet
Service architecture	Semi-distinct	Ad hoc	Distinct
Integrated control and management	No	Yes	Yes
Service reliability	High	Low	High
Service creation	Complex	Ad-hoc	Systematic
Ease of use of services	Medium	High	High
Evolvability/modularity	Low	Medium	High
Time to market of services	Long	Short	Short
Architecture openness	Low	High	High

5.1 Service Domain Provides Open Software Platform

The service architecture of NGN seeks to provide a universal open software platform on which a large variety of services can be architected in a data-centric multi-provider environment. In broad terms, a service is defined as a software application that provides a useful and well-defined functionality to the user. A service can be realized by a set of interacting software components distributed across multiple computing elements.

The user of the service can be any human or machine entity. One can classify services into a few major categories in order to gain insights on the architectural issues that impact service domain requirements in NGN:

- The highest classes of services offered by the NGN are interactive communications services. These include real-time communications services involving multiple parties interacting in real-time and using multiple media. Multiple qualities or grades of these services must be supported. Because of their real-time performance and dynamic control requirements, these services are likely to become the most complex set of services NGN has to offer its customers.
- The second major class of services can be broadly labeled as information/data services. These services may be thought of as the evolution of modern Internet services: browsing, information retrieval, online directories, scheduled information delivery, e-commerce, advertising, and others. They also include a rich set of remote access and control services, such as telemetry, monitoring, security, network management, and other data-oriented services. The evolution of this class of services includes not only the rudimentary versions that exist today, but also major improvements to enhance response time and reliability, and provide advanced billing, QoS, and policy enforcement options.
- The third class of services that next generation service providers inevitably need to enable is delivery of content. Typically, such content delivery is for purposes of entertainment and/or education. These services can be offered on-demand, nearly on-demand, on a broadcast or multicast basis, or on a deferred delivery basis for use at a later time. The various flavors of on-demand and/or multicast services (video on demand, high-quality music on demand, etc.) can pose interesting technical challenges from the point of view of scalability. The NGN's service domain architecture must address these challenges and provide an efficient and economical infrastructure to support them.
- Finally, another class of services that has to do with management of other services. These services may or may not have revenue-generating potential by themselves, but they are as useful and necessary as others. This class includes services such as subscription, customer provisioning, customer network management, and customer service management. The dominant mode of accessing these services is likely to be a Web-based interface. Other services in this class include configuration management, performance monitoring, billing and accounting management, service security management, policy enforcement, and similar services. The users of some of these services may be internal customers (e.g., IT personnel).

According the service categories mentioned above, the notion of a session becomes fundamental to NGN's service domain architecture. Three distinguishable types of service sessions exist. An access session is the starting point of interaction

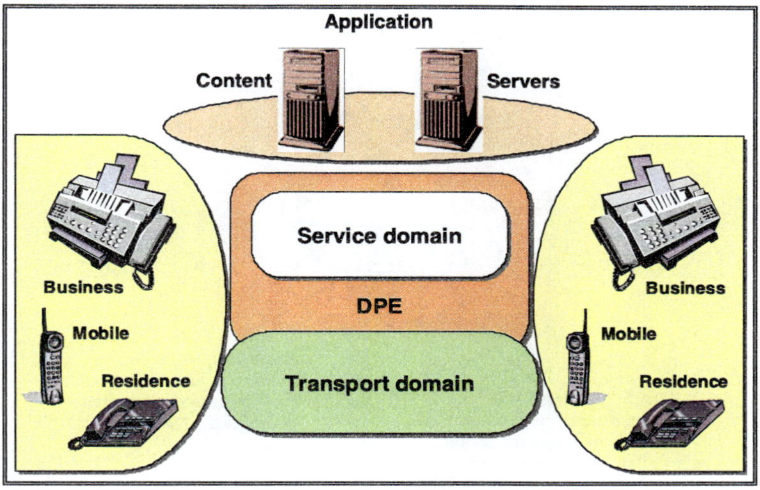

Fig. 7 NGN domains.

between the user of a service and the service provider. Authentication and subscription related functions would take place within the access session. A usage session is the actual context in which service execution takes place and constitutes the utilization phase of the service. A communication session provides an abstract view by the service domain of connection-related resources necessary for the delivery of a particular service. A communication session is typically paired with a usage session. The actual connection-related resources are provided by the transport domain through a transport session [21, 22].

5.2 Transport Domain Supplies Connectivity for Service Domain

The transport domain provides the connectivity requested by the service domain by using an underlying packet-based transport infrastructure. Such connectivity is characterized by communication associations that meet the requirements of QoS, scalability, traffic engineering, reliability, and evolvability. The transport layer provides all functions generally attributed to the lower four layers of the open systems interconnections (OSI) model.

A communication session in the service domain maps to a transport session in the transport domain. The service domain can initiate two types of transport sessions. A default transport session gives a customer access to the network. Activities such as Web-browsing, notification, and sending and receiving email can occur in such a session, and do not require the service domain to establish a new session specific to such activities. An enhanced transport session would go beyond the default transport session to support some combination of QoS guarantees, policy, and billing. It is within an enhanced transport session, mapped from a communication session, that the service domain can exert control over establishing and managing connections and/or associations [17, 23, 34].

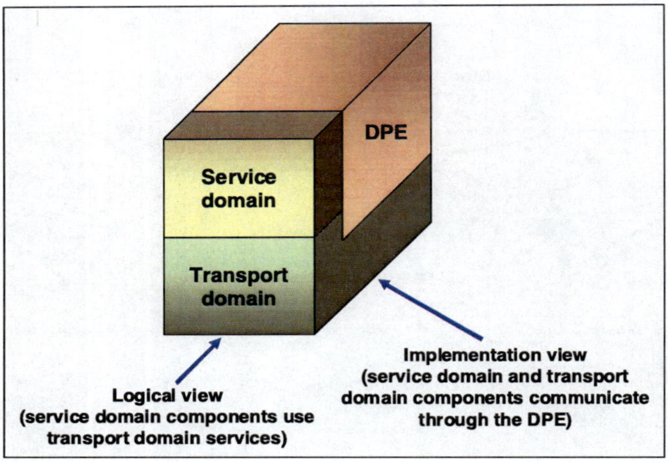

Fig. 8 The relations between domains in NGN

5.3 Distributed Processing Environment as Software Framework for Development and Deployment of Distributed Applications

The Distributed Processing Environment (DPE) provides a software infrastructure to support the development and deployment of distributed applications in the NGN. The DPE enables service components and transport components to interact with each other and relieves them of having to deal with and solve the difficult problems of distribution and communication on a per-service basis. Components communicate with other components as though all components were resident on the same local host. Thus, service developers are by and large relieved of explicit concerns about details and complexities associated with distributed processing, communication protocol design, concurrency control, fail-over strategy, platform redundancy and reliability, and so on.

Fig. 8 shows both the logical and the implementation views of the relationship between various domains in NGN [13, 24, 25].

6 NGN Applications

Next generation networks bring consistency between the traditional call applications and the information world thanks to the use of a unique IP-based transport plane and decoupling between the transport, control and application layers. Next generation applications can be distinct, falling into the following categories:

- *connectivity applications* - reflect peer-to-peer relations, which are more widely defined than a call since they include real-time multimedia, voice, video and data aspects.

- *information applications* - include mail capabilities, as well as streaming, presence, instant messaging, geographic location, virtual office, communities' capabilities and behaviors, and even business-oriented processes.
- *interactions* - applications that can range from simple, such as the click-to-dial services which launch a connectivity application from a data process (e.g. yellow pages or presence) or the sending of a multimedia message, to complex interactions, such as content delivery processes that can be a mix of both worlds, depending on the type of content (e.g. push or pull messages, streaming with or without request for a connectivity application) or call center processes [26, 27].

6.1 Next Generation Application Model

A definition of next generation applications can only be reliable if it is based on a general model, which must ensure an appropriate balance between the network-centric and terminal-centric concepts. This definition is based on two main assumptions:

- Applications must be consistent from the user's point of view. Even if they are provided by various suppliers and third parties, they must be of carrier grade. For example, registration and authentication must be the same for all applications; similarly, a common community definition must apply to all relevant applications.
- The operator must act as a broker between subscribers and application providers. This brokerage must benefit the applications by providing them with a full set of capabilities. For example, adding a presence capability or a location capability to an application can make it richer and more user-friendly than a standalone application.

The operator is the owner of at least a minimum core set of capabilities to meet the above requirements. This core set must be modular enough not to restrain innovative or differentiating applications. In addition, it must include all the capabilities needed to ensure consistency from the viewpoints of the subscribers and the operator's business model.

Finally, the operator must be able to develop rapidly data and connectivity applications that take advantage of the wide range of capabilities, as well as protocols and underlying data world technologies and services. Many of the services evolve continually and therefore have to be enhanced, field tested, refined, etc. The application creation and run-time environments are thus of prime importance to the operator. The technology, openness and ability to evolve are major success factors [26 ÷ 28].

The diversity of applications and the above model lead to the following main architectural trends:

- acceptance and behavior of applications with regard to the subscriber profile and the consistency between interactions is a key component of the service control node.

- subscribers data is organized in dedicated servers.
- basic connectivity service, known as the serving call state control function, is part of the service control; enhanced connectivity applications are supported by this service node or other application servers [26, 29].

6.2 Application Service-Enabling Platform

The future networks open new opportunities for providing advanced value-added services and applications and have led to the development of the concept of a so-called application service enabling platform (Fig. 9). Based on this concept, the added value in broadband networks is provided at the network edge, reaching out to the customer's premises, and assumes the ubiquity of IP. The challenge is to position broadband access networks as key components in the new service era, the major inputs and drivers for which are coming from the application service providers (ASP) and content providers that are deemed to provide value over broadband networks.

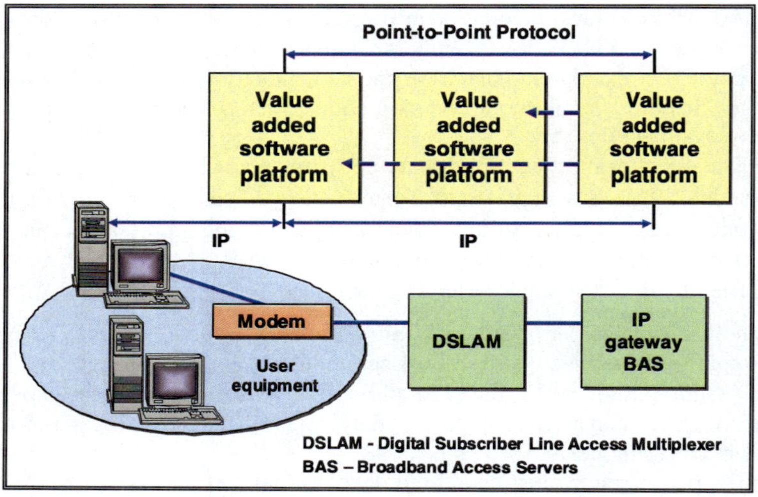

Fig. 9 Value-added software platform example

The various application-level protocols constitute a first category of enablers for this platform. Any new differentiating application service will be composed of two major component classes: client-server and peer-to-peer.

A second category of enablers are common functional components, which are likely to be reused by most applications and content. Presence, a function that maintains information as to where each user is, is a classical example. Session control and accounting is another such component.

In addition, while the protocols typically refer to fixed-line broadband DSL access, the architecture of Fig. 9 also applies largely to local multipoint distribution services (LMDS) and universal mobile telecommunications system (UMTS) networks [30 ÷ 32].

7 Disclosure and Evaluation of Services Enabling Shared/ Collective Intelligence

It is difficult to predict upcoming NGN applications and services in detail. However, it is possible to infer the types of service characteristics and capabilities that can be important in the NGN environment by examining current service-related industry trends.

7.1 NGN Service Features

The main task of traditional network service providers has been to offer the mass market basic transport of information between end users, with various value-added capabilities. These services tended to involve narrowband voice calls, with a single point-to-point connection per call. However, this view of services has been rapidly changing as the world's economies are becoming increasingly reliant on information as a basic resource.

As multiple carriers, service providers, equipment vendors, and other business entities all become involved in providing services to end users, federated network and business systems become increasingly important. The primary goal is to enable users to get the information content they want - in any media/format, over any facilities, anytime, anywhere, and in any volume. Based on the above-mentioned trends, the following provides a summary of several service characteristics that are important in an NGN environment:

- *ubiquitous, real-time, multi-media communications* - high-speed access and transport for any medium, anytime, anywhere, and in any volume;
- *personal intelligence distributed throughout the network* - includes applications that can access users' personal profiles (e.g., subscription information and personal preferences), learn from their behavior patterns, and perform specific functions on their behalf (e.g., intelligent agents that notify users of specific events or that search for, sort, and filter specific content);
- *network intelligence* - includes applications that know about, allow access to, and control network services, content, and resources. It can also perform specific functions on behalf of a service or network provider (e.g., management agents that monitor network resources, collect usage data, provide troubleshooting, or broker new services/content from other providers);
- *personal service customization and management* - involves the users' ability to manage their personal profiles, self-provision network services, monitor usage and billing information, customize their user interfaces and

the presentation and behavior of their applications, and create and provision new applications;
- *intelligent information management* - helps users manage information overload by giving them the ability to search for, sort, and filter content, manage messages or data of any medium, and manage personal information (e.g., calendar, contact list, etc.) [2, 27].

7.2 Typology of NGN Services

A variety of services, some of which are already available while others are still at the conceptual stage, have been linked to NGN initiatives and considered candidates for NGN implementations. While some of these services can be offered on existing platforms, others benefit from the advanced control, management, and signaling capabilities of NGNs. Although emerging and new services are likely to be the strongest drivers for NGNs, most of the initial NGNs profits may actually result from the bundling of traditional services. Thus, bundled traditional services pay for the network, whereas emerging services fuel its growth. NGNs can enable a broad collection of service types, including:

- *specialized resource services* - provision and management of multipoint conferencing bridges, media conversion units, voice recognition units;
- *processing and storage services* - provision and management of information storage units for messaging, file servers, terminal servers, OS platforms;
- *middleware services* - naming, brokering, security, licensing, transactions;
- *application-specific services* - business applications, e-commerce applications, supply-chain management applications, interactive video games;
- *content provision services that provide or broker information content* - electronic training, information push services;
- *interworking services* - for interactions with other types of applications, services, networks, protocols, or formats;
- *management services* - to maintain, operate, and manage communications/computing networks and services.

Furthermore, the following services are important drivers in the NGN environment, in terms of how pervasive they can be, what profit margins they are likely to generate, and how much they can benefit from an NGN type of environment:

- *unified messaging* - supports the delivery of voice mail, email, fax mail, and pages through common interfaces. Through such interfaces, users can access, as well as be notified of, various message types (voice mail, email, fax mail, etc.), independent of the means of access (i.e., wireline or mobile phone, computer, or wireless data device).
- *information brokering* - involves advertising, finding, and providing information to match consumers with providers. For example, consumers could receive information based on pre-specified criteria or based on personal preferences and behavior patterns.

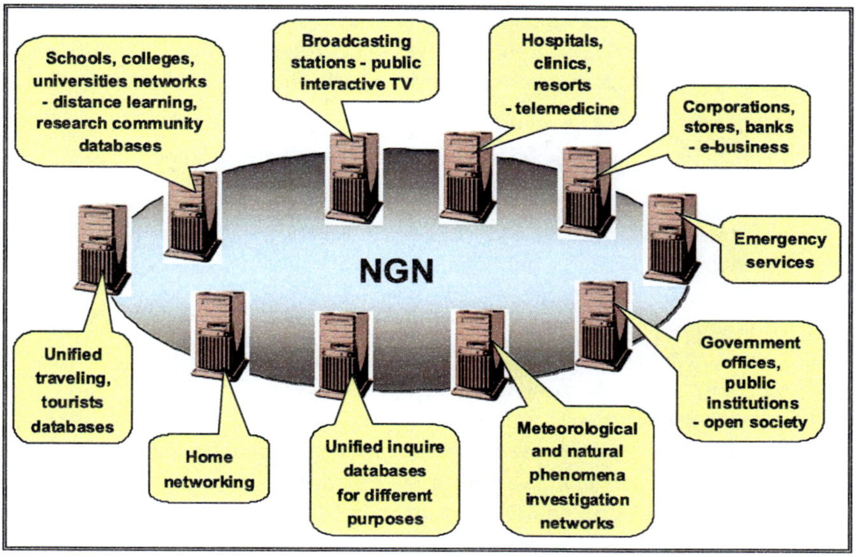

Fig. 10 Example of NGN services categorization

- *e-business* - allows consumers to purchase goods and services electronically over the network. This can include processing the transactions, verifying payment information, providing security, and possibly trading (i.e., matching buyers and sellers who negotiate trades for goods or services). Home banking and home shopping fall into this category of services. This also includes business-to-business applications (e.g., supply-chain management and knowledge management applications);
- *call center services* - a subscriber can place a call to a call center agent by clicking on a Web page. The call could be routed to an appropriate agent, who could be located anywhere, even at home (i.e., virtual call centers). Voice calls and e-mail messages could be queued uniformly for the agents. Agents would have electronic access to customer, catalog, stock, and ordering information, which could be transmitted back and forth between the customer and the agent.
- *interactive gaming* - offers consumers a way to meet online and establish interactive gaming sessions (e.g., video games);
- *distributed virtual reality* - refers to technologically generated representations of real-world events, people, places, experiences, etc., in which the participants in and providers of the virtual experience are physically distributed. These services require strong coordination of multiple, diverse resources.

- *home manager* - with the advent of in-home networking and intelligent appliances, these services can monitor and control home security systems, energy systems, home entertainment systems, and other home appliances [21, 28, 33].

Also, the NGN services are divided using other principles than in modern networks (Fig. 10). One of the possible NGN services categorization is as follows:

- school, college, university networks – distance learning, research community databases;
- broadcasting stations – public interactive TV, audio/video on-demand;
- hospitals, clinics, resorts – telemedicine, e-health;
- corporations, stores, banks – e-business, m-business;
- government offices, public institutions – open society, e-government;
- emergency services, disaster recovery;
- home networking, teleworking, social networks;
- unified traveling, tourists databases, virtual enterprise/services;
- unified inquire databases for different purposes, e-logistics;
- meteorological and natural phenomenon's investigation networks, and so on [26, 34].

8 Business Model for Service Engineering in NGN

Within a service engineering approach (Fig. 11), business modeling is used at two distinct levels: at the planning level (as input to business design and service planning) and at the requirements level (as input to service design, acquisition and management). We evaluate here the development of business models for planning purposes. Furthermore, we want to establish why it is important to build up a business model before new project planning.

Geoffrey Moore provides a few strategic thoughts in his recent book "Dealing with Darwin" [35]. Moore argues that IT needs to support the business strategy by providing different kind of systems and services to support different capabilities, according to how each capability is seated within the business strategy. Moore proposes a set of issues that assist in establishing the enterprise strategy. He sets out a few questions that have to be answered before every business modeling begins:

- What is the most critical thing/goal in every business/enterprise?
- Does every team member know that it is the most critical thing/goal for his enterprise?
- And what percentage of time does every team member spend on it?

This approach also underlines why business modeling is essential. It is impossible to get the standard functional decomposition for any generic industry that can be implemented without further thought. Of course, if a standard industry model exists, it can be used as a starting point, but what is important - e.g., core to certain business strategy- is what can be inscribed onto this model.

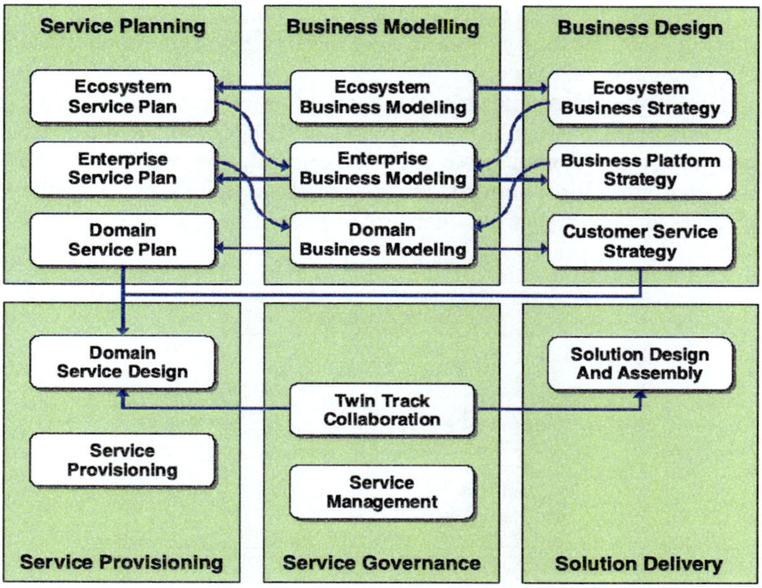

Fig. 11 Model of Service Engineering.

In a top-down process, the business modeling can be done at several different levels. It can be modeled as a whole ecosystem, for a single enterprise, or for a domain within an enterprise. The domains themselves may be defined as part of the service planning activity.

At the higher levels, the primary purpose of business modeling is to understand the requirements for business design and service planning. This corresponds to what was known as a "Business Strategy Planning" in older methodologies. At the lower levels, the primary purpose of business modeling is to understand the detailed requirements for service design [10, 36, 37].

8.1 Modeling Methodology - Strategy, Processes, and Outcomes

On the whole, the purpose of a business modeling methodology is to obtain different kinds of outcomes at different levels. For instance, at the planning level, desired outcomes are the business goals and the outcomes that affect the success of the business, e.g., key uncertainties or risks. Thus, it is important to identify the capabilities that manage these high-level outcomes, including monitoring and control capabilities; and to understand how each capability contributes to the top-level business goals.

The classification of capabilities depends partly on the link to business outcomes/value. It also depends on the knowledge associated with each capability.

Correct decoupling between capabilities depends on encapsulating the knowledge embedded in each capability. This is a version of the well-known architectural principle - separation of concerns.

Going further to address the details, a standard service provides a fixed one-size-fits-all service to each user in isolation. Network services provide the capability for users to connect and communicate with one another. The service network represents an opportunity for a network provider to move upwards. A variable service network gives an opportunity for a network provider to move sideways (Fig. 12).

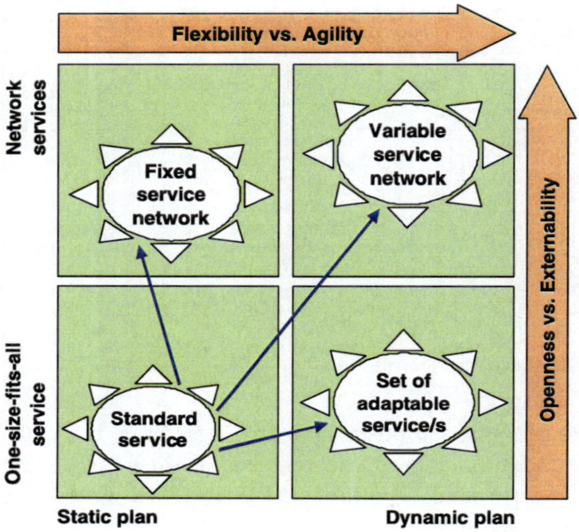

Fig. 12 Stack strategy and capabilities matrix, conceptual/general view.

Thus, it may not be necessary to specify how an enterprise generates the idea for a service; there may be many alternative triggers for identifying a service opportunity. But, what is important is that an enterprise has a systematic way of evaluating and implementing such ideas, e,g., has developed an efficient business model and strategy [10, 35 ÷ 37].

In this context, the NGN services planning and development process is responsible for the definition of rules for network planning, installation and maintenance, application of new technology and supplier strategy, development and acceptance of new network types, and description of standard network configurations for operational use. Also, this process is responsible for designing the network capability to meet a specified service or need at the desired cost and for ensuring that the network can be properly installed, monitored, controlled and billed. It is also responsible for ensuring that enough network capacity will be available to meet the forecasted demand. Based on the required network capacity, orders are issued to suppliers or other network operators (Fig. 13). A design of the logical network configuration is provided to network provisioning process/function.

The input triggers to this process have to be as follows:

- new service description from service planning and development;
- new network technology from supplier;
- capacity plan from service planning and development;
- capacity request from network provisioning, inventory management and maintenance and restoration.

The output triggers of it:

- orders to suppliers and/or other network operators;
- work orders to network inventory management or a network constructor;
- configuration requirements to network provisioning;
- performance goals to network data management;
- maintenance rules to network maintenance and restoration;
- work orders to network inventory management;
- deployment plans to service planning and development.

The output data, to be generated within the process:

- purchasing, installation, performance and maintenance rules, including standard network configurations and routing restrictions/requirements;
- network capabilities (including performance goals);
- planned network capacity;
- planned logical network configuration;
- deployment plans.

And, potential NGN services planning and development process responsibilities:

- develop and implement procedures;
- set-up framework agreements with suppliers;

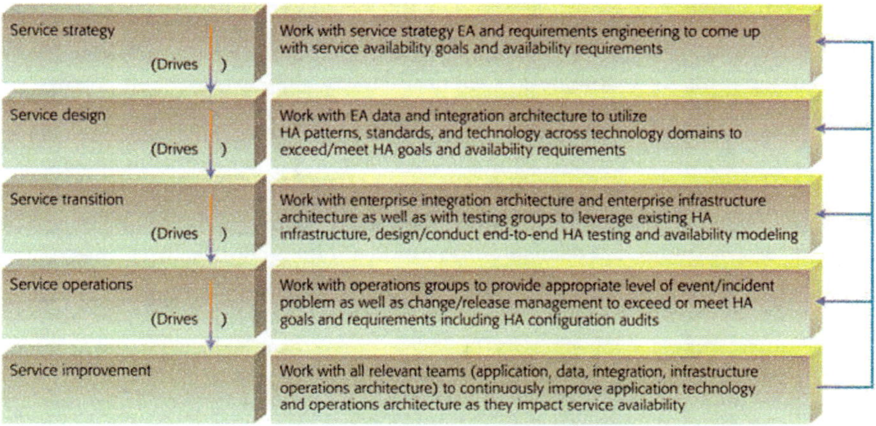

(EA- Enterprise Architecture, HA – High Availability)

Fig. 13 NGN Services Planning and Development Process [38].

- develop new networks and architectures, determine network capabilities, based on network technology and architecture;
- plan required network capacity;
- plan the mutation of network capacity (including destruction of obsolete networks);
- issue orders to suppliers and other network operators;
- plan the logical network configuration;
- plan the required physical site facilities [25, 36, 39].

8.2 Challenges and Support in Business Decisions

The business and technology decision makers already place a high priority on providing optimized communication between remotely located knowledge workers and their teams. Without a doubt, the modern organization is already overprovided with regard to communications devices.

Numerous industrial studies have established that on average, there are more than six communications devices and almost five communications applications per employee at an enterprise. At the same time, the "quality of interaction" expectations have increased as the use of Web chats and Web conferences, video conferences, and multimedia contact centers has grown.

Nevertheless, the crucial problem - how to create and sustain an effective communication environment for mobile personnel and distributed workgroups persists. In order to solve this problem, modern organizations have overcome two major communications challenges:

- increase visibility into the availability, schedules, or presence status of primary decision makers and primary players in a timely way;
- improve collaboration: spontaneously, productively, cost-effectively, and on an ongoing basis, regardless of their location or the device being used; to produce tangible results and business benefits.

The failure to address these challenges inflicts real drawbacks, e.g., business projects are delayed. And ultimately, both profitability and customer and channel relationships are put at risk, more specifically:

- communication-caused delay and disruption is a pervasive business problem, the multiplicity of communications devices has not yet solved perhaps the most obvious problem in the distributed and mobile workplace: how to access an important decision maker on the right device the first time.
- communications complexity affects long-term productivity, business process reform, and financial performance - poor communication affects strategic initiatives such as lean or just-in-time manufacturing, supply chain optimization, and customer relationship management.
- decision support outcomes suffer from the inability to access and collaborate effectively with primary players. The knowledge workers need ways and means to resolve urgent questions across an entire project lifecycle.

Upstream decisions affect future decision making and even the ultimate viability or relative success of the project itself.

- resources are improperly used or misallocated because of the complexity of communication. Mindful of the critical importance of maintaining customer loyalty and building brand equity, businesses invest heavily in both developing human talent and enabling technologies to, for example, better manage customer inquiries about a new product or service. But, the excessive use and lack of integration of communications tools and applications have undermined that strategic objective, and can squander those investments. In the example, if the customer does not receive the requested information in a timely manner, the sale is canceled. And, over time, the organization cannot effectively grow its business or improve customer satisfaction ratings and loyalty.

The developments in communications devices, and the challenges caused by them, have created a need among enterprises for an effective communications system that enables them to streamline business processes; reach the right person the first time; make communications more personal, collaborative, and mobile; and improve profitability.

The NGN unified communications can enable people to find peers or decision makers using a single telephone number or Internet address. They integrate email, instant messaging, and calendaring applications with communications devices and applications, telephony wired and wireless; voice messaging; and audio, video, and Web conferencing. Unified communications applications support advanced presence-sensitivity and find-me capability, and media independence. They can also provide voice access to applications and data [22, 40].

8.3 Significance of Applications Intelligence in NGN Business/Service Modeling and Engineering

It is clear that the enterprises develop applications to support business processes. Business leaders have commissioned large investments into the networking infrastructures to support the automation of their business processes.

Thus, the application intelligence has become available just in time, as new application delivery models have emerged, representing unknown and unforeseen traffic patterns and network loads. Structured IT applications are converging with Web services, making it difficult to distinguish mission critical from recreational applications.

Telecom networks have traditionally supported applications by providing a transport service. More than ten years ago, this transport service was optimized through primitive network ports and IP address-based quality of service mechanisms in an effort to prioritize one set of applications over another. Technologies such as traffic shaping or rate limiting were introduced to control applications up and down. While these tools and techniques were powerful, they are not systemic or supported across an entire corporate network, leaving large gaps or blind spots to application performance.

Next Generation Networks can connect all IT resources and therefore become a strategic business platform. An important platform attribute is broad and deep application recognition and performance support. It is fundamental that a broad portfolio of underlying network technology, which enables increased application performance and security, is resident within the network fabric. An application network service can optimize an application life cycle process from application development, test, roll out, production and maintenance. For example, in data centers, networks increasingly understand content and message structure, enabling them to make decisions and perform network operations such as content-based routing and load balancing, optimizing application performance delivery.

To deliver and optimize application performance, network-based application services need to be end-to-end, with the network recognizing applications and providing service at ingress and egress points. The overall network will respond not just to prioritizing applications, but will also deliver a set of underlying services to ensure that application performance is of the highest order. This means that networks will recognize applications, signatures and protocols at the point of entry, classify them, mark them with pre-defined priority which network devices understand and enforce them as the application flows across the network independent of its final location, be it to a data center, an end-point, over the wide area, through the campus, or to a branch office. This is application intelligence and it wraps around an entire network.

Next, application intelligence offers a different approach to campus network design by automating application performance management. Where there are multiple different applications flowing across the network, they can all be logical, meaning that a network is made up of "n" logical networks with application intelligence managing these logical entities.

Furthermore, application intelligence can manage the connection between applications, whether the connection is strong or weak, or no connection at all exists. It can allow the application to view the network as though it is matching the network's capabilities with its own unique requirements.

And finally, application intelligence allows every application to obtain its fair share of resources, bandwidth, QoS, and Service Level Agreement (SLA) in the presence of all other applications [7, 12, 28].

9 Conclusions and Implications for Future Work

The vision of information and communication anytime, anywhere, and in any form is coming into focus as major players begin to position themselves for a radical transformation of their network and service infrastructures. It has become increasingly clear that a prerequisite for realizing such a vision is the convergence of the current multiple networks - each employing different transport and control technologies - into a unified, multiservice, data-centric network offering services at different qualities and costs on open service platforms. The evolution toward such a vision is undeniably influenced by the growing trends of deregulation in the telecommunications industry on one hand, and the rapidly evolving technological convergence between distributed computing and communications on the other. The fundamental driving forces for NGN can be categorized as follows [18, 32]:

- *environmental factors* - reflect changes that have been happening in the Information Communication Technologies (ICT) business environment over the past two decades. During this period, the global telecommunications industry as a whole has been gradually moving away from the model of state-owned and/or regulated monopolies to that of a competitive industry operating in an open market.
- *service/market factors* - reflect the continuously expanding set of capabilities and features customers in various markets demand to satisfy their constantly evolving set of personal and professional needs. For example, mobility of various kinds has become a paramount requirement. Other market-driven needs include ready access to information, easy-to-use communication capabilities involving more than one medium, greater and more granular end-user control over services, and progressively higher quality content delivered for purposes of entertainment and/or education. Service and market factors, even as they get modulated by other factors, are unquestionably the ultimate drivers for services architecture evolution, because services are what customers use and pay for.
- *technology factors* - include all the technological enablers a service provider, in partnership with its vendors, can take advantage of in the process of architecting and composing its range of services. In modern information society, technology factors have a lot to do with shaping customer expectations, thereby modulating service/market and environmental factors. The impact of the spectacular rise of the Internet on the convergence of distributed computing and communication technologies has underlined the critical importance of technology drivers in elevating and reshaping customer expectations, and pushing out the restrictive envelope of regulatory constraints [16, 27, 29].

These factors, operating against the backdrop of some mega-trends - including the growing diversity and complexity of new services, the increasing variety and power of end-user devices, and the competitive push to minimize time-to-market - unmistakably underline the urgency of fundamental transformation in communication networks and service architectures towards NGN [13, 36].

Consequently, the evolutional strategy proposed and explored in this chapter enables the development of a proficient concept of transformations in the way networks are planned, paying particular attention to the business needs of diverse service providers and network operators.

Furthermore, we have observed that communication application development platforms are rapidly being aligned with major IT and telecommunication platforms based on Web services and SOA (Service-oriented Architecture). Enterprise IT departments are the early adaptors of Web services as a means to write communications-enabled business process. Web services and SOA are powerful forces in the IT and Telecom communities, since they address the entire supply chain of an enterprise by abstracting the multitude of APIs into a standard set of programming interfaces with SOA governed business applications. Besides, IP telephony providers can offer real-time communications into an SOA construct by using

Web services as the main programming interface into real-time communication application servers.

Assuming that the service providers need a new approach to service creation and delivery that addresses their business requirements and bridges the technical gaps between closed, proprietary telecom software development and open Web-based software development environments, we would like to focus on these open issues in our future work.

References

1. Estes, G.H.: NGN: Preparing for Tomorrow's Services. Alcatel Telecommunications Review, 2nd Quarter (2001)
2. Carbone, P., Romagnino, S.: Extreme value from next-generation applications and services. The Nortel Technical Journal (5) (March 2007)
3. Ecma International. Enterprise Communication in Next Generation Corporate Networks (NGCN) involving Public Next Generation Networks (NGN). Technical Report ECMA TR/91, 1st Edition (December 2005)
4. KATE-KOM. Next Generation Service Delivery v1.0. White Paper from KATE-KOM (2005), http://www.kate-kom.com
5. EMC. Managing Next-Generation Networks: Service Assurance for new IP Services, Technology Concepts and Business Considerations. White Paper from EMC Corporation (September 2006)
6. Uebele, R., Verhoeyen, M.: Strategy for migrating voice networks to the next generation architecture. Alcatel Telecommunications Review, 2nd Quarter (2001)
7. Becchina, W., Ciccarelli, L., Giotis, C., Kenny, J.: Reinventing the enterprise with communications-enabled applications. The Nortel Technical Journal (5) (March 2007)
8. Achterberg, R.A., Hanrahan, H.E.: A Framework for Evolution to Next Generation Networks. In: The IEEE Intelligent Networks Workshop – IN 2000, pp. pp. 0_1 - 0_11 (2000)
9. Hao, Q.M.: Toward a Unified Service Delivery Process for Next-Generation Services. Bell Labs Technical Journal 12(4), 5–20 (2008)
10. Gurbani, V., Sun, X.H., Brusilovsky, A.: Inhibitors for Ubiquitous Deployment of Services in the Next-Generation Network. IEEE Communications Magazine 43(9), 116–121 (2005)
11. Ecma International, Next Generation Corporate Networks (NGCN) – General. Technical Report ECMA TR/95, 1st Edition (June 2008)
12. Augustine, S.: Managing Agile Projects. Prentice-Hall PTR, Englewood Cliffs (2005)
13. Modarressi, A.R., Mohan, S.: Control and Management in Next-Generation Networks: Challenges and Opportunities. IEEE Communications Magazine 38(10), 94–102 (2000)
14. ITU-T. Principles for the Management of Next Generation Networks. ITU-T Rec. M.3060/Y.2401 (March 2006)
15. Alcatel-Lucent. Delivering Innovative Services With the Alcatel-Lucent Service Delivery Environment (SDE) - a Service Delivery Environment Featuring Common Enablers for the Delivery of Personalized and Blended Services Across Multiple SDPs. White Paper (June 2007)

16. ETSI. Telecommunications and Internet Converged Services and Protocols for Advanced Networking (TISPAN) NGN Management, Operations Support Systems Architecture. ETSI TS 188 001, v1.2.1 (March 2006)
17. Kryvinska, N., van As, H.R., Brusilovskiy, S.: Packet Intelligent Networks based on a Potential Signaling System No.8 Targeting towards the Next Generation Business Model. In: St. Petersburg Regional International Teletraffic Seminar 2002, St. Petersburg, Russia, pp. 12–22 (2002)
18. Levine, S.: Service Delivery Platforms: A Deployment and Market Assessment. International Data Corporation (IDC), Doc. No. 06C5055 (December 2006)
19. Camp, K.: The Definitive Guide To Converged Network Management. Realtimepublishers (2006)
20. IBM IBV. Services over IP Delivering new value through next-generation networks. IBM Corporation, IBM Institute for Business Value (IBV) (2005)
21. Ericsson. Service Delivery Platform - Efficient Deployment of Services. White Paper, 284 23-3083 Uen Rev A (October 2006)
22. Lofstrand, M., Carolan, J.: Sun's Pattern-Based Design Framework: The Service Delivery Network. Sun Microsystems, Sun BluePrints™ OnLine, Part No 819-4148-10, Revision 1.0, 10/3/05 (September 2005)
23. Stratus Technologies. How to Transform Voice over Broadband Opportunity into Sustainable Profitability. Stratus Technologies (June 2007)
24. AT&T. Tuning Applications for IP'. An AT&T Survey and White Paper in Cooperation with the Economist Intelligence Unit, AT&T Knowledge Ventures, 07/12/07 AB-1116 (2007)
25. Tarzey, B., Bamforth, R., Nosseir, S.: Managing 21st Century Networks A world of convergence. An independent study by Quocirca Ltd. Quocirca Insight Report (January 2007)
26. Frelot, O., Taeymans, J., Bonnet, G.: Introduction to next generation applications. Alcatel Telecommunications Review, 2nd Quarter (2002)
27. Daniel, C., Walker, S.: Service Solutions in Next-Generation Networks. Multiservice Switching Forum, MSF Technical Report, MSF-TR-ARCH-003-FINAL (April 2003), `http://www.msforum.org`
28. Lippis: Application Intelligence: A New Network Service. Nicholas John Lippis III Publisher, The Lippis Report (April 2007)
29. Avaya. Avaya Communication Architecture. White Paper 03/06, LB1842 (March 2006)
30. Vanderstraeten, H.: Myriad: an application service-enabling platform. Alcatel Telecommunications Review, 3nd Quarter (2001)
31. Gurbani, V., Sun, X.H.: Terminating Telephony Services on the Internet. IEEE/ACM Trans. Net. 12(4) (August. 2004)
32. ABI Research. 267 Million Residential Voice over IP Subscribers Worldwide by 2012. ABI Research, Press Release (January 31, 2007)
33. Crimi, J.C.: Next Generation (NGN) Services. White paper, Telcordia Technologies (2001)
34. Kryvinska, N., van As, H.R.: From Call Control towards Service Control in Next Generation Networks. In: Proceedings of the International Conference on Software, Telecommunications and Computer Networks (Softcom 2002), Split, Croatia, pp. 29–33 (2002)

35. Moore, G.A.: Dealing with Darwin - How great companies innovate at every phase of their evolution. Penguin, USA (January 2006)
36. Kryvinska, N., Strauss, C.: A Strategy towards Next Generation Service Delivery Network - Architectural Planning and Design, Business Perspective. In: Proceedings of the Fifth IEEE Advanced International Conference on Telecommunications AICT 2009, Venice, Italy, May 24-28, pp. 410–415 (2009)
37. Veryard, R.: BSP for SOA - Business Strategy Planning for the Service Economy. CBDI Forum Limited, CBDI Journal (May 2006)
38. Radhakrishnan, R., Mark, K., Powell, B.: IT service management for high availability. IBM Systems Journal 47(4) (2008)
39. TeleManagement Forum. Network Management Detailed Operations Map, Evaluation Version 1.0, GB908. TM Forum (March 1999), `http://www.tmforum.org`
40. Cisco. Enhancing Business with Smarter, More Effective Communications. Cisco White Paper, C11-332778-01 3/07 (2007)

Glossary of Terms and Acronyms

API	Application Programming Interface
CPE	Customer Premises Equipment
DPE	Distributed Processing Environment
IN	Intelligent Network
IP	Internet Protocol
IT	Information Technologies
LMDS	Local Multipoint Distribution Services
NGN	Next Generation Network
PSTN	Public Switched Telephone Network
QoS	Quality of Services
SIP	Session Initiation Protocol
TDM	Time Division Multiplexing
UMTS	Universal Mobile Telecommunications System

Chapter 19
Utilizing Next Generation Emerging Technologies for Enabling Collective Computational Intelligence in Disaster Management

Nik Bessis, Eleana Assimakopoulou, Mehmet E. Aydin, and Fatos Xhafa

Abstract. Much work is underway within the broad next generation emerging technologies community on issues associated with the development of services to foster synergies and collaboration via the integration of distributed and heterogeneous resources, systems and technologies. In this chapter, we discuss how these could help coin and prompt future direction of their fit-to-purpose use in various real-world scenarios including the proposed case of disaster management. Within this context, we start with a brief overview of these technologies highlighting their applicability in various settings. In particular, we review the possible combination of next generation emerging technologies such as ad-hoc and sensor networks, grids, clouds, crowds and peer to peer with intelligence techniques such as multi-agents, evolutionary computation and swarm intelligence for augmenting computational intelligence in a collective manner for the purpose of managing disasters. We then conclude by illustrating a relevant model architecture and by presenting our future implementation strategy.

Nik Bessis
School of Computing and Mathematics, University of Derby, Derby, United Kingdom
e-mail: n.bessis@derby.ac.uk

Nik Bessis · Eleana Asimakopoulou · Mehmet E. Aydin
Department of Computer Science and Technology, University of Bedfordshire, Luton, United Kingdom
e-mail: eleana.asimakopoulou@googlemail.com,
{nik.bessis,mehmet.aydin}@beds.ac.uk

Fatos Xhafa
Departament de Llenguatges i Sistemes Informàtics, Universitat Politècnica de Catalunya, Barcelona, Spain
e-mail: fatos@lsi.upc.edu

N. Bessis, F. Xhafa (Eds.): Next Generation Data Technologies for CCI, SCI 352, pp. 503–526.
springerlink.com

1 Introduction

The use of collaborative systems has evolved over the years through developments in distributed computational science in a manner, which provides applicable intelligence to their problem-solving capabilities. Within this context, data integration has long been discussed in distributed computing literature reviews. Many concerns have been encountered, as most of the data systems addressed by individual systems and their applications are both heterogeneous and geographically distributed.

The data integration concern is mainly due to the different contexts and purposes for which the data systems were originally built. In other words, the main concern resides on the view that data sources have been originally produced for purposes other than their integration [1]. Currently, another aspect is that data is available and accessible to and from a wider audience, and thus, we suggest that data must support a many-to-many exploitation type of relationship with their owners and/or on-demand users who are also geographically distributed. Thus, data is now utilized in purposes other than what was it originally produced for [13].

In turn, the ability to make data systems and their stores interoperable remains a crucial factor for the development of these types of systems [55]. One of the challenges for such facilitation is that of data integration, which aims to provide seamless and flexible access to information from multiple autonomous, distributed and heterogeneous data sources through a query interface [45]. Moreover, the combination of large dataset size, geographic distribution of users and resources, and computationally intensive analysis results in complex and stringent performance demands that, until recently, have not been satisfied by any existing computational and data management infrastructure [28]. Rather, various technologies have been developed to address the issue of collaboration, data and resource sharing.

Most of these have emerged with the view of producing frameworks and standards to fully or partially – yet purposefully – support data integration processes within heterogeneous distributed environments. Emerging paradigms and their associated concepts highlighting their benefits include but are not limited to P2P, Grids and Cloud computing. Their goal is to enable an approach relevant to collective resource utilization and thus, enhance multi-user participation and collaboration in functioning as a coherent unit through the use of a Cyber Infrastructure (CI). That is, to purposefully work together, collaborate and solve a well-defined problem of mutual interest from a multi-disciplinary perspective. As such, they typically enable the provision of shared and often real-time access to, centralized or distributed resources, such as applications, data, toolkits and sensors.

In this chapter, we start off with an overview highlighting how these next generation emerging technologies fit into the broader picture of IT and also present their current applicability in various settings (Section 2). In Section 3, we discuss the notion of intelligence within the contexts of computational collective setting and collective computational intelligence. In Section 4 we briefly present a disaster management scenario as the means to highlight the emergency management stakeholders' requirements. In Section 5, we build up on previous works via the use of a model architecture and by discussing our implementation effort. Finally, in Section 6 we conclude with future work.

2 Overview of Next Generation Emerging Technologies

In this section we briefly describe most important emerging technologies and paradigms that serve as a basis for collective intelligence platforms.

2.1 Grid Computing

Grid computing have been described as the infrastructure and set of protocols to enable the integrated, collaborative use of distributed heterogeneous resources including high-end computers (nodes), networks, databases, and scientific instruments owned and managed by multiple organizations [28]. The concept of Grid technology has emerged as an important area differentiated from open systems, clusters and distributed computing [11]. Specifically, open systems remove dependencies on proprietary hardware and operating systems, but in most instances are used in isolation. Unlike conventional distributed systems, which are focused on communication between devices and resources, Grid technology leverages of computers connected to a network, making it possible to compute and to share data resources. Unlike clusters, which have a single administration and are generally geographically localized, Grids have multiple administrators and are usually dispersed over a wide area. But most importantly, clusters have a static architecture, whilst Grids are fluid and dynamic with resources entering and leaving. In brief, Grid can be viewed as a dynamic, enabling paradigm supporting synchronous and asynchronous resource utilization in a c-cube mode (communication, co-operation and collaboration). It has been purposefully developed for solving well-known scientific problems (mainly by academic researchers) of e-Science family [12]. Not less important, the cross-administrative domain, poses challenges and legal issue to integration, access and use of resources.

One of the most important standards that have been emerged within the Grid community is the Open Grid Services Architecture (OGSA), an informational specification that aims to define a common, standard and open architecture for Grid-based applications. The need to integrate databases into the Grid has also been recognized [39] in order to support science and business database applications [2]. Significant effort has gone into defining requirements, protocols and implementing the OGSA-DAI (Data Access and Integration) specification as the means for users to develop relevant data Grids to conveniently control the sharing, accessing and management of large amounts of distributed data in Grid environments [2; 7]. Ideally, OGSA-DAI as a data integration specification aims to allow users to specify 'what' information is needed without having to provide detailed instructions on 'how' or 'from where' to obtain the information [45]. An important merit of the OGSA-DAI model is that all components of the environment can be virtualized. It is the virtualization of Grid services that underpins the ability to map common service semantic behavior seamlessly on to native platform facilities. These particular characteristics extend the functionality offered by Web Services and other conventional open systems. In turn, the OGSA standard defines service interfaces and identifies the protocols for invoking these services. The potential range of OGSA services is vast and currently includes data and information services, resource and service management, and core services such as name resolution and discovery, service domains, security, policy, messaging, queuing, logging, events, metering and accounting.

On the other hand, Web Services aim to provide a service-oriented approach to distributed computing issues, whereas Grid arises from an object-oriented approach. That is to say, Web Services typically provide stateless, persistent services whereas Grids provide state-full, transient instances of objects [11]. In fact, emergence of Web Services with Grid computing has resulted in a service-oriented architecture for the Grid. An important merit of this (grid) model is that all components of the environment can be virtualized, a feature, which points to what is currently known as Cloud computing.

2.2 Cloud Computing

Various approaches and definitions of Cloud computing exist [54; 23; 10; 46]. All conclude that a Cloud is comprised from Grid, virtualization and Utility computing notions. [18] defines a Cloud as a type of parallel and distributed system consisting of a collection of inter-connected and virtualized computers. They are dynamically provisioned and presented as one or more unified computing resources based on service-level agreements established through negotiation between the service provider and consumers.

Grids have been developed for solving scientific problems and thus security, reliability and use by non-academics were initially far away from being the primary concerns. On a similar vein, we could also point out that Grids are the first generation of this type of paradigm and thus, it would be completely unfair to expect a fully functional paradigm which would fully meet real-world business requirements. The fact that Clouds are based on Grids demonstrates Grids' sustainable robustness, as well as business value and prospects. Thus, taking a broader view we can define a Cloud as a re-factored business-oriented Grid model. Users forming the Cloud can access resources, solve problems such as in Grids, but in a well-defined robust commercialized context; offering a more structured, scalable and personalized management control; as well as by being charged with a cost [12]. In brief, one can conclude that the goal of Grids and Clouds is to purposefully utilize resources (data, computational power, software, toolkits, expertise, etc) that is available from/to Virtual Organizations (VO) partners so they can more effectively solve mainly scientific (Grid) or commercial (Cloud) problems.

2.3 Pervasive Computing

Lately, Pervasive computing as a new paradigm aims to enable resource computation and utilization in a far more mobile or environmentally-embedded manner. Pervasive computing embeds computing and information technologies into our environments by integrating them seamlessly into our everyday lives [53]. Pervasive computing has many potential real-world applications ranging from health to environmental monitoring systems. It is quite common to involve a number of devices including mobile phones, PDAs, sensors and computers. Lately, Situated Computing as an emerging paradigm deals with computing devices having the autonomous ability of adapting, detecting, interpreting and responding to the user's environment. Readers are pointed to [30] who gives a solid background of the fundamentals for Situated Computing. Situated Computing makes use of concepts from situated cognition [21]. Thus, where and when someone is, it matters, and that the state s/he is in affects what s/he does. The fundamental difference is between encoding all knowledge prior to its use and allowing the knowledge to be developed and grounded in the interaction between the tool and its environment. The effect of this is to provide a computational system such as a tool with experience based on its interaction with its environment. That experience is then used to guide future actions. The effect of this grounded experience is to provide the tool with the capability to respond differently when exposed to the same environment again depending on the experiences it has had between these two exposures. The objective knowledge within the tool remains unchanged, only the knowledge that is the result of the interaction of the tool with its environment is changed. This provides the basis for computational systems to learn and change their behavior based on their experiences. The learning is not necessary to improve the performance of the system rather it is designed to customize it to its user [30].

2.4 *Crowd Sourcing*

More recently a new paradigm called Crowd Sourcing (also known as Crowd Computing or Citizen Science) has been introduced. Some studies have proven the potential worth of so-called "crowd-sourced" mobile phone data [42; 12]. Some of these pilot studies have shown that mobile phones and mobile sensors can be used by ordinary "citizens" to gather data that could be useful in various settings. [42] has also coined the term "citizen science" for solutions that seek to leverage collective citizen-based collection. However, participatory data collection activities of this kind and their subsequent aggregation and analysis by decision makers pose significant opportunities and challenges.

2.5 *Sensor Technologies*

Finally, another technology that attracts current attention is wearable sensors. These come in many forms: wrist watches, rings, smart clothes (shirts, shoes etc.), spectacles, plasters, or implanted devices (e.g. subcutaneous); and are known in different contexts or just with different usage under several generic names: wearable or body sensors or ambulatory monitoring, and extended into body sensor networks or body area networks (BANs). [29] reports successful modeling for "Intra-Body Communication" (IBC), as 'a short range "wireless" communication technique [that] relies on the conductive property of human tissue to transmit the electric signal [within the] human body'. In contrast, the disposable digital plaster, which won the electronic category in the 2007 Institution of Engineering Technology (IET) Innovation Engineering Award, "sticks to a patient's chest and has an ultra low power wireless smart sensor in a silicon chip attached to the plaster, which monitors in real-time a range of vital signs like ECG, body temperature, respiration and physical activity" [47]. The device began clinical trials in November 2009 [33]. Applications of body sensors are now widespread: health care, military applications, athlete training, law enforcement, tracking professional truck driver's vital signs for fatigue, motion capture in the wider sense of biosensors, environmental monitoring and industrial process control. The signals that body sensors measure include physiological vital signs such as heart rate, temperature, and other biological signs such as sweat production or glucose levels.

3 Computational Collective Intelligence versus Collective Computational Intelligence: Are the Two the Same?

In our view, the concept of collective intelligence refers to the intelligence achieved and managed collectively by multiple independent homogeneous or heterogeneous actors. Thus, the focus should be on how to develop a satisfactory level of intelligence in a collective way. This requires being related to cognition addressing individual intelligence, to coordination for pointing out how to create collective behavior and to collaboration for clarification of what information to share and exchange towards collectivity. Thus, collective intelligence could be seen as a synergy achieved among individual entities with having some sort of intelligence, sharing and exchanging individual information and intelligence to accomplish missions, that could not been achieved by each individual participant alone.

Computational intelligence is quite a mature concept and it refers to a certain level of intelligence achieved by machines using computational methodological procedures. It is a subset of artificial intelligence, which embraces heuristic approaches, fuzzy logic etc. These methodologies are implemented into computational entities/objects to solve particular problems. The methodologies with which computational intelligence is achieved can vary from fuzzy logic to swarm intelligence. For example, a typical computational intelligence is computational collective intelligence.

Computational collective intelligence is the notion of computationally creating collective intelligence. The core of this concept is collective intelligence created and managed in a computational manner. Usually, the notion of collective intelligence is borrowed from nature, in which real creatures make decision collectively, share intelligence, attend emerging circumstances via group intelligence etc. Thus, the main concept behind computational collective intelligence is to seek analogies between real and artificial (computational) processes in modeling the collaboration and coordination processes towards the development of a collective intelligence with a motivation to solve large scale and/or complex problems. Computational collective intelligence does not necessitate individual intelligence for each unit taking part of collective action, where the intelligence is expected to be the concluding result. Therefore, the coordination is managed among the individuals towards a level of intelligence, where there is no prerequisite cognition per individual participant.

In contrast, collective computational intelligence is a new computational approach. It has also brought in attention of researchers for modeling and solving complex and large-size problems. It mainly associates with collectivity in computational intelligence, which is a reasonably mature concept in problem solving. It is related to computational collective intelligence approach but certainly these are not the same. Collective computational intelligence seeks ways in which various computational intelligence techniques collaborate towards collective behavior as well as intelligence. The main distinction is that all three underlying principles of collective intelligence, cognition, coordination and collaboration, play their complete roles in a way that each individual participant has certain level of intelligence, which corresponds to cognition, collaboration with other peer individual participant through a particular coordination procedure. Collective computational intelligence approaches are mainly based on multi agent systems and swarm intelligence.

Collective intelligence remains as the main ground of both computational collective intelligence and collective computational intelligence. That is to say, genetic/evolutionary algorithms/programming, swarm intelligence and multi-agent systems can be considered as the main computational methods to create any sort of collective intelligence.

3.1 Evolutionary Computation

Evolutionary Computation (EC) is very mature subfield of artificial intelligence acting as the umbrella name of the family of genetic/evolutionary algorithms/programming. It embraces all varieties of genetic algorithms, genetic programming, evolutionary algorithms and programming. It is mainly based on population of solutions evolved towards targeted status with various operating functions called operators. Since the individuals take part of collectivism managed by EC are simple and static solutions, there is no way to assume a priori intelligence as part of cognition required for collectivism. However, intelligence is expected to be a result through the process in which collaboration is achieved using selection operators as coordination method. In this sense, any version of EC algorithm can act as a collective intelligence technology in a way that the collectivism is only managed in the way of computational collective intelligence. On the other hand, collective computational intelligence can be achieved if the individuals – as contributors to the collectivism – bring up a certain level of a priori intelligence.

3.2 Swarm Intelligence

Swarm intelligence is also a subfield of artificial intelligence, where an intelligent behavior can emerge as the outcome of the self-organization of a collection of simple entities, organisms or individuals. Simple organisms that live in colonies such as ants, bees, bird flocks etc., have long fascinated many people for their collective intelligence that is manifested in many of the things that they do. A population of simple units can interact with each other as well as their environment without using any set of instruction(s) to proceed, and compose a swarm intelligence system.

The swarm intelligence approaches are to reveal the collective behavior of social insects in performing specific duties; it is all about modeling the behavior of those social insects and using these models as a basis upon which varieties of artificial entities can be developed. Within this context, problems could be solved by models, which exploit the problem solving capabilities of social insects. The motivation is to model the simple behaviors of individuals and the local interactions with the environment and neighboring individuals in order to obtain more complex behaviors that can be used to solve complex problems. These are mostly referring to optimization problems [22; 34; 50].

Bee colonies-based algorithms are recently developed swarm intelligence algorithms, which are inspired of the social behavior of the natural bee colonies. This family of algorithms has been successfully used for various applications such as modeling on communication networks [27], manufacturing cell formation [43], training artificial neural networks [44]. There is a rather common opinion on that bee colony algorithms are more successful in continuous problems than combinatorial problems. The main idea behind a simple bee colony optimization algorithm is to follow the most successful member of the colony in conducting the search. The scenario followed is that once a bee found a fruitful region, then it performs the waggle dance to communicate to the rest of the colony.

Once any member of the colony realizes that there is a waggle dance performance by a peer fellow it then moves to that member's neighborhood to collect more food. Inspiring of this natural process, bee colony optimization algorithms are implemented for efficient search methodologies borrowing this idea to direct the search to a more fruitful region of the search space. That would result a quicker search for an appropriate solution to be considered as a next nearby optimum solution. For further information readers are pointed to [43; 44] and [27].

Particle swarm optimization (PSO) is another population-based approach inspired of social behavior of bird flocking and fish schooling. PSO inventors were implementing such scenarios based on natural processes as explained below to solve the optimization problems [34]. Suppose the following scenario: a group of birds are randomly searching for food in an area, where there is only one piece of food available and none of them knows where it is, but they can estimate how far it would be. The problem here is "what is the best way to find and get that food". Obviously, the simplest strategy is to follow the bird known as the nearest one to the food. In PSO, each single solution, called a particle, is considered as a bird, the group becomes a swarm (population) and the search space is the area to explore. Each particle has a fitness value calculated by a fitness function, and a velocity of flying towards the optimum, food. All particles search across the problem space following the particle nearest to the optimum. PSO starts with initial population of solutions, which is updated on an iteration-by-iteration basis. For more information readers are pointed to [34; 35; 20; 24].

Ant colony optimization (ACO) became the general term of algorithms inspired of the collectivism in real ant colonies for food hunting and collection. These are mainly used for solving combinatorial optimization problems such as the traveler salesman problem [48] or routing problem in various networks [25]. It is well known that ant individuals use their personal knowledge and experience in taking action as well as the information shared by other peer ants. Once a particular ant has found a more useful way to go through, it immediately leaves some sort of substance, called pheromone, to let other peer ants to trace through. The pheromone is used to guide the search and let the ants cooperate as a team/colony to find high quality solutions [49; 16; 25].

3.3 Multi-Agent Systems

This is a well-known and pretty mature collective intelligence approach with which a set of agents aims to act individually and collaboratively for solving problems. The idea is to develop the models in a distributed manner and build a certain level of coordination to let them efficiently collaborate in solving the problems using their distributed intelligence. Multi agent systems are designed to ease the use of artificial intelligence techniques in a more efficient way in which various independent individual agents are implemented separately to make various versions of a single technique or various related techniques available to tackle the same problem source. This is because it is much easier to implement and develop sole algorithms then more complicated ones. Each of the agents harvests its own intelligence out of the algorithm equipped in it. Therefore, every individual agent applies the cognition principle of collective intelligence in this way. This leads to the 'coordination of the agents corresponding to the coordination principle of collective intelligence' problem. Tackling this problem universally remains a challenging issue despite of many approaches proposed with being domain-specific [41; 51; 36]. The collaboration is to harvest the synergy among the collection of the agents, where each one contributes proportionally to their own intelligence. The collaboration is delivered with letting each agent contribute once an efficient coordination is managed in one way or another.

A typical instance of multi agent systems can be the meta-heuristic agents. The meta-heuristic agents are identified to describe multi agent systems equipped with meta-heuristics to tackle hard optimization problems. Meta-heuristics have mostly been implemented as standalone applications in an ordinary sense and examined under such circumstances. Few multi agent systems implementing meta-heuristics are introduced and overviewed with respect to their performances. For further information readers are pointed to [8] and [32].

The idea of multi agent systems is to build up intelligent autonomous entities, which form up teams and solve problems in harmony. The agents equipped with meta-heuristics aim to solve hard and large-scale problems with their own intelligent search skills. Since standalone heuristic search usually face with local minima, ideas such as hybrid algorithms etc. have received intensive attention to overcome such shortcomings. On the other hand, the idea of multi agent agency eases building collaboration among various methods and approaches in a form of collaborating independent computational entities. Ideally, multi agent systems are to be executed on distributed systems due to their computational cost. Once swarms of agents are created that will be further expensive with respect to computational cost. In this respect, further information on how a PSO can be distributed to solve problems is available from [52].

The agents equipped with meta-heuristics become entities with in-built cognition facilities of intelligent search. Coordinating them will harmonize their various search skills towards a synergy, where each agent will contribute with their own skills for refreshing and diversifying the search mechanism. That will help the search to be rescued of prospective local optimum solutions. [8] and [9] propose a PSO-based swarm intelligence approach to coordination of meta-heuristic agent teams to solve combinatorial optimization problems. The idea is to run an ordinary PSO algorithm, which proposes swarms of enabled search agents homogeneously or heterogeneously with simulated annealing. The results seem promising. The implementation is reported that a particular grid programming tool [38] is used to overcome the computational cost.

4 Using Next Generation Emerging Technologies in Disaster Management

Here we discuss the projected applicability of the aforementioned technologies for augmenting intelligence in disaster management situations.

4.1 Motivation

There is clear evidence demonstrating the impact – on the society – from disasters [5]. In the EU alone, 494 disasters occurred between 2000 and 2007, claiming over 79,000 lives. The economic cost of these disasters is estimated at €103 billion, or approximately €15 billion per year. The statistics are even more sobering at global level. Since 1975, the annual number of disasters worldwide has increased from 75 to 400. There is also growing recognition among EU Member States of the imperative to work together on the prevention of, preparedness for and timely response to disasters occurring on their territories [26].

In managing a disaster, it is apparent that a number of teams and individuals from multiple, geographically distributed organizations (such as medical teams, civil protection, police, fire and rescue services, health and ambulance services, etc) will be required to communicate, co-operate and collaborate – in real time – in order to take appropriate decisions and actions [31; 40; 3]. 'The need for information exchange during an emergency situation is present; however it can be very diverse and complex' [19]. [19] also report that 'there are frequent quotes regarding the lack and inconsistent views of information shared in emergency operations'. There are also many small communities that 'do not have the resources, personnel and expertise to develop a set of requirements to assist them in managing their activities as they pertain to emergency response' [17]. Moreover, recent emergency management approaches are also characterized as inefficient because of their 'unstructured poor resource management and centralized nature with fixed hierarchical instructions'. Many scholars in the field also point out that for the management of emergency response operations, a number of ICT and relevant collaborative computer-based systems have been developed. However, report findings from National Resources Services [37] suggest that sustained efforts should be made with respect to data and resource archiving, sharing and dissemination. [37] refers to it as the 'hazards and disaster research informatics problem that is not unique to this research specialty, or other fields but it demands immediate attention and resolution'.

Naturally, information collection is one of the most crucial issues when managing disasters. This is due to the need for decisions to be done on a timely fashion, as well as based on correct and up to date information [6; 11; 5]. Our view here is that managing disasters should not be seen as a single sub-system but reflect, rather, systemic outcomes that result from the combined interaction of multiple sub-systems at different levels, where a sub-system may refer to a single communication means or to an event. It is our view that there should be a pervasive approach in investigating, acting on, controlling and managing the vast complexity of events and information related challenges occurred during any of the four phases of disasters (mitigation, preparedness, response and recovery). Thus, our motivation and challenge here is to pose the question: how are we going to enable disaster managers with the facility of exploring combined collections in a meaningful manner or simulation results that are defined across a broad range of ad-hoc, spatial and/or temporal scales?

4.2 A Disaster Management Scenario

We present here a previously published [13] and fictional yet typical disaster management case scenario, which is used throughout the remainder of our chapter.

An earthquake of a magnitude between 6.0 and 7.0 of the Richter scale occurs in an urban area x. The area is highly populated and characterized by multi-storey buildings, such as blocks of offices, malls and other public buildings. Specifically, there are three building types, b1, b2 and b3 in this area. The first type of buildings, b1 (area x1, a low-density populated area) refers to medium resistant buildings as they have been constructed using previous seismic construction regulation and their resistance is up to 6.5 of the Richter scale. These buildings have also experienced a number of earthquakes during their lifetime. The second type of buildings, b2 are located in area x2 that is a highly populated area. These buildings have been constructed under the current seismic regulations (up to 10.0 of the Richer scale), and thus, they refer to as high resistant buildings. Finally, in the highly populated area x2, there are few old buildings of b3 type, which have been constructed without following any building regulation. We assume that the occurrence of the earthquake caused a disastrous situation, as some of the buildings have collapsed and some people (victims: v) have been injured and trapped. Further to this, a number of secondary phenomena follow the occurrence of the main hazard, such as electricity failures, fires and a series of aftershocks. The area's civil protection department has organized the emergency operation in order to respond to the disaster. According to the area plans and to the emergency calls that reach the emergency services, operational units (OU) including rescue teams, engineers, health officials and medical doctors have been sent on site to locate, support and rescue earthquake victims. The members of these individuals and teams have to work as a unit and to report back to their operation centre's about their status and progress. OU members and experts have to find ways to reach trapped victims within the collapsed buildings. This process is dangerous, as the stability of the affected structural elements cannot be easily assessed. Further to this, the fact that aftershocks with different magnitudes and without lead-time occur in the area makes these attempts more difficult and dangerous. For example, imagine that while members of an OU-1 are inside an affected multi-storey block of offices an aftershock occurs. This in turn results in some of the already affected structural elements of the building collapsing. Our assumption leads to a realistic scenario whereby some OU-1 members are injured and trapped inside the building alongside the originally trapped victims. Other OUs (e.g. OU-2 or OU-n) and the operation centre do not know the condition of OU-1 members: if they are alive, seriously injured, as well as their exact condition and location. The scenario yields even more uncertainties, increased workloads, pressures and problems, as other OUs have to locate and rescue their OU-1 colleagues, help assist in rescuing victims meant to be rescued by the OU-1 team as well as deliver their original rescue plan (issued to them prior to the aftershock) without compromising more lives. Rescuing OU-1 members is considered a top priority as these now-victim members are valuable personnel with significant immediate value and inreplacable expertise in rescue operations.

The aforementioned scenario highlights several challenges. For example, [4; 13] disaster managers require to know where people are and be able to measure the present and projected impact: there is no real benefit in sending operational units to a place if there is no one actually there; equally, there is great benefit in sending in operational units to a place not considered at a great risk if someone injured is there. As someone may realize, it is easy to construct many plausible scenarios where knowledge can be collected from various emerging technologies to the benefit of the disaster managers and the society. That is to say, access to shared information about the number, whereabouts and health of people in an area struck by a disaster will significantly enhance the ability of disaster managers to respond timely to the reality of the situation. The aforementioned challenge is so vast and multifaceted that it is clearly insufficient to address all of them here.

4.3 Stakeholders' Requirements

Primary research findings [3] indicate that emergency management stakeholders include the civil protection, police, fire and rescue services, health and ambulance services, engineering sector, utility companies, local authorities, central government, relief bodies armed forces, monitoring, research and observatory centers and humanitarian organizations.

Interviews [3] with emergency managements stakeholders demonstrated clear evidence of the following requirements. In particular, they wish having the ability to/of:

- Allow decision makers to request, access and, assess information from various sources (including resources, external experts and instrumentation) related to a situation under alert;
- Various sources to collect information about a situation;
- Store information in one or more repositories;
- Authorized decision makers to plan and decide an appropriate action plan to tackle the situation based on what is available on them;
- Alert authorized decision makers if there is a situation that requires attention (including when an action plan decided is considered incorrect, incomplete or even if more resources are required);
- Send the job plan to relevant and available resources (including operational units);
- Resources or operational units to use allocated resources to take action once job plan has been received;
- Resources to take a job on demand;
- Operational units to report back of the job status including cases when more resources are required;
- Authorities and resources to set up a code of practice in the form of a set of policies (including ethics and body of law).

Table 1 Mapping Indicative Requirements to Indicative Technologies.

Indicative Questions	Indicative Answers	Indicative Technologies	Indicative Methods	Indicative Challenges
Where data is residing (capture)?	Data come from various sources and platforms. Environment is distributed and heterogeneous. Data sources can also fed from various instruments such as sensors.	P2P, push & notifications, pervasive, web 2.0, situated, crowd sourcing, wireless communications and mobile devices, ad-hoc networks.	Native or gateway engines including but not limited to Oracle, mySQL, Access, XML documents, FTP, HTTP, emails, etc.	Social networking, content and context aware data merging and integration methods.
Is data integration required (transmission, retrieval)?	Yes. Remote access from hierarchical, relational, object and flat files. Integration algorithms including extract from source, match, map and move to the target. On demand and on the fly integration is required.	P2P, push & notifications, pervasive, web 2.0, situated computing, crowd sourcing, wireless communications and mobile devices, ad-hoc networks.	Web services, APIs, OGSAI-DAIS, RFID, etc. Support for various drivers such as CGI, ODBC, JDBC, etc. Transfer from/to E-mail, URL, FTP, etc.	Social networking, autonomic, semantics, trust, reliability, reputation, security, in-compatibility, data anomalies, policies, etc.
Are there any analyses and/or simulation modeling involved (manipulation)?	Yes. Decision support systems, intelligent agents and artificial intelligence.	Complex event processing, models, cases, scheduling, monitoring, etc.	Markov, Bayesian, multi-variation, optimization, interpolation, etc.	Data and text mining, collective computational intelligence.
Are these computationally intensive (store)?	Yes. Remote access is required.	Clusters, grid and cloud computing.	Virtualization, and replication, batch processing, firewalls, etc.	Outsourcing, sustainability, scalability, load balancing, Queuing, etc.
Do results are outputted and/or communicated in different ways (display)?	Yes. Data could be displayed in various devices including PCs, GIS, mobiles, mashups, maps, etc.	Pervasive and situated computing, P2P, web 2.0, telepresence, etc.	Ajax, XML, mobile agents, location based services, transcoding, content adaptation, etc.	Context awareness, standards, collaborative technologies, personalization, RSS, etc.
What is the end-user's IT related skill-set?	Varies from novice to highly experienced. Non-human users are also included.	Agents, context aware technologies agents, etc.	Learning theory, Swarm intelligence, q-learning, remote profile discovery.	Ambient intelligence, adaptive interfaces, HCI, virtual worlds, tagging, etc.

Table 1 illustrates the stakeholders' requirements whilst Figure 1 illustrates an attempt to draft a relevant technology roadmap [14]. However, it is our view that the aforementioned outcomes stand as a preliminary activity for identifying the stakeholders, their requirements and success criteria. Thus, further engagement with emergency management stakeholders is required. The proposed engagement will fully update our pilot findings in which a commonly agreed shared vision amongst stakeholders can be developed. It is also believed that the proposed approach could lead to the identification of good practices in the sector, the identification of common problems and challenges and development of a definitive roadmap describing what specific needs are to be solved.

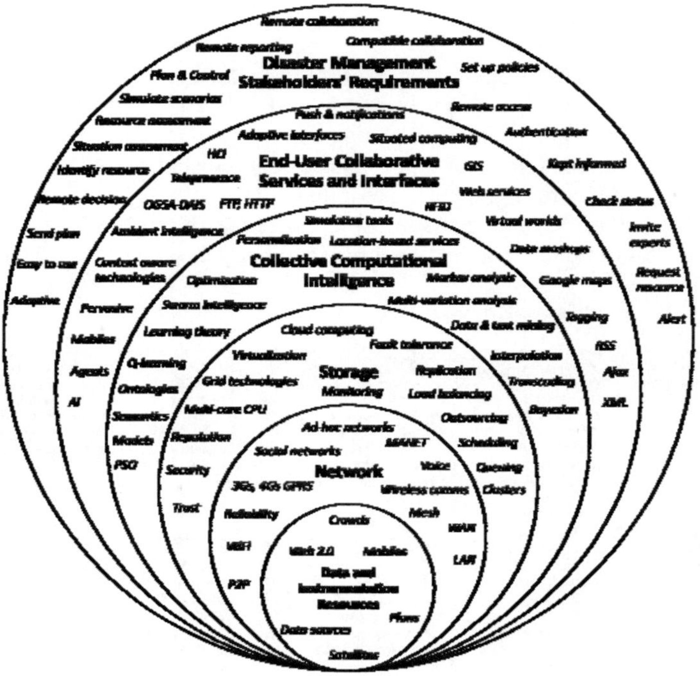

Fig. 1 A Draft Technology Roadmap to Enable Collective Computational Intelligence in Disaster Management (adapted from [14]).

5 Model Architecture

In this section we illustrate a model architecture and details our current implementation strategy.

5.1 A Technical Model Architecture

To start with, Figure 2 illustrates our previously published low-level model architecture [13]. This shows the flow of interactions between computational devices capable of sensing the environment and establishing an ad-hoc mobile network. In brief, the flow takes into account that during a disaster we could usefully leverage various distributed emerging technologies to visualize the status or conditions of victims who have trapped in a structurally damaged building.

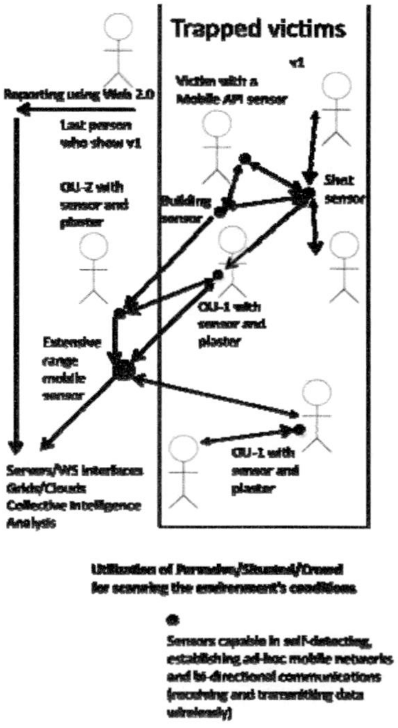

Fig. 2 A Low-level Flow (adapted from [14]).

For example, let us assume that every member of an operational rescue unit wears a plaster that records data about individual health condition, as well as devices that sense the environment. We also assume that trapped rescue team members who are in good enough condition to do so could disperse one or more sensors so they can start collecting relevant data about the environment over a range for which their own sensor and plaster could not function and/or detect. We also assume that buildings could have installed sensors and finally, we assume that victims could have installed sensor APIs on their mobile devices. However, we do appreciate that the latter APIs would be limited in data transfer as well as in detecting and capturing a variety of vital signs.

Arguably, a trapped person's sensor could detect other sensors available in the environment; and this would create and establish a limited ad-hoc network. This will enable communication between mobile APIs, sensors and plasters with the view to transferring data across networks residing outside the building. We suggest the use of Grids and Clouds for data processing and storage, as well as the use of collective computational intelligence tools including complex event processing for their meaningful analysis. Complex event processing works on the event – condition – action (ECA) logic and it is particularly useful for analyzing the complexity of multi-criteria that could lead to trigger an alert. For example, certain levels of combined smoke and temperature could imply a fire or, a complete change of sensors' positioning in a building could imply a change in the building structure.

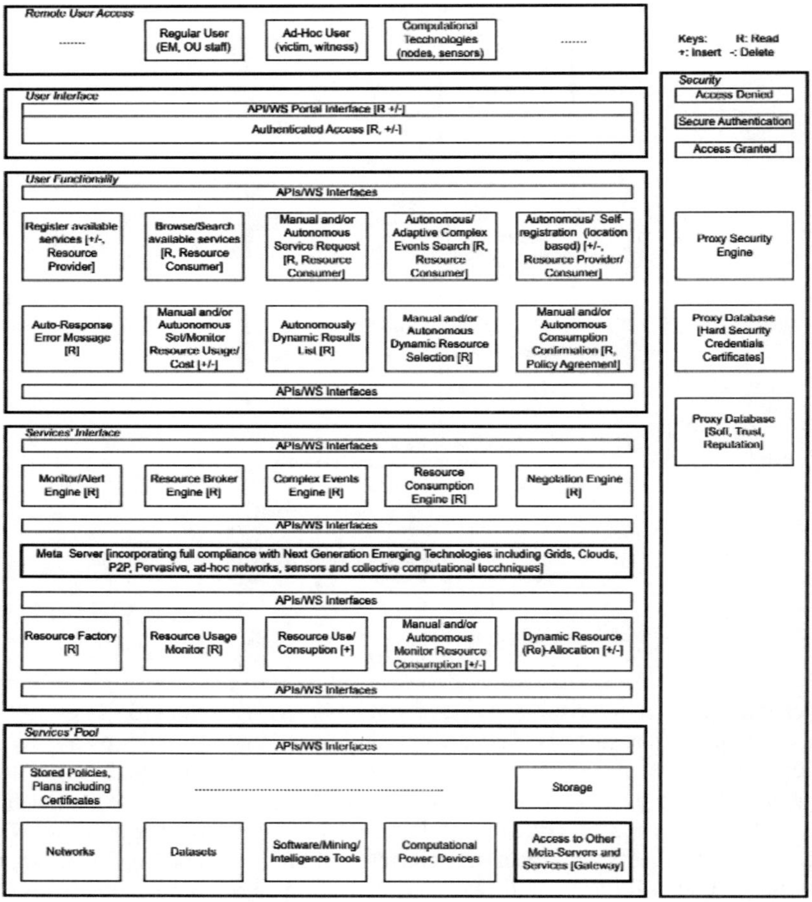

Fig. 3 A High-Level Technical Model Architecture for Augmenting Collective Computational Intelligence for Disaster Management

Figure 3 illustrates a more detailed model architecture demonstrating how the functionality available and the aforementioned next generation emerging technologies relate and impact in realizing, making sense of and ultimately enabling a more informed decision-making based on the actual situation rather than a speculative analysis. Specifically, the model appreciates that each member from the Virtual Organization (VO) community may have a different domain of specialization, which requires taking into account when managing disasters and occupational hazards. In other words, like the technologies that have been developed with the view of complementing each other, limitations of individual members and their infrastructure may be satisfied from any other member. Since neither everybody nor any technology can perform all tasks, a group encompassing different resources, support technologies and individuals may be utilized in a manner, which will collectively cover a much larger domain. Due to the complexity involved we have not made links between functions and services.

In terms of the functionality available, the model appreciates the need for accessing the portal interface remotely and that there are various remote users who could access it including but not limited to sensing devices that capture a situation, emergency management users, ad-hoc users such as victims or other who have witness a victim or an emergency. Users get access to the portal after a successful authentication control. Authentication takes decisions on the basis of both security standards (PKI, X509, etc) and softer issues such as trust and reputation as there is a need to ensure the reputation of a service requestor and/or provider. Following the authentication procedure users can register their resources using some metadata descriptions, which can be stored in a factory for their future harvesting. Users may also request for resources in either manual or autonomous manner. For example, a wearable sensor such as a plaster monitoring someone's status could raise an alert, which requires a resource's attention and/or consumption. Following the search procedure (manual or else) involving a broker the resource availability becomes live. A policy based negotiation between resource provider and requestor is performed prior to the resource allocation. A monitoring function is used to dynamically re-allocate resources when these become unavailable for some reason. A complex event engine is also available as to ensure that combination of parameters may lead to alerts. As resource consumption may come with some cost, a resource consumption function is embedded to ensure that usage is within the set limit and that resources are used in the most efficient and effective manner from both the consumer and provider ends. It is important to remind that each function or service support multi-instances regardless if they are shown as single instances. Finally, the pool refers to all possible resources, which could be consumed locally or else. Resources include but not limited to Clusters, Grids, Clouds, P2P, Crowd sourcing devices, Web 2.0 feeds, sensors, networks, mining tools and computational collective intelligence tools, policies, plans and maps.

5.2 Learning Birds for Information Gathering: A Future Implementation Scenario

Let us assume there is an area in which massive information is required to be gathered as a matter of emergency in order for decision makers to take appropriate actions, i.e. facilitation of rescuing services.

A typical area is presented in Figure 4, where both urban and rural parts are included. One way to fulfill the aforementioned requirement would be the use of a number of mobile vehicles interconnected and organized in certain way to scan the area in a distributed manner. The team of the mobile vehicles should be devised with a collective computational intelligence system to organize the physically distributed and dispersed vehicles so that the entire team behave in harmony and collaboration for collecting information as accurate as possible.

We suggest a collective computational intelligence implementation through learning agents called learning birds. These shall be able to move around the area as a team, scan the area as a team and learn from one another during the phase of information gathering. One of the first challenges is to learn how to collaborate in developing collective behaviors. To achieve this, they will need to develop and thus, reach a certain level of learning ability by exchanging information.

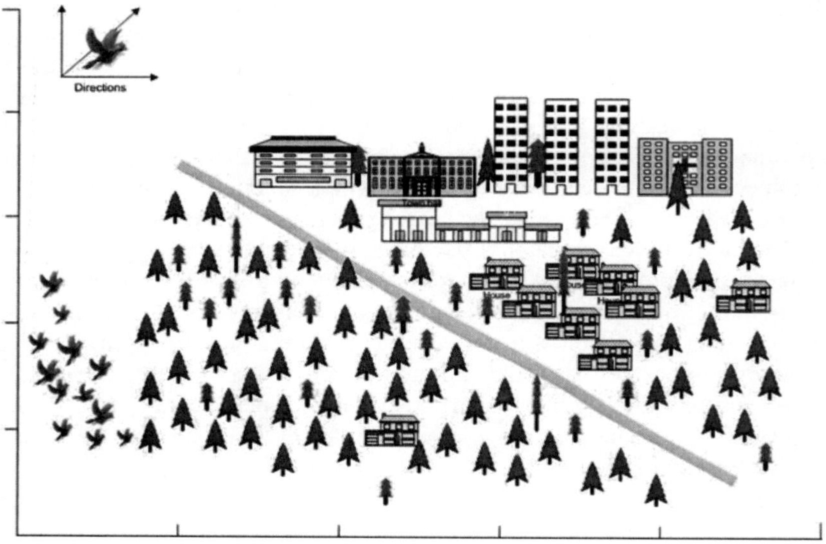

Fig. 4 A Typical Scenario Area where Scanning by Learning Birds is required.

The proposal here is that a swarm intelligence algorithm is required to be implemented with the purpose of gathering information autonomously. Thus, each member of the swarm shall be designed as an intelligent agent in order to satisfy the role of learning community of birds as a team. A reinforcement-learning algorithm, namely Q learning algorithm, should be embedded into a swarm intelligence algorithm, namely particle swarm optimization algorithm, for the purpose of implementing swarms of learning birds. Each bird taking part of the swarm shall be implemented as a particle of the swarm, where the position vector shall be used to indicate the position of the bird based on the Euclidean space whilst the velocity one is ignored. This is because moving from one position to another is made subject to Q learning algorithm not to the velocity vector, where the behaviors of the birds are assessed and reinforced through. The forward movement is defined based on the direction and amount of distance to be taken per step. Once a direction chosen and the size of the step to move forward will be taken a reinforcement function assesses and then creates a reward for approving the decision made if applicable. The goal is to let the birds gain experience while move around scanning the environment for the information gathering. They should try to keep in contact with not being too far neither too close to one another. Therefore, the step-size and the direction remain essential functions for optimization purposes. Our view is that right at start, they should share their positioning data so they can start detect how far they should be from the rest of the members of the swarm. Next, they should update themselves at regular intervals with data about the current situation of the environment and also about their positioning, as they may need to change their trajectory. Currently, we are under the process of developing such an algorithm.

6 Conclusions

In this chapter, we have discussed a visionary opportunity among various emerging technology-based paradigms including Grid, Cloud, Crowd, Pervasive and Situated Computing, to be integrated for a collective computational intelligence model for disaster management where a smart approach would be a significant advantage. To achieve this, we reviewed the possible combination of the aforementioned next generation emerging technologies with intelligence techniques such as multi-agents, evolutionary computation and swarm intelligence for augmenting computational intelligence in a collective manner for the purpose of managing disasters.

This chapter, therefore, has addressed our motivation and a notable issue. That is to say, how to enable disaster managers with the facility of exploring combined collections in a meaningful manner that are defined across a broad range of ad-hoc, spatial and/or temporal scales. To that end, the chapter has discussed our current implementation efforts and also presented a relevant technical model architecture that illustrates the use of these technologies enabling an augmented collective computational intelligence approach in disaster management. Further to these, our future implementation strategy aims to develop a high-level technical roadmap and start on simulating and implementing a number of technical services based on realistic scenarios.

References

1. Ahmed, E., Bessis, N., Norrington, P., Yue, Y.: Managing Inconsistencies in Data Grid Environments: A Practical Approach. International Journal of Grid and High Performance Computing, IGI (2010) (in press)
2. Jackson, M., Krause, A., Laws, S., Magowan, J., Paton, N., Pearson, D., Sugden, T., Watson, P., Westhead, M.: The design and implementation of grid database services in OGSA-DAI. Concurrency and Computation: Practice and Experience 7(2-4), 357–376 (2005)
3. Asimakopoulou, E.: A Grid-Aware Emergency Response Model for Natural Disasters, PhD Thesis, Loughborough University (2008)
4. Asimakopoulou, E., Bessis, N.: Advanced ICTs for Disaster Management and Threat Detection: Collaborative and Distributed Frameworks. IGI Publishing (2010) ISBN: 978–1615209873
5. Asimakopoulou, E., Bessis, N., Varaganti, R.: The Implementation of a Personalised Forest Fire Evacuation Data Grid Push Service. In: International Conference in Disaster and Reduction, IDRC, Davos, May 30– June 3 (2010)
6. Asimakopoulou, E., Bessis, N., Varaganti, R., Norrington, P.: A Personalized Forest Fire Evacuation Data Grid Push Service – The FFED-GPS Approach. In: Asimakopoulou, E., Bessis, N. (eds.) Advanced ICTs for Disaster Management and Threat Detection: Collaborative and Distributed Frameworks, pp. 279–295. IGI Publishing (2009) ISBN: 978-1615209873
7. Atkinson, M, Dialani, V, Guy, L, Narang, I, Paton, N, Pearson, P, Storey, T, Watson P (2003) Grid database access and integration: requirements and functionalities. Report, `http://www.ggf.org/documents/GFD.13.pdf` (retrieved August 17, 2008)
8. Aydin, M.E.: Metaheuristic agent teams for job shop scheduling problems. In: Mařík, V., Vyatkin, V., Colombo, A.W. (eds.) HoloMAS 2007. LNCS (LNAI), vol. 4659, pp. 185–194. Springer, Heidelberg (2007)
9. Aydin, M.E.: Coordinate metaheuristic agents with swarm intelligence. Journal of Intelligent Manufacturing (2010a)
10. Aydin, M.E.: Collaboration of heterogenous metaheuristic agents. In: Proceedings of ICDIM 2010, pp, pp. 540–545 (2010b)
11. Bessis, N.: Model Architecture for a User tailored Data Push Service in Data Grids. In: Bessis, N. (ed.) Grid Technology for Maximizing Collaborative Decision Management and Support: Advancing Effective Virtual Organizations, pp. 235–255. IGI Publishing (2009) ISBN: 978-1-60566-364-7
12. Bessis, N.: Using Next Generation Grid Technologies for Advancing Virtual Organizations. In: Bessis, N. (ed.) Keynote Talk in the International Conference on Complex, Intelligent and Software Intensive Systems (CISIS 2010), Krakow, Poland, xlvii - xlvii (February 2010)
13. Bessis, N., Asimakopoulou, E., French, T., Norrington, P., Xhafa, F.: The Big Picture, from Grids and Clouds to Crowds: A Data Collective Computational Intelligence Case Proposal for Managing Disasters. In: Proceedings of 5th IEEE International Conference on P2P, Parallel, Grid, Cloud and Internet Computing (3PGCIC-2010), 1st International Workshop on Emerging Data Technologies for Collective Intelligence (EDTCI-2010), Fukuoka, Japan, pp. 351–356 (November 2010) ISBN: 978-07695-4237-9

14. Bessis, N., Asimakopoulou, E., Xhafa, F.: A next generation emerging technologies roadmap for enabling collective computational intelligence in disaster management. International Journal of Space-Based and Situated Computing 1(1) (2011)
15. Bessis, N., Brown, A., Asimakopoulou, E.: A Mathematical Analysis of a Disaster Management Data-Grid Push Service. International Journal of Distributed Systems and Technologies, IGI 1(3), 56–70 (2010) ISSN: 1947-3532
16. Blum, C., Dorigo, M.: The hyper-cube framework for ant colony optimization. IEEE Transactions on System, Man, and Cybernetics, Part B, 1–12 (2004)
17. Bui, T., Lee, J.: An Agent-Based Framework for Building Decision Support Systems, Decision Support Systems. The International Journal 25(3) (1999)
18. Buyya, R.: Cloudbus Toolkit for Market-Oriented Cloud Computing (2008), `http://www.buyya.com/papers/Cloudbus-Keynote2009.pdf`
19. Carle, B., Vermeersch, F., Palma, C.R.: Systems Improving Communication in Case of a Nuclear Emergency. In: Conference of the International Community on Information Systems for Crisis Response Management (ISCRAM 2004), Brussels, Belgium, May 3-4 (2004)
20. Chen, A., Yang, G., Wu, Z.: Hybrid discrete particle swarm optimization algorithm for capacitated vehicle routing problem. Journal of Zhejiang University Science A 7(4), 607–614 (2006)
21. Clancey, W.: Situated Cognition. Cambridge University Press, Cambridge (1997), `http://cs.gmu.edu/~jgero/publications/2003/03oGerooCAADRIA03.pdf`
22. Colorni, A., Dorigo, M., Maniezzo, V., Trubian, M.: Ant system for job-shop scheduling. Belgian Journal of Operations Research, Statistics and Computer Science (JORBEL) 34(1), 39–53 (1994)
23. De Assuncao, M.D., di Costanzo, A., Buyya, R.: Evaluating the cost-benefit of using cloud computing to extend the capacity of clusters. In: Proceedings of the 18th ACM international Symposium on High Performance Distributed Computing, HPDC 2009., Garching, Germany, pp. 141–150. ACM, New York (2009)
24. Dong, C., Qiu, Z.: Particle swarm optimization algorithm based on the idea of simulated annealing. International Journal of Computer Science and Network Security 6(10), 152–157 (2006)
25. Dorigo, M., Stützle, T.: Ant Colony Optimization. MIT Press, Cambridge (2004)
26. European Union, Reinforcing the European Union's Disaster Response Capacity, (2010), `http://ec.europa.eu/governance/impact/planned_ia/docs/28_echo_eu_disaster_response_capacity_en.pdf`
27. Farooq, M.: Bee-inspired protocol engineering: From nature to networks. Springer, Berlin (2008)
28. Foster, I., Kesselman, C., Tuecke, S.: The anatomy of the grid: enabling scalable virtual organisations. International Journal of Supercomputer Applications 15(3), 200–222 (2001)
29. Gao, Y.M., Pun, S.H., Du, M., Mak, P.U., Vai, M.I.: Simple electrical model and initial experiments for intra-body communications. Proceedings of the IEEE Eng. Med. Biol. Soc., 697–700 (2009)
30. Gero, J.S.: Situated Computing: A New Paradigm for Design Computing (2006), `http://citeseerx.ist.psu.edu/viewdoc/summary?`, doi:10.1.1.91.4545

31. Graves, R.J.: Key Technologies for Emergency Response. In: Conference or the International Community on Information Systems for Crisis Response (ICSCRAM 2004), Brussels, Belgium, May 3–4 (2004)
32. Hammami, M., Ghédira, K.: COSATS, X-COSATS: Two Multi-agent Systems Cooperating Simulated Annealing, Tabu Search and X-Over Operator for the K-Graph Partitioning Problem. In: Khosla, R., Howlett, R.J., Jain, L.C. (eds.) KES 2005. LNCS (LNAI), vol. 3684, pp. 647–653. Springer, Heidelberg (2005)
33. ICL (2009) Digital 'plaster' for monitoring vital signs undergoes first clinical trials. News release. Imperial College London, Faculty of Medicine. November 02 (2009), http://www1.imperial.ac.uk/medicine/news/20091102_digitalplaster
34. Kennedy, J., Eberhart, R.C.: Particle swarm optimization. In: Proceedings of IEEE International Conference on Neural Networks, Perth, Austrailia (1995)
35. Kennedy, J., Eberhart, R.C.: A discrete binary version of the particle swarm optimization. In: Proceedings of IEEE Conference on Systems Man and Cybernetics, Pisctaway, NY, USA (1997)
36. Kolp, M., Giorgini, P., Mylopoulos, J.: Multi-agent architectures as organizational structures. Autonomous Agents and Multi Agent Systems 13, 3–25 (2006)
37. National Research Council (NRC), Facing Hazards and Disasters: Understanding Human Dimensions. National Academy Press, USA (2006)
38. Nguyen, T.A., Kuonen, P.: Programming the grid with POP C++. Future Generation Computer Science 23(1), 23–30 (2007)
39. Nieto-Santisteban, M.A., Gray, J., Szalay, A.S., Annis, J., Thakar, A.R., O'Mullane, W.J.: When database systems meet the grid. Technical Report. Microsoft Research, Microsoft Corporation (2004)
40. Otten, J., Heijningen, B., Lafortune, J.F.: The Virtual Crisis Management Centre. An ICT Implementation to Canalise Information! In: Conferenceof the International Community on Information Systems for Crisis Response (ISCRAM 2004), Brussels, Belgium, May 3-4 (2004)
41. Panait, L., Luke, S.: Cooperative multi-agent learning: The state of the art. Autonomous Agents and Multi-Agent Systems 11, 387–434 (2005)
42. Paulos, E.: Designing for Doubt: Citizen Science and the Challenge of Change, Engaging Data. In: Proceedings of 1st International Forum on the Application and Management of Personal Information, MIT, Cambridge, USA (2009)
43. Pham, D.T., Afify, A., Koc, E.: Manufacturing cell formation using the Bees Algorithm. In: Pham, et al. (eds.) IPROMS 2007 Innovative Production Machines and Systems Virtual Conference, Cardiff, UK (2007)
44. Pham, D.T., Otri, S., Ghanbarzadeh, A., Koc, E.: Application of the bees algorithm to the training of learning vector quantisation networks for control chart pattern recognition. In: Proceedings of the Information and Communication Technologies (ICTTA 2006) pp, Syria, pp. 1624–1629 (2006)
45. Reinoso Castillo, J.A., Silvescu, A., Caragea, D., Pathak, J., Honavar, V.G.: Information extraction and integration from heterogeneous, distributed, autonomous information sources – a federated ontology – driven query-centric approach. In: Paper presented at IEEE International Conference on Information Integration and Reuse, http://www.cs.iastate.edu/~honavar/Papers/indusfinal.pdf(retrieved August 17, 2004)
46. Schubertt, L., Jeffery, K., Neidecker-Lutz, B.: Expert Group Report, The Future of Cloud Computing: Opportunities for European cloud computing beyond, European Commission, Belgium (2010)

47. Smith, C.: A little plaster goes a long way (2007), http://www3.imperial.ac.uk/newsandeventspggrp/imperialcollege/newssummary/news_11-12-2007-9-27-28
48. Stützle, T., Dorigo, M.: ACO algorithms for the traveling salesman problem. In: Miettinen, K., Makela, M., Neittaanmaki, P., Periaux, J. (eds.) Evolutionary Algorithms in Engineering and Computer Science: Recent Advances in Genetic Algorithms, Evolution Strategies, Evolutionary Programming, Genetic Programming and Industrial Applications, John Wiley & Sons, Chichester (1999)
49. Stützle, T., Hoos, H.H.: Max-min ant system. Future Generation Computer Systems 16(8), 889–914 (2000)
50. Tasgetiren, M.F., Liang, Y.C., Sevkli, G., Gencyilmaz, M.: Particle swarm optimization algorithm for makespan and total flowtime minimization in permutation flowshop sequencing problem. European Journal of Operational Research 177(3), 1930–1947 (2007)
51. Vazquez-Salceda, J., Dignum, V., Dignum, F.: Organizing multi-agent systems. Autonomous Agents and Multi-Agent Systems 11, 307–360 (2005)
52. Wang, X., Ma, J.J., Wang, S., Bi, D.W.: Distributed particle swarm optimization and simulated annealing for energy - efficent coverage in wireless sensor networks. Sensor 7, 628–648 (2007)
53. Weiser, M.: The Computer for the Twenty-First Century. In Scientific American 265(3), 94–104 (2001)
54. Winton, L.J.: A Simple Virtual Organization Model and Practical Implementation. In: Winton, L.J. (ed.) Proceedings of the 2005 Australasian workshop on Grid computing and e-research, vol. 44, pp. 57–65 (2005)
55. Wohrer, A., Brezany, P., Janciak, I.: Virtalisation of heterogeneous data sources for grid information systems.(2004), http://www.par.univie.ac.at/publications/other/inst_rep_2002-2004.pdf

Glossary of Terms and Acronyms

ACO	Ant Colony Optimization
API	Application Programming Interface
BAN	Body Area Network
CI	Cyber Infrastructure
DAI	Data Access and Integration
EC	Evolutionary Computation
ECA	Event – Condition – Action
EU	European Union
FTP	File Transfer Protocol
GIS	Geographical Information Systems
HCI	Human-Computer Interaction
HTML	Hyper Text Markup Language
HTTP	Hyper Text Transfer Protocol
IBC	Intra-Body Communication

ICT	Information Communication Technology
IET	Institution of Engineering Technology
IT	Information Technology
JDBC	Java Database Connectivity
ODBC	Object Database Connectivity
OGSA	Open Grid Services Architecture
OU	Operational Unit
PC	Personal Computer
PDA	Personal Digital Assistant
PKI	Public Key Infrastructure
PSO	Particle Swarm Optimization
P2P	Peer-to-peer
RFID	Radio Frequency IDentification
RSS	Really Simple Syndication
SQL	Structured Query Language
XML	Extensible Markup Language
VO	Virtual Organizations
URL	Uniform Resource Locator

Chapter 20
Emerging, Collective Intelligence for Personal, Organisational and Social Use

Sotiris Diplaris, Andreas Sonnenbichler, Tomasz Kaczanowski, Phivos Mylonas, Ansgar Scherp, Maciej Janik, Symeon Papadopoulos, Michael Ovelgoenne, and Yiannis Kompatsiaris

Abstract. The main objective of this chapter is to present novel technologies for exploiting multiple layers of intelligence from user-contributed content, which together constitute Collective Intelligence, a form of intelligence that emerges from the collaboration and competition among many individuals, and that seemingly has a mind of its own. User contributed content is analysed by integrating research and development in media analysis, mass content processing, user feedback, social analysis and knowledge management to automatically extract the hidden intelligence and make it accessible to end users and organisations. The exploitation of the emerging Collective Intelligence results is showcased in two distinct case studies: an Emergency Response and a Consumers Social Group case study.

Sotiris Diplaris · Ioannis Kompatsiaris
Informatics & Telematics Institute, Thermi, Thessaloniki
e-mail: {diplaris, ikom}@iti.gr

Andreas Sonnenbichler · Michael Ovelgoenne
Karlsruhe Institute of Technology, Germany
e-mail: {andreas.sonnenbichler, michael.ovelgoenne}@kit.edu

Tomasz Kaczanowski
Software Mind S.A., Krakow, Poland
e-mail: tomasz.kaczanowski@softwaremind.pl

Phivos Mylonas
National Technical University of Athens, Greece
e-mail: fmylonas@image.ntua.gr

Ansgar Scherp · Maciej Janik
University of Koblenz-Landau, Germany
e-mail: {scherp, janik}@uni-koblenz.de

Symeon Papadopoulos
Informatics & Telematics Institute, Thermi, Thessaloniki, Greece and Department of Computer Science Aristotle University of Thessaloniki, Greece
e-mail: papadop@iti.gr

N. Bessis, F. Xhafa (Eds.): Next Generation Data Technologies for CCI, SCI 352, pp. 527–573.
springerlink.com

1 Introduction

Due to advances in communications, mobile devices and Web technologies, it is nowadays easy for users and organisations to generate and share content, individually or within communities. Social media sharing properties, such as Flickr, Facebook and PicasaWeb host billions of images and video, which have been annotated and shared among friends, or published in groups that cover a specific topic of interest. The fact that users annotate and comment on the content in the form of tags, ratings, preferences etc and that these are applied on a daily basis, gives this data source an extremely dynamic nature that reflects events and the evolution of community focus. Although current Web 2.0 applications allow and are based on annotations and feedback by the users, these are not sufficient for extracting this "hidden" knowledge, because they lack clear semantics and it is the combination of visual, textual and social context, which provides the ingredients for a more thorough understanding of social content. Therefore, there is a need for scalable and distributed approaches able to handle the mass amount of available data and generate an optimized 'Intelligence' layer, also called Collective Intelligence, that would enable the exploitation of the knowledge hidden in the user contributed content. There already exist a number of approaches, which use user-contributed content in order to provide useful information for various applications. For example, mobile location information and uploaded content is used to monitor online traffic and generate traffic patterns in [84], connect citizens in Boston [76], share nature experience [9], discover travel patterns and provide travel advice [11] [6], or communicate problems in a city [8]. The MIT Center for Collective Intelligence[1] is hosting a series of projects aiming at harnessing and using Collective Intelligence. These include the Climate Collaboratorium [53], which deals with climatic changes and the Collective Prediction effort, which tries to make accurate predictions about future events such as product sales, political events, and outcomes of medical treatments [43]. Collective Intelligence is also used in healthcare [65], while studies are conducted on its applications in today's organizations [10]. However, the main characteristic of such applications is that they are mostly based on collecting well-structured contributions through specific applications, on shallow statistical processing of the contributions and their visualization. Very few focus on analysis and on dealing with unstructured large-scale data, where an important source of knowledge is hidden.

Large-scale user contributions and more specifically, tags, which can be used to extract Collective Intelligence, suffer from a number of limitations, such as polysemy, lack of uniformity, and spam, thus not presenting directly an adequate solution to the problem of content organization. Therefore, how to manage, index and search for this content effectively and efficiently is becoming an increasingly important research topic. There have been many approaches dealing with the relevant tasks of tag refinement (to refine the unreliable user-provided tags) and automatic annotation or tagging, especially for user-contributed photos. In [78] the Content-based Annotation Refinement (CBAR) method re-ranks the tags of an image, reserving the top ones as the refined results. A more elaborated method [49] refines and enriches

[1] http://cci.mit.edu/index.html

tags based on the visual and semantic consistency residing in social sites. Other algorithms perform automated tagging of an untagged image, either by building classifiers for individual semantic labels [48] [21], or by learning relevance models between images and keywords [40] [2]. The most frequently used method for automatic tagging is the semi-supervised graph-based level propagation technique [86], where a graph is constructed to model the relationship among individual images in terms of visual similarity. Such approaches, although they make use of the available content collections and their connections, they are trying to improve the quality of each independent user contributed content item and therefore they do not directly address the objective of extracting the hidden Collective Intelligence, which can be used in applications other than annotation and retrieval.

Existing approaches towards extracting Collective Intelligence usually build upon restricted combinations from the available social media attributes. For example, in [44] geo-locations and tag information are used in order to generate representative city maps. In [66] tags and visual information together with geo-location are used for objects (e.g. monuments) and events extraction. Tags from Flickr images and timestamp information are used in [35] to form a chronologically ordered set of geographically referenced photos and distinguish locals from tourist travelling. The description of city cores can be derived automatically, by exploiting tag and location information [37]. The approach is able of distinguishing between administrative and vernacular uses of place names, thus avoiding the potential for confusion in the dispatch of emergency services. But besides these combinations, user generated content can be viewed as a rich multi-modal source of information including attributes such as time, favorites and social connections. For example, beyond harnessing content and the surrounding tags or text, limited effort has been made to include the social patterns into the media analysis. The important aspect of fusion of modalities and different sources is currently lacking in existing Collective Intelligence applications. In this chapter novel techniques for exploiting these multiple layers of intelligence from user-contributed content are presented, which together constitute Collective Intelligence, a form of intelligence that emerges from the collaboration and competition among many individuals. The Collective Intelligence technologies to be described were developed in the context of the FP7 EU project WeKnowIt: Emerging, Collective Intelligence for personal, organisational and social use[2]. User contributed content is analysed by integrating research and development in visual content analysis for localisation (Media Intelligence), tag clustering and Wikipedia ontology-based categorization (Mass Intelligence), analyzing social structures and user communities access rights (Social Intelligence) and event representation (Organisational Intelligence). The exploitation of the emerging Collective Intelligence results is showcased in two distinct case studies: an Emergency Response and a Consumers Social Group case study.

The chapter structure comprises an overall of nine sections. After this Introduction, the following section details the current state-of-the-art for each intelligence layer. The next four sections describe the developed technologies in each

[2] http://www.weknowit.eu

intelligence layer individually, namely the Media, Mass, Social and Organisational Intelligence layers. The seventh section presents the integration of the different technologies within one framework, namely the Integrated Collective Intelligence Framework (ICIF), which is exploited in the Emergency Response and Consumer Social Group[3] use cases. The eighth section describes these two application scenarios where the presented techniques have been used together in order to leverage Collective Intelligence. The section also includes user evaluation results for the developed demonstrators. The last section contains conclusions and possibilities on building on top of the achieved Collective Intelligence results.

2 Background

The presented approach towards Collective Intelligence builds on two aspects: mass content availability provided by a lot of users and availability of analysis techniques and results from different layers. Collective Intelligence methods can be classified based both on the number of the input modalities or layers that they employ in the analysis and the number of users contributing to the data. More specifically, the different Intelligence layers that contribute to Collective Intelligence can be classified to the following:

Media intelligence is the intelligence originated from digital content items (images, video, audio, text) and contextual information analysis, either provided by the user or pre-existing, and their merging. For this purpose, intelligent, automated content analysis techniques are used for different media to extract knowledge from the content itself. Since the amount of data is large and noisy, machine learning, data mining and information retrieval methods are used. Also the methods are able to fuse information from different sources/modalities, contextual information (e.g. time, location, and EXIF metadata), personal context (profile, preferences, etc.) and social context (tagging, ratings, group profiles, relevant content collections etc.).

Mass intelligence analyzes user feedback. Mass analysis enables input information clustering and ranking as well as information and event categorization. Also, bursts of information can be detected that may indicate potential events (emergency) and trend analysis and prediction. Facts and trends are recognized and modelled by interpreting user feedback on a large scale. For instance, a single road being blocked in a storm may not be very critical, but all access roads being blocked towards a hospital centre may be very critical in the case of an emergency.

Social Intelligence is the exploitation of information about the social relations between members of a community. Nearly everything humans do, they do in a social context because they communicate, collaborate or in some other way interact with other people. Information about the various types of social relations may be represented in communication networks, friendship networks or organization charts.

Organisational Intelligence allows support of decision making through workflows exploiting the generated knowledge and taking into account existing procedures within an organisation. This is quite a departure from traditional methods

[3] http://weknowit.research.yahoo.com/csg/

where knowledge is produced by the individual knowledge worker and collected and integrated manually in knowledge based systems or organisational repositories.

2.1 Advances in Media Intelligence

Following current web socializing and multimedia hanging user trends, most users nowadays upload, describe, geo-tag and localize their personal photos, based on previous personal or community content. Undoubtedly, the growth of such image collections and change of user behaviours created the need for intelligent algorithms, able to analyse mass heterogeneous multimedia content. In the following we focus on content-based retrieval systems whose aim is to identify the same object under various viewpoints within a large database and where in most cases local features extraction is no longer sufficient. Perhaps the most popular features used for object categorization are the SIFT features [50]. A typical example of using local patches around key-points for the retrieval of objects is [58]. A very well-known category of approaches is commonly known as "bag-of-words" model, which shares significant similarities with text retrieval approaches. This model has been adopted by Sivic and Zisserman [75]. Towards the need of lowering down this complexity, Chum et al. [22] propose a technique called Locality Sensitive Hashing and a random sketch of the set of visual words present in the image is used as an image representation. Another approach that uses a hashing scheme is presented in [36], where sets of feature vectors are indexed under their partial correspondences in sub-linear time. Checking the spatial consistency between the query image and the top retrieved images is adopted in the works [64] and [66], using the well-known RANSAC parameter estimation algorithm, introduced by Fischler and Bolles [28]. Geometric hashing has been introduced by Wolfson and Rigoutsos [83] for matching geometric features against a database.

2.2 Advances in Mass Intelligence

Collaborative Tagging is nowadays a standard feature of content sharing web applications enabling users to: (a) upload new, or bookmark existing content and, (b) annotate it by means of free-text keywords (tags). Created folksonomies in such Social Tagging Systems constitute a direct encoding of the views of a large number of users on how content items should be organized through a flexible annotation scheme (tagging). Detected tag clusters reveals relations between tags perceived by users as related to each other. To date, tag clustering has been dealt with either by conventional clustering algorithms, such as k-means [34] and Hierarchical Agglomerative Clustering [20], or by use of community detection methods [18]. Conventional clustering schemes are frequently troubled by two shortcomings: (a) the need for providing the number of clusters as input to the algorithm, and (b) their computational complexity. The new tag clustering scheme was designed with the above limitations in mind. Created folksonomies and large-scale ontologies in addition to discovery of tag communities, can be used as significant resources in categorization. Accumulated knowledge helps to overcome the shortcoming of the classical approaches, like Support Vector Machines [77], Naïve Bayes [46], or Latent Semantic Analysis [24], as they all require training set of pre-classified documents. When a training set is not available, alternative approach should use available knowledge such as named entities, relationships between them and ontology schema to perform classification. In such approach named entities and relationship between them can be successfully used for term disambiguating and vocabulary unification or calculation of semantic relatedness [19]. Additionally, descriptions of neighbouring entities can enrich information about a classified document [31].

2.3 Advances in Social Intelligence

The usage of social intelligence proofed to be useful in different areas. Winerman [81] shows that first information in crisis situation is often not provided by professionals but average citizens. She investigates how to exploit this knowledge to create official community-response grids. Besides analysing the pure content of the provided information it can be helpful to use complementary data about sources, distribution and routing of information. Social network analysis provides tool sets. New ways of information acquisition require methods and tools to route, store and operate these data. Existing methods in access control (to govern these new ways of collaboration) do not really fit the new demands. Standards like XACML [69] cannot support self-organizing communities with distributed access rights management. Another example lacking self-organization is EPAL [16]. To provide intelligent information routing the identification of social groups and structures is a key requirement. As online social networks can be of large size, fast and scalable algorithms are necessary. Few methods are capable of identifying groups in networks with millions of nodes like the Label Propagation Algorithm [67].

2.4 Advances in Organisational Intelligence

Events are understood as the occurrences in which humans participate. They may be very complex and a variety of aspects need to be considered. Models of events exist in various domains like the Eventory [79] system for journalism, the Event Ontology [68] as part of a music ontology framework, the ISO-standard of the International Committee for Documentation on a Conceptual Reference Model [25] for cultural heritage, the event markup language EventML [3] for news, the event calculus [56] for knowledge representation, the Semantic-syntactic Video Model [26] and Video Event Representation Language (VERL) [29] for video data, and the event model E [80] for event-based multimedia applications. From this related work we have derived that event aspects such as time and space, objects and persons involved, as well as mereological, causal, and correlative relationships between events, and interpretations of events have to be considered. The Event-Model-F [73] we have developed provides full support for these requirements. It advances the current state of the art by its full support for causality, correlation, and interpretation of events. The ER Log merging and management (WERL) application makes use of the Event-Model-F and is related to the general domain of C4I software, which stands for Command and Control Systems and Components. Among the numerous solutions available in the market that target at a wide range of applications such as military operations and surveillance, WERL is most closely related to ER incident management solutions. Three well-known solutions are: the Atlas incident management system (Aims) [5], the Emergency Command System [7], and the Bristol City Council map-based application [14]. The main focus of such software solutions is the support for information sharing and communication, as well as task and asset management. However, they lack the reusability and shareability of information that WERL offers thanks to its Event-Model-F-based representation, as well as the semantic enrichment capabilities of WERL.

3 Media Intelligence

In spite of the multitude of current intelligent, automated visual content analysis activities, there is still a lack of appropriate outlets for presenting high quality research in the prolific and prerequisite field of multimedia content retrieval. Thus, the goal of developing intelligent, automated content analysis and retrieval techniques for different media to extract knowledge from the content itself remains a major research task; still image retrieval turns out to be one of the most exciting and fastest growing research areas in the field of multimedia technology that aids towards the taming of those needs.

In principle, designing Collective Intelligence with a user-centred approach requires the involvement of users from the very beginning, as it is fundamental to understand the reality of what are people doing, how, when, and why. A user-centred design approach often works by trying to answer typical questions like who are the users, which are the user tasks and goals, what information do the users need and so on. Therefore, the undoubtedly growth of multimedia content collections and the change of user behaviours have created the need for fast, robust and efficient Media Intelligence algorithms, able to analyse large-scale diverse and heterogeneous visual content. More specifically, the popularity of social networks and web-based personal image collections has resulted to a continuously growing volume of publicly available photos and videos. Users are uploading, describing, tagging and annotating their photos, whereas recently they also geo-tag the location they were taken. In addition, a heavily increasing percentage of people are using the Internet to find and provide additional information with respect to past or forthcoming events or actions.

In this framework, the main scope of Media Intelligence tackles the area of such visual content interpretation. Media Intelligence suggests a content-based retrieval approach and focus is being given on a visual retrieval and localisation framework of content applicable to large web collections that may become extremely useful in pre-, during or post-travelling user activities. Motivated by this observation, a fast and robust retrieval system is being proposed [41], based on the popular bag-of-words model. More specifically, Speeded-Up Robust Features (SURF) have been selected to capture the visual properties of digital images and a visual vocabulary is created, through which images are efficiently represented. Geometric constraints on the image features are then taken into account, facilitating more accurate retrieval in comparison to traditional approaches. The performance of the proposed method is evaluated through the development of a web-based image retrieval application, that yields a geographic position estimation about a query image, exploiting already geo-tagged datasets. The latter aids users to identify so-called Points of Interest (POIs) and famous landmarks (i.e. in a "what to see?" concept), as well as additional background info about them together with interesting activities and events (i.e. in a "what to do?" concept).

In the following, we present the VIRaL tool[4], a web-based tool used to identify and localise similar multimedia content under different viewpoints, applicable to any functionality that involves a still image search and retrieval task. The main research principle of VIRaL exploits the fact that typical metadata usually contain a free text description together with some representative user-generated tags. In some cases, some metadata are related to the geographical position of the image taken (a.k.a. geo-tags). A geo-tag consists of the actual geographic coordinates, i.e. the longitude and the latitude, and is either extracted automatically through GPS or is manually defined by the user.

Initially, in order to represent the visual content of any given digital still image, a set of interest points is selected and visual features are extracted locally from their surrounding area. Since the goal is to choose scale invariant interest points, their localisation is carried out on a Gaussian scale-space. Using SURF features has been proven to achieve high repeatability and distinctiveness, whereas their extraction speed is very fast, when compared e.g. with SIFT features [50]. An example of the extracted SURF features is depicted in Figure 1. To further understand the notion of a visual vocabulary, one should consider it as an equivalent to a typical language vocabulary, with an image corresponding to a part of a text. In the same way that text may be decomposed to a set of words, an image can also be decomposed to a set of visual words. Then, in order to compare two images, their corresponding visual words may be compared. Thus, it is interesting to create a visual vocabulary in such a way that parts of images could be meaningfully assigned to visual words. Figure 2 depicts two regions of interest extracted from two different images, which correspond to the same visual word. The visual vocabulary is presented as the Voronoi cells of the clustered space of visual words. We should note here that due to their polysemy, visual words cannot be as accurate as natural language words.

To create the visual vocabulary, a clustering process is followed. More specifically, the well-known K-means clustering algorithm [52] is applied on the SURF descriptors corresponding to a very large number of points of interest. If the number of the points to be clustered is significantly large, clustering using the K-means algorithm becomes a very slow task. For example, clustering of 5M of points (which are typically extracted from 10K of images) requires a few days of processing. However, to efficiently deal with large scale retrieval problems, the size of the

[4] http://viral.image.ntua.gr

Fig. 1 SURF features extraction.

vocabulary should be in the order of a few tenths of thousands of visual words [22] [39]. Thus, in order to rapidly create an appropriate vocabulary, the clustering process is performed on a smaller subset, carefully selected to contain the most representative images. Finally, after constructing the visual vocabulary, each image has to be represented with a description that captures its relation to all the words of it. The process of querying an image database without and with a visual vocabulary is depicted in Figure 3.

In the first case the comparison of the local descriptors is performed immediately for two images and after exhaustive comparisons in the whole database, the closest regions are found. In the latter case, for every image of the database all points

Fig. 2 Regions of interest extracted from two different images that correspond to the same visual word.

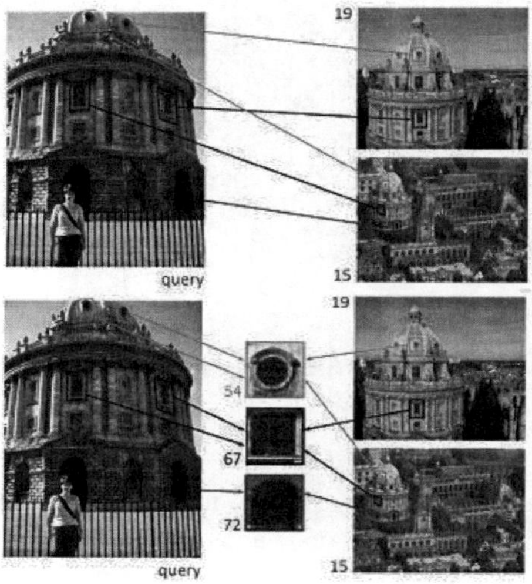

Fig. 3 Querying an image database without and with a visual vocabulary.

have been assigned to appropriate visual words of the visual vocabulary. Thus, for a new query, its points have to be assigned to the closest visual words of the vocabulary. After this process, two images are considered to be similar if their points are assigned to similar visual words.

Fig. 4 Correspondence of images.

When a user query reaches the system, then the local low-level features are extracted from the query image and the model vector is computed. Then the similarity of the query model vector with all database model vectors is computed using an inverted-file structure to speed up the process, and the N most similar images, the images with the highest such value, are either returned to the user as similar, or become candidates for geometric consistency check using the RANSAC algorithm [42] [41]. Finally, regarding geo-tag estimation, if a user issues a query containing a landmark image against a large database of geo-tagged images, then most probably the top-retrieved results will contain the actual landmark that the query image depicts. Those correctly retrieved images are expected to have near identical geo-tag values. Little variance is expected since geo-data can be defined by users and also because the same building may be photographed by different distances (using appropriate camera lenses). However, the estimated geo-tag for the initial query image is expected to be within the larger consistent subset of the result images (Figures 4, 5 and 6).

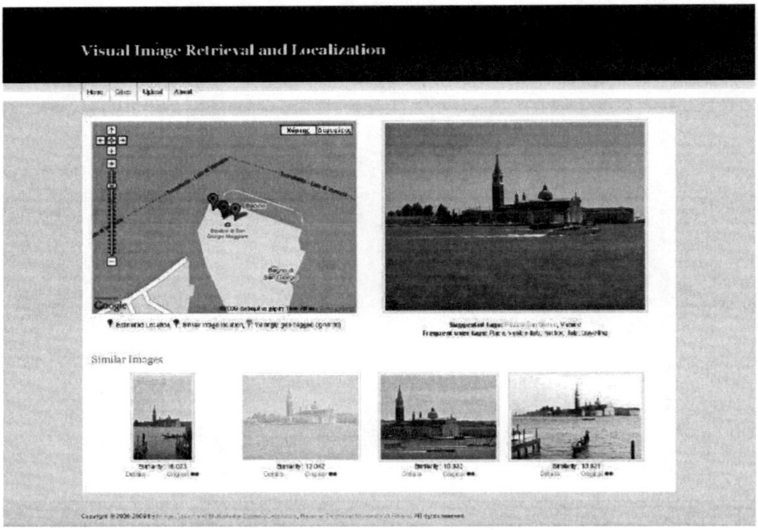

Fig. 5 Geo-tag estimation.

Fig. 6 Consistent subset of result images.

The above described methodology has been evaluated on a challenging 1M urban image dataset, namely European Cities 1M[5]. It consists of a total of 1.037.574 geo-tagged images from 22 European cities, which is a subset of the dataset used in the VIRaL tool. In order to acquire detailed information on the evaluation process, the reader is encouraged to consult [41] or [17].

4 Mass Intelligence

The vast amount of available and produced information in the current web requires new approaches to categorization and clustering in order to help users in efficient navigation and information finding. With the analysis of tag networks and training-less ontology-based categorization, large-scale user generated content became the centre of interest for Mass Intelligence. Nevertheless, Mass Intelligence does not only concentrate on such information. It can consume both massive direct user input as well as aggregated inputs already processed by other intelligence layers. Processing of inputs or joining results of the other intelligence layers creates basis for the Collective Intelligence. The example approaches presented below also cooperate with other intelligences to create new values and solutions, like hybrid image clustering that joins tag community detection with clustering based on visual features created in Media Intelligence.

4.1 Community Detection on Tag Graphs

Folksonomies comprise three types of entities, namely users, resources and tags, as well as the associations among them [55]. Tag clustering involves a process that groups the tags in a way such that members of the same tag cluster are perceived by users as related to each other. Despite the subjectivity of users involved in judging the degree of relatedness between tags, tag clusters are expected to correspond to meaningful topic areas, which can be useful in a series of tasks [61], such as information exploration and navigation, automatic content annotation, user profiling, content clustering and tag recommendation. The proposed scheme builds upon the notion of (μ,ε)-cores introduced in [85]. The original algorithm, referred to as SCAN, suffers from two problems. First, it needs two parameters, namely μ and ε, to be provided as input. Second, it leaves a substantial number of nodes unassigned to clusters. As a result, its utility is limited in IR tasks such as tag recommendation. For that reason, our scheme conducts an efficient iterative search over the parameter space (μ,ε) in order to discover cores for multiple values of the parameters. Finally, the identified cores are expanded by maximizing a local measure of modularity [51] in order to increase the number of nodes that are assigned to communities and to allow for overlap among communities. The scheme is described in detail in [62] and its three steps are briefly explained below:

[5] http://image.ntua.gr/iva/datasets/ec1m/

Core set discovery. The definition of (μ,ε)-cores is based on the concepts of *structural similarity*, *ε-neighborhood* and *direct structure reachability*. The structural similarity between two nodes is defined as:

$$\sigma(u,w) = \frac{|\Gamma(u) \cap \Gamma(w)|}{\sqrt{|\Gamma(u)| \cdot |\Gamma(w)|}} \tag{1}$$

, where $\Gamma(u)$ is the structure of node $u : \Gamma(u) = \{w \in V | (u,w) \in E\} \cup \{u\}$. Then, the ε-neighborhood of a node is the subset of its structure containing only the nodes that are at least ε-similar with the node, i.e. have similarity equal to or higher than ε. A node is called a (μ,ε)-core if its ε-neighborhood contains at least μ nodes and a node belonging to the ε-neighborhood of such a core is said to be directly structure reachable from it. Eventually, community seed sets are extracted by identifying the (μ,ε)-cores of a network and attaching to each one of them the nodes that are structure reachable with them.

Parameter space exploration. One issue that is not addressed in [85] pertains to the selection of parameters μ and ε. Setting a high value for ε (the maximum possible value is 1) will render the core detection step very eclectic, i.e. few (μ,ε)-cores will be detected. Moreover, higher values for μ will also result in the detection of fewer cores (for instance, all nodes with degree lower than μ will be excluded from the core selection process). For that reason, we employ an iterative scheme, in which the community seed set selection operation is carried out multiple times with different values of μ and ε so that a meaningful subspace of these two parameters is thoroughly explored and the respective (μ,ε)-cores are detected. The scan of the parameter space starts from high μ and ε values, moves logarithmically towards lower μ values, then lowers ε by a small step and starts again from the high μ. It terminates as soon as it reaches the lowest meaningful (μ,ε) values.

Core set expansion. Once the community seed sets have been identified by the above process, a core set expansion step is carried out in order to enrich existing communities with more relevant nodes. It achieves this by starting a local exploration process with the goal of maximizing a local quality measure, namely subgraph modularity [51]:

$$M(S) = \frac{ind(s)}{outd(s)} \tag{2}$$

, where $ind(S)$ stands for the number of within-subgraph connections for subgraph S, and $outd(S)$ stands for the number of connections from subgraph nodes to the rest of the graph. Example of the tag community around tag 'computers' is presented in Figure 7.

In qualitative evaluation our method produced much more meaningful tag clusters compared to CNM [23], which contained few gigantic clusters and many small ones, and comparable clusters, but richer in terms of tag coverage than the ones produced by SCAN [85]. Qualitative evaluation was based on subjective assessment of the derived tag communities and an implicit evaluation by using the derived

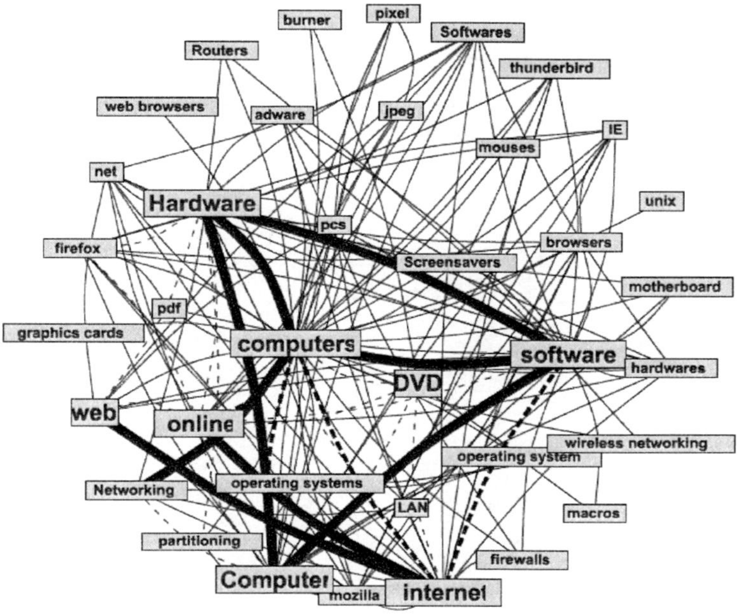

Fig. 7 Tag community around tag "computers".

clusters for tag recommendation and measuring the achieved performance on historical tagging data from three different tagging sources (Delicious, Flickr, and BibSonomy). Complete results are reported in detail in [62]. The process of extracting tag clusters from massive user tagging leads to promising results, however it is solely based on the statics of tag usage. As a result, the extracted tag clusters may often contain irrelevant tags (low precision) or miss related tags (low recall). Tag cluster precision can be improved by exploiting large-scale semantic resources, such as WordNet and Wikipedia, in the way presented in the following section. Cluster recall can be improved by propagating tag descriptions of photos to other photos that are visually very similar to them. Visual similarity can be established with the analysis tools provided by the Media Intelligence layer. In that way, a synergy between two intelligence layers is established in order to improve the quality of the analysis result. This Collective Intelligence driven idea resulted in a Hybrid Image clustering approach joining Media and Mass Intelligences. An image similarity graph was built encoding both the visual and the tag similarity between images of the collection. Experiments and evaluation of the hybrid image clustering are described in details in [63]. According to the performed user study results, used image clustering approaches are characterized by very high precision scores ($\geq 90\%$). Visual-only clusterings are characterized by superior precision ($\simeq 98\%$), but suffer from low recall. Tag-only clusterings behave in an IR-complementary way, yielding higher recall rates at lower precision. This allows creating of an image cluster that captures

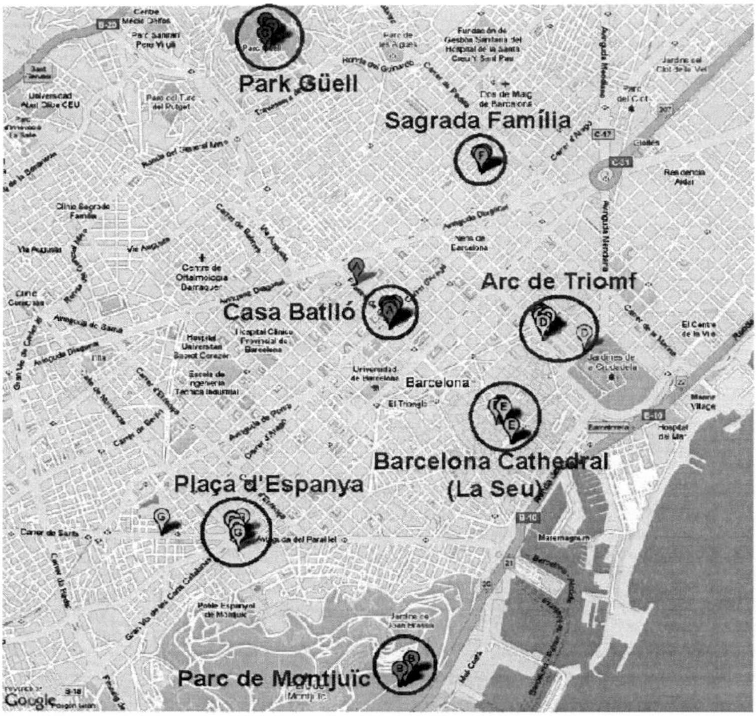

Fig. 8 Example Barcelona landmarks that were identified as image clusters by hybrid approach.

pictures of the same landmarks taken from very different spots. Evaluation was conducted on a set of 128,714 geotagged images located within the metropolitan area of Barcelona. Sample image cluster detected by the hybrid approach is presented in Figure 8.

The result of this Collective Intelligence instance can be further exploited by the Organisational Intelligence layer, namely the WeKnowIt Emergency Response Log merger and manager, for improved log entry indexing and retrieval.

4.2 Ontology-Based Classification

The method relies on the domain knowledge represented in the form of an ontology to perform the categorization task. It concentrates on the recognized named entities and relationships in the document text to measure the semantic similarity of the created thematic graph to the categories, defined as ontology fragments, and perform the categorization. In the proposed text categorization method the ontology effectively becomes the classifier. As a result, it overcomes the limitation of classical approach, and does not require a training set of pre-classified documents. Instead, it

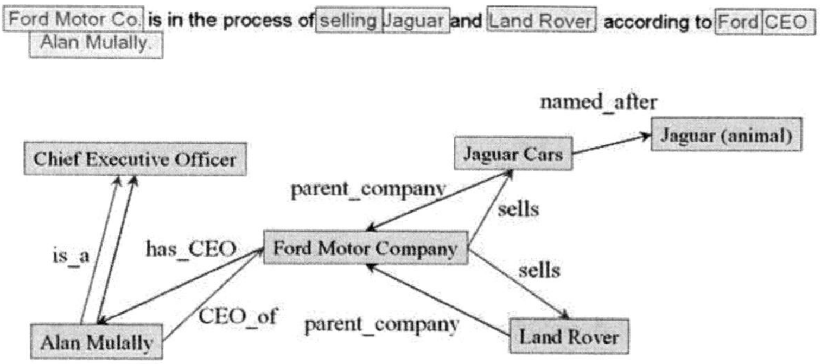

Fig. 9 Sentence with marked phrases and created semantic graph.

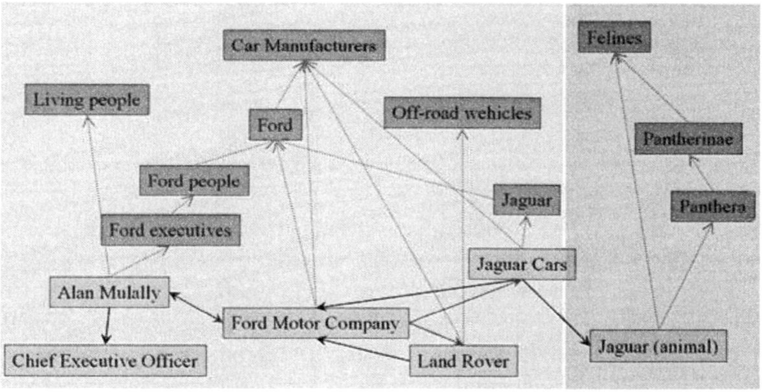

Fig. 10 Entities with category trees from ontology.

allows defining categories as fragments of ontological knowledge. Our categorization algorithm consists of three main steps: (1) construction of the semantic graph, (2) selection and analysis of the thematic graph, and (3) categorization of the selected thematic graph. The complete approach is presented in [38] and [4]. The approach uses ontology created from Wikipedia due to richness of represented domains, high number of interconnected entities, and included categorization scheme. Semantic graph construction requires identification of named entities (based on entity phrases and labels known to ontology), relationship extraction with shallow NLP and connectivity inducement that utilizes ontology as background knowledge. Analysis of a sample sentence is presented in Figure 9.

Thematic graph is selected after identification of created components in semantic graph and finding most authoritative entities using HITS algorithm [45]. The dominant thematic graph is further taken for categorization. It is based on discovered entities and initial categories assigned in ontology.

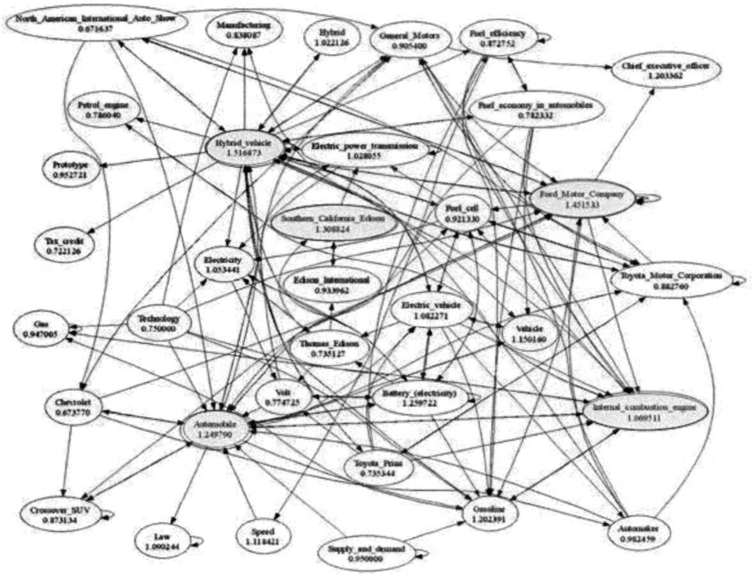

Fig. 11 Semantic graph from news document about hybrid vehicle announced by Ford.

Separate category trees for identified entities also can perform disambiguation function. As presented in Figure 10, entities matched from the sample sentence belong to two category trees: automotive (core) and animals (side category). Concentrating on core entities and dominating category tree, establishes categorization of the whole document. The document including presented sample sentence was categorized by the presented method as Wikipedia categories: "Hybrid vehicle" and "Automobile". The semantic graph created from the whole news document is presented in Figure 11. Most authoritative entities are highlighted.

In the performed experiments on news from CNN (www.cnn.com) RSS feeds and subset of the Reuters RCV1 corpora [47] we compared our method with Naïve Bayes from BOW implementation [54] and SVM from WEKA package [82]. The proposed method reached over 85% accuracy without the need of training set.

5 Social Intelligence

The analysis of the structures of social networks reveals information on general properties of the relations, on the hierarchical composition of a network and on the roles and positions of people. One example for the general analysis of network properties we conducted is the analysis of the communication on Question & Answer platforms. This analysis revealed the topic-specific answer behaviour. The knowledge about the different structures of communication explained the varying performance of expert identification algorithms. Expert identification is used to guide users to domain experts for one or several topics or to provide users information on the trustworthiness of fellow users. Gaining an understanding of how people interact or relate to other persons is the first step in extracting knowledge from social networks. For example the knowledge of 'natural groups' of similar people is a key issue for many real-world applications. If the members of a person's social group are known, new social applications become feasible. One of those applications is the social emergency alert service (EAS) which activates members of the social group of a person in need in a case of emergency [32] [60]. This service promises to facilitate faster help than the status quo. The social emergency alert system consists of software clients for smart phones and a server component. By analysing the users' communication (voice and text) social groups can be identified and used for broadcasting help request. A feasibility assessment showed that there is a good chance to receive help from a near-by friend [32].

Social networks can be very large. The networks of popular online communities like MySpace and Facebook have hundreds of millions of users. Therefore, scalability is one of the major problems for network analysis techniques. A community identification algorithm that is able to process huge networks is the randomized greedy modularity clustering algorithm (RG) we proposed in [59]. Modularity is a measure for the quality of a graph clustering (i.e. the decomposition of the set of nodes into non-overlapping groups) introduced by Newman and Girvan [57]. RG is an algorithm that tries to identify groups by maximizing the modularity of a graph partition. This algorithm is explicitly designed for natural networks like social networks that have many structural similar substructures. The agglomerative hierarchical algorithm exploits these redundancies by using a randomization strategy that finds with a high probability but little effort nodes that belong together to iteratively build a near optimal dendrogram. Processing a network with about 5 million nodes and 40 million edges takes less than 10 seconds on standard desktop computers. Another algorithm we developed is optimized for the processing of huge dynamic datasets [30]. The algorithm is based on restricted random walks that are incrementally updated. The incremental update process avoids that for every change of the data set all data needs to be reprocessed. The EAS is a good example for Collective Intelligence: It makes use of network data from the different layers of intelligence to reason about social groups.

5.1 Protecting Virtual Communities

Analysing online communities to support their members is one element of social intelligence. Another task is, to create a safe environment where people can interact. For almost any IT system, access control is an important issue. Especially for modern, state-of-the-art social networking platforms like Yahoo![6], Facebook[7] and Flickr![8], a flexible, reliable, manageable and easy way of access control is important. Personal data and media as well as social interactions have to be protected. Users

[6] http://www.yahoo.com, last accessed 30.6.2010

[7] http://www.facebook.com, last accessed 30.6.2010

[8] http://www.flickr.com, last accessed 30.6.2010

want to define in detail, who can access their media, e.g. holiday photos. They also want to differentiate access with regard to the social groups they belong in a convenient way. The same principle can be applied to two use case scenarios, Emergency Response Scenario and a Consumer Group Scenario. Both scenarios relay on media objects like images, videos, text documents and audio streams. It is obvious, that an access rights concept is necessary to allow or disallow access to these resources. In traditional approaches, access rights are mainly modelled by 3-tuples of the form (user, permission, object). Access is granted, if the current system state matches one of such 3-tuples in the access rights facts ("facts"). Giving an example, the 3-tuple fact (alice, read, pic1.jpg) allows a user with an account named Alice to access a file pic1.jpg with read access. More generally, the 3-tuple (U,P,O) is an element in the space $U \times P \times O$. U hereby is the set of users or subjects, sometimes also called principals. Subjects are entities, typically users or programs, executing in a system on behalf of a real world user [71]. P is the set of permissions. Typical elements of this set are read, write, execute and delete. The semantic of an access right is usually given to it by the developers of the application or software system. O represents the set of objects the access right applies to. Typical objects can be files and devices. It is important to notice, that user accounts can be subjects as well as objects, depending whether they act active (do something) or passive (something is done with them). Two conclusions are obvious: By this design, a very fine-grained facts modelling is possible. Secondly, it is obvious, that it is practically impossible to manage a more complex system with several thousand objects and lots of users by defining these triples individually. The access rights facts base would be huge, and any change has a quite high probability to affect several access rights which all must be changed manually. This leads to a high error rate by access rights not correctly set. To avoid this, two major approaches have been introduced. First, several elements are grouped in sets. For example, all user accounts of a department are grouped assigned to one group. The same can be applied to groups of access rights and objects. This grouping then allows defining facts on them and no longer on individual elements. Many approaches using this technique can be found [74] [33]. Another improvement are the concept of authoritative roles and access control lists (ACL). Roles combine permissions and objects (e.g. "read file A", "write all files in dir B"). Roles become an intermediate layer, a container holding objects and permission combinations. Roles are assigned then to users, allowing them to perform operations described by their assigned roles. Usually the role approach is combined with grouping. Roles allow to abstract access rights from individuals by modelling rights in abstract roles. If a role of a user changes his new access rights can be modified easily by changing his roles. ACLs do work similar as roles, but use a different perspective: ACLs combine users and access rights to access control lists, which are then assigned to objects. We see immediately, that this is the same approach as roles, applied this time on objects and not users. The common RBAC model [27] (Role-based Access Model), for example, is an example of the first extension. In RBAC access rights - the two latter elements of the triple - are collected in so called roles. In our example a very simplistic role could be defined by role1, allowing to read Pic1.jpg. Roles are then assigned to users. Obviously, the RBAC and the triple

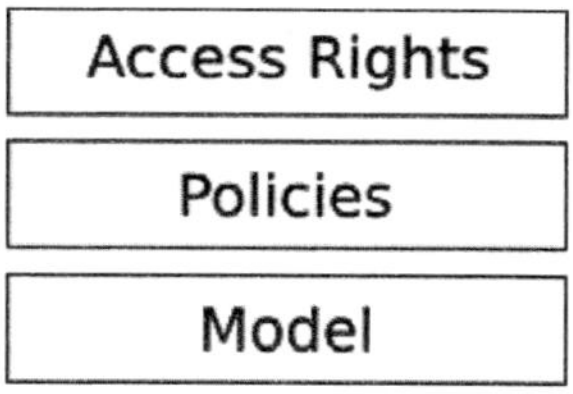

Fig. 12 Layers of an Authorization System.

model have the same expressive power. RBAC collects permissions and objects in roles and assigns them to users, while the basic triple model directly models triples. It is important to notice, that these approaches do not extend the functional expressive power of an access rights model. The same result could still be retrieved by the triples model. The extensions focus on usability and manageability. They relay all on the same *authorization model*, thus the (u, p, o) triple.

5.2 The Community Design Language

We suggest a different approach allowing defining not only policies but defining and modifying the underlying authorization model (Figure 12).

The Community Design Language (CDL) is a formal language to define (1) the model, (2) the policies and, (3) deduce access rights. The model describes the possible conditions which can be used to formulate policies in. Policies are the high-level definition of the conditions which grant or deny access. The access rights are the deduced combinations of entities which allow or deny access. For a better understanding, we provide an example: In the model it is defined, which circumstances and conditions may be used to formulate policies. To model a "traditional" *u-p-o* model, the conditions user, permission and object must be defined. If additionally the access time shall be introduced as possible policy condition, it is defined in the model. Let us define for this example, users, permissions, objects and access time in the model. A policy defines - based on the model - if access shall be granted. In our example, all users of the group scientist may read all files through the intranet. This is a policy. The deduced access rights is then the fact, that Alice (as member of the group scientists) can read the file pic1.jpg from the network intranet. This example is presented in the Community Design Language in Table 1.

6 Organisational Intelligence

Organisational Intelligence is the ability of an organisation to understand and to leverage knowledge that is relevant to its goals and purpose. As a consequence, it is the goal of Organisational Intelligence to bring the right piece of knowledge, at the right time to the right person, in order to support decision making to best accomplish the organisations purpose. For instance, in emergency response, typically several professional entities are involved such as emergency hotline, police

Table 1 This table shows the example presented in the text in the formal language CDL.

Model definition:
CREATE SETS users, permissions, files, networks;
CREATE SETS usergroups;
CREATE RELATION useringroups (users, usergroups);
Define policies:
CREATE ACCESSCONDITION ac:
([users].usergroups IN ”scientists”, [permissions] IN ”read”, [networks] IN ”intranet”);
Fill facts in database:
CREATE ELEMENTS users: { Alice }, permissions: { read }, files: { pic1.jpg }, networks: { intranet },
usergroups: { scientists };
CREATE LINK useringroups: {(Alice, scientists)};
Perform access check:
CHECK ACCESS (users=Alice, permissions=read, files=pic1.jpg, network=intranet);

department, fire department, and emergency control center. All these entities need to exchange event descriptions like the one above. However, they typically use different systems and applications with their own proprietary data models for events. Using the formal Event-Model-F [73] instead, these systems can commonly represent and effectively communicate event descriptions. The Event-Model-F bases on the foundational ontology DOLCE+DnS UltraLight (DUL)[9] and provides a set of ontology design patterns to represent the different relations of events as derived from the related work in Section 2. The participation of objects in events is implemented by the participation pattern. This pattern also provides for modelling the absolute time and location of events and objects. The mereology pattern, causality pattern, and correlation pattern implement the structural relationships between events. The mereology pattern allows representing composition of events along temporal, spatial, and temporal-spatial relations of events and objects. The documentation pattern provides for annotating events, e.g., by media or sensory data. It can be seamlessly linked with other ontologies like the Multimedia Metadata Ontology [70] for precisely describing digital media data. Finally, the interpretation pattern supports different event interpretations.

6.1 Event Log Merger Application

The Event-Model-F is used in the ER Log merging and management (WERL) application. WERL addresses the problems arising in reviewing and searching through the logs that are produced by different members of the ER personnel. The disparate log entries are automatically merged and represented on the basis of Event-Model-F. Furthermore, semantic information is extracted from their text in order to enable a concise view of the ER log content. Commonly, a log file contains information

[9] http://wiki.loa-cnr.it/index.php/LoaWiki:DOLCE-UltraLite

pertaining to the documented incident and the log creator, as well as a set of time-stamped log entries, each of which documents a message communicated between some members of the ER personnel and an associated action. Thus, two granularities of events are defined, the high-level emergency incident (e.g. a fire incident in a factory) and the specific actions taken by the ER personnel, which are considered as sub-events of the high-level incident (composition pattern). Depending on the granularity of the event, different documentation properties are attached to it (documentation pattern). Furthermore, log entries are described based on their location and temporal attributes as well as the involvement of specific individuals to them (participation pattern). In order to derive several of the aforementioned log attributes (location, person names, etc.), the recorded log text is undergoing a semantic enrichment process. As training data is not available to learn the extraction patterns and inconsistencies are observed in the log entries, in terms of linguistic and syntactic style, the extraction processes does not rely on natural language patterns, but applies a knowledge-intensive approach. This requires that quality resources (gazetteers/taxonomies) are available, containing the desired named-entities likely to be found in the logs. For instance, for emergency incidents there is a requirement to identify fine-grained locations (i.e. at the street level). In addition to the extraction of location and person names, prominent key phrases and ER-specific terms are extracted from the text. The semantically enriched log entries are surfaced to the professional users through the WERL front-end. The front-end of WERL provides online filtering capabilities for facilitating the interactive exploration of the available log entries. A snapshot of the application main screen is provided in Figure 13. At the top, a slider-based time filter is available that enables the examination of a particular time interval of the incident. In addition, standard full text search capabilities are provided for retrieving only the subset of log entries that are relevant to the input query. Most importantly, there is a series of four semantic filters that summarise the main entities found in the log files by the text annotation component of the system. Thus, it is possible to view only the log entries that are related to a particular location, person name, or significant keyword or ER acronym. The presentation of all identified semantic entities in these lists can provide the ER user with a quick overview of the semantic content of the log file. Another significant feature of WERL is the presentation of provenance information and the possibility to filter based on the provenance of log entries. Beside each log entry there is a marker indicating its origin. At the bottom of the log entry list, there is an associated legend, which can also be used for filtering based on the log entry provenance. In that way, it is possible to inspect only the log entries produced by a particular log creator, thus gaining insight into his/her perspective of the incident. In larger scale incidents, involving many members of ER personnel coming from different organisations (e.g. fire department, police), it is expected that more sophisticated provenance mechanisms will be necessary, e.g. provenance by organisation, unit, role in the organisation, etc. Finally, WERL provides a map-based view of the log entries based on the automatic identification of fine-grained location information from their text.

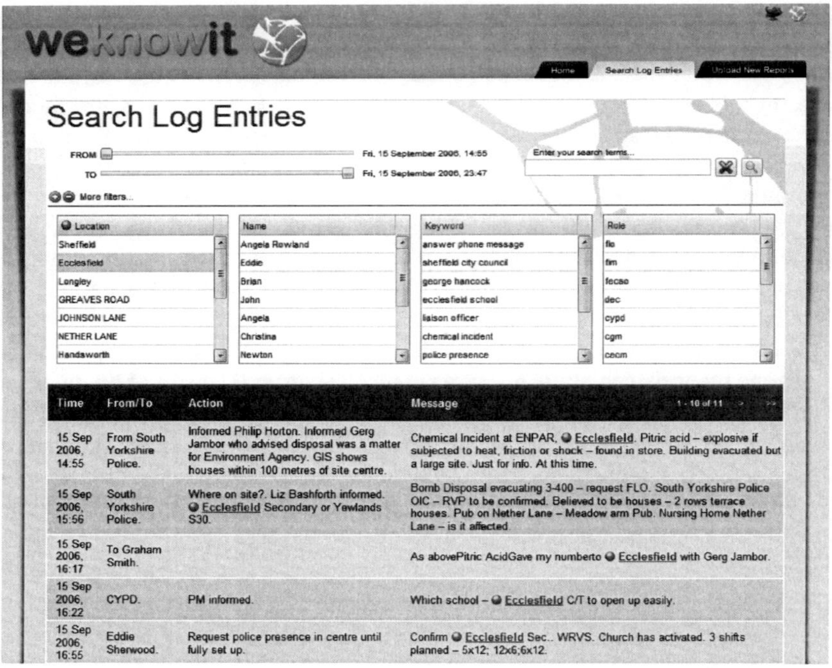

Fig. 13 Snapshot of the WERL front-end main screen.

6.2 Sharing Event Descriptions with SemaPlorer

Besides the use of the Event-Model-F in the log merger application it is also used in the SemaPlorer++ application, an extension of the SemaPlorer [72] application, for creating and sharing event descriptions. A domain specific ontology on emergency incidents provided by the Sheffield City Council (SCC) has been the developed and is used within the SemaPlorer++ application. The SemaPlorer application allows its users to interact with events on the map view as shown in Figure 14. An ontology browser is located on the left hand side with an emergency response ontology loaded. The emergency events defined in the SCC ontology are listed in a tree-structure of the ontology browser. It enables the user to easily create event descriptions by clicking on a concept in the ontology, i.e., clicking on the emergency event IAEP_Major_Industrial_Fire in the ontology browser representing a major industrial fire, and dragging and dropping it on the map. Once the user has placed an event description on the map, an instance of the event participation pattern of the Event-Model-F with the information about the event is created. It comprises the time and location of the event and a default object participating in the event. In addition, an instance of a specialization of the event documentation pattern is created and connected with the event participation pattern. This specialization of the documentation pattern allows storing additional information about the event such as a

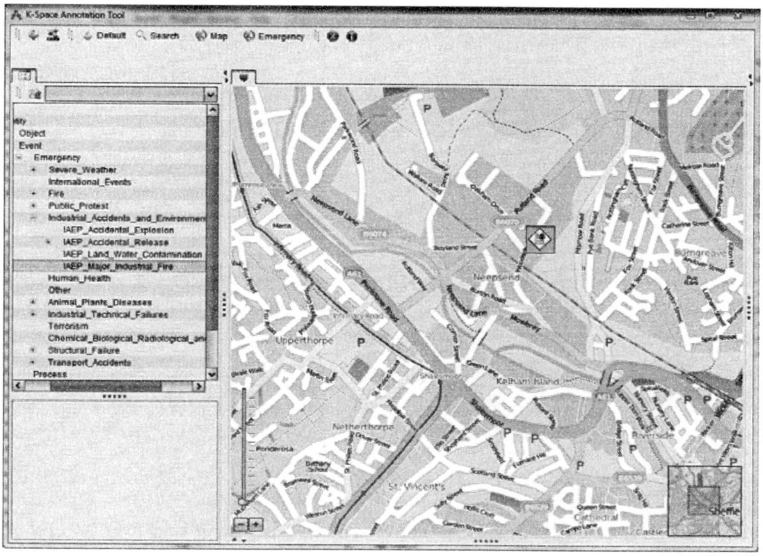

Fig. 14 Screenshot of the SemaPlorer++ Application for Creating and Sharing Emergency Response Event Descriptions.

title, written location name, description, and documenting pictures. The example in Figure 14 shows an icon on the city map of Sheffield. It represents a major industry fire that happened in a bakery in Sheffield.

6.3 Application of the Event-Model-F to Tourism and Sports

With the Event-Model-F, we can create and exchange sophisticated descriptions of real world events. It has been formally specified in Description Logics [1] using the Web Ontology Language [12]. We have verified its validity by using the build-in reasoning tools of the ontology engineering tool Protégé [13]. Besides the use of the Event-Model-F in the two applications above for emergency response, it can also be applied to arbitrary other domains that need to represent events as occurrences in which humans participate. Different examples for applying the Event-Model-F include soccer games and tourism and are available at: http://west.uni-koblenz.de/eventmodel. The soccer example models an entire soccer game, i.e., the first halftime and second halftime. Different events happen during the game such a foul and goal. The soccer example makes full use of all patterns of the Event-Model-F, namely participation, causality, correlation, composition, and interpretation. It further shows how a domain specific ontology can be embedded and used to describe the events happening during a soccer game. A tourism example models a two-day weekend trip. Three people are participating in this trip. On the first day, there is a sub-event dinner. On the second day are two sub-events, a visit to a museum and a sight. The tourism example applies different patterns of the Event-Model-F such as participation and composition.

7 Integrated Collective Intelligence Framework

The technologies described in the previous sections can be used alone. But its cooperation brings real value to the end users. To this end a new Collective Intelligence methodological approach is introduced which is able to combine the different intelligent layers, and exploit their interactions and synergies in order to effectively harness Collective Intelligence in the integration level. From a technical perspective, the role of Integrated Collective Intelligence Framework (ICIF) is to achieve the synergy effect ("Collective Intelligence") by combining the results of work from the different services into a cohesive system. In order to achieve this, the following technical tasks have been fulfilled:

- specification and implementation of the overall software and hardware architecture,
- preparation of a set of commonly used objects and implementation of a storage component capable of storing them,
- integration of services (software components) based on the common model and API.

7.1 Collective Intelligence Methodology

As can be seen in the intelligence layers chapters, the presented Collective Intelligence techniques in most cases exploit links, references and relations among different content items contributed by the users, thus differentiating from the legacy large scale data analysis techniques. Typical examples of such techniques are Flickr-based visual analysis, tag clusters extraction from massive user tagging and the Wikipedia-based community detection methods. The integration of such different techniques originating from different intelligence layers could potentially leverage Collective Intelligence within diverse usage scenarios. However, the proposed Collective Intelligence approach moves a step further; instead of a mere concatenation of the different layers intelligent methods, it imposes a pairwise combination of different intelligence layers within the architecture of some of the developed techniques. Multi-modal analysis is often exploited to enhance the results in each intelligence layer. For example, Mass Intelligence tag clustering results are improved by using Media Intelligence visual analysis features when building graph clusters. As a result the produced clusters are evaluated as more coherent, since they incorporate cross-domain knowledge. Furthermore, the added value of Collective Intelligence is also evident in the integration level, where the different techniques are combined to produce better results in each case. Geo-tagging through visual and tag analysis yields better localisation results, while WERL can achieve improved log entry indexing and retrieval when incorporating the Collective Intelligence tag clustering method.

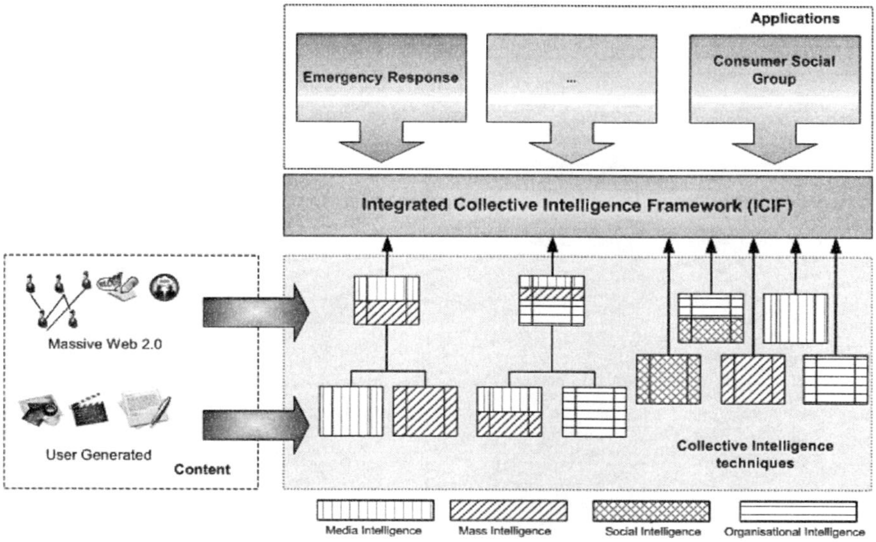

Fig. 15 Collective Intelligence Methodology.

Figure 15 depicts the proposed Collective Intelligence approach, which is able to produce enhanced results by:

- exploiting large-scale user contributed content
- combining different layers in building Collective Intelligence techniques
- fusing results from different intelligent layers

7.2 Architecture and Integration

The architecture of the ICIF is based on the idea of loosely-coupled components. Cooperation of the components is realized via the registry of the OSGi[10] framework. The connections between the components can be specified programmatically or declaratively (using enterprise integration patterns). Some functionality provided by the ICIF are exposed via REST API allowing external applications to use ICIF's services. Taking into account the massive calculations performed by some components (e.g. visual analysis), a scalable architecture has been prepared. An Enterprise Service Bus (ESB) component is responsible for passing messages between services. Additionally, some "heavyweight" components can run on different machines and be accessed remotely. All components of the system are provided as OSGi bundles and integrated within the ICIF architecture. Figure 16 presents how the services, described in previous sections are grouped into (conceptual) modules and deployed within the ICIF platform. The ICIF is written in Java and uses many open-source enterprise ready technologies including Fuse ESB (integration

[10] http://www.osgi.org/

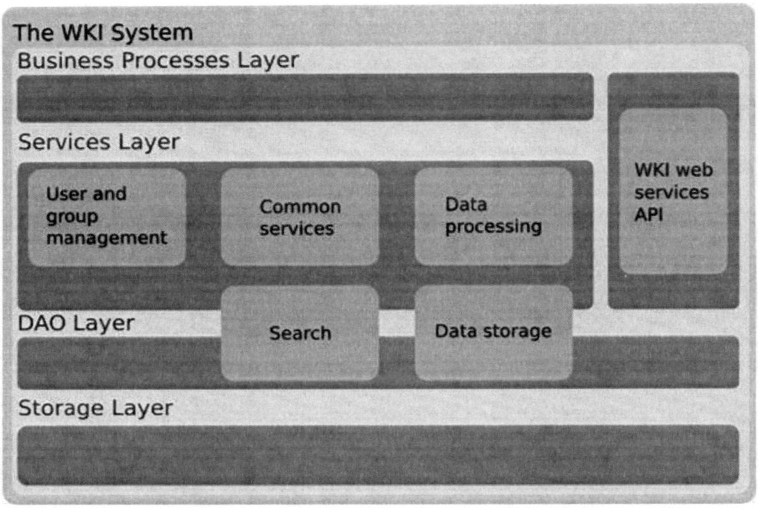

Fig. 16 Layers and groups of modules within the WKI System.

Table 2 Services of the WeKnowIt System

Group	Services Examples
User and group management	WP4_CommunityDesignLanguage, WP5_GroupManagement
Common services	WP6_CommonsModel
Data processing	WP2_TextClassification, WP2_TextClustering, WP2_VisualAnalysis, WP2_SpeechIndexing, WP3_SpamDetectionService, WP3_LocalTagCommunityDetectionService
Search	WP6_DataStorage, WP2_SearchInSpeech
Data Storage	WP6_DataStorage

platform), Apache Camel (integration framework), Apache Karaf (OSGi framework), Apache CXF (web services framework), Spring (Java/J2EE application platform), Lucene (fulltext indexing and search engine) and other[11]. All of them are mature, and proved their stability in enterprise projects. No vendor-specific products are used, and open communication standards are utilized. Currently, more than 15 services (of varying granularity), including the ones presented in this chapter and others developed in the framework of the WeKnowIt project, are integrated within the ICIF. Table 2 provides exemplary services belonging to each of the aforementioned conceptual groups. Note that each group can contain services from different

[11] http://fusesource.com/products/enterprise-servicemix/, http://camel.apache.org/, http://karaf.apache.org/, http://cxf.apache.org/, http://www.springsource.org/, http://lucene.apache.org/

Intelligence Layers. For example group "Data processing" includes services from Media and Mass Intelligence. Some of the services are accessible on their own, while other are combined in order to perform multiple operations in response to one call of the client applications. For example, a simple act of file upload triggers a chain of actions within the ICIF platform. First, a mime-type of a file is determined. Provided that the uploaded file is an image the following services are executed:

- WP4CommunityDesignLanguage service determines user rights to upload a file and stores permissions of the newly uploaded file after it was successfully processed,
- WP2VisualAnalysis service determines the geo-location of a picture and returns tags that are likely associated with it,
- WP2TagNormalization service matches the tags to a domain ontology concepts and adds some annotations to metadata,
- WP6DataStorage stores the file with all generated metadata.

As a result, a file, along with extracted and generated metadata, is stored within the system thanks to which various applications can benefit from it.

7.3 Hybrid Storage

The storage of the ICIF (called WKI DS) is capable of storing both files and business objects (entities that are exchanged between the services of the system). It can be divided into three types of storages:

- file storage (Hadoop DFS),
- triple store (Jena), and
- object storage (NeoDatis).

All these storages are realized using open-source technologies. The pluggable architecture allows the actual implementation of each storage to be changed. For example, it is possible to replace Jena with some other triple store, if such need arises. The idea behind the redundant storages for business objects (triple store and object database) is the following. On the one hand, the applications built on top of the ICIF require flexible access to data. This requirement is satisfied by the SPARQL endpoint of the triple store. On the other hand, a typical web application most frequently performs a number of operations during a certain period of time for which the triple stores are yet not designed. In order to make many operations (i.e. simple CRUD operations) perform much faster, the second type of storage - object database - is used. There are two obvious problems with this approach though. First, there is redundancy of data that is duplicated among both storages. At the same time, there is necessity of mapping between objects and triples. Regarding data redundancy, it is handled internally by the WKI DS component, which stores some additional information that allows for matching entities from object storage with graphs from triple store (this information is stored in separate relational database that is used only internally). The API of the WKI DS does not allow for operations on single triples in order to keep content of both storages synchronized. Insertions, updates

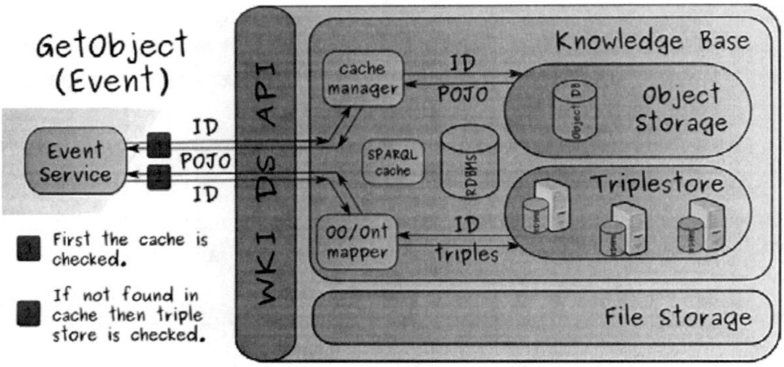

Fig. 17 Retrieval of business object from the WKI DS hybrid storage.

and removals of objects are taken care of by the internal mechanism, which guarantees the consistency of data in both storages. The mapping of business objects to triples is performed by an internal mechanism that is implemented using Jenabean library. The objects that the mapping mechanism can transform into triples (and the other way round) belong to a Common Model, which is a set of objects created based, on the common ontology that is used throughout the framework. The WKI DS storage can be fine-tuned to meet the needs of an application. The object storage can be used as a full copy of triple store, but can be also configured to store only a part of its data (thus serving as a cache). Figure 17 presents this scenario. It is worth to notice that the mapping from triples to objects is completely transparent to the client ("Event service" in this example). In fact, the client is even unaware of the dichotomous structure of the WKI DS and operates solely on Java objects. The existence of two storages allows for further improvements of the application's performance apart from benefiting from fast access to business objects guaranteed by the object database. An additional effort is being put into making the system even more efficient. Transformation of some of the SPARQL queries directed at the triple store into native queries of the object storage may result in a significant performance boost. The first proof of concept (implemented using a fixed set of queries from the Berlin SPARQL benchmark[12]) is very promising, but the whole idea still needs a lot of polishing.

7.4 Development Environment

Integration of work of independent development teams require good communication and coordination of efforts. It can be improved by the usage of proper open-source tools[13]:

[12] http://www4.wiwiss.fu-berlin.de/bizer/BerlinSPARQLBenchmark/

[13] http://www.mantisbt.org/, http://nexus.sonatype.org/, https://hudson.dev.java.net/, http://subversion.tigris.org/, http://www.sonarsource.org/, http://webdav.org/

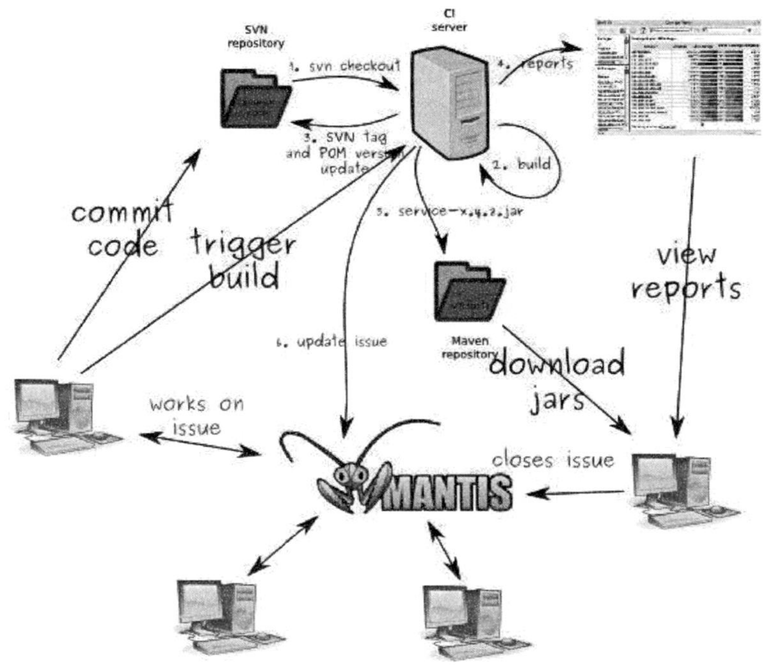

Fig. 18 Development environment of the Integrated Collective Intelligence Framework.

- Source Code Repository (SVN),
- Continuous Integration server (Hudson) with code quality checking tools (Sonar),
- Artifacts Repository (Nexus),
- Bug Tracker (Mantis),
- WebDav server.

Figure 18 presents a typical flow of development activities and role of different tools involved - from SVN checkout, through bug report, up to patch commit.

7.5 Summary

The ICIF platform brings into life Collective Intelligence by providing a runtime environment that can be used to combine services and achieve synergy effect. The first prototypes of an Emergency Response and a Consumer Social Group scenario prove this approach feasible. The hybrid storage can be used with different applications that require both fast business objects access and execution of complex SPARQL queries. Its unique capabilities open new perspectives for business applications that are powered by triple stores.

8 Applications

The techniques described in the previous sections have been integrated in two different scenarios in order to depict the feasibility of harnessing and producing Collective Intelligence. In an Emergency Response (ER) scenario (Figure 19), upon an emergency event (e.g. fire, flood, etc.) a user logins to the application and is able of capturing the event and contextualize it with metadata, e.g. tags. The Media Intelligence VIRaL localisation method is then used to automatically add location information to the image, by analysing massive input from Flickr. Thus, even if GPS is not activated or available, the uploaded image can be enriched with geo-localisation information. In the Emergency Responders side, Mass Intelligence techniques are applied in the user contributions, which are also enriched by relevant content from Web 2.0 sources. The community detection technique is applied on the received tag volume, so as to induce clusters of the received image data. Moreover, an ER domain ontology is used for the classification of the available tags. Meanwhile, if a user of the application is in danger, she is able of activating her connections in her social group, by utilizing the Social Intelligence Emergency Alert service. Finally, after the event, in the headquarters of the Emergency Planning team, the personnel is able of monitoring and reviewing the team's reports and actions by the use of the Organisational Intelligence ER log merger. In overall, the different analysis layers that are used in the different stages of the event contribute altogether to the leveraging of Collective Intelligence, which in turn yields better handling of such Emergency Response events. In a different scenario, the presented techniques can analyse user contributed content from various sources to help travellers discover essential information about what to see and do on travel or one-day cultural trip events. More specifically, by making use of automatically extracted Collective Intelligence

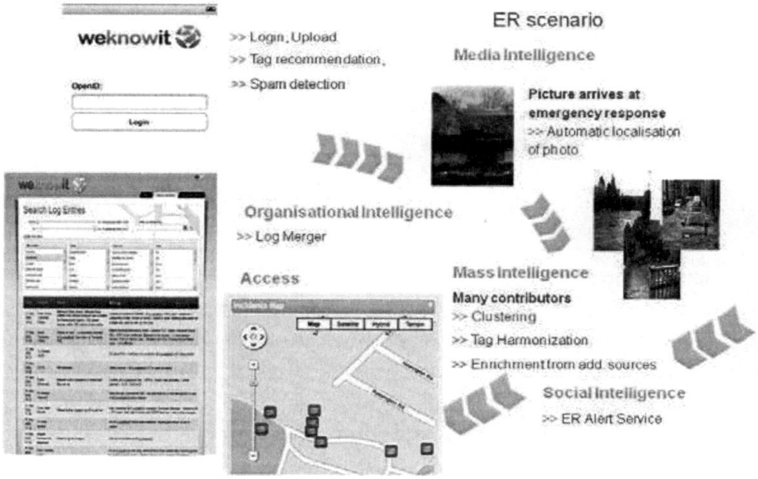

Fig. 19 Emergency Response scenario.

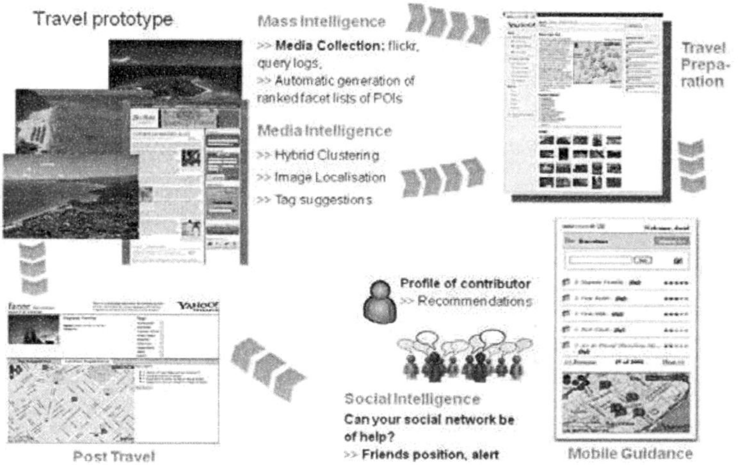

Fig. 20 Consumer Social Group scenario.

results, it assists users in a travel exploration experience by identifying points of interest (POIs), ranking and prioritizing them (according to the most popular places and users profiles) and by presenting them along with additional background information aggregated from different sources. The scenario contains three main parts: a Travel Preparation, a Mobile Guidance, and a Post travel stage (Figure 20). During the travel preparation part, users need a tool, which is able to provide them with information about the different candidate places to be explored and visited. The Travel Preparation tool, in the form of a web application, provides the relevant information that users need to prepare their travel, e.g. information about the locations, multimedia content, and points of interest. This information is aggregated with content coming from different sources (e.g. Wikipedia), is ranked according to trends to add value to user's experience, and is presented to the user by the use of Mass Intelligence clustering techniques. Figure 21 depicts a snapshot from the online available travel preparation tool[14]. In the second part, Mobile Guidance, users perform the planned trip. With the features offered by mobile devices and the developed application, users are able to access relevant information about their physical environment and search for new events or points of interest. They are also able of using Social Intelligence techniques to find the position of their social connections, or even notify them in case of emergency. Users can also take pictures and record videos of the places they are visiting, which are likely to correspond to the ones chosen in the travel preparation phase. These data, generated by users, such as images and videos, comments, ratings and notes, can then be added into the system where all this content is stored and shared with other users, enriching the repository of information for other users of the system. Finally, in a post travel application, the user

[14] http://weknowit.research.yahoo.com/csg/

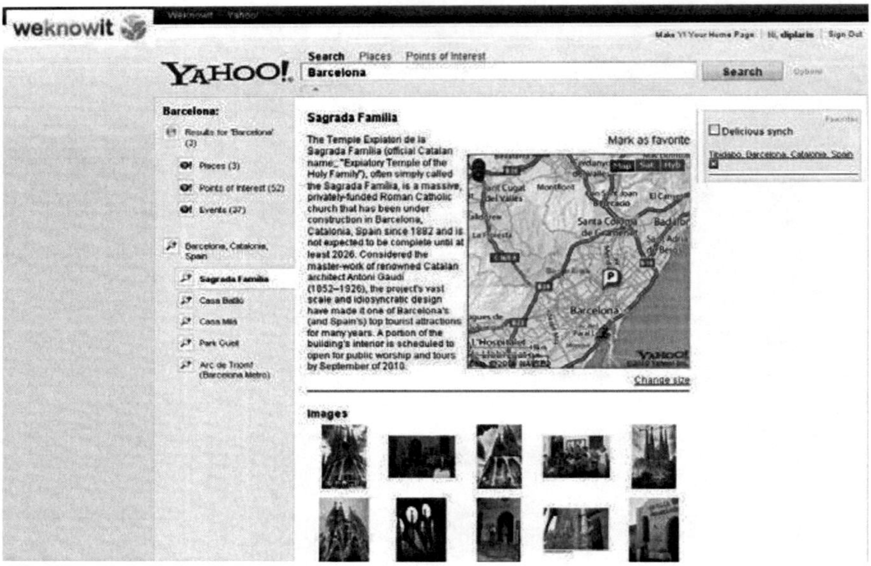

Fig. 21 Travel preparation tool snapshot.

can exploit the Media Intelligence image localisation technique, as well as the Mass Intelligence tag suggestion methods to enrich metadata in her data collection and also automatically geo-locate her photos.

8.1 Evaluation

The Emergency Response and Travel scenarios have been developed in the form of demonstrators. As the two demonstrators are harnessing Collective Intelligence in a different manner, different evaluation approaches and metrics are exploited for each case accordingly. For the ER demonstrator two interfaces were evaluated during the evaluation runs - the access interface, used by citizens and ER experts to explore incidents and the upload interface, used to provide information to the system. The demonstrator was evaluated by two groups. Six Emergency Response experts evaluated the intelligent access interface to the Emergency Response demonstrator and a smaller group of experts evaluated the intelligent upload interface to the demonstrator. Additionally, 12 citizens evaluated both the upload and access interfaces. All evaluations were run in the city of Sheffield. The desktop and mobile components of the travel demonstrator were evaluated separately. Two evaluation runs were carried out for each one of the demonstration modules, comprising 15 and 21 end users for the desktop module, while 4 and 22 end users evaluated the mobile guidance module accordingly. The evaluations were split between lab exercises and field trials. Users were given tasks typical of the usage of the respective demonstrators. Following exposure to the demonstrators, the user response to the prototypes was evaluated through questionnaires or discussed in consensus meetings. These materials were analysed in order to assess the added value of Collective Intelligence and to identify recommendations for the future enhancement of the demonstrators. The evaluations addressed several dimensions: usability, complexity, efficiency, responsiveness, satisfaction etc. Feedback was also collected in a non-structured format in order to get explicit recommendations on how the demonstrators can be improved.

8.1.1 Evaluation of Emergency Response Demonstrator

The evaluation focused on both whether target users were able to make use of the application to carry out the essential tasks and whether their performance in these tasks was better than the current status quo. To assess the functionality of the ER demonstrator, users were asked to carry out the tasks they would typically be involved in and assess the interface on the basis of these tasks. Thus, citizens were asked to upload an image and to use the intelligent access interface to determine information related to an incident they had previously witnessed in the city. The time taken, title, description and tags applied to the images were noted and the response to the interface was gathered. The ER experts were asked to primarily evaluate the access interface, although a secondary evaluation assessed the functionality of the upload interface. The task for the ER experts was to gather information about and distinguish between three events which were co-occurring in the city. In both the access cases a further goal of the evaluation was to assess how the Collective Intelligence implemented by WeKnowIt impacted on the ability of citizens and ER experts to make sense of the information provided to them. Therefore, evaluations were carried out with three different interfaces in order to measure this impact, namely a) information gathered from incoming calls from the general public, b) raw data in the form of a simple web interface which allowed the users to see the images and comments in a serial manner, and c) the ER demonstrator. A number of images and comments were collected corresponding to the severity of the incident. ER experts were then asked to use the interfaces to build an understanding of each of the incidents, their location and their corresponding location. The citizen participants had access to the same data and interfaces but were instead asked to find out more information about one of the incidents (one with medium severity) on the basis of a small amount of information. To measure the performance (as opposed to the efficiency) of the participants, a simple scoring system was used. The answers given by the participants were compared to the information used to generate the evaluation data on a one-to-one basis and given a score of 1 if the participant gave the same answer and 0 otherwise. The scoring was made for each incident, for the location, severity and type of incident. Thus each answer was scored out of 9 (9 meaning that the answer was completely correct and 0 meaning that the answer was completely incorrect). The scoring was carried out by a single person without reference to the condition to prevent biasing. Figure 22 shows the mean scores per condition for the access evaluations. As Figure 22 shows, the mean scores were generally high. As expected the score arising from the comments condition alone were lowest and the scores achieved in the ER demonstrator condition were marginally higher than those

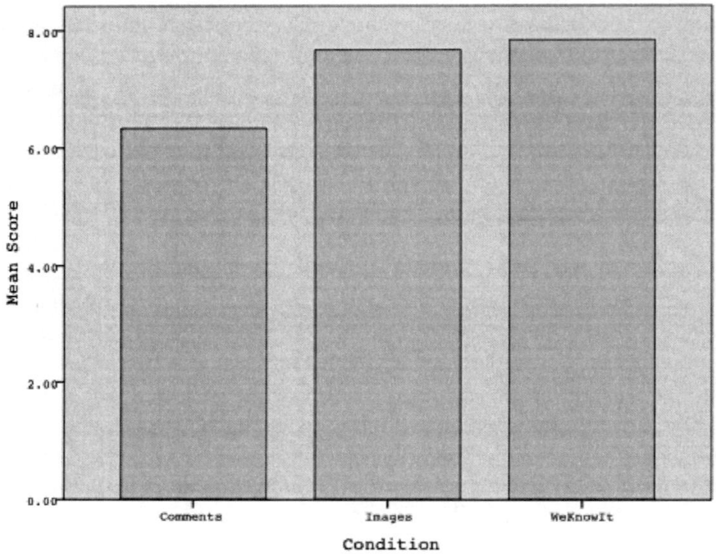

Fig. 22 ER - Mean Score by Condition for Experts.

for the images (7.8 for ER demonstrator and 7.7 for Images). Given that the information present in the two interfaces is the same, this result is not surprising. The scores for the comments are lower; the near equality of the scores for the images and ER demonstrator conditions is reflective of the power that the image has for this user group. Figure 23 shows the average score achieved with each interface for the citizens. The difference in performance between the conditions was less pronounced in the citizens case than that for the ER experts, though this is largely due to the simpler task that the citizen was asked to perform. Again, however, it can be seen that the ER demonstrator receives the highest score overall and that the citizens found that the images added some value to their interpretation of the incident. Scores were also computed for the usability and efficiency of the interfaces, using post-condition questionnaires. Due to space limitations the results are not presented here. The interested reader can find all ER evaluation results in [15]. Overall both groups of participants were positive about the ER demonstrator application. They were able to complete the tasks required of them both in terms of uploading information, accessing information and also interpreting and interacting with the information in order to build an understanding of what incidents were happening in the city.

8.1.2 Evaluation of Consumer Social Group Demonstrator

Regarding to the travel desktop prototype, the evaluation process was carried out with two different groups. One group, consisted by some of the WeKnowIt

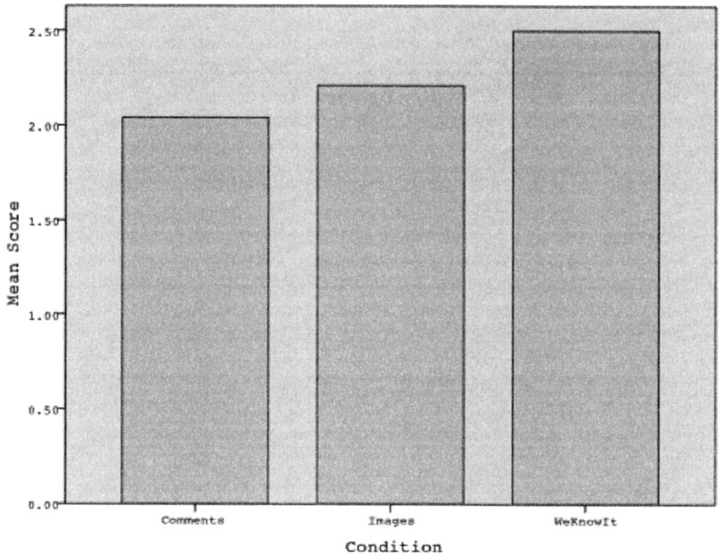

Fig. 23 ER - Average score by condition for Citizens.

consortium members, had prior knowledge of the Collective Intelligence approach taken in the demonstrator and served as a provider of information that can be used in future enhancements of the prototype. The second group, on the other hand, were completely external users who showed their opinions for a product and not for a research prototype. Their opinion is especially important to observe how a tool like the travel desktop demonstrator would impact as a real product for planning a trip, aiming to help users to get the most complete view of places they want to visit. For evaluation purposes, a generic Survey platform was implemented with the aim to provide an easy way to perform the evaluation from remote locations. The demonstrator was evaluated in terms of usability and satisfaction, while implicit user click feedback was collected to observe the usage of the tool and thus improve and refine workflows. Moreover, the effectiveness of its Collective Intelligence services was evaluated through qualitative questionnaires. The SUS Usability and Satisfaction questionnaire was used, which allows for the usage of a standard methodology, which outputs a score on a scale from 0 to 100, the greater the score the more usability and satisfaction (Figure 24). The average SUS score obtained was 68.92 over 100 for the experienced group and 60.66 over 100 for external users, which is interpreted as a decent overall usability score with room for future improvement. The difference in the scores for the two user groups is not statistically significant ($P > 0.85$ using two-tailed unpaired t-test). The interested reader can find more detailed results of the evaluation in [15]. In terms of using the evaluation to assess the

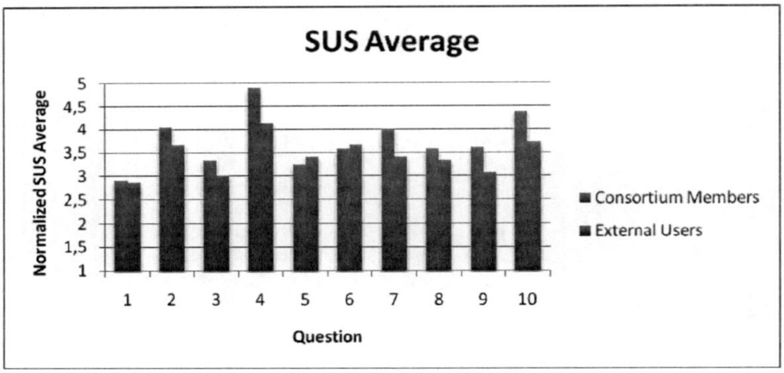

Fig. 24 Desktop travel application - SUS Average for both groups. Normalized values per question.

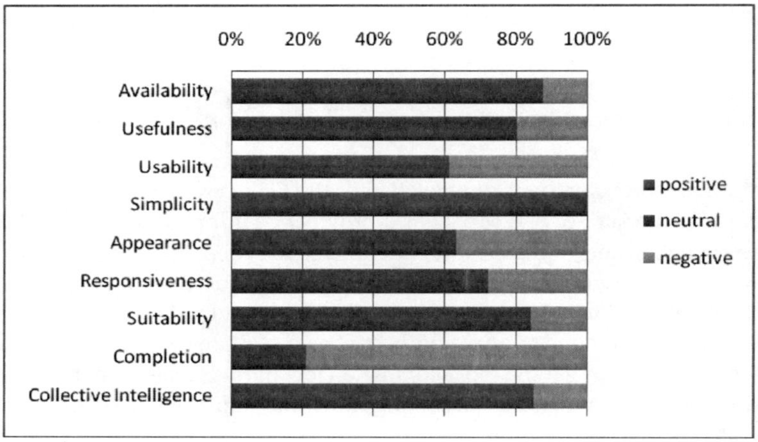

Fig. 25 Mobile guidance evaluation outcome.

quality of the Collective Intelligence services, it can be concluded that the search functionalities of the services (exploiting content from different sources) and the POIs clustering features were of highest importance to the end users. Concerning the mobile part of the travel scenario demonstrator, two field test evaluations were performed, each implemented by different groups of evaluators. The first evaluation was performed in Madrid by a group of four people who perform a two-hour trip in Madrid. The second evaluation has been performed in Barcelona by a group of 21 persons, who evaluated the demonstrator during a 5-hours field test. The Mobile Guidance demonstrator suggested to the users different routes, according to their preferences. The user feedback was collected by means of a questionnaire. The evaluation addressed the following user needs: touristic information and recommendations, field navigation guidance and group communication, which were supported by Collective Intelligence automatically extracted results. The items of functionality considered for evaluation were: get recommendations of Points of interest; search for places; search for Points of Interest; search for friends; get detailed information about places; get detailed information about Points of interest; get image gallery about Points of interest. The following dimensions were explored in the evaluation: availability, usefulness, usability, simplicity, appearance, responsiveness, suitability, completion and contribution of Collective Intelligence to the user experience. The evaluation outcome is displayed in Figure 25. Most dimensions received positive feedback, which means complete or partial agreement on that dimension. Most evaluators were positive on the ability to exploit Collective Intelligence in the travel scenario using the Mobile Guidance application.

9 Conclusions

We have presented state-of-the-art technologies from the media, mass, social and organisational intelligence layers which exploit massive Web 2.0 and user generated content. The methods can be used either solely or in combination, so as to provide an enriched level of intelligence, the so called Collective Intelligence. The integration of the different technology layers which analyse large amount of user generated data in complementary approaches has led to the leveraging of an integrated Collective Intelligence framework, a software platform incorporating analysis services and functionalities from these diverse modalities.

Regarding the exploitation of single layer techniques, in media intelligence the developed VIRaL tool presents an integrated approach on visual image retrieval and localization. We show how the bag-of-words model can be extended by adding geometrical consistency into it and how geo-tags may be exploited in order to allow localization and POIs identification. The nature of this research and the hot topic it comprises within the Web 2.0 framework allows us to identify numerous extensions among the future goals of this work. The most important ones are the expansion of the utilized datasets to larger sets of publicly available, user-generated images, improvements on the algorithmic geometrical consistency model, extension of the presented content matching algorithm in order for it to be used for fast and robust processing of mass amount of non-geotagged images and additional integration of external social web sources as ad-hoc services.

In the mass intelligence layer, methods developed for ontology-based classification of documents create the basis for ontological navigation within thematically related documents. Analysis of semantic graphs created from the input documents can reveal overlapping information that constitutes the core of a topic, and complementary information covered only in a document subset. Such distinction between major and minor topics covered by document set will facilitate more fine-grained topical navigation within the given set of related documents. This approach can be used to create a comprehensive description of a topic and facets to efficiently navigate within it.

The developed community detection methods are a valuable tool for studying the mesoscopic structure of graph-based data, e.g. folksonomies, which represent associations among users, content items and tags. The study of the derived communities can be valuable for a series of tasks such as content indexing, tag clustering and tag recommendation. The efficiency of the developed methods make them suitable for tackling large-scale analysis problems.

In the context of the combination of the different intelligence layers, new Collective Intelligence techniques can be created that leverage the information of several sources as well as services that combine technologies from different research areas. For example, image similarities can be used to weight the networks of their photographers. New opportunities for social network analysis and recommendation systems arise. On the organisational intelligence side, the ontology support for collaboratively creating and sharing semantic POIs can be transferred and used as modelling basis to support collaborative user activities in other applications and domains. Due to its domain independent design, the Event-Model-F can be applied in various other domains. First examples of using the Event-Model-F for an application in the domain of history of art and research of scientific art pieces have shown that it can be easily specialized towards domain-specific requirements. The events represented using the Event-Model-F can also be associated with documentary support from the Web 2.0 such as images and Flickr and tweets on Twitter. Finally, one can conduct reasoning on the Event-Model-F by leveraging domain-specific knowledge such as in emergency response.

From the integration point of view, the produced Data Storage component for the ICIF is a generic solution suitable for storing data of different applications. It can be tailored to store business objects of any application. The WKI DS can be used to enrich typical business applications with SPARQL querying capabilities.

Apart from the possibilities on building on top of the results of each intelligence layer, the most important outlook of the presented Collective Intelligence approach comes from the fact that the aggregated benefits of Collective Intelligence acquired through the various Intelligence Layers are, in particular, realised by both the end users and the organisations when uploading, searching for, browsing and consuming the content. Therefore, each combination of technologies does not only contribute to the generation of Collective Intelligence resulting from each technique separately, but the integration of these varied techniques results in an overall added value which exceeds the aggregate of the individual techniques benefits.

Acknowledgements. This work was sponsored by the European Commission as part of the Information Society Technologies (IST) programme under grant agreement n215453 - WeKnowIt.

References

1. The Description Logic Handbook: Theory, Implementation and Applications. Cambridge University Press, Cambridge (2003)
2. Multiple Bernoulli relevance models for image and video annotation, vol. 2 (2004), http://ieeexplore.ieee.org/xpls/abs_all.jsp?arnumber=1315274
3. Iptc, eventml (2008), http://iptc.org/
4. Wikipedia in Action: Ontological Knowledge in Text Categorization, doi:10.1109/ICSC 2008.53 (2008)
5. Aims: Atlas incident management system (2010), http://www.atlasops.com/products/aims.php
6. Dopplr (2010), http://www.dopplr.com/
7. Emergency command system (2010), http://www.emergencycommandsystem.com
8. Fixmystreet (2010), http://www.fixmystreet.com/
9. ispot, your place to share nature (2010), http://ispot.org.uk/
10. Mit center for collective intelligence, distributed collaboration project (2010), http://cci.mit.edu/research/collaboration.html
11. Mobnotes (2010), http://www.mobnotes.com/
12. Owl 2 web ontology language (2010), http://www.w3.org/TR/owl2-overview/
13. The protege ontology editor and knowledge acquisition system (2010), http://protege.stanford.edu/
14. Using geography can help you to meet your flood management responsibilities (2010), http://bit.ly/fc3GQX
15. Weknowit project deliverable d7.5.1: Consumer and emergency response use case first evaluation report (2010), http://www.weknowit.eu/deliverables
16. Anderson, A.H.: A comparison of two privacy policy languages: Epal and xacml. In: Proceedings of the 3rd ACM Workshop on Secure Web Services, SWS 2006, pp. 53–60. ACM, New York (2006), doi:http://doi.acm.org/10.1145/1180367.1180378
17. Avrithis, Y., Kalantidis, Y., Tolias, G., Spyrou, E.: Retrieving landmark and non-landmark images from community photo collections. In: Proceedings of the International Conference on Multimedia, MM 2010, pp. 153–162. ACM, New York (2010), doi:10.1145/1873951.1873973
18. Begelman, G., Keller, P., Smadja, F.: Automated Tag Clustering: Improving search and exploration in the tag space (2006), http://www.pui.ch/phred/automated_tag_clustering/
19. Bloehdorn, S., Hotho, A.: Text classification by boosting weak learners based on terms and concepts. In: Proceedings of the Fourth IEEE International Conference on Data Mining, ICDM 2004, pp. 331–334. IEEE Computer Society, Washington, DC, USA (2004)
20. Brooks, C.H., Montanez, N.: Improved annotation of the blogosphere via autotagging and hierarchical clustering. In: Proceedings of the 15th International Conference on World Wide Web, WWW 2006, pp. 625–632. ACM, New York (2006), doi:http://doi.acm.org/10.1145/1135777.1135869
21. Chang, E., Goh, K., Sychay, G., Wu, G.: Cbsa: Content-based soft annotation for multimodal image retrieval using bayes point machines. IEEE Transactions on Circuits and Systems for Video Technology 13, 26–38 (2003)

22. Chum, O., Philbin, J., Zisserman, A.: Near Duplicate Image Detection: min-Hash and tf-idf Weighting. In ACM British Machine Vision Conference 2, 1
23. Clauset, A., Newman, M.E.J., Moore, C.: Finding community structure in very large networks (2004), doi:10.1103/PhysRevE.70.066111
24. Deerwester, S., Dumais, S.T., Furnas, G.W., Landauer, T.K., Harshman, R.: Indexing by latent semantic analysis. Journal of the American Society for Information Science 41(6), 391–407 (1990)
25. Doerr, M., Ore, C.E., Stead, S.: The cidoc conceptual reference model: a new standard for knowledge sharing. In: Tutorials, Posters, Panels and Industrial Contributions at the 26th International Conference on Conceptual Modeling. ER 2007, vol. 83, pp. 51–56. Australian Computer Society, Inc, Darlinghurst (2007)
26. Ekin, A., Tekalp, A.M., Mehrotra, R.: Integrated semantic-syntactic video modeling for search and browsing. IEEE Transactions on Multimedia 6, 839 (2004)
27. Ferraiolo, D.F., Kuhn, D.R., Chandramouli, R.: Role-Based Access Control. Artech House, Inc., Norwood (2003)
28. Fischler, M.A., Bolles, R.C.: Chap. Random Sample Consensus: a Paradigm for Model Fitting with Applications to Image Analysis and Automated Cartography. In: Readings in computer vision: issues, problems, principles, and paradigms, pp. 726–740. Morgan Kaufmann Publishers Inc, San Francisco (1987)
29. Francois, A.R., Nevatia, R., Hobbs, J., Bolles, R.C.: Verl: An ontology framework for representing and annotating video events. IEEE Multimedia 12, 76–86 (2005), doi:http://doi.ieeecomputersociety.org/10.1109/MMUL.2005.87
30. Franke, M., Geyer-Schulz, A.: An update algorithm for restricted random walk clustering for dynamic data sets. Advances in Data Analysis and Classification 3(1), 63–92 (2009)
31. Gabrilovich, E., Markovitch, S.: Overcoming the brittleness bottleneck using wikipedia: enhancing text categorization with encyclopedic knowledge. In: Proceedings of the 21st National Conference on Artificial Intelligence, vol. 2, pp. 1301–1306. AAAI Press, Menlo Park (2006)
32. Geyer-Schulz, A., Ovelgoenne, M., Sonnenbichler, A.: Getting Help In A Crowd - A Social Emergency Alert Service. In: International Conference on e-Business 2010 (ICETE ICE-B), Athens, Greece, pp. 207–218 (2010)
33. Geyer-Schulz, A., Thede, A.: Implementation of hierarchical authorization for a web based digital library. In: 3rd International Conference on Cybernetics and Information Technologies, Systems, and Applications, pp. 139–144 (2006)
34. Giannakidou, E., Koutsonikola, V., Vakali, A., Kompatsiaris, Y.: Co-clustering tags and social data sources. In: The Ninth International Conference on Web-Age Information Management WAIM 2008, pp. 317–324 (2008), doi:10.1109/WAIM.2008.61
35. Girardin, F., Calabrese, F., Fiore, F.D., Ratti, C., Blat, J.: Digital footprinting: Uncovering tourists with user-generated content. IEEE Pervasive Computing 7, 36–43 (2008), doi:10.1109/MPRV.2008.71
36. Grauman, K.: Pyramid match hashing: Sub-linear time indexing over partial correspondences. In: CVPR (2007)
37. Hollenstein, L., Purves, R.: Exploring place through user-generated content: Using Flickr to describe city cores. Journal of Spatial Information Science 1(1), 21–48 (2010)
38. Janik, M., Kochut, K.: Training-less Ontology-based Text Categorization. In: Workshop on Exploiting Semantic Annotations in Information Retrieval (ESAIR 2008) at the 30th European Conference on Information Retrieval, ECIR 2008 (2008)

39. Jegou, H., Douze, M., Schmid, C.: Hamming embedding and weak geometric consistency for large scale image search. In: Forsyth, D., Torr, P., Zisserman, A. (eds.) ECCV 2008, Part I. LNCS, vol. 5302, pp. 304–317. Springer, Heidelberg (2008), `http://dx.doi.org/10.1007/978-3-540-88682-2_24`
40. Jeon, J., Lavrenko, V., Manmatha, R.: Automatic image annotation and retrieval using cross-media relevance models (2003)
41. Kalantidis, Y., Tolias, G., Avrithis, Y., Phinikettos, M., Spyrou, E., Mylonas, P., Kollias, S.: Viral: Visual image retrieval and localization. Multimedia Tools and Applications, 1–38 (2010), doi:10.1007/s11042-010-0651-7
42. Kalantidis, Y., Tolias, G., Spyrou, E., Mylonas, P., Avrithis, Y.: Visual image retrieval and localization. In: 7th International Workshop on Content-Based Multimedia Indexing, Greece (2009)
43. Kemp, C., Shafto, P., Berke, A., Tenenbaum, J.B.: Combining causal and similarity-based reasoning. nips (2006)
44. Kennedy, L., Naaman, M., Ahern, S., Nair, R., Rattenbury, T.: How flickr helps us make sense of the world: context and content in community-contributed media collections. In: Proceedings of the 15th International Conference on Multimedia, MULTIMEDIA 2007, pp. 631–640. ACM, New York (2007), doi:http://doi.acm.org/10.1145/1291233.1291384
45. Kleinberg, J.M.: Authoritative sources in a hyperlinked environment. J. ACM 46, 604–632 (1999), doi:http://doi.acm.org/10.1145/324133.324140
46. Lewis, D.: Naive (bayes) at forty: The independence assumption in information retrieval. In: Nédellec, C., Rouveirol, C. (eds.) ECML 1998. LNCS, vol. 1398, pp. 4–15. Springer, Heidelberg (1998), doi:10.1007/BFb0026666
47. Lewis, D.D., Yang, Y., Rose, T.G., Li, F.: Rcv1: A new benchmark collection for text categorization research. J. Mach. Learn. Res. 5, 361–397 (2004)
48. Li, J., Wang, J.Z.: Automatic linguistic indexing of pictures by a statistical modeling approach. IEEE Trans. Pattern Anal. Mach. Intell. 25, 1075–1088 (2003), doi:10.1109/TPAMI.2003.1227984
49. Liu, D., Hua, X.S., Wang, M., Zhang, H.J.: Retagging social images based on visual and semantic consistency. In: Proceedings of the 19th International Conference on World Wide Web, WWW 2010, pp. 1149–1150. ACM, New York (2010), doi:http://doi.acm.org/10.1145/1772690.1772848
50. Lowe, D.G.: Distinctive image features from scale-invariant keypoints. Int. J. Comput. Vision 60, 91–110 (2004), doi:10.1023/B:VISI.0000029664.99615.94
51. Luo, F., Wang, J.Z., Promislow, E.: Exploring local community structures in large networks. In: Proceedings of the 2006 IEEE/WIC/ACM International Conference on Web Intelligence, WI 2006, pp. 233–239. IEEE Computer Society, Washington, DC,USA (2006), doi:http://dx.doi.org/10.1109/WI.2006.72
52. MacQueen, J.B.: Some methods for classification and analysis of multivariate observations. In: Le Cam, L.M., Neyman, J. (eds.) Proc. of the Fifth Berkeley Symposium on Mathematical Statistics and Probability, vol. 1, pp. 281–297. University of California Press, Berkeley (1967)
53. Malone, T.W., Klein, M.: Harnessing collective intelligence to address global climate change. Innovations: Technology, Governance, Globalization 2(3), 15–26 (2007), doi:10.1162/itgg.2007.2.3.15
54. Mccallum, A.K.: BOW: A toolkit for statistical language modeling, text retrieval, classification and clustering (1996), `http://www.cs.cmu.edu/~mccallum/bow/`

55. Mika, P.: Ontologies are us: A unified model of social networks and semantics. In: International Semantic Web Conference, pp. 522–536 (2005)
56. Mueller, E.T.: Chapter 17 event calculus. In: van Harmelen, V.L.F., Porter, B. (eds.) Handbook of Knowledge Representation. Foundations of Artificial Intelligence, vol. 3, pp. 671–708. Elsevier, Amsterdam (2008), doi:10.1016/S1574-6526(07)03017-9
57. Newman, M.E.J., Girvan, M.: Finding and evaluating community structure in networks. Phys. Rev. E 69(2), 026,113 (2004), doi:10.1103/PhysRevE.69.026113
58. Niste'r, D., Stewe'nius, H.: Scalable recognition with a vocabulary tree. In: CVPR, pp. 2161–2168 (2006)
59. Ovelgoenne, M., Geyer-Schulz, A., Stein, M.: A randomized greedy modularity clustering algorithm for community detection in huge social networks. In: 4th International Workshop on Social Network Analysis and Mining (SNA-KDD 2010), Washington, DC, USA (2010)
60. Ovelgoenne, M., Sonnenbichler, A.C., Geyer-Schulz, A.: Social emergency alert service - a location-based privacy-aware personal safety service. In: Proceedings of the 2010 Fourth International Conference on Next Generation Mobile Applications, Services and Technologies, NGMAST 2010, pp. 84–89. IEEE Computer Society, Washington, DC, USA (2010), doi:http://dx.doi.org/10.1109/NGMAST.2010.27
61. Papadopoulos, S., Kompatsiaris, Y., Vakali, A.: A graph-based clustering scheme for identifying related tags in folksonomies. In: Bach Pedersen, T., Mohania, M.K., Tjoa, A.M. (eds.) DAWAK 2010. LNCS, vol. 6263, pp. 65–76. Springer, Heidelberg (2010)
62. Papadopoulos, S., Vakali, A., Kompatsiaris, Y.: Community detection in collaborative tagging systems. In: Pardede, E. (ed.) Community-built Database: Research and Development. Springer, Heidelberg (2010)
63. Papadopoulos, S., Zigkolis, C., Tolias, G., Kalantidis, Y., Mylonas, P., Kompatsiaris, Y., Vakali, A.: Image clustering through community detection on hybrid image similarity graphs. In: 2010 International Conference on Image Processing (ICIP 2010), Hong-Kong, September 26-29 (2010)
64. Philbin, J., Chum, O., Isard, M., Sivic, J., Zisserman, A.: Object retrieval with large vocabularies and fast spatial matching. In: IEEE Conference on Computer Vision and Pattern Recognition, CVPR 2007, pp. 1–8 (2007)
65. Prelec, D.: A Bayesian Truth Serum for Subjective Data. Science 306(5695), 462–466 (2004), doi:10.1126/science.1102081
66. Quack, T., Leibe, B., Van Gool, L.: World-scale mining of objects and events from community photo collections. In: Proceedings of the 2008 International Conference on Content-based Image and Video Retrieval, CIVR 2008, pp. 47–56. ACM, New York (2008), doi:http://doi.acm.org/10.1145/1386352.1386363
67. Raghavan, U.N., Albert, R., Kumara, S.: Near linear time algorithm to detect community structures in large-scale networks. Physical Review E 76(3), 036,106 (2007), doi:10.1103/PhysRevE.76.036106
68. Raimond, Y., Abdallah, S.: The event ontology (October 2007), `http://motools.sf.net/event`
69. Rissanen, E.: Extensible access control markup language (xacml) version 3.0 committee draft 03 (2010), `http://docs.oasis-open.org/xacml/3.0/xacml-3.0-core-spec-cd-03-en.pdf`

70. Saathoff, C., Scherp, A.: Unlocking the semantics of multimedia presentations in the web with the multimedia metadata ontology. In: Proceedings of the 19th International Conference on World Wide Web, WWW 2010, pp. 831–840. ACM, New York (2010), doi:http://doi.acm.org/10.1145/1772690.1772775
71. Sandhu, R.S., Samarati, P.: Access control: Principles and practice. IEEE Communications Magazine 32, 40–48 (1994)
72. Schenk, S., Saathoff, C., Staab, S., Scherp, A.: Semaplorer-interactive semantic exploration of data and media based on a federated cloud infrastructure. Web Semant. 7, 298–304 (2009), doi:10.1016/j.websem.2009.09.006
73. Scherp, A., Franz, T., Saathoff, C., Staab, S.: F–a model of events based on the foundational ontology dolce+dns ultralight. In: Proceedings of the Fifth International Conference on Knowledge Capture, K-CAP 2009, pp. 137–144. ACM, New York (2009), doi:http://doi.acm.org/10.1145/1597735.1597760
74. Sikkel, K.: A group-based authorization model for cooperative systems. In: Proceedings of the Fifth Conference on European Conference on Computer-Supported Cooperative Work, pp. 345–360. Kluwer Academic Publishers, Norwell (1997)
75. Sivic, J., Zisserman, A.: Video Google: A Text Retrieval Approach to Object Matching in Videos. In: IEEE International Conference on Computer Vision, vol. 2, pp. 1470–1477 (2003), doi:10.1109/ICCV.2003.1238663
76. Torres, L.H.: Citizen sourcing in the public interest. Knowledge Management for Development Journal 3(1), 134–145 (2007)
77. Vapnik, V.N.: The nature of statistical learning theory. Springer-Verlag New York, Inc., New York (1995)
78. C. Wang, F. Jing, L. Zhang, H.-J. Zhang: Content-based image annotation refinement. In: CVPR (2007)
79. Wang, X.j., Mamadgi, S., Thekdi, A., Kelliher, A., Sundaram, H.: Eventory – an event based media repository. In: Proceedings of the International Conference on Semantic Computing, pp. 95–104. IEEE Computer Society, Washington, DC, USA (2007), doi:10.1109/ICSC.2007.33
80. Westermann, U., Jain, R.: Toward a common event model for multimedia applications. IEEE MultiMedia 14, 19–29 (2007), doi:10.1109/MMUL.2007.23
81. Winerman, L.: Social networking: Crisis communication. Nature 457(7228), 376 (2009), doi:10.1038/457376a
82. Witten, I.H., Frank, E.: Data Mining: Practical Machine Learning Tools and Techniques, 2nd edn. Morgan Kaufmann Series in Data Management Systems. Morgan Kaufmann, San Francisco (2005), http://bit.ly/ihBvSG
83. Wolfson, H.J., Rigoutsos, I.: Geometric hashing: an overview. IEEE Computational Science and Engineering 4(4), 10–21 (1997), doi:10.1109/99.641604
84. Work, D.B., Blandin, S., Tossavainen, O.P., Piccoli, B., Bayen, A.M.: A Traffic Model for Velocity Data Assimilation. Applied Mathematics Research Express 2010(1), 1–35 (2010), doi:10.1093/amrx/abq002
85. Xu, X., Yuruk, N., Feng, Z., Schweiger, T.A.J.: Scan: a structural clustering algorithm for networks. In: KDD 2007: Proceedings of the 13th ACM SIGKDD International Conference on Knowledge Discovery and Data Mining, pp. 824–833. ACM, New York (2007)
86. Zhou, D., Bousquet, O., Lal, T.N., Weston, J., Schölkopf, B.: Learning with local and global consistency. Advances in Neural Information Processing Systems 16 (2004)

Glossary

ACL	Access Control Lists
CBAR	Content-based Annotation Refinement
CDL	Community Design Language
CNM	Clauset, Newman and Moore
CRUD	Create, Read, Update and Delete
DOLCE	Descriptive Ontology for Linguistic and Cognitive Engineering
DUL	DOLCE+DnS UltraLight
EAS	Emergency Alert Service
EPAL	Enterprise Privacy Authorization Language
ER	Emergency Response
EXIF	Exchangeable Image File Format
HITS	Hyperlink-Induced Topic Search
ICIF	Integrated Collective Intelligence Framework
NLP	Natural Language Processing
OSGi	Open Services Gateway initiative
POI	Point of Interest
RANSAC	RANdom SAmple Consensus
RBAC	Role-Based Access Control
REST	Representational State Transfer
RG	Randomized Greedy modularity clustering algorithm
SCAN	Structural Clustering Algorithm for Networks
SIFT	Scale-Invariant Feature Transform
SPARQL	SPARQL Protocol and RDF Query Language
SURF	Speeded-Up Robust Features
SUS	System Usability Scale
SVM	Support Vector Machines

VERL	Video Event Representation Language
VIRaL	Visual Image Retrieval and Localization
WEKA	Waikato Environment for Knowledge Analysis
WERL	WeKnowIt ER Log merging and management
WKI DS	WeKnowIt Data Store
XACML	eXtensible Access Control Markup Language

Chapter 21
Mobile Sensing Technologies and Computational Methods for Collective Intelligence

Daniel Olguín Olguín, Anmol Madan, Manuel Cebrian, and Alex (Sandy) Pentland

Abstract. This book chapter is a review of mobile sensing technologies and computational methods for collective intelligence. We discuss the application of mobile sensing to understand collective mechanisms and phenomena in face-to-face networks at three different scales: organizations, communities and societies. We present an overview of the state-of-the art in individual behavior recognition from sensor data. We discuss related work on group behavior recognition such as face-to-face interaction, social signaling, conversation detection, and conversation dynamics. We also present a brief overview of pattern recognition methods in social network analysis for the automatic identification of groups and the study of social network evolution. We describe a sensor-based organizational design and engineering system for computational collective intelligence applications in organizations. We also provide two example applications of collective intelligence and modeling user behavior at the community scale. Finally, we investigate the impact that these new sensing technologies may have on the understanding of societies, and how these insights can assist in the design of smarter cities and countries.

1 Introduction

Collective intelligence is the shared intelligence that emerges from the collaboration of individuals (Scarlat & Maries, 2009). It can also be defined as the ability of a group to solve problems more effectively than any of its individual members (Heylighen, 1999). Thus, we can think of computational collective intelligence as the ability to achieve higher group performance with the help of computational tools and methods.

To date, research on human interactions has relied mainly on one-time, self-reported data on relationships. New technologies, such as video surveillance, e-mail, and mobile phones, offer a moment-by-moment picture of interactions over extended periods of time, providing information about both the structure and content of relationships. This has given rise to an emerging field of "computational

Daniel Olguín Olguín · Anmol Madan · Manuel Cebrian · Alex (Sandy) Pentland
The Media Laboratory, Massachusetts Institute of Technology

N. Bessis, F. Xhafa (Eds.): Next Generation Data Technologies for CCI, SCI 352, pp. 575 – 597.
springerlink.com

social science" that leverages the capacity to collect and analyze data with an unprecedented breadth and scale (Lazer, et al., 2009). Social science and computer science researchers have begun to apply methods from different disciplines in an attempt to understand the nature of the complex organizational and social network footprints left behind by the use of personal electronic devices such as mobile phones and RFID cards. It is now possible to understand human behavior and social interaction at different levels, ranging from individuals and small groups to large organizations and even entire nations. Even though well known statistical methods and social network analysis have been applied to large datasets generated by pervasive communication media, new analytical tools and methods need to be created in order to deal with the complexity of these systems.

These new sensing and modeling technologies help us understand what local properties enable collective intelligence at the community scale. Ideas, opinions, behaviors, and information play a fundamental role in collective intelligence and group decision-making—but how do they diffuse through face-to-face interactions? We show that mobile sensing approaches help understand the formation of strong and weak local ties, and the behaviors that characterize them in the context of communities. We also find that these interactions predict the adoption of different types of behaviors in these communities. Understanding collective behaviors will allow us to design methods to maximize community-level benefit in the future.

The following questions apply at the scale of cities, countries, and societies in general: Is the way in which individuals interact, intentionally or unintentionally, designed to maximize global benefit? Or does it result in a fundamentally non-egalitarian stratification of society, where a small number of individuals inevitably dominate? Our ability to observe and record interactions between individuals in large populations has improved dramatically with modern technological innovations, but using this data to model cooperation and collaboration between individuals and its global effect on the entire population is still a difficult task. To shed light on these questions we review recent advances in social modeling that allow us to quantify an individual's value in society as a mathematical function of a set of choices, and the collective potential of a population as the expected value of an individual over time (Lazer & Friedman 2007; Bramoullé & Kranton, 2007; Cebrian et al., 2010; Lahiri & Cebrian, 2010). These theories assume that individuals try to selfishly improve their societal value by adopting the choices of their neighbors, constrained by the actual observed interaction topology and order. As a result, we are also able to investigate how far societies are from an optimal regime of functioning, and therefore help in the design of more functional cities and countries.

2 Mobile-Sensing of Collective Behavior in Organizations

Our research group at the MIT Media Laboratory has demonstrated that wearable technology can be used to characterize face-to-face interactions, measure individual and collective patterns of human behavior, and automatically map out a company's de facto organizational chart (Choudhury & Pentland, 2003; Pentland, 2006; Olguin-Olguin et al., 2009b). This capability can be an extraordinary resource for studying group behavior, group performance and team formation processes.

We have also developed several tools for analyzing voice patterns and quantifying social context in human interaction, as well as several socially aware platforms that objectively measure different aspects of social context, including non-linguistic social signals measured by a person's tone of voice, movements or gestures. We have found that non-linguistic social signals are particularly powerful for analyzing and predicting human behavior, sometimes exceeding even expert human capabilities (Pentland et al., 2005; Pentland, 2008; Madan et. al. 2006).

2.1 Sensing and Modeling Individual Behavior

By recognizing human behavior from sensor data at the individual and group levels, and combining pattern recognition methods with dynamic social network analysis, we can create a general framework for modeling group dynamics. At the individual level, researchers have applied pattern recognition methods to several aspects of human behavior such as primitive motor activities (i.e. standing, walking, running, etc.), complex or high-level activities (i.e. cooking, washing dishes, etc.), body posture, facial expressions, hand gestures, and displacement patterns (i.e. location tracking).

Previous work in human activity recognition using motion sensors (e.g. accelerometers, gyroscopes, vibration sensors, etc.) has shown that it is possible to classify in real time several postures and activities, however most of the early work on human activity recognition from sensor data has focused on the identification of a specific activity in a particular scenario, such as sitting on a chair or walking. More recently, there has been increasing interest on modeling more complex patterns of behavior over extended periods of time (Oliver & Horvitz, 2005).

For example, Van Laerhoven and Cakmakci (2000) augmented a pair of pants with an accelerometer to register what the wearer was doing and anticipate his behavior. DeVaul and Dunn (2001) developed a two-layer model that combined a Gaussian mixture model with first-order Markov models to classify a range of activities including: sitting, walking, biking, and riding the subway. A single three-axis accelerometer placed on the torso was used. Mantyjarvi, et al. (2001) used Principal Component Analysis (PCA), Independent Component Analysis (ICA), and a multilayer perceptron for activity classification in subjects wearing two accelerometers, one on each side of the hip. Lee and Mase (2002) proposed a method to determine a user's location, detect transitions between locations, and recognize sitting, standing, and walking behaviors. In (Kern & Schiele, 2003), the acquisition of context information from audio and acceleration sensors is investigated. They developed a model that uses personal and social interruptability of the user to decide both whether or not to notify the user and to decide which notification modality to use.

Bao and Intille (2004) evaluated several algorithms to classify twenty different physical activities from data acquired using five 2-axis accelerometers following a semi-naturalistic data collection protocol. Sukthankar and Sycara (2005) proposed a full-body motion capture system that recognizes human behavior from a set of military maneuvers, based on the subject's motion type and proximity to landmarks. Low-level motion classification is performed using Support Vector

Machines (SVMs) and Hidden Markov Models (HMMs). Lester, et al. (2006) suggested that a wearable activity recognition system should have the following properties: (i) data should only be needed from a single body location (it doesn't need to be the same for every user), (2) it should work across individuals and require personalization only to enhance its recognition capabilities, and (3) it should be effective even with a subset of the sensors and data features.

Chung and Liua (2007) presented a Hierarchical HMM (HHMM) for behavior understanding from video streams in a nursing center. The proposed approach infers elderly behaviors through three contexts: spatial, activities, and temporal. Liao, et al. (2007) have used conditional random fields (CRFs), a form of undirected graphical models and boosting to identify a person's indoor activities, such as using a computer, having a meal, or watching TV; using accelerometer, audio and light sensor data. By directly modeling the conditional distribution over hidden states given the observations, CRFs make no assumptions on the dependency structure between observations.

2.2 Sensing and Modeling Group Behavior

At the group level, we are interested in automatically identifying face-to-face interactions, conversations, and conversation dynamics. A wide range of studies has shown that hand-coded analyses of communication in teams can predict performance (Foltz & Martin, 2009). These studies have looked at the frequency, patterns and content of communication. For instance, an analysis of the communication patterns of aircrews in flight simulation experiments revealed significant differences between successful and unsuccessful crews (Bowers, et al., 1998). In some cases, high-performing teams communicate with higher overall frequency than low-performing teams, but in other cases, this finding has not been supported.

According to Foltz and Martin (2009), to develop a human performance model, one needs to find out if, and the degree to which, a relationship between communication and performance exists. Computational models must accurately measure features in communication that relate to measures of team performance. To create such model, recent advances in the fields of computational cognitive models (i.e. latent semantic analysis, or LSA, social network analysis, and pattern recognition techniques (i.e. clustering, classification, generalization) can be leveraged.

Oliver, et al. (2004) proposed the use of Layered HMMs (LHMMs) to classify different office activities (i.e. phone conversation, face-to-face conversation, distant conversation, presentation, etc.). In their model, there is a hierarchy with multiple HMMs, each corresponding to a certain concept (for example, audio signals). These HMMs take as observations either the features computed from the raw signals or the inferential results from the previous level. In LHMMs, each layer of the architecture is connected to the next layer via its inferential results.

Gatica-Perez (2006) discusses work on automatic analysis of face-to-face multiparty conversations from multisensory data that has appeared in the literature spread over several communities, including signal processing, computer vision, multimodal processing, machine learning, human-computer interaction, and ubiquitous computing. The author proposes a categorization of conversational group activities on the basis of temporal scale and group size. The proposed categories are: addressing (i.e., who speaks to whom at every time), turn-taking patterns (i.e. floor control, discussions, monologues), and group trends (i.e. interest levels, dominance, and influence). In (Gatica-Perez, 2009) the author extended his review to more than a hundred different works addressing the computational modeling of interaction management, internal states, personality traits, and social relationships in small group conversations. His review focuses on small groups, non-verbal behavior, computational models, and face-to-face conversations.

Other aspects of human behavior that can be automatically measured are the subconscious signals displayed during social interactions. According to Pentland (2008), these "honest" signals can be measured by analyzing the timing, energy, and variability of speech and body movement patterns. He describes four different types of honest signals in humans: influence (the extent to which one person causes the other person's pattern of speaking to match their own pattern), mimicry (the reflexive copying of one person by another during a conversation), activity (speaking time and energy), and consistency (low variability in the speech signal). The pattern of signaling behavior and social roles largely determines the pattern of communication within an organization. Consequently, the dynamics of group interaction can be inferred from the pattern of communication. For instance, dominant, high-influence individuals cause the pattern of communication to flow through them, making them more central in the organization.

2.3 Social Network Analysis

A social network consists of a set of actors (or nodes) and the relations (or ties) between these actors. Actors may be individuals, groups, organizations, or entire communities, and relations may span across or within levels of analysis. These relational variables are defined and measured at the dyadic level and can include a wide variety of social and physical ties, each of which may have a number of different basic properties (Wasserman & Faust, 1994).

Pattern recognition methods have also be applied in social network analysis. Clustering techniques have been used to identify communities and study their evolution over time (Mishra, et al., 2007). Interest often focuses on finding clusters of actors or ties, and the number of groups in the data is typically unknown. For example, Handcock, et al. (2007) proposed a latent position cluster model, under which the probability of a tie between two actors depends on the distance between them in an unobserved Euclidean "social space", and the actors' locations in the latent social space arise from a mixture of distributions, each corresponding to a cluster.

An important property found in many networks is community structure, in which network nodes are joined together in tightly knit groups, between which there are only looser connections. Girvan and Newman (2002) proposed a method for detecting such communities, built around the idea of using centrality indices to find community boundaries. The traditional method for detecting communities is hierarchical clustering; however, the authors propose an alternative approach: Instead of trying to construct a measure that tells us which edges are most central to communities, they focus instead on those edges that are least central, the edges that are most "between" communities. Communities are constructed by progressively removing edges from the original graph. By removing these edges, groups can be separated from one another and the underlying community structure of the graph can be revealed.

Bagrow and Bollt (2005) proposed another method of community detection. This method is local in the sense that a community can be detected within a network without requiring knowledge of the entire network. A community could be loosely described as a collection of vertices within a graph that are densely connected to the rest of the graph. Palla, et al. (2007) developed a new algorithm based on clique percolation that allows to investigate the time dependence of overlapping communities on a large scale and to uncover basic relationships characterizing community evolution. Kossinets and Watts (2006) analyzed a dynamic social network comprising students, faculty, and staff at a large university, in which interactions between individuals were inferred from time-stamped e-mail headers recorded over one academic year and were matched with affiliations and attributes. They found that network evolution is dominated by a combination of effects arising from network topology itself and the organizational structure in which the network is embedded. They emphasize that understanding tie formation and related processes in social networks requires longitudinal data on both social interactions and shared affiliations.

Some of the most recent analytical developments are in the form of exponential random graph models (Robins, et al., 2007), which allow for modeling of complex patterns of dependencies at different levels of analysis. Exponential random graph models define the probability of an observed network as a function of different structural characteristics such as density, reciprocity, or cliquing; these models allow for simple research questions, such as whether there are homophily effects in networks, or more complex research questions like whether the tendency toward various hierarchy-related structures differ across groups that use different strategies to complete a team task (Slaughter, et al., 2009). Another interesting application of pattern recognition is the use of graphical models in dynamic social network analysis (Goldenberg, 2007).

2.4 Sensor-Based Organizational Design and Engineering

The basic goal of organizational research has been to discover what kinds of organizational designs or structures will be most effective in different situations, as well as to identify variables that will enable researchers to make consistent and valid predictions of what kinds of organizational structures will be most effective in said situations (Tushman & Nadler, 1978).

Olguin-Olguin and Pentland (2010) have proposed a sensor-based organizational design and engineering approach that combines behavioral sensor data with other sources of information such as e-mail, surveys, and performance data in order to design interventions aimed at improving organizational outcomes. The proposed approach includes the following steps:

1. Capturing the interactions and social behavior of employees, managers and customers using wearable and/or environmental sensors. Other sources of information that can be incorporated into the system are any form of digital records (e.g. e-mail, chat, phone logs).
2. Performing data mining and pattern recognition to extract meaningful information from these data.
3. Combining the extracted information with performance data (e.g. sales, tasks, timing) and finding relationships between objective measurements and performance outcomes.
4. Generating feedback in the form of graphs, interactive visualizations, reports, or real-time audio-visual feedback for employees, managers and/or customers in order to improve organizational performance and customer satisfaction.
5. Designing and implementing organizational interventions based on behavior simulation and prediction.
6. Continuous measurement and performance assessment.

A sensor-based system for organizational design and engineering consists of environmental and wearable sensors, computers, and software, that continuously and automatically measure individual and collective patterns of behavior, identify organizational structures, quantify group dynamics, and provide feedback to their users. The purpose of such system is to improve productivity, efficiency, and/or communication patterns within an organization. The proposed system is composed of one or more wearable sensing devices functioning in a wireless sensor network, one or more radio base stations, a computer system, and several data processing algorithms. The system may include some or all of the following:

- Environmental sensors that monitor the current conditions of the workplace (temperature, light, movement, activity, sound, video, etc.) and that can be used as base stations.
- Wearable sensors that employees carry around and that measure human behavior (social interaction, activities, location, etc.). These can be mobile devices such as cell phones, PDAs, or electronic badges that collect data, communicate with a database (via Ethernet or wirelessly) to retrieve information, and provide feedback to their users.
- Software that automatically identifies relevant keywords in documents, web pages, e-mail, and instant messaging communication.
- A database that stores all the information collected by the environmental, wearable and software sensors (who-knows-what, who-knows-who, and where-is-who).
- Simulation and data mining algorithms.
- Feedback and visualization mechanisms.

The organizational re-engineering process consists of short cycles of measurement-feedback-intervention-measurement until significant improvements have been reached. The first measurement phase may last a few weeks or up to a few months. The feedback phase can happen in real time (while sociometric data is being collected), or after the first measurement phase. Interventions have to be implemented soon after the feedback phase and the second measurement phase can carried out a few days or weeks after the intervention has been put into practice. The second measurement phase is confirmatory step and the entire cycle can be repeated again. Figure 1 shows a diagram of a sensor-supported organizational re-engineering process.

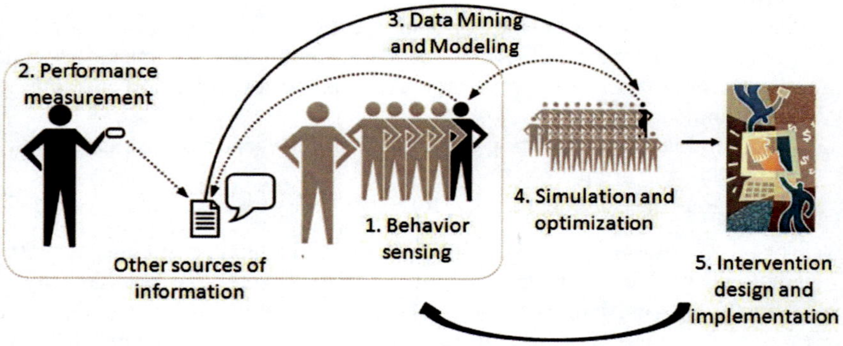

Fig. 1 Sensor-supported organizational re-engineering process

We have used *sociometric badges* in sensor-based organizational systems to detect face-to-face interactions, conversations, body movement, and proximity to others (Olguin-Olguin, 2007). A *sociometric badge* is capable of extracting speech features without recording the content of conversations in order to maintain privacy, and of wirelessly transferring data to a central server. We have used them in several organizations to capture face-to-face communication patterns and study the relationship between collective behavior and performance outcomes, such as productivity and job satisfaction (Olguin-Olguin et al., 2009a, 2009b).

The design of the *sociometric badges* was motivated by the fact that a large number of organizations already require employees to wear RFID name tags that identify them and grant them access to several locations and resources. These traditional RFID name tags are usually worn around the neck or clipped to the user's clothing. With the rapid miniaturization of electronics, it is now possible to augment RFID badges with more sensors and computational power that allow us to capture human behavior without requiring any additional effort on the user's side. By capturing individual and collective patterns of human behavior with *sociometric badges* and correlating these behaviors with individual and group performance, it is possible to identify successful vs. unsuccessful teams, high performing teams, and predict group outcomes. The added value for the users is the feedback that they can receive about their daily behaviors and interactions with others, and how these behaviors affect their individual and group performance.

3 Mobile-Sensing of Collective Behavior in Communities

To optimize collective intelligence in the context of communities, it is important to understand how information, ideas, behaviors and opinions propagate, and how individuals influence this diffusion process. The proliferation of social web applications on the Internet has generated copious amounts of data about how people behave and interact with each other in online communities—and these data, in turn, are being extensively used to model the role of social interactions in our lives. However, many characteristics of our lives about are expressed only in real-world, face-to-face interactions. In order to model such collective behavior, we need fine-grained data about face-to-face interactions between people—who talks to whom, when, where, and how often.

Social scientists have traditionally relied on survey questionnaires to model these interactions. However, these approaches are not scalable and inaccurate- it is impossible to use these methods with fine resolution, over long timescales (e.g. months or years), or for a large number of people, (e.g. thousands or millions). In a survey of informant accuracy literature, Bernard et al. (1984) have shown that recall of social interactions in surveys is typically 30-50% inaccurate.

With automated social sensing tools we now accurately capture face-to-face interactions – measuring who talks to whom, how often, and so forth – and this more accurate data helps us understand the circumstances under which new ideas and opinions spread from one person to another: For example, we can now explore questions like: Does regular co-location or frequent communication imply greater social influence? What is the relative role of different types of communication and interactions? How do interactions in the workplace or in other contexts translate into different types of social influence? Is one context more powerful than the other?

The design goals and research methodology for such collective intelligence platforms are explained in the next section. We then provide two example applications of collective intelligence and modeling user behavior at the community scale, first in the context of social relationships and then in the context of political opinions.

3.1 Mobile Platforms for Collective Intelligence at the Community Scale

To generate a meaningful representation of human interactions in communities, it is important to devise platforms that can capture different aspects of these interactions, like the strength and nature of ties (or relationships), the overall social network structure (which often varies with the communication modality), the dynamics of network ties and interactions (evolution, reciprocity), emergent communities, and homogeneity of behaviors of individuals. A mobile platform designed for long-term use within communities should meet the following design goals:

- Designed for comfortable, long-term, continuous use, ranging from several weeks to years

- Meant to be used the user's, always-on primary device or mobile phone, such that day-to-day usage is natural and not cumbersome.
- Inbuilt security, encryption and removal of personal identifiers—such devices collect sensitive personal data and it has to be treated carefully.
- Incorporate user feedback and display— so that the user can visualize his/her data, evaluate application performance and provide cognitive support where automatic pattern recognition methods are inaccurate. Several platforms also use experience sampling at opportune periods for active learning or reinforcement learning techniques.

Various researchers have designed such collective intelligence platforms based on commodity mobile phones, or worked with mobile operators to generate user interaction datasets. Eagle and Pentland (2009) used mobile phone Bluetooth proximity, call data records and cellular-tower identifiers to detect the social network structure and recognize regular patterns in daily user activity. With human location traces, Gonzalez (2008) showed that call detail records could be used to characterize temporal and spatial regularity in human mobility patterns better than random walk or Levy flight simulations. Other examples of the use of mobile phones to map human interaction networks include the CENS participatory sensing project at UCLA (Abdelzahar 2007) and the mHealth project at Dartmouth (Avancha 2009). Our group at the MIT Media Lab has deployed mobile phones as automated collective intelligence sensors in the context of undergraduate communities, to capture face-to face interactions and phone communication patterns amongst residents for an entire year (Madan & Pentland 2009, 2010).

The following types of sensor hardware and electronic communication have been used to model interactions in communities:

- Location sensing: Cellular towers identifiers, global positioning system (GPS) receivers, 802.11 WLAN Access Points
- Proximity sensing: Bluetooth transceivers, Infra-red (IR) sensors
- Phone Communication: call-data records (CDRs) and SMS logs
- Electronic Communication: Email headers, network data from social networking sites like Facebook.

In the next section, we provide two examples of sensing and modeling collective intelligence in communities, one in the context of viral media and the second in the context of political opinions.

3.2 Collective Behavior in Communities – Viral Media

Leskovec (2007) found that viral media recommendations have a long-tail distribution, and that social influence plays a stronger role for niche products. Salganik et. al. (2008) measured social influence online with music in eight simultaneous worlds and found highly unpredictable and completely different rankings of music tracks in each world, unrelated to the actual quality of music –evidence of social influence of previous listeners.

But what role do face-to-face interactions play in the spread of viral media? A mobile platform with the previously mentioned capabilities was deployed residents of three floors of a similar undergraduate dormitory for one month. In addition to social interaction data, how the participants played and shared viral music on the mobile devices was also logged on a central server. Communication features (e.g., total communication, off-peak communication, incoming vs. outgoing calls) and 802.11 WLAN co-location features were extracted from the raw interaction data.

The phone communication features and music sharing behaviors were significantly correlated with the self-reported relationships amongst the residents, and can be used to discriminate between close friends and casual acquaintances. While total communication counts were positively correlated with both friends and acquaintance relationships, off-peak communication and SMS communication features were positively correlated only with the 'friend' relationships, as shown in Figure 3.2-1. The overall classification accuracy for identifying close friends (versus non-friends) with communication features alone was found to be 87% with communication features alone, and 90% with both communication and music sharing features.

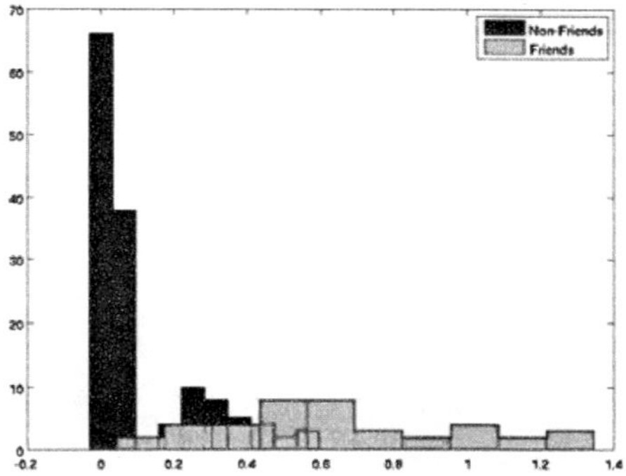

Fig. 2 Histogram of 'friend' relationships vs. values predicted using the location, communication and sharing features. X-axis values are computed as a linear function of the raw features and the Y axis represents the number of dyads in each bin. Friends can be visually separated from other classes by drawing a vertical line at x=0.2 or fitting Gaussian for each class.

The overall classification accuracy for identifying close-friends (versus non-friends) from communication features alone was found to be 87% with communication features alone, and 90% with both communication and music-sharing features. A cost-sensitive Bayesian Network model with 5-fold cross-validation was used for classification, as classes were unbalanced.

The viral music sharing observed in the community had two components—about 70% of total music shared was between mutually acknowledged close friends (i.e., strong social ties) and the remaining 30% of music shared was between relative strangers or weak ties (i.e., students in the same classes or residence floor).

The communication and co-location features were significantly correlated with music sharing behaviour, and the correlations were even stronger if only strong ties are considered. Dyadic music sharing behaviour shows a higher correlation with automatically captured communication and location features than with self-reported relationships—this result highlights the power of automated collective intelligence sensing technologies, and their resulting improvement over survey tools. It may be possible to better predict viral music propagation amongst weak ties with other forms of interactions, like mapping email or Facebook networks. The Author-Recipient-Topic (ART) model and Latent Dirchlet Allocation (LDA) (McCallum 2007) are examples of approaches that have been used to identify roles, relationships and group membership from email interactions.

What about the ability of participants to influence other's viral music preferences? One approach to measuring social influence is the latent-state influence model, proposed by Dong (2007), a tractable approximation for hidden Markov modelling of multiple interacting stochastic processes, where the model parameters for an increasing number of chains are reduced, by introducing influence parameters to summarize the coupling between interacting chains (also referred to as 'alpha' parameters in literature). The forward backward algorithm for estimating latent states and the maximum likelihood algorithm for estimating model parameters can be derived from the equivalence between the influence model and corresponding Hidden Markov model.

When the influence model is fitted to the viral media dataset, each participant is represented as Markov chain (with latent states) and observations in the model are their respective music consumption. The influence parameters then capture the social influence dynamics between the individuals, as shown in Figure 3.2-2 for sixteen interacting participant-chains.

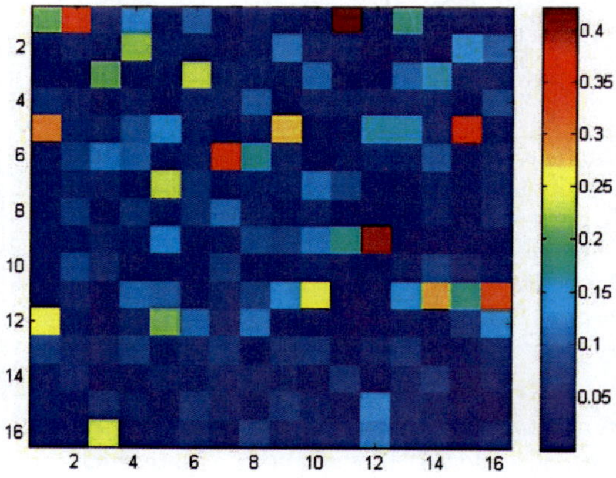

Fig. 3 Social influence matrix (influence or alpha parameters) for 16 participants based on their music consumption. The observed variable for each chain is the number of times the three most popular tracks are played by the participant. The time-step for all chains in 1-day, with data from 30 days used in the model. The model was trained used Expectation-Maximization. Two latent states are assumed per chain and represent the level of 'activation' for the participant. The influence parameters represent inter-chain dynamics for the entire 30-day period.

3.3 Collective Behavior in Communities – Political Opinions

A similar mobile platform for collective intelligence was deployed with the undergraduate residents of a university residence hall for an entire academic year in 2008 (Madan et. al. 2010). The US Presidential elections were held during the same period, and the data collected included the last three months of the election campaigns of Barack Obama (Democrat candidate) and Senator John McCain (Republican candidate). The political opinions of the undergraduate students were surveyed monthly during this period along different dimensions: political party preference (scale from strong Republican to strong Democrat), interest in politics (scale from 'no interest' to 'high interest'), and liberal or conservative (scale from 'very conservative' to 'very liberal').

Using mobile phone interaction features based on phone communication and physical proximity, it is possible to estimate individual exposure to different types of opinions in this community. Due to the geographic location of the university, a majority of the students showed higher exposure to democratic and liberal opinions. These patterns of exposure reveal patterns of *dynamic homophily*, where individuals show increased exposure to like-minded individuals around external political events during the 2008 election period (e.g. political debates and election day). Measured exposure to other nodes in the network can be used to predict future opinions for individuals ($r=0.8$, $p < 0.0001$). it was found that these mobile phone based features and dynamic exposure increases the explained variance by up to 30%, with stronger effects for freshmen (first-year) students in the community.

4 Mobile-Sensing of Collective Behavior in Societies

A robust empirical observation in economics is that population density is positively correlated with productivity (Ciccone, 1996; 2002). Recent research has shown that the relationship between human agglomerations and sociological quantities goes beyond pure economic productivity: variables such as creative output, crime, and disease are also super-linear in the size of the city (Bettencourt, 2007). That is, if y represents the number of new patents produced in a city, and is its population N, super-linear scaling implies that $y = cN^{\alpha}$, with $1 \leq \alpha \leq 2$ and a constant c.

This empirical regularity offers an important target for science to aim at. Understanding this relationship might enhance our understanding of the creativity of society, with the concomitant potential to shed light on the policy implications for economic growth. There are, however, multiple empirical challenges in studying this empirical relationship. Distinguishing the causal direction of this relationship is the first conundrum. Productivity might be high because of agglomeration effects. Alternatively, agglomeration might be a consequence, not a cause, of high productivity. Enhancing this inferential challenge is the difficulty in examining the intervening processes that might connect productivity and city population size.

Specifically, as economists studying this question have observed, distinguishing these alternative explanations is almost impossible if one cannot observe the real process by which individuals interact and transfer knowledge to their peers, a factor that determines total factor productivity (Jacobs, 2970, 1984; Glaeser, 1992). Most of this spontaneous exchange of information happens in the physical proximity which makes the mining of face-to-face interactions crucial (P. Hinds, 2002; Wu et al., 2008). If density is the exogenous variable we want to study, then we need a technology able to sense the information exchange at the finest possible spatial granularity.

There are strong reasons to believe that at least part of the explanation for this empirical regularity lies in the systematic difference in network structure of cities. Arbesman et al. (2009), building on studies on the hierarchical structure of some networks (Leskovec, 2005; Aaron, 2008), hypothesize that cities might be structured as trees as well, i.e. that large scale human communities will be decomposable into overlapping groups, which may be hierarchically decomposable into subgroups, etc. The authors then use a random tree to generate the connections between the inhabitants of a city, and assume that the probability of linking two individuals decreases exponentially with their distance in the three.

As a conclusion of this, that larger the tree, the more weak ties (i.e. links between far apart individuals in the social network, with little or no social overlap) may link its inhabitants. It has been long conjectured (Granovetter, 1973), and finally supporting evidence has been provided (Macy, 2009), that information about economic opportunities is more likely to come from these weak ties (socially distant acquaintances) than from clustered relationships (friends). Using all these ingredients, Arbesman et al. derive equations for the growth of innovation as a function of the size of the tree, yielding exponents similar to the observed ones, and conclude that this mechanism is the reason for the super-linear scaling.

With the advent of Reality Mining (Eagle & Pentland, 2009) it is now possible to perform an empirical testing of Arbesman et al.'s proposal. For this, we can now to combine all reality mining data we can gather about the inhabitants of corporations and cities with social scientific data collected by governments. An example of this is sort of social scientific data is The European Urban Audit Project, which provides urban statistics for 258 cities across 27 European countries. It contains almost 300 statistical indicators presenting information on matters such as demography, society, the economy, the environment, transport, the information society and leisure.

4.1 From Social Interaction Data to Collective Intelligence

The study of how people interact in successful corporations is providing managers with better tools to allocate human resources, organize work, and be more efficient in general. Similarly, natural science research into social insect behavior is helping computer scientists design robust distributed control and optimization algorithms. Cooperation and competition within natural populations are the bedrock of complex social structures, but although our technological ability to observe the dynamics of interactions within these populations has improved, modeling and quantifying the global sociological effects of these dynamics remains a difficult problem. There has also been little work in quantifying how far a population of entities is from its optimal regime of functioning in terms of global wellness of the entire population.

This problem is fundamentally connected to an important question in social science that concerns the interplay between individual and collective success in social networks: how does a person's interactions with other people affect their social position? Furthermore, how do the collective effects of these local interactions globally influence society at large? In the social sciences, this has long been the focus of a "positivist" line of thinking, which defines social progress as the changing of society towards an ideal state, generated by individual contributions and aggregated by collective interactions. It is also the question that computational social science is investigating today, albeit in a strictly quantitative sense.

We are starting to leverage the unprecedented opportunities offered by the recent availability of large amounts of social interaction data, such as e-mail and phone call records, and ideas from optimization theory and social network analysis, to analyze real populations in terms of the questions just posed. The result is a novel framework that allows us to characterize populations from social interaction data in a mathematically robust way, based on the population's intrinsic ability for local interactions to produce a positive global outcome over time. We describe this as the collective potential of a population, and define it in terms of the impact of interaction topology and order on its collective potential.

Our framework for computing the collective potential of a population is based on the notion of a hypothetical *collective potential curve*. Each individual in a population has a dynamic state at any given time, which we model as a real-valued function of multiple, interacting binary choices that the individual has made. The interactions between these binary choices, and thus the overall state function, may be made *arbitrarily complex*, with the end result being that the "value" of each individual within the population is expressed as a single, continuous value. We also assume that each individual seeks to increase their state value by interacting with their neighbors over time, and adopting some of their neighbors' more beneficial choices, i.e., when imitating the neighbor's choices would result in the selfish positive outcome of increased state value.

The collective potential curve of a population is then defined as the trajectory of the expected state value in the population over time, with a key contribution of our method being that the expectation is computed over all possible state functions in a computationally tractable way. The collective potential curve therefore represents the efficiency of constructive, collaborative processes in a population over time, or how efficiently the structure and dynamics of social interactions can foster positive global change in the population through selfish local interactions. Although an expectation over all possible state functions cannot be computed analytically, it can be shown that in practice, computational simulation estimates of the collective potential curves of large, real populations converge very quickly, usually in a few dozen iterations, even with populations of millions of individuals.

The collective potential curve presents some interesting avenues for the analysis of populations from dynamic social interaction data. Diffusion processes that take place in social networks have been studied in the context of epidemiological modeling, and more recently in "viral marketing" scenarios, where word-of-mouth recommendations drive the adoption of a product. Since the collective potential curve models a form of diffusion through a dynamic network over time, it allows us to explicitly compare the effect of interaction order and topology on the *efficiency* and *speed* of diffusion, and also to compare the dynamics of different populations which have been controlled for size and other external factors. Our definition of the collective potential curve of a population has its roots in a number of research areas. The global dynamics of individuals following binary choice models has been studied extensively in mathematical sociology. There has also been interest in how the structure of the network impacts the rates of diffusion of information.

Our contribution is to model a population by means of an arbitrarily complex state function that operates on a multitude of choices made by each individual. We use a simple model of interactions between individuals—determined by real interaction data—and vary the complexity of the state function, computing an expectation over all possible functions in a countably infinite class of state functions. The sociological idea of a society progressing towards an ideal state through interactions between individuals is modeled by a form of *collective optimization*. The specific type of collective optimization we use is a modified form of a simple genetic algorithm, which bears similarities to parallel genetic algorithms with spatially distributed populations and mating topologies. The generation of arbitrarily complex state functions is based on a class of synthetic functions called *Hyperplane Defined Functions*, initially devised as difficult test cases for genetic optimization methods.

4.2 Collective Potential in Dynamic Networks

Our framework quantifies a number of sociological principles in as simple a way as possible. Each individual is represented by a binary state vector that encodes a set of choices it has currently made, without specifying the form or function of each choice. By allowing the state vectors to grow arbitrarily long, we can encode any number of choices. A global objective function operates on the state vectors and assigns each one an objective score, as a measure of value for its choices. Although the presumption of a global measure of worth for all individuals might violate some sociological principles, we compensate by allowing the objective function to be arbitrarily complex.

Individuals in the population seek to increase their own worth, which they achieve by interacting with other individuals. We assume a simple model of interactions between individuals, where the topology and order of interactions between individuals are governed by recorded data. For example, in a dataset of phone call records, an undirected or mutual interaction between two individuals takes place when they call each other. Similarly, if the dataset consists of e-mail records, an e-mail sent between two addresses qualifies as a directed interaction from the sender to the recipient, i.e., the sender gains no advantage in sending the e-mail, but the recipient might. Furthermore, communications networks are by no means the only type of data that can be used. Interaction networks in the recent past have been derived from physical proximity determined by Bluetooth sensing devices, wearable badges, radio tracking collars on wild animals, and bibliographic databases of co-publication patterns, among others.

During each interaction, a random subset of choices (state) is temporarily exchanged between the pair of interacting individuals. This exchange becomes permanent if the value of the state of either individual increases (unless the interactions are directed, in which case only the recipient's state can change). Although this is a very simple model of interactions between individuals, it is surprisingly flexible when paired with an appropriate objective function. Furthermore, by holding the interaction model constant, the only variable in our model is the objective function.

The collective potential curve of the population is defined as the rate of increase in the expectation of the objective value of individuals in a population over time. The expectation is computed over all possible objective functions, which encompass a large class of collective behavior models. Thus, our method is essentially parameter free. This allows us to measure how effective a population is at spreading positive processes, which may be choices, ideas, information, or any of a number of other diffusion processes.

5 Conclusions

In this paper, we presented an overview of state-of-the art pattern recognition methods for individual and group behavior recognition using wearable and environmental sensors. Research on individual human behavior recognition has focused on classifying motor activities such as standing, walking, or running, as well as higher level activities. Research at the group level has focused on face-to-face interaction detection, conversation detection, and conversational dynamic modeling. However, most of the research has used supervised learning and classification methods due to the inherent difficulty of implementing fully unsupervised human behavior learning and classification systems, as well as the difficulty of collecting data on physical social interactions.

Another interesting area of research where pattern recognition methods can be applied is social network analysis. We presented a brief overview of the research in social network analysis for the automatic identification of groups and the study of social network evolution. On the one hand, there has been an enormous body of research in the area of longitudinal social network analysis, and many experiments involving large amounts of data collected using mobile phones and other digital forms such as e-mail and internet databases have been conducted. On the other hand, there has also been a lot of research on individual and group behavior recognition of face-to-face social interactions using wearable and environmental sensors but with very few experimental subjects. We believe it is a natural extension to bring together these two streams of research in order to study and model human social behavior and physical interactions from data collected using wearable and environmental sensors.

The use of pervasive wearable and environmental sensors has made it possible to collect large amounts of data on physical social interactions. While labeling large amounts of data is unrealistic and time consuming, applying unsupervised learning methods to this type of data is possible. This would allow us to accurately identify and characterize group behavior (i.e. face-to-face interactions, meetings, conversations, etc.) in large groups of people. By applying dynamic models (dynamic Bayesian networks and dynamic social network analysis) to the automatically identified social interactions it will be possible to model interactions at different time scales, from millisecond-level conversational dynamics, to larger time granularities such as day-to-day face-to-face interactions.

Collecting, processing, and analyzing data in real time is one of the greatest challenges. The amount of data that is generated from everyday's use of digital technology grows at a faster rate than it can be stored, processed, and analyzed to extract useful value from it. Therefore, it is of great importance that real-time data collection processes be developed in order to extract valuable information without having to store raw data. This could also help resolve the privacy issue since sensitive data would not have to be stored or shared across different entities.

Being able to accurately predict the creation and extinction of ties in a social network created from data on physical interactions would have a huge impact in research areas such as the study of team formation, organizational dynamics, the evolution of friendship networks, and the spread of disease, among many others.

It is still unclear how privacy issues should be addressed in order to preserve the integrity of individuals and organizations. Consumers already share private information with companies. It is surprising how some people are willing to give private information in exchange for rewards, and sometimes even without being completely informed about the way this information will be handled. We can only expect to obtain more contributions from people when they know the advantages and understand the data that is collected.

Companies must abide to ethical principles when using these data for their own benefit. A better incentive system that gives added value to the users must be put into place in order to gain approval from the majority of the population to use data collected from their digital interactions and digital transactions. The use of anonymous data should be enforced and analysis at the group level should be preferred over that at the individual level. Most researchers agree that studying general trends and aggregated patterns of behavior is much more interesting than looking at the individual level.

Emperical evidence (Madan & Pentland 2009, 2010) shows that social interactions in communities predict the adoption of political opinions and the spread of viral media. Unlocking how ideas and opinions diffuse in our communities, will allow us to design better individual participation and feedback systems for collective intelligence. Similar to companies and employee data, privacy is a very important concern with consumer data and modeling. For socially beneficial applications of these technologies, it is important that consumers own their interaction data and have a say in how it is used.

Cebrian et al (2010) computed collective potential curves based on real social interaction data by means of a robust, parameter-free, estimate of the capacity of a population to increase their collective wellness in a given time period. They empirically investigated the impact that the network topology and the order of interactions have on the collective potential of a population. These results are compatible with known results from population genetics and evolutionary computation, namely that networks with random topology asymp- totically yield the highest collective potential, and small levels of perturbation in the timing of the interactions help to prevent inbreeding of good solutions, and tend to speed-up the collective potential growth rate. It is interesting to note that under our model, having a large number of contacts with other individuals and a large neighborhood of contacts leads to a high final fitness, but the converse is not necessarily true, i.e., certain individuals consistently achieve a high final fitness value in spite of having smaller and more infrequent circles of contacts.

Experimental results also show a number of interesting features that warrant further sociological analysis and experiments. For example, shuffling the order of interactions in some dataset has a more pronounced effect on the lift in collective potential than in others dataset. It is known that the first dataset was drawn from has a higher level of regional deprivation than the second. Since shuffling the interaction order breaks down temporal clusters, it is interesting to ask whether sociological factors such as an inherently low level of communication diversity in the deprived region is responsible for this phenomenon.

In a continuation study Lahiri et al. (2010) presented a case study on the Enron e-mail dataset, modeling probabilistic information flow between people as they exchange e-mails. The results indicated that a small section of vertices in the dataset are privileged in terms of the structure and dynamics of the network, and consistently receive more information than other vertices, regardless of how much they start with. We also showed that this phenomenon in the Enron dataset is not a trivial result that can be explained by simple network properties. The relationship between a vertex's non-trivial network properties (e.g., PageRank or betweenness centrality), and its level of 'information value' is a relationship that warrants further study. It would also be interesting to see if this phenomenon, like the commonly observed skew in degree distribution, exists across different networks. They noted note that this new diffusion model utilizes a very basic form of the genetic algorithm. Like the profusion of different types of Genetic Algorithm themselves, it is simple to extend the model presented here to model even more complex phenomena. For instance, the incorporation of mutation into the model loop would encompass a richer class of diffusion models, as would the introduction of more complex state replacement rules after crossover, instead of the retain-the-best method.

In future work, we envision two directions for this research. The first deals with validating and explaining the sociological implications of our findings based on computational collective intelligence. Some possible questions are to ask if there are commonalities in the local network properties of individuals at each strata of society, or if the collective potential curves are correlated with global network properties such as hierarchy or community structure. A second line of research would be to investigate the nature of the genetic stochastic process we have described in this paper, in terms of its capabilities as a collective behavior model, its convergence, and other possible uses.

References

Abdelzaher, T., Anokwa, Y., Boda, P., et al.: Mobiscopes for Human Spaces. IEEE Pervasive Computing, 20–29 (2007)

Avancha, S., Baxi, A., Kotz, D.: Privacy in mobile technology for personal healthcare. ACM Computing Surveys (2009)

Aaron, C., Moore, C., Newman, M.: Hierarchical structure and the prediction of missing links in networks. Nature 453(7191), 98–101 (2008)

Arbesman, S., Kleinberg, J., Strogatz, S.: Superlinear scaling for innovation in cities. Physical Review E 79(1), 16115 (2009)

Bagrow, J.P., Bollt, E.M.: Local method for detecting communities. Phys. Rev. E. 046108 (2005)

Bao, L., Intille, S.S.: Activity Recognition from User-Annotated Acceleration Data. In: Proceedings of the 2nd International Conference on Pervasive Computing, pp. 1–17 (2004)

Bettencourt, L., Lobo, J., Helbing, D., Kuhnert, C., West, G.: Proceedings of the National Academy of Sciences 104, 7301 (2007)

Bernard, H.R., Killworth, P., Kronenfeld, D., Sailer, L.: The Problem of Informant Accuracy: The Validity of Retrospective Data. Annual Review of Anthropology 13, 495–517 (1984)

Cebrian, M., Lahiri, M., Oliver, N., Pentland, A.: Measuring the Collective Potential of Populations From Dynamic Social Interaction Data. IEEE Journal of Selected Topics in Signal Processing 4(4), 667–686 (2010)

Lahiri, M., Cebrian, M.: The genetic algorithm as a general diffusion model for social networks. In: Proceedings of the AAAI Conference on Artificial Intelligence (2010)

Choudhury, T., Pentland, A.: Sensing and Modeling Human Networks Using the Sociometer. In: 7th International Symposium on Wearable Computers, October 21-23 (2003)

Chung, P.-C., Liua, C.-D.: A daily behavior enabled hidden Markov model for human behavior understanding. Pattern Recognition, 1572–1580 (2007)

Ciccone, A., Hall, R.: Productivity and the density of economic activity. The American Economic Review 86, 54–70 (1996)

Ciccone, A.: Agglomeration effects in Europe. European Economic Review 46(2), 213–227 (2002)

DeVaul, R.W., Dunn, S.: Real-Time Motion Classification for Wearable Computing Applications. Technical Report. MIT Media Laboratory (2001)

Dong, W., Pentland, A.: Modeling Influence Between Experts. In: Huang, T.S., Nijholt, A., Pantic, M., Pentland, A. (eds.) ICMI/IJCAI Workshops 2007. LNCS (LNAI), vol. 4451, pp. 170–189. Springer, Heidelberg (2007)

Eagle, N., Pentland, A., Lazer, D.: Inferring Social Network Structure Using Mobile Phone Data. In: Proceedings of NAS, vol. 106, pp. 15274–15278 (2009)

Gonzalez, M., Hidalgo, C., Barabasi, A.-L.: Understanding Human Mobility Patterns. Nature 453, 779–982 (2008)

Foltz, P.W., Martin, M.J.: Automated Communication Analysis of Teams. In: Salas, E., Goodwin, G.F., Burke, S. (eds.) Team Effectiveness in Complex Organizations, pp. 411–431. Taylor & Francis Group, New York (2009)

Gatica-Perez, D.: Analyzing group interactions in conversations: a review. In: Proceedings of the International Conference on Multisensor Fusion and Integration for Intelligent Systems, pp. 41–46. IEEE, Los Alamitos (2006)

Gatica-Perez, D.: Automatic nonverbal analysis of social interaction in small groups: A review. Image and Vision Computing (2009)
Girvan, M., Newman, M.E.: Community structure in social and biological networks. In: PNAS, pp. 7821–7826 (2002)
Glaeser, E., Kallal, H., Scheinkman, J., Shleifer, A.: Growth in Cities. Journal of Political Economy 100, 1126–1152 (1992)
Goldenberg, A.: Scalable Graphical Models for Social Networks. PhD Thesis. Pittsburgh, PA, USA: Carnegie Mellon University (2007)
Granovetter, M.: The strength of weak ties. American Journal of Sociology 78, 1360–1380 (1973)
Handcock, M., Raftery, A., Tantrum, J.: Model-based clustering for social networks. Journal of the Royal Statistical Society: Series A (Statistics in Society) 170(2), 301–354 (2007)
Heylighen, F.: Collective Intelligence and its Implementation on the Web: algorithms to develop a collective mental map. Computational and Mathematical Organization Theory 5(3), 253–280 (1999)
Hinds, P.: Distributed Work. MIT Press, Cambridge (2002)
Jacobs, J.: The economy of cities. Vintage Books (1969)
Jacobs, J.: Cities and the wealth of nations: Principles of economic life. Random House (1984)
Kern, N., Schiele, B.: Context-Aware Notification for Wearable Computing. In: Proceedings of the 7th International Symposium on Wearable Computing, pp. 223–230 (2003)
Kossinets, G., Watts, D.J.: Empirical Analysis of Evolving Social Networks. Science, 88–90 (2006)
Bramoullé, Y., Kranton, R.: Public Goods in Networks. Journal of Economic Theory 135(1), 478–494 (2007)
Lahiri, M., Cebrian, M.: The genetic algorithm as a general diffusion model for social networks. In: Proc. of the 24th AAAI Conference on Artificial Intelligence, Atlanta, Georgia (2010)
Lazer, D., Friedman, A.: The Network Structure of Exploration and Exploitation. Administrative Science Quarterly 52(4), 667–694 (2007)
Lazer, D., Pentland, A., Adamic, L., Aral, S., Barabási, A.-L., Brewer, D., et al.: Computational Social Science. Science 323, 721–723 (2009)
Lee, S.W., Mase, K.: Activity and Location Recognition Using Wearable Sensors. In: Pervasive Computing, pp. 24–32 (2002)
Leskovec, J., Kleinberg, J., Faloutsos, C.: Graphs over time: densification laws, shrinking diameters and possible explanations. In: Proceedings of the Eleventh ACM SIGKDD International Conference on Knowledge Discovery in Data Mining, p. 187 (2005)
Leskovec, J., Adamic, L., Huberman, B.: The dynamics of viral marketing. ACM Transactions on the Web, 1–1 (2007)
Lester, J., Choudhury, T., Borriello, G.: A practical approach to recognizing physical activities. In: Fishkin, K.P., Schiele, B., Nixon, P., Quigley, A. (eds.) PERVASIVE 2006. LNCS, vol. 3968, pp. 1–16. Springer, Heidelberg (2006)
Liao, L., Choudhury, T., Fox, D., Kautz, H.: Training Conditional Random Fields Using Virtual Evidence Boosting. In: Proceedings of the International Joint Conference on Artificial Intelligence, Hyderabad, India, pp. 1–6 (2007)
Macy, M., Eagle, N., Claxton, R.: Network Diversity and Economic Development. Science 328(5981), 1029 (2010)

Madan, A., Pentland, A.: Modeling Social Diffusion Phenomena Using Reality Mining. In: AAAI Spring Symposium on Human Behavior Modeling, Stanford University, CA (2009)

Madan, A., Pentland, A.: Social Sensing of Political Opinions (2010) (in submission)

McCallum, A., Wang, X., Corrada-Emmanuel, A.: Topic and Role Discovery in Social Networks with Experiments on Enron and Academic Email. Journal of Artificial Intelligence Research 30, 249–272 (2007)

Mantyjarvi, J., Himberg, J., Seppanen, T., Center, N.: Recognizing human motion with multiple acceleration sensors. In: IEEE International Conference on Systems, Man, and Cybernetics, pp. 747–752 (2001)

Mishra, N., Schreiber, R., Stanton, I., Tarjan, R.E.: Clustering social networks. In: Bonato, A., Chung, F.R.K. (eds.) WAW 2007. LNCS, vol. 4863, pp. 56–67. Springer, Heidelberg (2007)

Olguin-Olguin, D.: Sociometric Badges: Wearable Technology for Measuring Human Behavior. Master's Thesis. Cambridge, MA, USA: Massachusetts Institute of Technology (2007)

Olguin-Olguin, D., Gloor, P.A., Pentland, A.: Wearable Sensors for Pervasive Healthcare Management. In: 3rd International Conference on Pervasive Computing Technologies for Healthcare, London, UK, April 1-3, pp. 1–4 (2009a)

Olguin-Olguin, D., Waber, B., Kim, T., Mohan, A., Ara, K., Pentland, A.: Sensible Organizations: Technology and Methodology for Automatically Measuring Organizational Behavior. IEEE Transactions on Systems, Man, and Cybernetics-Part B: Cybernetics 39(1), 43–55 (2009b)

Olguin-Olguin, D., Pentland, A.: Sensor-Based Organisational Design and Engineering. To appear in: International Journal of Organisational Design and Engineering (2010)

Oliver, N., Horvitz, E.: A Comparison of HMMs and Dynamic Bayesian Networks for Recognizing Office Activities. In: Ardissono, L., Brna, P., Mitrović, A. (eds.) UM 2005. LNCS (LNAI), vol. 3538, pp. 199–209. Springer, Heidelberg (2005)

Oliver, N., Garg, A., Horvitz, E.: Layered representations for learning and inferring office activity from multiple sensory channels. Computer Vision and Image Segmentation 96, 163–180 (2004)

Palla, G., Barabasi, A.L., Vicsek, T.: Quantifying social group evolution. Nature, 664–667 (2007)

Pentland, A.: Automatic Mapping and Modeling of Human Networks. Physica A: Statistical Mechanics and its Applications 378(1), 59–67 (2006)

Pentland, A.: Honest Signals: How they Shape our World. The MIT Press, Cambridge (2008)

Ravasz, E., Barabasi, A.: Physical Review E 67, 26112 (2003)

Robins, G., Snijders, T., Wang, P., Handcock, M., Pattison, P.: Recent developments in exponential random graph (p*) models for social networks. Social Networks 29(2), 192–215 (2007)

Salganik, M., Dodds, P., Watts, D.: Experimental Study of Inequality and Unpredictability in an Artificial Cultural Market. Science 311 (2006)

Scarlat, E., Maries, I.: Towards an Increase of Collective Intelligence within Organizations Using Trust and Reputation Models. In: Nguyen, N.T., Kowalczyk, R., Chen, S.-M. (eds.) ICCCI 2009. LNCS (LNAI), vol. 5796, pp. 140–151. Springer, Heidelberg (2009)

Slaughter, A.J., Yu, J., Koehly, L.M.: Social Network Analysis: Understanding the Role of Context in Small Groups and Organizations. In: Salas, E., Goodwin, G.F., Burke, S. (eds.) Team Effectiveness in Complex Organizations, pp. 433–459. Taylor & Francis Group LLC, Abington (2009)

Sukthankar, G., Sycara, K.: A cost minimization approach to human behavior recognition. In: Proceedings of the 4th International Joint Conference on Autonomous Agents and Multiagent Systems, The Netherlands, pp. 1067–1074 (2005)

Van Laerhoven, K., Cakmakci, O.: What shall we teach our pants? In: Proceedings of the 4th International Symposium on Wearable Computers, pp. 77–83 (2000)

Wasserman, S., Faust, K.: Social Network Analysis: Methods and Applications. Cambridge University Press, London (1994)

Chapter 22
ICT and Dataveillance

Darryl Coulthard and Susan Keller

Abstract. Dataveillance, the collection, storage and mining of data and images, is increasing and new emerging technologies seem to inevitably contribute to ever more dataveillance. In this chapter, we outline the key social drivers for dataveillance and illustrate some of the roles emerging technology plays in dataveillance. We then turn to the question of the relationship of technology to its use and how non-neutral outcomes eventuate. Why does new technology seemingly lead to dataveillance rather than empowerment of the citizen, worker and consumer? To unravel this, we develop a social ecological model of technology cooption. In this model, we show how technology cooption is contested at each stage of technology development. Further, we show that the outcome of such contestation is the non-neutrality of the technology. The technology cooption model provides a middle range theory for empirical analysis by identifying the key elements of technology cooption and their proposed links and the role of the stakeholders in such cooption.

1 Introduction

Digitization of information provides undreamt of production and access to information. It would seem to be a world where everyone is a winner. Consumers are empowered by access to information and choice, popular sentiments about government policy can be quickly ascertained from blog sites and opinion surveys, and businesses can produce and market more efficiently and effectively. As a consumer, there is the promise of convenience and empowerment. Personal technology such as the iPhone promises the consumer seamless and effortless access to information about products, the ability to monitor their health, listen to music and orient themselves geographically and of course to purchase products [1, 29]. For the government and corporation, the information collected by current and emerging technologies can be used to improve services and offer a more 'customized' response. Not only is increased efficiency and effectiveness of the production process and supply chain promised, but also an unprecedented understanding of consumer and political behavior [56].

Darryl Coulthard · Susan Keller
School of IS, Deakin University, Melbourne
e-mail: {darryl.coulthard,susan.keller}@deakin.edu.au

N. Bessis, F. Xhafa (Eds.): Next Generation Data Technologies for CCI, SCI 352, pp. 599 – 623.
springerlink.com

Indeed, in the age of information a concern for information flow, its collection, storage, mining and use seems anachronistic. Scott McNealy of Sun Microsystems' "There is no such thing as privacy – get over it" encapsulates this view. In-formation has become a commodity and it has a use value that can only be realized by its flow [46]. Information does indeed seem to 'demand' its collection, integration and transmission. Personal and business computers contain readily available and transmissible documents, emails, transaction details or purchases and images. Servers and ISPs routinely store emails, log access to the Internet and record web sites visited. Internet information and social software providers routinely keep profiles of their users. Access to buildings and rooms are increasingly electronic with access routinely logged. Camera surveillance and image recognition software can monitor, store and recall car or personal movements. Software and digital recordings of customer interactions can monitor worker performance. Our actions seem never to have been so monitored or potentially open to scrutiny. The rise of 'dataveillance' [13] 'surveillance society' [43] and the 'superpanopticon' [65] have been well described by authors such as [1, 10, 50, 72]. However, the next generation data technologies that enable collective computational intelligence will also enable the capture, consolidation, analysis, mining and interpretation of distributed data in ways that will make current capabilities seem primitive. While the technical possibilities may excite IT professionals, and the practical possibilities may excite governments and corporations, the privacy implications for civil society are serious and deserve considered debate.

This chapter is an exploration of how dataveillance has come about, and barring intervention, how it will continue to rise. First we will discuss the rise of dataveillance and the broad social reasons that have been put forward. As we shall show, these explanations tell a great deal about what has happened but not how these changes have been enacted or emerge in our daily life as worker of the government or corporation, consumer and citizen. Such social explanations do not show the linkages between the technology, the citizen/consumer and the government/corporation.

How new technology interacts with people, businesses and organizations to have particular social and organizational outcomes is a key theoretical problem in Information Systems. Some outcomes will be desired, some may lead to changes in the distribution of power, and others may be in some way undesirable or rejected. Most will have unanticipated outcomes. One of the central tasks we believe of Information Systems is to examine how, from the set of possible out-comes, particular outcomes are realized. It is to wrestle with Kranzberg's 'laws of technology'[38] and in particular the first law: "Technology is neither good nor bad; nor is it neutral" and to work out how 'non-neutral' or biased [36] outcomes eventuate. It is also important to the discipline to understand as far as possible what those outcomes are. Too often, we believe, Information Systems has restricted itself to the immediate, mostly business outcomes of efficiency and effectiveness, and their related technical outcomes. As a profession it has a moral responsibility we believe to bring to attention such negative outcomes that it identifies and to promote ways of ameliorating or avoiding them.

This chapter is divided into two major sections. First we shall briefly outline the rise of dataveillance, provide some examples of emerging technologies that can be used for dataveillance and outline the targets of dataveillance. Secondly, we shall then introduce a model that helps to identify the interaction between the properties of the technology and its use that lead to non-neutral outcomes such as dataveillance. We suggest that dataveillance is an outcome of a social ecology where different actors and interests design the properties of information technology and appropriate the affordances of those properties to meet their local needs. Finally, we argue that this social ecological model is a general model for technology use and provides a useful approach to empirical investigation.

2 The Rise of Dataveillance

While surveillance has always been with us, data surveillance is a relatively recent phenomenon. Dataveillance is a term coined by Clarke [13] to describe "the systematic monitoring of people's actions or communications through the application of information technology". Clarke's seminal paper drew attention to the integration of personal information from multiple sources and the emergence of technologies that allowed the data to be exploited in new ways. The technologies in use at the time included front-end verification of transactions and individuals, cross-system checks against individuals, profiling and data matching [13]. These technologies are now commonplace and new emerging technologies promise even greater dataveillance capacity. The possibility of combining data originating from different times and places to infer intelligence that was never intended for disclosure, dwarf the imaginations of our most dystopian novelists and science fiction writers.

The second half of the 20th century has seen a substantial increase in the use of technology to aid surveillance including the use of video, audio, electronic tagging, monitoring of email and web usage, computer matching and profiling, data mining and mapping as well as network analysis and simulation [50]. We will use the term ‘surveillance creep’ and ‘dataveillance’ interchangeably to describe the general phenomena.

Progressively, surveillance has moved from physical observations of the ‘party of interest’ to the digitization of the data-flows that are produced from everyday living in the modern world. Capturing, analyzing and monitoring these data-flows is economically efficient because the process can be automated. In addition, the watched are mostly unaware of the degree of dataveillance or its importance. Dataveillance is largely hidden and rarely discussed or acknowledged [14].

3 Examples of Enabling Technologies of Information Integration and Surveillance

In this section, we review three current and growing technology trends to highlight the potentialities of surveillance, data-mining, profiling, and identification, that they provide.

3.1 Transaction Data Streams

Transaction data streams arise from the interaction between entities. An entity may be a credit card number, a person, a company, a telephone number or an IP address. Streams of data between entities include data from Automatic Teller Machine (ATM) transactions, credit card purchases, electronic fund transfer orders, and buy and sell orders for stock market trading [12]. The volume of data from such transactions is so great that the traditional approach of analyzing the data after it has been stored in a database is not practical.

There are various mechanisms for analyzing the data but in concept the idea is to tap the data as it enters the database in order to build up summaries or 'signatures' of the behavior of individual entities [19]. These signatures evolve over time in response to changes in the data streams. Once a signature is created it can be used to monitor unusual patterns in transactions for fraud detection and also to build up information of customer behavior [40]. The profile provides a summary of characteristics of the entity. Figure 1 shows a high level conceptual view of how signatures are developed on the fly from transaction data.

For instance, a signature may contain characteristics of the buying behavior on a particular credit card number such as average number of daily transactions, size of transactions, time of transactions, and geographic region of where transactions take place. The profile can be used in fraud detection by comparing all new transactions to the usual behavior embodied in the signature.

Transaction data streams are also mined for marketing purposes [40]. For example, a business can use the information to identify which customers are most likely to buy particular goods and services at particular times. The key point is that signatures reflect current customer-level behavior and thus are ideal when the goal is to monitor or influence customer behavior in close to real-time [40]. Since signature databases typically maintain only one summary record per customer they can be easily merged with other databases to provide more in-depth analysis of customer behavior if required (see Figure 1) [40].

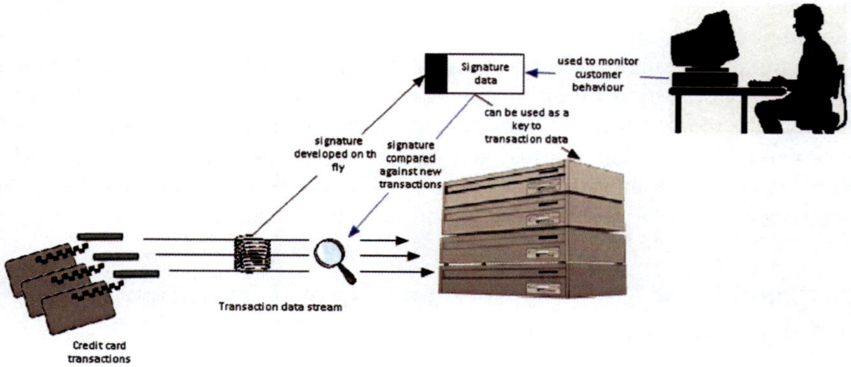

Fig. 1 Transaction data streams and the creation of signatures

Thus unbeknownst to most consumers, corporations have built up summaries of spending habits as well as having the exact details of expenditure. This may lead to not only the possibility of unwelcome marketing and security risks of the data collected, but also to the corporation having a far more intimate knowledge of its customers that could disadvantage the consumer vis-à-vis the corporation.

3.2 Measurement Data Streams

Measurement data streams are high volume data streams produced through network traffic, sensor data and from so-called clickstreams[12]. Much of this data is not stored at all but processed in memory as it arrives in real or near real-time [62].

Clickstreams are produced by users clicking on keyboards or targets such as web-links. The analysis of clickstreams can be used to monitor how visitors are using a website, for software testing, market research, or monitoring employee activity. Internet Service Providers (ISPs) store data on visitors' use of a website and can resell it. In 2006, AOL Research, created a public uproar after posting a compressed text file on one of its websites that contained the search keywords of over 650,000 AOL customers over a three month period. The file was taken down three days later but by that time it had been widely distributed on the Internet, and the nature of the Internet is such that it is impossible to tell how many copies of the file are still in existence. While individuals' names had been replaced with numbered codes it was possible to indirectly indentify particular individuals because people sometimes search on their own names, addresses, credit cards and even social security numbers. Indeed, the New York Times followed the trail of the search data of Internet user number 4417749 and with her permission, identified her as Thelma Arnold, a then 62-year-old widow who lived in Lilburn, Georgia, US. While Thelma's search data was fairly prosaic some AOL members' data was, at the very least, embarrassing. For instance, AOL user 17556639 search data, according to [51] included items such as:

how to kill your wife
pictures of dead people
photo of dead people
car crash photo

While many Internet users understand that they can clear traces of their search data from their own browsers most do not know that those same searches are being saved by ISPs and mined for marketing purposes and that their anonymous data is not that anonymous. Indeed, a recent study based on the 2000 US census data found that 63% of the US population can be uniquely identifiable by just three attributes - gender, ZIP code and full date of birth [33].

3.3 Data Mashups

Data mashup is a term used to describe the creation of new intelligence based on combining different sources of data. For instance, a web mashup might use

Google maps, local restaurant information, and information from the Health Department to provide a map of local restaurants and information on their food handling history.

In fact, the largest proportion of mashups involve maps, but photo, video, shopping and news mashups are also popular[1]. Mashup developers can use application programming interfaces (APIs) supplied by sites such as Google, Flickr, YouTube, Twitter, Amazon, eBay, Facebook to source data. These APIs have been provided specifically to encourage the development of mashups. However, where no APIs exist, some mashup developers resort to screen scraping tools to acquire information. Unless a web-site monitors for such activity, screen scraping tools can automatically harvest information from existing web pages without the knowledge of the content providers (which may be users of social networking sites).

Web mashups lead to a number of thorny privacy issues especially mashups that aggregate information supplied by Internet users in different contexts. This was highlighted by the recent launch of a site called 'Please Rob Me'. The mashup merged social networking feeds from Twitter with information from FourSquares a geo-location site that encourages users to share their location. The 'Please Rob Me' site developers wanted to point out the dangers of people sharing their whereabouts in real-time because this broadcasts the fact that they are not at home[2]. In a similar vein, a computer consultant created a mashup using book wish-lists from Amazon and Google Maps to demonstrate the ease in which data mashups could erode people's privacy. The book wish-lists contained users' full names, city and state; enough information to find their street address using Yahoo People Search. The consultant, armed with street addresses then produced a map showing where people who liked 'subversive' books lived [55]. One can easily imagine the same process being used to locate other people of interest to government agencies or vigilante groups – perhaps homosexuals, unionists or home-schoolers. Perhaps more likely is the possibility of mashups of this type being created for businesses interested in targeted marketing. A location map showing all individuals within a certain geographic area with an interest in babies and children as indicated by book wish-lists would be of great interest to businesses in that market.

Having surveyed a few examples of the technologies that enable dataveillance we now turn our attention to the targets of dataveillance.

4 The Targets of Dataveillance

There are three principal targets for dataveillance: the citizen, the consumer and the employee. These may at times overlap or interconnect. For our purposes the 'citizen' is used broadly to denote the individual as a political actor or subject.

For most of us, the term dataveillance immediately conjures up the totalitarian image of 'Big Brother' with its emphasis on state security and coercive control

---

[1] http://www.programmableweb.com/apis/directory/1?sort=mashups

[2] http://pleaserobme.com/why

and monitoring over our everyday life. We think of totalitarian regimes monitoring its people for potential subversion and democratic activities and these seem far removed from the democratic impulses of Western governments and our day-to-day experience: we do not feel monitored in such a way and the restrictions on freedom of expression appear small. The monitoring that is in place, is argued in parliaments, limited by privacy laws, and justified by the threat to security or reduction in crime. The threat from the Western State seems far-fetched and paranoid on the one hand, or is of limited concern to the law-abiding citizen on the other. Underpinning this is a form of legitimacy of the Western state and a benign or trusting view of the intentions of the state and the limits to its power [34, 42].

4.1 The State and Dataveillance

The terrorist attack on the US in 2001 has focused attention and provided support for increased surveillance by the State to defend itself and its citizens against terrorist and criminal threat. In the post 911 environment, most countries have implemented a wide range of surveillance measures on its citizens and visitors. Illustrations of these measures include the FBI Carnivore and the multi-state Echleon system, national laws and practices to erode email and mobile telephony privacy and police practices allowing integration of electronic and other information to profile or identify suspects.

The pressure and temptation of state security and police to use dataveillance to detect and prevent crime and atrocity must be simply enormous. Recently, police in Australia used travel card data on the Brisbane transport system to identify and interview people on a particular bus as part of a murder investigation [5]. Crime detection, safety and security often make powerful arguments for dataveillance by police, and pleas for privacy and strong controls on police use of dataveillance appear weak in comparison [2].

Terrorism and crime in an uncertain world is only the tip of the iceberg of state dataveillance. Dataveillance as Giddens [32] and Foucault [25] have shown has been an integral part of the emergence of the nation state and the imperative of military competition between states and with subjecting its citizens to central government control. Lyon [44] refers to this view as the 'Big Brother' perspective. To this perspective can be added the 'Soft Sister' perspective that employs and considers the kind of dataveillance necessary to make the welfare state possible [7:407-425, 16].

The state and its bureaucracy are instrumentalist and consequentialist in its approach to public policy. As a bureaucracy it is instrumentalist, seeking the most efficient and effective means to the policy ends of its political masters. Where the bureaucracy itself is apolitical and therefore 'agnostic' concerning the ends its masters seek, the ends are all the same to the bureaucrat. From this point of view, dataveillance aids effective and efficient policy delivery.

Secondly, consequentialist thinking predominates in bureaucracies and governments. Consequentialism, or more specifically utilitarianism is the view that happiness is the fundamental yardstick of policy and the happiness of the many can outweigh the unhappiness of the few. The collation and meshing of security data more efficiently may well lead to better security and safety of citizens from crime and terrorism and arguments against government intrusion or surveillance seem weak and overblown by comparison, especially in the context of what is seen as a benign state [42].

Finally, there is the collation of data for planning and policy purposes. Data is collected and used to promote and plan public goods. Health promotions, traffic safety and occupational health and safety campaigns are some of the most visible. There is a point however where such data and policy advance an overbearing "Big Sister" or "Nanny State" that aims to socially reengineer harmful or undesirable behaviors out of its citizenry, all for 'their own good'.

4.2 Dataveillance and the Employer

Closer to our everyday life is the use of dataveillance for the monitoring of work [49, 76]. What is most striking is the acceptance of the control the employer has over the employee and the ability of the employer to collect and monitor the work of the employee, should they so wish. Organizations also have operational effectiveness and efficiency imperatives [64]. Information collection and integration is routinely used to manage employee efficiency and effectiveness and also routinely to identify and reduce service costs. Employee performance appraisals, key performance targets and increased surveillance of the employee is a general trend [73].

Here the key underpinning principle is again that of instrumental rationality: what organizational arrangements, what performance targets, what monitoring maximize the ends in the contemporary understanding of the managers and shareholders? The level of dataveillance on the employee is subject to the understanding of its effects on performance of the individual and business. In such circumstances the employee has already given up various rights as a citizen and we are left with the hope that being nice to people brings better results than being nasty. Unfortunately for us, as Sennett [69] argues, providing an uncertain future currently produces on balance the most effective performance.

Industrial democracy seems to be unheard of [18] as is the use of technology to empower its workers in the sense of increased autonomy, enriched work and increased participation in decision making.

4.3 Dataveillance and the Consumer: Business Intelligence

Whereas the dangers and powers of dataveillance in the political and industrial areas are more or less understood and delimited and at times contested, the effects of dataveillance on the consumer is less well known and we will consider these effects in greater detail and in particular consider the effects of consumer dataveillance in the ongoing creation of the consumer in the consumer society.

The routine collection of information of the consumer or prospective consumer is unprecedented. Through loyalty schemes, customer relationship management programs and point of sale data, businesses can identify and profile who buys what, what else they buy, when and how often. This information can be used to customize marketing and to organize the shop layout, merchandise location, and sales [see for example 24, 44]. Supply chain initiatives such as "Efficient Customer Response" (ECR) and "Collaborative Planning, Forecasting and Replenishment" (CPFR) form an integral part of the management of the consumer and the market.

Initiatives such as ECR, CPFR have led many to argue that dataveillance is the governance of the consumer [7, 8, 27, 63]. Just as a Big Sister or a Nanny State may attempt to manage or socially engineer the citizen, dataveillance opens up the opportunity to socially engineer the consumer. Andrejevic [1] shows how interactive technologies are being used to better understand, manage and produce markets.

Much of the discussion of the protection of the citizen, employee and consumer is taken up with the discussion of privacy [42]. The right to privacy is typically viewed as the main conceptual (and legal) defense against dataveillance and monitoring and other forms of intrusions into one's 'private life'. Presumably, the privacy of the employee can be conveniently limited by the fact that they are engaged in a form of public life. It can be readily seen that the rights to privacy seem ready to be over-ridden by the State with arguments of the greater public good. It might similarly be argued that the greater market efficiencies and convenience that dataveillance provides justifies the erosion of privacy rights [42].

5 Life World Uses of Information and Dataveillance

Up to this point we have been considering the use of dataveillance from the perspective of the State and the business. However we must also consider how citizens and consumers take up enabling technologies and how they wittingly or un-wittingly disclose information and permit dataveillance on their activities.

The Internet and other information and telecommunication technologies have provided the opportunity for people to communicate and share information in novel ways. Websites, Facebook, Myspace and similar social networking sites, blogs, emails and wikis provide the opportunity for people to disclose or, to use the most widely used term, share information about themselves.

Thus for many of us, the picture we have of these new technologies is not one of surveillance but one of empowerment. This has not been lost on the advertisers and manufacturers. The overriding view of information technology, almost incessantly promulgated and advertised by ICT businesses and also by businesses that perceive a competitive advantage through ICT adoption, is that ICT, or as Andrejevic (2007) terms it 'interactivity or 'iCulture[3] ', is empowering. iCulture promises unhitherto dreamt of immediate and convenient access to information and communication where the limits of their use is only limited by the imagination of the user.

The potential for empowerment is possibly real enough lending credence to the claims of the advertisers and ICT businesses. There is no doubt that eShopping

[3] The lower case 'i' refers to interactivity, the interaction and information exchange between the consumer and business.

can be more convenient and allows the consumer to compare pricing and information on goods on offer. The use of the mobile phone and hand held devices can provide information on fitness routines, carbon footprints of activities and consumption, the location of nearby restaurants and so on. The Internet provides a huge range of information, educational material and opportunity to discuss, learn, and make contact with others.

Underpinning much of the hype and fanfare of interactivity, Andrejevic [1] points out, is the use of the old left critique of the media as an oligopolistic disseminator of biased or selected information or limiter of choice. This has been drafted into the argument that ICT empowers by placing the consumer of media in charge of the editorial and even the production process. Interactivity, by contrast, puts the 'you in control' of what you watch and when you watch it. Interactivity also allows people through blogs, Facebook, twitter, wikis, YouTube and so on to contribute content. Ubiquitous connectivity allows people to stay in touch, seek information, log their travels, never get lost and monitor their children. The technology promises so much empowerment. The consumer is in control with the certainty and safety of the technology.

This provides a left wing or rebellious gloss to ICT as an empowering and iconoclastic tool that shatters conventions and politics. The 1983 advertisement for the introduction of the Macintosh computer draws on and epitomizes this theme. In the advertisement, a young women in color runs into a room of gray, identically clad and shaven people that recalls the clothing and shaved heads of the people in George Miller's dystopia THX 1138. Chased by police, she throws a sledgehammer at the telescreen of 1984's talking head, Big Brother[4] . Similarly, the 1995 Microsoft Windows 95 'Start me up'[5] campaign, featured the song by the rebellious 'Rolling Stones' while the viewer is treated to the empowering nature of the technology.

Early adoption of a technology has probably never been this cool. Coming from Silicon Valley, California, it does seem to have 'flowers in its hair' and indeed some of the early adopters such as Howard Rheingold, did see ICT as the harbinger of new hippy world[6]. The reality of Silicon Valley with Los Alamos, the massive military presence in the area, the financial support of industrial military triangle and the cheap Mexican labor is carefully screened and forgotten.

The image of the ICT user is one of being empowered, discerning, knowledgeable and cool. Those who lag behind are not generally seen as Luddites possibly because the Luddites were indeed true rebels. The laggards are, at least in the Australian advertisements[7], generally older, less attractive and less intelligent, and he advertisements play on the insecurities of parents and their children's education

---

[4] http://www.youtube.com/watch?v=OYecfV3ubP8 Accessed 30 April 2010

[5] http://www.youtube.com/watch?v=5VPFKnBYOSI&feature=fvst Accessed 30 April 2010

[6] The logo for the Apple Corporation is an apple with a bite taken out of it. It clearly suggests an "Adam and Eve" rebellion against a controlling God and that they have eaten of the fruit of the tree of knowledge. In the aforementioned '1984' advertisement apple logo was rainbow colored, the colors of the hippie.

[7] E.g. the Telstra series of advertisements for broadband such as the following: lhttp://www.youtube.com/watch?v=DvlWQyvEI38 Accessed 30 April 2010

and that even if they have no need for such technologies, their children do. Adopting shows that you are "with it", affluent, in control, young (at heart) and imaginative. Possessing such gadgetry may suggest something like owing a sports car.

5.1 Dangers of Social Software

What is poorly understood is how the information willingly or inadvertently revealed or shared can embarrass, disadvantage or harm the revealer. Such harm can include ridicule and bullying, stalking and the use of such sites by employers or prospective employees to check on employees or screen candidates. This is partly due to social networking being new and its status as a form of intimacy and revelation being unclear. People are now warned to use such sites as public spaces. They are also commercial and public spaces as Hodgkinson [35] points out. Facebook routinely compile data they collect from Facebook sites and sell this to advertisers. They dissect our relationships and our wants and desires.

There is also the misuse of information and personal details that may arise should the integrated information be lost or stolen. Geddes [30] for example, explains that the level of personal detail remaining in a mobile telephone, even after most information has been deleted by the user, is substantial and could easily be used for identity theft, stalking and credit card fraud.

While some may argue that information posted on a social networking site is public information and therefore open property, the problem is determining what public and private mean in the Internet age [45, 59, 71]. There is a difference between public and publicized, and between discrete information and information that has been collated and published. Making something which is public, more public would feel like an invasion of privacy to most people [9]. In providing information to web sites, users do so for a particular purpose and the information is provided in a particular context.

5.2 Uncertainty and Dataveillance

It is a well-established observation that we live in a time of great uncertainty [4, 6, 32, 47, 69]. By this is meant that we live in a post traditional order where old social structures, positions and customs have crumbled and fragmented. Employment and careers are less certain and less long term and increased individualization means that the self is more vulnerable to the vicissitudes of outcomes.

Bauman [4] argues that this uncertainty has lead people to value security more than previous generations. The dystopian visions are now not so much 1984 and Brave New World of hard or soft totalitarianism, or Big Brother and Soft Sister respectively, as "I am Legend" or Cormac McCarthy 's "The Road", where the individual or fragmented family is pitted alone against a world of degenerate human beings. Our fear is the falling apart of society and failure of security, not with totalitarianism. The rise of terrorism and attacks on Western targets adds further anxiety and predisposal to security measures.

There is also faith and certainty in technology. One of the characters in the film I, Robot expresses this sentiment well when asked why she trusts the robots. She states that unlike humans they are 'safe'. Romanyshyn [67] expresses this faith as 'technology is the magic of the modern world and every man, woman and child, however humble their circumstance, can be a practitioner of its art'. Lyotard [46] argues that we use technology, and information technology in particular, as a form of certainty in an uncertain world. There is a mechanical certainty to it.

We therefore at this point have two aspects, firstly a more favorable attitude towards security – we need it more than ever in an uncertain world and secondly, the form of security itself is technological. Fear of dataveillance, of the state or its misuse by corporations according to Bauman [4] is arguably less feared than what it might have been in previous generations.

6 Empowerment and Dataveillance and Technology Cooption

From our overview, dataveillance strengthens the power of the state and citizen, the corporation over the consumer, and the employer and employee. It would seem that, contrary to popular views of ICT as empowering, there is a very real danger of increased control by the state, by businesses and by employers. This is both surprising and alarming and raises important political issues for civil society. It represents the infiltration of the life world of the citizen and the making of the consumer in the image of the business analyst and marketer.

The forgoing provides some overview of what is happening and it also helps to explain why surveillance creep and dataveillance is occurring. Our overview also foreshadows the concepts of conflict of power and contest of interests between state and citizen, corporation and consumer, which is often lacking in Information Systems research. Indeed it could be said that the overview provides an overview of the different interests at stake.

However, our interest as Information Systems researchers is how these interests are played out in the development and use of the technology. In the first instance, a purely social or political explanation such as our overview is not sufficiently fine grained and fails to get to the details of when and how the technology is used in these ways by Government and corporation employees and by citizens and consumers. Secondly, it ignores or at the very least it obscures the involvement of the technologies themselves and how they are appropriated.

The challenge we set ourselves was to develop a conceptual model that identifies the key elements of the process of this technology cooption. Technology cooption, stresses the co-construction of the technology use and appropriation by the designer, owner of the designed artifact, and the user. The more orthodox terms of technology diffusion, acceptance or adoption implies a certain given-ness of the technology; something that is taken up or left. An artifact is designed and produced and the user adopts or fails to adopt the artifact. What we wish to include, is that how an artifact is adopted may be unintended by the designer or owner of the design, and future design may incorporate or constrain ways of using the artifact. Finally, cooption also points to the notion that by use and appropriation of an artifact, the context of use changes, enabling more, less or different use opportunities but also transforming our lives and changing our selves. In sum, cooption stresses the interplay between the various agents imagining, developing, using and restricting the technology, the interaction of the potentialities of the technologies with the agents themselves and the resulting social/organizational outcomes.

The simple model of technology cooption introduced here illustrates how the properties of new information technology and interactivity, and the agents involved in particular, lead to 'surveillance creep' and increased management and measurement by the state, corporation and employer over the citizen, consumer and employee. The model is what sociologists describe as a theory of the middle range [53]. As shall be described, it is underpinned and receives much of its dynamic from our understanding of the broader social currents and attempts to explain how these broader trends interact with agents engaged in the making and using of technology and the affordances those technologies offer.

7 Dataveillance and Technology Cooption

Our initial question was that given the promise of empowerment for the citizen, consumer and worker of information technology, why does it appear, according to a near consensus of commentators, to be doing the exact opposite [1, 10, 13, 43, 50, 65, 72]. Why is there surveillance creep? Why, in the words of Scott McNealy of Sun Microsystems is there 'no such thing as privacy on the Internet' and why should we 'just get over it'. Is it an inevitability emerging from the new technologies? Do the new, interactive technologies have an inherent bias towards dataveillance and surveillance creep[36]?

As we have seen, one of the main research streams for surveillance creep has been cataloguing its rise and the social and business pressures leading to increased dataveillance. Another strand has been to examine the ethical and legal issues to determine whether there is a legal or ethical defense against surveillance creep. This is usually discussed in terms of privacy. Lindsay [42] in his extensive review of this, suggests not. Lindsay among others e.g., [15] suggest that privacy is a flawed and often inchoate concept where the concerns of the individual can often be readily dismissed by appeals to the commonweal (e.g. security and economic benefits). There is in a sense, nothing to stop surveillance creep.

Finally there is the 'temptation theory' of technology. The temptations provided by the technologies, particularly in the face of weak ethical or legal opposition is simply too great (see for example, [61, 74]).

With the partial exception of temptation theories, none of these theories and approaches, to our knowledge, have attempted to identify how the technology itself is coopted and put into service of dataveillance. They tend to explain what has happened and why but not the processes by which the technologies themselves are taken up, the various ends they are put to, and how this produces surveillance creep. At best, we get an overall view of how the consumer is duped into unequal bargains or how the citizen is unaware of the reaches of surveillance[23]. This particular area, we believe is not only under theorized but that it is the task of Information Systems to study this particular area – the relationship of technology to the agents involved and how the adoption or cooption of the technology occurs. What is it that makes the technology simply too 'tempting' not to exploit?

Our work is to focus on how surveillance creep occurs, or in this introduction of our social ecological model, how technology in our everyday life proves 'tempting' and is coopted for multifarious and potentially conflicting purposes by the agents involved and how such cooption of technology leads to surveillance creep.

7.1 A Social Ecological Model for Technology Cooption

Kranzberg's first law of technology states that: "Technology is neither good nor bad; nor is it neutral"[38]. By this, Kranzberg means that technology interacts with the social ecology in ways that are difficult to foresee or control. By social ecology [38] or information ecology [20, 57] we mean a 'system of people, practices, values and technologies in a particular local environment' [57]. The term 'ecology' is appealing as a metaphor because it emphasizes a process of dynamic complex interactions, conflicts and struggles that lead to particular ecological outcomes. It looks at how things come about. As an ecology it also provides a nexus between the physical properties of technology and their human design and use.

The same technology can be 'good' or 'bad' depending on the context and indeed whether a short-term or long-term perspective is taken. Technology is designed, and as such, it is a reflection of a society's values and power relationships. Embodied in the design are decisions about who will use the technology, how they will use it and to what purpose. However, technology may be used and customized by people in ways not envisaged by the designers [48]. This raises the question of what it is about a particular technology that enables such unplanned and unintended uses. A useful approach for thinking about this issue is the concept of affordances. In simple terms, an affordance is an action possibility that involves the interaction between an actor's capabilities and the real, objective or physical properties of an environment [68] or in this context the physical properties of technology.

We contend that there are five key elements that effect the cooption of a technology and that produce a particular social or behavioral outcome such as surveillance creep. This forms a design and use cycle, which is provided in Figure 2.

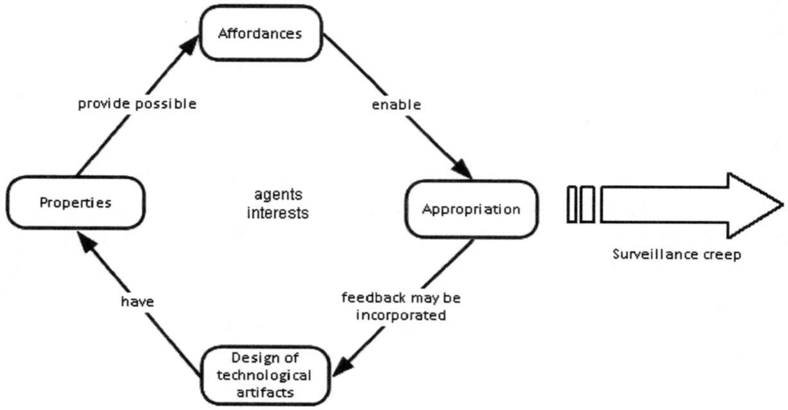

Fig. 2 Social ecological model for technology cooption

The circular arrows and the elements (design, properties, affordances, appropriation, agent interests) describe the design and use cycle. It is circular to highlight the ongoing process of design and use. A particular IT artifact is designed with physical properties. These properties provide affordances that are taken up by agents. These affordances, which may or may not have been explicitly designed, are then appropriated by various agents to address their interests. This is how the technology is used, how it is adopted and adapted by the user. This in turn leads to new design that might design out undesired features or enhance desired ones. A similar approach to considering how technology is coopted is the work of Carroll [11]. While Carroll [11] focuses on appropriation as part of the design process, our interest is how appropriation occurs, especially the interests of the agents that impact this process.

In the centre of the circle are the interests of the agents – the users, designers, and owners of the technologies. As shall be described further below the interests of the various agents, ranging from designers, businesses, governments, citizens, employees and consumers will differ and at times conflict according to what they see as an affordance, what they appropriate and use, and what they see as desirable designs.

From this social ecology of interests, design, properties, affordances and appropriations, social outcomes are produced as a result of the playing out of the possibilities in design and properties between the different interests of the agents. We also suggest that it is an essentially contested process and one that is generally an unequal one. It is contested because agents use technologies according to their own interests, and it is unequal as the owners of the design of the technology are in a far better position to realize their interests.

The model allows the researcher to drill down into each of the elements for greater examination while having some understanding of where the element fits into the larger social ecological context. It also shows how each element impacts other elements and contributes to social outcomes. We believe that how technological artifacts interact with human agency to be a central question of Information Systems. The model also places issues of interest, power and conflict as being central. How technology is coopted is potentially a contested phenomenon.

In the next section, we illustrate how our model sheds light on ICT surveillance creep.

7.2 The Social Ecology Model and the Rise ICT Dataveillance

7.2.1 Design

While design means different things to different people, Winograd [75] argues that through the various definitions of design there runs a common thread 'linking the intent and activities of a designer to the results that are produced when a designed object is experienced in practice'. Artifacts thus reflect the intentionality of the designer which consciously or unconsciously reflects society's values [58]. And since much technology is designed by corporations to pursue their interests, designs will favor these aims rather than broader social concerns such as privacy.

For example, when Netscape invented cookies in 1994, no effort was made to address privacy issues; the browser did not provide cookie management tools nor were cookies even mentioned in the documentation [70]. It was only the public uproar after the Financial Times [37] ran a story about the cookies that Netscape began a redesign that gave users the option to turn cookies off.

Shah and Kesan [70] conclude that as a society we should not expect that firms will uphold values for the greater good if these values are in conflict with the profit motive. While this point is arguable, what is certain is that technology is designed and can thus be engineered in ways to intentionally destroy privacy, for example to capture consumer information or to produce audit trails [26]. Conversely, privacy enhancing technologies could be designed. For instance, designers could engineer technologies that withhold or do not gather identifying information [26]. Technologies are designed to open up certain possibilities and close down others and this outcome may be intentional or unintentional. Lack of privacy and surveillance may be wittingly or unwittingly built into the design. Perhaps most commonly, technology is designed blindly without serious thought towards privacy or other social implications and may nevertheless provide privacy eroding affordances.

7.2.2 Properties of ICT

A designed artifact by its nature exhibits various properties. In this context, we define a property as a real or objective attribute of the technology. Properties are characteristics of the technology and while a technology may have many properties, the ones of interest are those that support the goals of the agent. It is this interaction between the properties of the technology and the goals of the agent that give rise to affordances. Technologies may be built on other technologies with resulting properties. For instance, data streams are built on a combination of high-speed communication networks, powerful computer processors, Database Management Systems, inexpensive storage devices, and sophisticated software that can build signatures dynamically. Together these technologies provide data streams with properties including:

- Instantaneous transmission of data
- Instantaneous collection of data
- Storage of vast amounts of data
- Processing of records as they arrive
- Global dissemination of information

It is an agent's interaction with the properties of an artifact that gives rise to various ICT affordances, in other words an affordance is an emergent action possibility, it is something that the technology can be brought to do that is within the boundary of consciousness or arc of intentionality of the agent [22].

7.2.3 Affordances

The word affordance was coined by ecological psychologist James Gibson [31] and refers to the action possibilities an environment can offer an actor. The affordance concept was popularized by Norman [60] in his book 'The Design of Everyday Things'.

Dreyfus [22] drawing on Gibson [31] and Merleau-Ponty [52] provides a useful approach to understanding affordances. Dreyfus argues there are three aspects to an affordance or what makes a thing an action possibility. Firstly, the physical properties of the object must meet or address, at the very least, minimum human physical capabilities. Secondly, action possibilities emerge as a result of the general skills that a human being may possess. Finally, affordances arise because of the stock of cultural skills within a community. Dreyfus uses the example of the chair. A chair affords sitting for Westerners because, one, it is a physical shape that permits the action and humans get tired standing. Secondly, sitting can be learned, and finally it forms a cultural practice in Western society.

The concept of affordance provides the nexus between the physical property and use. In the context of social interaction, affordances have been used to describe the material properties of the environment that affect how people interact [28, 39]. This is not to say that social behavior cannot be accounted for in terms of 'social conventions' and 'communities of practice', but the ecological psychology approach examines how social activities are embedded in and influenced by the physical environment [28].

Our interest here is how the properties of an ICT system influence and interact with social activities leading to non neutral or 'biased' outcomes [36, 38]. For instance, we can examine how privacy or the lack of privacy is embedded in and influenced by the properties of the technology.

Continuing our data stream example, the previously listed properties can provide certain actors with emergent dataveillance action possibilities such as:

- Real-time (or nearly real-time) ability to monitor the behavior of individuals through the capture of transaction data.
- Ability to create knowledge without the consent or even the awareness of the individual (for example, clickstreams, sensor monitoring, network traffic).

Which affordances are perceived of the properties of a designed artifact depends to a large degree on a particular actor's needs and goals. For the individual, Yahoo's mashup tool, Yahoo Pipes, can afford the creation of a webpage that assembles, in one place, information of interest from a variety of sources. For Yahoo, the mashup tool can afford a means of understanding the interests of a large number of individuals; information that has value for marketers. This leads many to observe, free Internet services are paid for using micropayments of our personal data [1, 10, 17, 23].

7.2.4 Appropriation

Technologies are encoded with forms of use, some intended by the designer and some not [48]. The intended forms of use may be reinforced by advertising but the properties of the artifact may allow it to be used in unintended ways. For example, cassette recorders were designed to play pre-recorded tapes but the recording property provided an affordance that meant the device was widely used for recording from records [48].

In using mobile phones, teenagers often avoid connection costs by the practice of pranking (calling and hanging up before the receiver answers). Using the property that allows the caller's number to be displayed, the mobile phone affords teens a modern-day smoke signal. The same caller-id property may provide a call-screening affordance for other phone users.

The process by which users adapt the technology to meet their needs is called appropriation [11], and we argue that it is the affordances of a technology, its 'intentional arc' that allow it to be appropriated by particular actors for particular purposes [22]. Some technological appropriations may be taken up and included in future designs of the technology while other appropriations will be closed out. Indeed, Bar et al.[3] argue that the appropriation process is primarily a political battle for power over the configuration of the technology and hence who can use it and how it can be used.

7.2.5 Agent Interests

Much of the business and IS literature, if not most, views product development cycles like our model as a consensual model where everyone is a 'winner' and the customer the winner above all. The implicit assumption or belief is that the business designs and appropriates a system for the benefit of the consumer. There is not only a belief that in the long run the customer is sovereign and those businesses that meet the needs and wants of the consumer will ultimately triumph but also that every step along the way, the strategies and tactics businesses employ, serve the interests of the customer.

It is however, quite plain that the consumer is not necessarily 'king' but is often just as much an adversary as are other businesses in strategic and tactical battles for business supremacy or survival. After all, businesses survive and run on profit and the consumer is merely a means to that end. As Porter [64] pointed out, the use of ICT for efficiency and effectiveness, production and distribution, is a business imperative not a strategy. Strategy, among other things, involves a combination of getting the customer to buy more of your product and purchasing it in a manner that is advantageous to the business. Porter [64] adds, that long-term advantage is doing this in a way that is difficult for other businesses to copy.

Put this way, there is conflict between the interests of the consumer and the business. In turn this sets up a conflict of interest between how new technologies are appropriated and designed. It is at this point that features can be designed in or out and particular affordances are appropriated that may not necessarily be in the interest of the consumer/citizen.

According to the surveillance literature most technological applications point to a Mephistophelean deal between the corporation and the consumer (for example, [1, 10, 23]). The corporation promises convenience and customization and obtains the consumers' data, their shopping soul. The corporation, in terms of the knowledge and understanding they receive concerning consumer behavior and response, is far greater than the convenience promised. In the words of Andrejevic [1]: "We are invited to actively participate in staging the scene of our own passive submission – and to view such participation as a form of power sharing" (p. 15).

This forms the crux of the conflict between the interests of the consumer and the corporation. The ICT promises empowerment but actually delivers surveillance and management of consumer behavior. It is a short step from here to consider the design of the systems to produce such surveillance that the ICT affords. Significantly most of the control of such design is in the hands of the corporation. Looked at from the consumers' point of view the empowerment promised in choice and interactivity is immense. To illustrate this conflict, consider the following examples. It is easy enough to imagine that the iPhone could be used to scan items to determine their carbon footprint, nutrition details and the labor practices of the business: what might be called the global impact of the consumer item. The phone could have an application that undertakes the necessary trade-offs, including price, to decide on which item to purchase. That would be empowering. Possibly even more empowering would be an iBot that remotely interrogated retailer databases and undertook comparisons of the cost and global impact of the delivered item to the home. However such a system, while technically feasible would raise enormous resistance among food retailers and manufacturers.

The objections to such a system would include the fear that such a system would reduce demand if people knew the true cost or 'global impact' of the items. The use of the iPhone might lessen impulse or unplanned purchases and the iBot certainly would destroy such impulses completely: the chocolate bars placed temptingly beside the milk would not seduce the iBot algorithm. Quite simply the databases required for such a system are unlikely to be forthcoming as there is little organized demand for such systems.

It is much more likely that systems incorporating these elements would be designed and appropriated by retailers for their strategic and tactical ends. Such a system could be used to interrogate a database on limited grounds, e.g. calorie intake of a food item. This interrogation would be collected and compared to aggregate purchase data. Varying information, product placement, special deal and promotions could be compared on purchase. The interrogation would also provide an opportunity for cross promotions and up-selling. Such a system would be very empowering to the retailer. Of the two systems, it is far more likely that this latter system would be designed and implemented, if it hasn't been already.

7.3 The Outcome of Technology Cooption: Surveillance Creep

The net effect of this social ecology process and the struggle or conflict between the various agents is surveillance creep, rather than consumer (or citizen or worker) empowerment. This is shown by the arrow in Figure 2. It could lead to other social outcomes. For example, some applications and technologies may lead to consumer empowerment rather than surveillance creep. The development of open source software and the peer-to-peer file sharing networks are two examples that led to greater user empowerment. The most important point however, is that the outcome is a product of the struggles of the agents over the appropriation of the affordances and the design of the technology.

Surveillance creep is at least in part an outcome of these quotidian local and specific struggles. It involves business and employees of businesses organizing and appropriating the technology to reach the strategic and tactical ends of the business and the motivations of managers and employees to those ends: satisfaction with a job well done, prospects for promotion, loyalty to the business, the technical interests in the work and staving off potential business failure and job loss. In comparison, the consumer is far less organized. At this point we propose that the overwhelming and general evidence for surveillance creep lies in the lack of power and organizing capacity of the consumer vis-à-vis the corporation which allows businesses to collect, store and manage data in the pursuit of specific corporate goals.

Surveillance creep does appear natural, as an unplanned outcome; it generally is. If it were not so, such creep would rightly be seen as an 'unnatural' political exercise. Only when each of the elements that constitute technology cooption is examined can we see how social processes and imperatives interact with the temptations of possibility to lead to surveillance.

It needs to be emphasized that surveillance creep is a net or overarching outcome of these local struggles over appropriation and design. Different technologies provide affordances that are 'biased' towards our simplified 'empowerment – surveillance' framework. The semi-porous or near boundless nature of the Internet provides a haven and organizing capability for non-socially approved activities such as pornography, hackers and alternative politics. The Internet as the wild west should not be under-estimated [66]. On the other hand, replacing an analogue key with a swipe card enables an employer to monitor office usage. Digitization affords monitoring in cultures that view such monitoring as a means to improve performance and employee discipline. Grounded in such a context and the relative strengths of employees and employers, the 'temptation' for monitoring proves 'natural'. The ability to monitor may be a 'bias' of the technology and supports Innis'[36] groundbreaking work on the bias of technology.

The model as depicted has not, however described effects exogenous to the local struggle other than those mediated by the agents. Our view is that this does not exhaust the influence of such exogenous factors. Agents draw upon the prevailing stock of knowledge, 'Weltanschauung' to inform their choices, motivations, and what works.

8 Conclusion

Dataveillance is on the rise. Emerging technologies are put into service of dataveillance. Such dataveillance may be used to increase security and provide better government services. Similarly, it can aid the corporation to increase work efficiency and to better understand the customer. Dataveillance however also opens the opportunity for the state, the employer and the corporation to manage, even socially engineer, the citizen, the worker and the consumer. The extent of these technologies to aid such management or increase control of the target of dataveillance is not widely known.

Our social ecological model provides an analytical tool for understanding how the introduction of information technology can lead to social outcomes such as surveillance creep or, potentially consumer empowerment. Dataveillance is not an inevitability. The social ecological model we believe demonstrates the truth of Kranzberg's first law and elaborates on its working. While the model and discussion presented has been limited to surveillance and the interests and positions of corporations and consumers, the model can be further developed to incorporate government/citizen and employer/employee positions and to generalize from surveillance to the bureaucratic management [41] and governmentality of the information age [54]. We believe that with development and refinement it can provide a general model for technology cooption.

The strength of our social ecological model is firstly that it views information design and appropriation as being essentially contested: the researcher must understand the differing motivations, interests and knowledge and also differing capabilities of agents. Secondly, by its focus on properties and affordances it provides the opportunities to examine how particular technologies are 'inherently biased'.

The social ecological model provides a middle range theory for empirical analysis by identifying the key elements of technology cooption and their proposed links. The researcher can investigate each point of the social ecology model and see how various properties are designed for, how such properties open the field of opportunity for action and then how they are appropriated by the various users and the conflicts between various stakeholders in the process. Of particular interest is what we suggest are the novelties of the model – (a) the introduction of affordance as the nexus of properties and appropriation, (b) that how a technology is ultimately appropriated or adopted is a matter of contestation of the agents involved and (c) the model sits within a social ecology where current everyday practices influence design, affordances and appropriation.

The next step is to test the model's utility in opening the design and appropriation cycle of an artifact or technological system in an empirical setting. For it to be successful it will need to demonstrate three key aspects. Firstly it must provide useful hypotheses and insights to be examined. To do so it needs to provide specific rather than general or sweeping insight. Good theory must offer empirical challenges and emerging fields of study. Secondly, it must continue to provide a plausible explanation of artifact appropriation – that the cycle and contestation we describe adequately reflect the practices of the agents involved. It is one thing to develop a logical model of appropriation but yet another to develop a model that describes actual practice. Finally, the model needs to be compared with other models of technological appropriation such as the Technology Acceptance Model (TAM)[21] to test which models are superior in terms of hypotheses generation, explanation and practice.

References

1. Andrejevic, M.: iSpy: surveillance and power in the interactive era. University Press of Kansas, Lawrence (2007)
2. Bagaric, M.: Privacy is the last thing we need. In: The Age, Fairfax, Melbourne (2007)

3. Bar, F., Pisani, F., Weber, M.: Mobile technology appropriation in a distant mirror: baroque infiltration, creolization and cannibalism. Prepared for discussion at Seminario sobre Desarrollo Económico, Desarrollo Social y Comunicaciones Móviles en América Latina Convened by Fundación Telefónica in Buenos Aires, April 20-21 (2007)
4. Bauman, Z.: Does ethics have a chance in a world of consumers? Harvard University Press, Cambridge (2008)
5. Baumgart, S.: Police watching where you Go. In: Bribane Times. Fairfax, Brisbane (2010)
6. Beck, U.: Risk Society: Towards a New Modernity. Sage, London (1992)
7. Beniger, J.R.: The control revolution: Technological and economic origins of the information society. Harvard University Press, Cambridge (1986)
8. Bennett, C.J., Raab, C.D.: The governance of privacy: Policy instruments in global perspective. Ashgate, Aldershot (2003)
9. boyd, D.: Facebook's Privacy Trainwreck: Exposure, Invasion, and Social Convergence. The International Journal of Research into New Media Technologies 14, 13–20 (2008)
10. Campbell, J.E., Carlson, M.: Panopticon.com: Online Surveillance and the Commodification of Privacy. Journal of Broadcasting & Electronic Media 46, 586–606 (2002)
11. Carroll, J.: Completing Design in Use: Closing the Appropriation Cycle. In: European Conference on Information Systems (ECIS). Association for Information Systems (2004)
12. Chaudhry, N.A.: Introduction to Stream Data Management. In: Chaudhry, N.A., Shaw, K., Abdelguerfi, M. (eds.) Stream Data Management, pp. 1–13. Springer, US (2005)
13. Clarke, R.: Information Technology and Dataveillance. Commununications of the ACM 31, 498–512 (1988)
14. Clarke, R.: While You Were Sleeping... Surveillance Technologies Arrived. Australian Quarterly 73, 10–14 (2001)
15. Cohen, J.E.: Examined Lives: Informational Privacy and the Subject as Object Stanford Law Review, vol. 52, pp. 1373–1438 (2000)
16. Cohen, S.: Visions of social control:crime, punishment, and classification. Polity Press:Blackwell, Cambridge Cambridgeshire Oxford, UK (1985)
17. Conti, G.: Googling considered harmful. In: Proceedings of the New Security Paradigms Workshop, pp. 67–76. ACM Press, New York (2006)
18. Cooley, M., Cooley, S.: Architect or bee?: the human/technology relationship. Langley Technical Services, Slough (1980)
19. Cortes, C., Fisher, K., Pregibon, D., Rodgers, A., Smith, F.: Hancock: A Language for Analyzing Transactional Data Streams. ACM Transactions on Programming Language and Systems 26, 301–338 (2004)
20. Davenport, T.H., Prusak, L.: Information ecology: mastering the information and knowledge environment. Oxford University Press, New York (1997)
21. Davis, F.D.: Perceived Usefulness, Perceived Ease of Use, and User Acceptance of Information Technology. MIS Quarterly 13, 319–340 (1989)
22. Dreyfus, H.L.: The Current Relevance of Merleau-Ponty's Phenomenology of Embodiment. The Electronic Journal of Analytic Philosophy (1996)
23. Fernback, J.: Selling ourselves? Critical Discourse Studies 4, 311–330 (2007)
24. Fishman, C.: The Wal-Mart Effect: How the World's Most Powerful Company Really Works–and How It's Transforming the American Economy. Penguin (2006)
25. Foucault, M., Sheridan, A.: Discipline and punish: the birth of the prison. Allen Lane, London (1977)

26. Froomkin, A.M.: The Death of Privacy? Stanford Law Review 52, 1461–1543 (2000)
27. Gandy, O.H.: The panoptic sort: A political economy of personal information. Westview Press, Boulder (1993)
28. Gaver, W.W.: Affordances for interaction: the social is material for design. Interactions 8, 111–129 (1996)
29. Geddes, L.: Rat in your cellphone. New Scientist 204, 34–37 (2009)
30. Geddes, L.: Rat in your cellphone. New Scientist, 34–37 (2009)
31. Gibson, J.J.: The Theory of Affordances. In: Shaw, R.E., Bransford, J. (eds.) Perceiving, acting, and knowing: Toward an ecological psychology, pp. 67–82. Lawrence Erlbaum Associates, Inc., Hillsdale (1977)
32. Giddens, A.: The nation state and violence. University of California Press, Berkeley (1987)
33. Golle, P.: Revisiting the uniqueness of simple demographics in the us population. In: WPES 2006: Proceedings of the 5th ACM Workshop on Privacy in Electronic Society, pp. 77–80. ACM, New York (2006)
34. Habermas, J.: Legitimation Crisis. Beacon, Boston (1975)
35. Hodgkinson, T.: Why you should beware of Facebook. In: The Guardian (2008)
36. Innis, H.A.: The bias of communication. University of Toronto Press, Toronto (1951)
37. Jackson, T.: This Bug in Your PC is a Smart Cookie. Financial Times, 15 (1996)
38. Kranzberg, M.: Technology and History: "Kranzberg's Laws". Technology and Culture 27, 544–560 (1986)
39. Kreijns, K., Kirschner, P.A.: The Social Affordances of Computer-Supported Collaborative Learning Environments. In: 31st ASEE/IEEE Frontiers in Education Conference, pp. T1F12–T11F-17. IEEE, Reno (2001)
40. Lambert, D., Pinheiro, J.C.: Mining a stream of transactions for customer patterns. In: Proceedings of the Seventh ACM SIGKDD International Conference on Knowledge Discovery and Data Mining. ACM, San Francisco (2001)
41. Lefebvre, H.: Everyday life in the modern world. Allen Lane, London (1971)
42. Lindsay, D.: An exploration of the conceptual basis of privacy and the implications for the future of Australian privacy law. Melbourne University Law Review 29, 131–178 (2005)
43. Lyon, D.: The electronic eye: the rise of surveillance society. Polity Press, Cambridge (1994)
44. Lyon, D.: Everyday Surveillance: Personal data and social classifications. Information, Communication & Society 5, 242–257 (2002)
45. Lyon, D.: Surveillance in cyberspace: the Internet, personal data, and social control. Queen's Quarterly, pp. 345–357 (2002)
46. Lyotard, J.-F.: The postmodern condition: a report on knowledge. Manchester University Press, Manchester (1984)
47. Lyotard J.-F., and Lyotard J.-F.: The postmodern condition : a report on knowledge. Manchester University Press, Manchester (1984)
48. Mackay, H., Gillespie, G.: Extending the Social Shaping of Technology Approach: Ideology and Appropriation. Social Studies of Science 22, 685–716 (1992)
49. Marx, G.: The Case of the Omniscient Organization. Harvard Business Review 90, 12–30 (1990)
50. Marx, G.T.: What's new about the "New Surveillance"? Classifying for change and continuity. Surveillance and Society 1, 9–29 (2002)
51. McCullagh, D.: AOL's disturbing glimpse into users' lives. In: CNET News (2006)
52. Merleau-Ponty, M.: Phenomenology of perception. Humanities Press, New York (1962)

53. Merton, R.K.: Social theory and social structure. Free Press, New York (1968)
54. Miller, P., Rose, N.: Governing the present: Administering economic, personal and social life. Polity, Cambridge (2008)
55. Montreal P. M.: 'Mashup' websites are a dream come true for hackers: privacy and security go out the window when websites are merged to make them more useful. In: New Scientist. pp. 28-30 (2006)
56. Mount, F.: Living with monsters. London Review of Books 32, 24–26 (2010)
57. Nardi, B.A., O'Day, V.: Information ecologies: using technology with heart. MIT Press, Cambridge (1999)
58. Nissenbaum, H.: How Computer Systems Embody Values. Computer, 118–119 (2001)
59. Nissenbaum, H.: Protecting Privacy in an Information Age: The Problem of Privacy in Public. Law and Philosophy 17, 559–596 (1998)
60. Norman, D.A.: The design of everyday things / Donald A. Norman Doubleday, New Yoir (1988)
61. O'Harrow, R.: No place to hide: The terrifying truth about the people who are watching our every move. Penguin Books, London (2006)
62. Olken, F., Gruenwald, L.: Data Stream Management: Aggregation, Classification, Modeling, and Operator Placement. IEEE Internet Computing 12, 9–12 (2008)
63. Parenti, C.: The soft cage: Surveillance in America from slavery to the war on terror. Basic Books, New York (2003)
64. Porter, M.E.: Strategy and the Internet. Harvard Business Review, 63–78 (2001)
65. Poster, M.: The mode of information: poststructuralism and social context. Polity Press in association with Basil Blackwell, Cambridge (1990)
66. Rheingold, H.: The virtual community: homesteading on the electronic frontier. Addison-Wesley Pub. Co., Reading (1993)
67. Romanyshyn, R.D.: Technology as symptom and dream. Routledge, London (1989)
68. Scarantinoyz, A.: Affordances Explained. Philosophy of Science 70, 949–961 (2003)
69. Sennett, R.: The corrosion of character: the personal consequences of work in the new capitalism. Norton, New York (1998)
70. Shah, R.C., Kesan, J.P.: Recipes for cookies: how institutions shape communication technologies. New Media & Society 11, 315–336 (2009)
71. Steeves, V.: If the Supreme Court were on facebook: evaluating the reasonable expectation of privacy test from a social perspective. Canadian Journal of Criminology and Criminal Justice 50, 331–347 (2008)
72. Webster, F., Robins, K.: I'll be watching you: Comment on Sewell and Wilkinson. Sociology 27, 243–252 (1993)
73. Wen, H.J., Gershuny, P.: Computer-based monitoring in the American workplace: Surveillance technologies and legal challenges. Human Systems Management 24, 165–173 (2005)
74. Winner, L.: Do artifacts have politics? Daedalus 109 (1980)
75. Winograd, T.: Bringing design to software. ACM Press, New York (1996)
76. Zuboff, S.: In the age of the smart machine: the future of work and power. Basic Books, New York (1988)

Glossary of Terms and Acronyms

- Affordance: A characteristic or quality of an artifact, or an environment, that allows an individual (with certain characteristics) to perform an action.

- Agent interests: Perceived wants or needs of a given individual, government or corporate entity.
- AOL: America Online, Inc Global. An internet services and media company.
- API: Application programming interface. Set of rules for accessing a particular software program or service.
- Appropriation: Commandeering of artifact and realizing the affordances that it provides for the agent's interest.
- Big Brother: Use of surveillance for state security and risk management.
- Clickstream: A stream of data resulting from the capture of the keystrokes generated by a computer user while Web browsing or using another software application.
- Consequentialist: The belief that the consequences of an action should be the basis for judging the morality of that action.
- Carbon footprints: The amount of green house gases produced as a result of the manufacture of a product.
- Data mining: The process of analyzing large datasets to find patterns and extract useful information.
- Data stream: A stream of data resulting from high volume transactions such as credit card purchases.
- Dataveillance: Neologism coined by Roger Clarke to refer to the integration of data collection management and mining with consumer, worker and citizen surveillance.
- Dystopian: A society characterized by human misery, often repressive and controlled by the state.
- Empowerment: The gaining of control over one's life or an increase in community control over its affairs.
- ICT: Information and Communication Technologies.
- IP: Internet Protocol. An IP address is a numerical label assigned to a computer on a network that uses the Internet Protocol for communication.
- ISP: Internet Service Provider.
- Loyalty scheme: Customer management and marketing programs that provide product, monetary and prestige incentives for repeat purchases.
- Mashup: New data or service produced by combining data, presentation or functionality from two or more sources.
- Privacy: The right to be let alone, free from observation, monitoring and data mining.
- Security: Concern for, or protection of, the safety of an agent from future or present loss.
- Social ecological model: schematic of the process of dynamic complex interactions, conflicts and struggles that lead to particular social and material outcomes.
- Soft Sister: Use of surveillance, monitoring and surveying of target populations usually by the state in order to better provide what the state sees as the interests of its population.
- Surveillance: Monitoring, observation and surveying of target population.

Index

D

G

Author Index